CW00702583

Exact and Approximate Controllability for Distributed Parameter Systems

The behavior of systems occurring in real life is often modeled by partial differential equations. This book investigates how a user or observer can influence the behavior of such systems mathematically and computationally. A thorough mathematical analysis of controllability problems is combined with a detailed investigation of methods used to solve them numerically, these methods being validated by the results of numerical experiments. In Part I of the book, the authors discuss the mathematics and numerics relating to the controllability of systems modeled by linear and nonlinear diffusion equations; Part II is dedicated to the controllability of vibrating systems, typical ones being those modeled by linear wave equations; finally, Part III covers flow control for systems governed by the Navier–Stokes equations modeling incompressible viscous flow. The book is accessible to graduate students in applied and computational mathematics, engineering, and physics; it will also be of use to more advanced practitioners.

ENCYCLOPEDIA OF MATHEMATICS AND ITS APPLICATIONS

All the titles listed below can be obtained from good booksellers or from Cambridge University Press. For a complete series listing

visit http://www.cambridge.org/uk/series/sSeries.asp?code=EOM

62 H. O. Fattorini *Infinite Dimensional Optimization and Control Theory*
63 A. C. Thompson *Minkowski Geometry*
64 R. B. Bapat and T. E. S. Raghavan *Nonnegative Matrices with Applications*
65 K. Engel *Sperner Theory*
66 D. Cvetkovic, P. Rowlinson, and S. Simic *Eigenspaces of Graphs*
67 F. Bergeron, G. Labelle, and P. Leroux *Combinational Species and Tree-Like Structures*
68 R. Goodman and N. Wallach *Representations and Invariants of the Classical Groups*
69 T. Beth, D. Jungnickel, and H. Lenz *Design Theory 1, 2nd edn*
70 A. Pietsch and J. Wenzel *Orthonormal Systems for Banach Space Geometry*
71 G. E. Andrews, R. Askey, and R. Roy *Special Functions*
72 R. Ticciati *Quantum Field Theory for Mathematicians*
73 M. Stern *Semimodular Lattices*
74 I. Lasiecka and R. Triggiani *Control Theory for Partial Differential Equations I*
75 I. Lasiecka and R. Triggiani *Control Theory for Partial Differential Equations II*
76 A. A. Ivanov *Geometry of Sporadic Groups I*
77 A. Schinzel *Polynomials with Special Regard to Reducibility*
78 H. Lenz, T. Beth, and D. Jungnickel *Design Theory II, 2nd edn*
79 T. Palmer *Banach Algebras and the General Theory of *-Albegras II*
80 O. Stormark *Lie's Structural Approach to PDE Systems*
81 C. F. Dunkl and Y. Xu *Orthogonal Polynomials of Several Variables*
82 J. P. Mayberry *The Foundations of Mathematics in the Theory of Sets*
83 C. Foias, O. Manley, R. Rosa, and R. Temam *Navier–Stokes Equations and Turbulence*
84 B. Polster and G. Steinke *Geometries on Surfaces*
85 R. B. Paris and D. Kaminski *Asymptotics and Mellin–Barnes Integrals*
86 R. McEliece *The Theory of Information and Coding, 2nd edn*
87 B. Magurn *Algebraic Introduction to K-Theory*
88 T. Mora *Solving Polynomial Equation Systems I*
89 K. Bichteler *Stochastic Integration with Jumps*
90 M. Lothaire *Algebraic Combinatorics on Words*
91 A. A. Ivanov and S. V. Shpectorov *Geometry of Sporadic Groups II*
92 P. McMullen and E. Schulte *Abstract Regular Polytopes*
93 G. Gierz et al. *Continuous Lattices and Domains*
94 S. Finch *Mathematical Constants*
95 Y. Jabri *The Mountain Pass Theorem*
96 G. Gasper and M. Rahman *Basic Hypergeometric Series, 2nd edn*
97 M. C. Pedicchio and W. Tholen (eds.) *Categorical Foundations*
98 M. E. H. Ismail *Classical and Quantum Orthogonal Polynomials in One Variable*
99 T. Mora *Solving Polynomial Equation Systems II*
100 E. Olivier and M. Euláliá Vares *Large Deviations and Metastability*
101 A. Kushner, V. Lychagin, and V. Rubtsov *Contact Geometry and Nonlinear Differential Equations*
102 L. W. Beineke, R. J. Wilson, and P. J. Cameron. (eds.) *Topics in Algebraic Graph Theory*
103 O. Staffans *Well-Posed Linear Systems*
104 J. M. Lewis, S. Lakshmivarahan, and S. Dhall *Dynamic Data Assimilation*
105 M. Lothaire *Applied Combinatorics on Words*
106 A. Markoe *Analytic Tomography*
107 P. A. Martin *Multiple Scattering*
108 R. A. Brualdi *Combinatorial Matrix Classes*
110 M.-J. Lai and L. L. Schumaker *Spline Functions on Triangulations*
111 R. T. Curtis *Symmetric Generation of Groups*
112 H. Salzmann, T. Grundhöfer, H. Hähl, and R. Löwen *The Classical Fields*
113 S. Peszat and J. Zabczyk *Stochastic Partial Differential Equations with Lévy Noise*
114 J. Beck *Combinatorial Games*
115 L. Barreira and Y. Pesin *Nonuniform Hyperbolicity*
116 D. Z. Arov and H. Dym *J-Contractive Matrix Valued Functions and Related Topics*
117 R. Glowinski, J.-L. Lions, and J. He *Exact and Approximate Controllability for Distributed Parameter Systems*

Exact and Approximate Controllability for Distributed Parameter Systems

A Numerical Approach

ROLAND GLOWINSKI
University of Houston

JACQUES-LOUIS LIONS
College de France, Paris

JIWEN HE
University of Houston

CAMBRIDGE UNIVERSITY PRESS
Cambridge, New York, Melbourne, Madrid, Cape Town, Singapore, São Paulo
Cambridge University Press
The Edinburgh Building, Cambridge CB2 8RU, UK

Published in the United States of America by Cambridge University Press, New York

www.cambridge.org
Information on this title: www.cambridge.org/9780521885720

First published 2008

Printed in the United Kingdom at the University Press, Cambridge

Library of Congress Cataloging in Publication data

Glowinski, R.
Exact and approximate controllability for distributed parameter systems : a numerical approach / Roland Glowinski, Jacques-Louis Lions, Jiwen He.
p. cm.
Includes bibliographical references and index.
ISBN 978-0-521-88572-0 (hardback : alk. paper)
1. Control theory. 2. Distributed parameter systems. 3. Differential equations, Partial–Numerical solutions. I. Lions, Jacques Louis. II. He, Jiwen. III. Title.
QA402.3.G56 2008
515′.642–dc22
2007042032

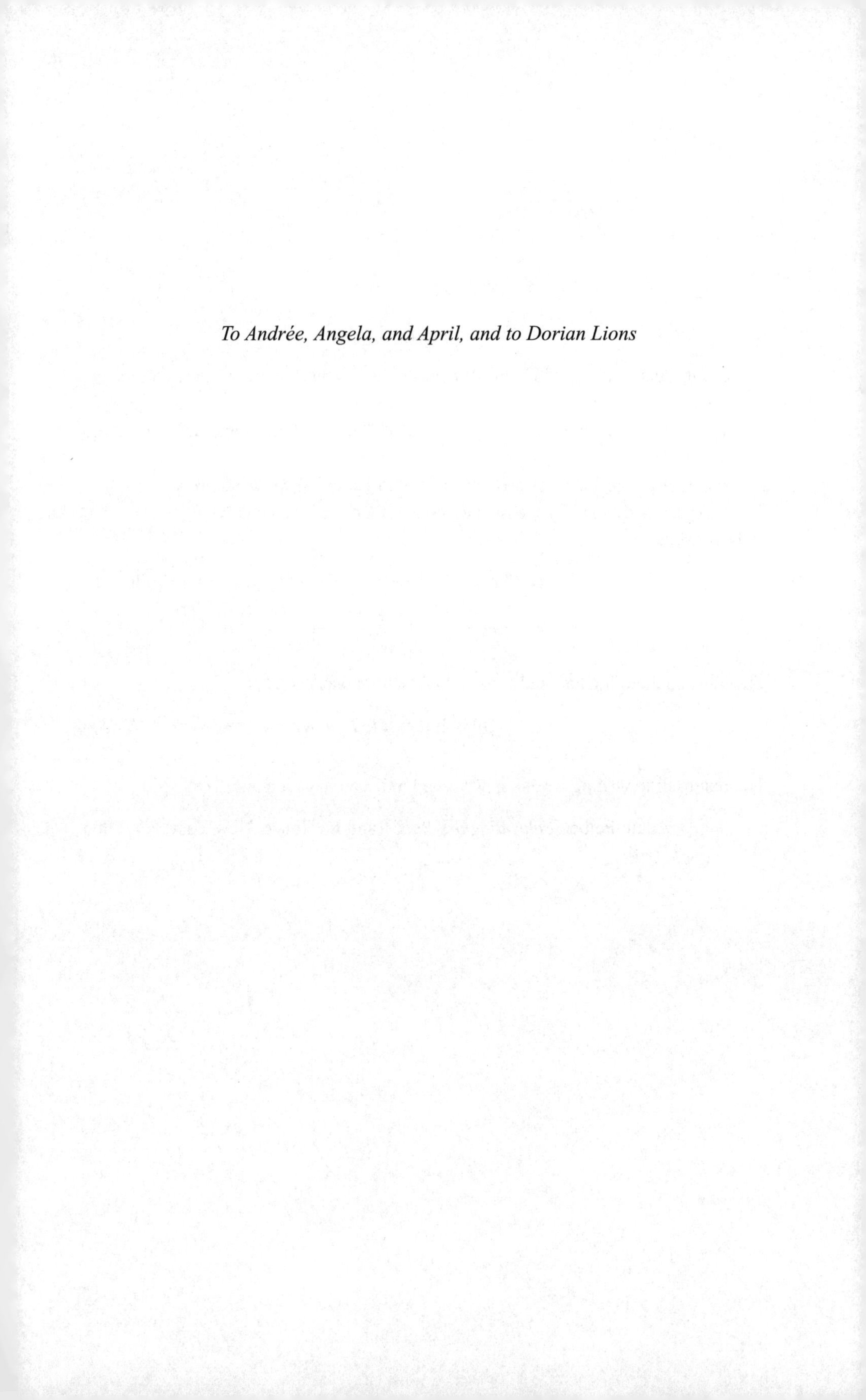

To Andrée, Angela, and April, and to Dorian Lions

LENS LARQUE-homonyms, with definitions.

1. Lencilorqua: a village of 657 inhabitants on Vasselona Continent, Reis, sixth planet to Gamma Eridani.
2. Lanslarke: a predacious winged creature of Dar Sai, third planet of Cora, Argo Navis 961.
3. Laenzle arc: the locus of a point generated by the *seventh theorem of triskoïd dynamics*, as defined by the mathematician Palo Laenzle (907–1070).
4. Linslurk: a mosslike . . .

Jack Vance, *The Face*. In *The Demon Princes*, Volume II,
Tom Doherty Associates, Inc., New York, NY, 1997

The most challenging course I took in high school was calculus.

Bill Clinton, *My Life*, Knopf, New York, NY, 2004

The real trick to writing a book is writing. Until you have a book.

Adam Felber, *Schrödinger's Ball*, Random House, New York, NY, 2006

Contents

Part II Wave Models

Part III Flow Control

Preface

During ICIAM 1995, in Hamburg, David Tranah approached Jacques-Louis Lions and myself and asked us if we were interested in publishing in book form our two-part article "*Exact and approximate controllability for distributed parameter systems*" which had appeared in *Acta Numerica 1994* and *1995*. The length of the article (almost 300 pages) was a justification, among several others, for such an initiative. While I was very enthusiastic about this project, J.L. Lions was more cautious, without being against it. Actually, his reservation concerning this book project was stemming from recent important developments on controllability related issues, justifying, in his opinion an in-depth revision of our article. Both of us being quite busy, the project was practically forgotten. As everyone knows in the Scientific Community, and elsewhere, Jacques-Lions passed away in June 2001, while still active scientifically. He largely contributed in making the *Control of Distributed Parameter Systems* a most important field where sophisticated mathematical and computational techniques meet with advanced applications. Therefore, when David Tranah renewed his 1995 suggestion during a conference of the European Mathematical Society held in Nice in February 2003, we thought that it would be a very nice way to pay to J.L. Lions the tribute he fully deserves. The idea was to respect as much as possible the original text, since it largely reflects J.L. Lions' inspired scientific vision, and also its inimitable way at making simple complicated notions. On the other hand, it was also agreed that additional material should be included to make the text more up to date. Most of these additions are concerned with *flow control*; indeed, for J.L. Lions, the control of flow modeled by the Navier–Stokes equations was a kind of scientific Holy Grail and we are most happy that he could witness the first real mathematical and computational successes in that direction, all taking place in the late 1990s.

The present volume is structured as follows:

- Motivations and some broad generalities are given in the Introduction.
- Part I is dedicated to the control of *linear* and *nonlinear diffusion models*; it contains Sections 1–5 of the *Acta Numerica* article, with additional materials such as the Neumann control of unstable advection–reaction–diffusion models, and a discussion of computer memory saving methods for the solution of time-dependent control problems by adjoint-equation-based methods. A short introduction to *Riccati-equation*-based control methods is also provided.

- Part II is concerned with the controllability of *wave equation* type models and of *coupled systems*. This material corresponds essentially to Sections 6 and 7 of the *Acta Numerica* article.
- Part III is the main addition to the original text; it is dedicated to the *boundary* control, by either rotation or blowing and suction, of *Newtonian incompressible viscous flow* modeled by the *Navier–Stokes equations*.

Since most of the additional material follows from investigations conducted jointly with Professor Jiwen He, a former collaborator of J.L. Lions, all the parties involved found it quite natural to have him as a coauthor of this volume.

Acknowledgments and warmest thanks should go first to David Tranah, Ken Blake, and Cambridge University Press for encouraging the publication of this augmented version of the *Acta Numerica* article, and also to Mrs Andrée Lions and Professor Pierre-Louis Lions for their acceptation of this project. The invaluable help of Dr H.L. Juárez (UAM-Mexico City) and of his collaborators (Bety Arce, in particular) is also acknowledged; they converted large parts of a text initially written in Word© to a LaTeX© file, a nontrivial task indeed considering the size of this volume.

Special thanks are due to S. Barck-Holst, M. Berggren, H.Q. Chen, J.M. Coron, J.I. Diaz, S. Gomez, M. Gorman, A.J. Kearsley, B. Mantel, R. Metcalfe, J. Périaux, T.-W. Pan, O. Pironneau, J.-P. Puel, A.M. Ramos, T. Rossi, D. Sorensen, J. Toivanen, and E. Zuazua for very helpful comments and suggestions concerning the additions to the original article (further acknowledgments may be found at the end of this volume; they concern the original *Acta Numerica* article).

We will conclude this preface with further thanks to Cambridge University Press for authorizing the reprinting of the above *Acta Numerica* article in Volume III of J.L. Lions, *Oeuvres Choisies*, SMAI / EDP Sciences, Paris, 2003, a three-volume testimony of the outstanding scientific contributions of Jacques-Louis Lions.

Guanajuato, Mexico — Roland Glowinski

Introduction

I.1 What it is all about?

We consider a system whose *state* is given by the solution y to a partial differential equation (PDE) of evolution, and which contains *control functions*, denoted by v.

Let us write all that in a formal fashion for the time being. The *state equation* is written as

$$\frac{\partial y}{\partial t} + \mathcal{A}(y) = \mathcal{B}v, \tag{I.1}$$

where y is a scalar- or vector-valued function.

In (I.1), $\mathcal{A}$ is a set of partial differential operators (PDOs), linear or nonlinear (at least for the time being). In (I.1), v denotes the *control* and $\mathcal{B}$ maps the "space of controls" into the "state space". It goes without saying that all this has to be made more precise. This will be the task of the following sections.

The PDE (I.1) should include *boundary conditions*. We do not make them explicit here. They are supposed to be contained in the abstract formulation (I.1), where v can be either applied *inside* the domain $\Omega \subset \mathbb{R}^d$, where (I.1) is considered (v is then a *distributed* control), or on the boundary Γ of Ω – or on a part of it (v is then a *boundary* control). If v is applied at points of Ω, v is said to be a *pointwise* control.

One has to add also *initial conditions* to (I.1): if we assume that $t = 0$ is the initial time, then these initial conditions are given by

$$y|_{t=0} = y_0, \tag{I.2}$$

with y_0 being a given element of the state space.

It will be assumed that, given v (in a suitable space), problem (I.1)–(I.2) (and the boundary conditions included in the formulation (I.1)) *uniquely defines a solution.* This solution is a function (scalar- or vector-valued) of $x \in \Omega$, $t > 0$, and of y_0 and v. We shall denote this solution by $y(v)$ $(= \{x, t\} \to y(x, t; v))$. Similarly, we shall denote by $y(t; v)$ the function $x \to y(x, t; v)$. Then, the initial condition (I.2) can be written as

$$y(0; v) = y_0. \tag{I.2*}$$

Remark I.1 The notions to be introduced below can be generalized to situations where the *uniqueness* of the solution to problem (I.1)–(I.2) is *not known*. We are thinking here of the Navier–Stokes equations (and related models) when the flow region Ω is a subset of $\mathbb{R}^3$ (and the *Reynolds number* is sufficiently large).

We can now introduce the notion of *controllability*, either *exact* or *approximate*.

Let $T > 0$ be *given* and let y_T (the *target* function) be a *given element* of the state space. We want to "drive the system" from y_0 at $t = 0$ to y_T at $t = T$, that is, we want to find v such that

$$y(T; v) = y_T. \tag{I.3}$$

If this is possible for *any* target function y_T in the state space, one can say that the system is *controllable* (or *exactly controllable)*. If – as we shall see in most of the examples – condition (I.3) is too strict, it is natural to replace it by the less demanding one

$$y(T; v) \text{ belongs to a "small" neighborhood of } y_T. \tag{I.4}$$

If this is possible, one says that the system is *approximately controllable*; otherwise, the system is *not controllable*.

Before giving precise examples, we want to say a few words concerning the motivation for studying these controllability problems.

I.2 Motivation

There are several aspects that make controllability problems important in practice.

Aspect # 1 At a *given time-horizon*, we want the system under study to behave *exactly* as we wish (or in a manner arbitrary close to it).

Problems of this type are common in Science and Engineering: we would like, for example, to have the temperature (or pressure) of a system equal, or very close, to a given value – globally or locally – at a given time. *Chemical Engineering* is an important source of such problems, a typical example in that direction being the design of *car catalytic converters*; in this example chemical reactions have to take place leading to the "destruction" at a given time-horizon (very small in practice) of the polluting chemicals contained in the exhaust gases (the modeling and numerical simulation of catalytic converter systems are discussed in, for example, Engquist, Gustafsson, and Vreeburg (1978), Friedman (1988, Chapter 7), and Friend (1993)).

Aspect # 2 For *linear* systems, it is known (cf. Russel (1978)) that exact controllability is equivalent to the possibility of *stabilizing* the system.

Stabilization problems abound, in particular in (large) composite structures – the so-called "multibody" systems made of many different parts which can be considered as three-, two-, or one-dimensional and which are linked together by *junctions* and *joints*. The modeling and analysis of such systems are the subject of many interesting studies. We want to mention here the contributions of P.G. Ciarlet and his collaborators (see, for example, Ciarlet, Le Dret, and Nzengwa, 1989, Ciarlet, 1990a,b), and those of Sanchez-Hubert and Sanchez-Palencia (1989), Lagnese, Leugering, and Schmidt (1992, 1994), J. Simo and his collaborators (see, for example, Laursen and Simo, 1993), Park and his collaborators (see, for example, Park, Chiou, and Downer, 1990 and Downer, Park, and Chiou, 1992).

Studying *controllability* is *one* approach to *stabilization* as shown in, for example, J.L. Lions (1988a).

Aspect # 3 (On *controllability* and *reversibility*): Suppose we have a system that *was* in a state z_1 at time $t = -t_0$, $t_0 > 0$, and that is *now* (that is, at $t = 0$) in the state y_0.

We would like to have the system *returning* to a state as close as possible to z_1, that is, $y_T = z_1$. If this is possible, it means some kind of "reversibility" property for the system under consideration. What we have in mind here are *environmental systems*; should they be "local" or "global" in the space variables?

Noncontrollable (sub)systems can suffer "irreversible" changes (cf. J.L. Lions, 1990 and Diaz, 1991).

We return now to the general questions of Section I.1, making them more precise before giving examples.

I.3 Topologies and numerical methods

The topology of the state space appears explicitly in condition (I.4). It is obvious that approximate controllability *depends* on the choice of the topology on the state space, that is, of the state space itself. Actually, *exact* controllability depends on the choice of the state space as well. The choice of the state space is therefore an obviously fundamental issue for the *theory*. We want to emphasize that it is also a fundamental issue from the *numerical point of view*. Indeed, if one has (as we shall see in several situations) exact or approximate controllability in a very general space (which can include elements that are not distributions but "ultra-distributions") but *not* in a classical space of smooth (or sufficiently smooth) functions, then the numerical approximation will *necessarily* develop singularities; "remedies" should be based on the knowledge of the topology where the theoretical convergence is taking place. We shall return on these issues in the following sections; actually, some of them have been addressed in, for example, Dean, Glowinski, and Li (1989), Glowinski and Li (1990), Glowinski, Li, and Lions (1990), Glowinski (1992a), where various *filtering* techniques are discussed in order to eliminate the numerical singularities mentioned above.

In the next section we shall address the following question (of general nature also), namely,

How to choose the control?

I.4 Choice of the control

Let us return to the general formulation (I.1), (I.2), (I.3), (or (I.4)). If there exists, *one* control v achieving these conditions, then there exist, in general, *infinitely many other* controls, vs, also achieving these conditions. Which one should we choose and how?

A most important question is: how to *norm* (we are always working in Banach or Hilbert spaces) the vs? This is related to the *topology* of the state space. It is indeed clear that the regularity (or irregularity!) properties of v and y in (I.1) are related. Let us assume that a norm $v \mapsto |||v|||$ is chosen. Once this choice is made, a natural formulation of the problem is then to find

$$\inf |||v|||, \tag{I.5}$$

among all those vs such that (I.1), (I.2), (I.3), or (I.4) take place.

Remark I.2 There is still some flexibility here since problem (0.5) still makes sense if one replaces $|||\cdot|||$ by a *stronger* norm. This remark will be of practical interest as we shall see later on.

Remark I.3 One can encounter questions of controllability for systems depending on "small" parameters. Two classical (by now) examples are

(i) *Singular perturbations.*
(ii) *Homogenization* which is important for the controllability of structures made of *composite materials*.

In these situations one has to introduce either *families* of norms in (I.5) or norms *equivalent* to $|||\cdot|||$, but which depend on the homogenization parameter.

I.5 Relaxation of the controllability notion

Let us return again to (I.1) and (I.2):

Condition (I.3) concerns the state y itself. In a "complex system" this condition can be (and will be in general) unnecessarily strong. We may want some *subsystems* to behave according to our wishes. We may also want some *average* values to behave accordingly, and so on. A general formulation is as follows:

We consider an operator

$$C \in \mathcal{L}(Y, \mathcal{H}), \tag{I.6}$$

where Y is the state space (chosen!) and where $\mathcal{H}$ is another Banach or Hilbert space (the *observation* space). Think, for instance, of C as being an *averaging* operator. Then, we "relax" condition (I.3) (respectively (I.4)) as follows:

$$Cy(T; v) = h_T, \quad h_T \text{ given in } \mathcal{H} \tag{I.7}$$

(respectively

$$Cy(T; v) \text{ belongs to a neighborhood of } h_T \text{ in } \mathcal{H}) \tag{I.8}$$

and consider (I.5) where v is subject to (I.7) (respectively (I.8)).

I.6 Various remarks

Remark I.4 For most examples considered in this book, the control function is either distributed (or pointwise) or of a boundary nature. It can also be a *geometrical* one. Namely, we can consider the domain Ω as variable or, to be more precise, at least a part of the boundary of Ω is variable, and we want to "move this part of the boundary" in order to drive the system from a given state to another one. In summary, we look for *controllability by a suitable variable geometry*. Problems of this type are discussed in Buschnell and Hefner (1990); they mostly concern *drag reduction* for viscous flow (see also Sellin and Moses, 1989); we will return on drag reduction in Part III.

Remark I.5 Recent events have shown the importance of *stealth technologies*. The related problems are very complicated from the modeling, mathematical, computational, and engineering points of view; several approaches can be considered (they do not exclude one another) such as active control, passive control via well-chosen coating materials and/or well-chosen shape, use of decoy strategies, and so on. Indeed, these methods can apply for planes and submarines as well. These problems justify a book by themselves and will not be specifically addressed here. We think, however, that various notions related to controllability, including the concept of *sentinels* can be most helpful in the formulation and solution of stealth related problems as shown in, for example, J.L. Lions (1989), (1990), and (1992a) (see also the references therein). It is also worth mentioning that the *exact controllability* based solution methods for the *Helmholtz equation at large wave numbers*, described in Chapter 7, have been motivated by stealth issues.

Remark I.6 We would like to conclude this introductory chapter by mentioning Zuazua (2007). This very recent and most remarkable publication contains an in-depth discussion of several of the controllability related topics discussed in the present volume.

Part I

Diffusion Models

1

Distributed and pointwise control for linear diffusion equations

1.1 First example

Let Ω be a bounded connected open set (that is, a bounded *domain*) in $\mathbb{R}^d$ (until recently, $d \leq 3$ in the applications, but *Finance modeling* and the *control of the Schrödinger equation* have brought situations where $d \geq 4$).

Remark 1.1 The "boundedness" hypothesis is by no means a strict necessity.

We shall also assume that $\Gamma = \partial\Omega$ is "sufficiently smooth," which is also not mandatory. Let $\mathcal{O}$ be an open subset of Ω. We emphasize here, at the very beginning, that $\mathcal{O}$ can be *arbitrary "small"*. The control function v will be with support in $\bar{\mathcal{O}}$; it is a *distributed control*. The *state equation* is given by

$$\frac{\partial y}{\partial t} + \mathcal{A}y = v\chi_{\mathcal{O}\times(0,T)} \quad \text{in } \Omega \times (0, T), \tag{1.1}$$

where $\chi_{\mathcal{O}\times(0,T)}$is the *characteristic function* of the set $\mathcal{O} \times (0, T)$ and where $\mathcal{A}$ is a *second-order elliptic operator*, with variable coefficients. *The coefficients of $\mathcal{A}$ can also depend on t.*

From now on, we shall denote by Q the space–time domain $\Omega \times (0, T)$.

Example 1.2 A typical operator $\mathcal{A}$ is the one defined by

$$\mathcal{A}y = -\sum_{i=1}^{d} \frac{\partial}{\partial x_i} \sum_{j=1}^{d} a_{ij} \frac{\partial y}{\partial x_j} + \mathbf{V}_0 \cdot \nabla y, \tag{1.2}$$

where, in (1.2), $\nabla = \left\{\frac{\partial}{\partial x_i}\right\}_{i=1}^{d}$ and where

(i) The coefficients $a_{ij} \in L^\infty(\Omega)$, $\forall i, j$, $1 \leq i, j \leq d$, and the matrix-valued function $(a_{ij})_{1\leq i,j\leq d}$ satisfies

$$\sum_{i=1}^{d}\sum_{j=1}^{d} a_{ij}(x)\xi_i\xi_j \geq \alpha \, \|\xi\|^2, \quad \forall \xi \left(= \{\xi_i\}_{i=1}^{d}\right) \in \mathbb{R}^d, \text{ a.e. in } \Omega, \tag{1.3}$$

with $\alpha \geq 0$ and $\|\cdot\|$ the canonical Euclidean norm of $\mathbb{R}^d$.

(ii) The vector $\mathbf{V}_0$ is *divergence-free* (that is, $\nabla \cdot \mathbf{V}_0 = 0$) and belongs to $(L^\infty(\Omega))^d$.
(iii) We have used the *dot product* notation for the canonical Euclidean scalar product of $\mathbb{R}^d$, that is,

$$\boldsymbol{\xi} \cdot \boldsymbol{\eta} = \sum_{i=1}^{d} \xi_i \eta_i, \quad \forall \boldsymbol{\xi} = \{\xi_i\}_{i=1}^d \text{ and } \boldsymbol{\eta} = \{\eta_i\}_{i=1}^d \in \mathbb{R}^d.$$

If the above hypotheses on the a_{ij}s and $\mathbf{V}_0$ are satisfied, then the *bilinear functional* $a(\cdot, \cdot)$ defined by

$$a(y, z) = \sum_{i=1}^{d} \sum_{j=1}^{d} \int_\Omega a_{ij} \frac{\partial y}{\partial x_j} \frac{\partial z}{\partial x_i} \, dx + \int_\Omega \mathbf{V}_0 \cdot \nabla y \, z \, dx, \tag{1.4}$$

is *continuous* over $H^1(\Omega) \times H^1(\Omega)$; it is also *strongly elliptic* over $H_0^1(\Omega) \times H_0^1(\Omega)$ since we have, from (1.3) and from $\nabla \cdot \mathbf{V}_0 = 0$, the following relation:

$$a(y, y) \geq \alpha \int_\Omega |\nabla y|^2 \, dx, \quad \forall y \in H_0^1(\Omega). \tag{1.5}$$

If $\mathbf{V}_0 = \mathbf{0}$ and if $a_{ij} = a_{ji}$, $\forall i, j$, $1 \leq i, j \leq d$, then the bilinear functional $a(\cdot, \cdot)$ is *symmetric*. Above, $H^1(\Omega)$ and $H_0^1(\Omega)$ are the *functional spaces* defined as follows:

$$H^1(\Omega) = \{ \phi \mid \phi \in L^2(\Omega), \ \frac{\partial \phi}{\partial x_i} \in L^2(\Omega), \ \forall i = 1, \ldots, d \}, \tag{1.6}$$

and

$$H_0^1(\Omega) = \{ \phi \mid \phi \in H^1(\Omega), \ \phi = 0 \text{ on } \Gamma \}, \tag{1.7}$$

respectively. Equipped with the classical *Sobolev norm*

$$\|\phi\|_{H^1(\Omega)} = \left(\int_\Omega \left(\phi^2 + |\nabla \phi|^2 \right) dx \right)^{\frac{1}{2}},$$

and with the corresponding *scalar product*

$$(\phi, \psi)_{H^1(\Omega)} = \int_\Omega (\phi \psi + \nabla \phi \cdot \nabla \psi) \, dx,$$

both spaces $H^1(\Omega)$ and $H_0^1(\Omega)$ are *Hilbert spaces*. Since Ω is *bounded*, the mapping

$$\phi \mapsto \left(\int_\Omega |\nabla \phi|^2 \, dx \right)^{\frac{1}{2}},$$

defines a norm over $H_0^1(\Omega)$ that is *equivalent* to the above $H^1(\Omega)$ norm, the corresponding scalar product being

$$\{\phi, \psi\} \mapsto \int_\Omega \nabla \phi \cdot \nabla \psi \, dx.$$

If we denote by $H^{-1}(\Omega)$ the *dual space* of $H_0^1(\Omega)$, then the above operator $\mathcal{A}$ is *linear* and *continuous* from $H^1(\Omega)$ into $H^{-1}(\Omega)$ and is an *isomorphism* from $H_0^1(\Omega)$ onto $H^{-1}(\Omega)$.

Back to (1.1), and motivated by the class of elliptic operators discussed in the above example, we shall suppose from now on that operator $\mathcal{A}$ is *linear* and *continuous* from $H^1(\Omega)$ into $H^{-1}(\Omega)$ and that it satisfies the following (*ellipticity*) property:

$$\langle \mathcal{A}\phi, \phi \rangle \geq \alpha \, \|\phi\|^2_{H^1(\Omega)}, \quad \forall \phi \in H_0^1(\Omega),$$

where, in the above relation, α is a *positive constant* and where $\langle \cdot, \cdot \rangle$ denotes the *duality pairing* between $H^{-1}(\Omega)$ and $H_0^1(\Omega)$. Operator $\mathcal{A}$ is *self-adjoint* over $H_0^1(\Omega)$ if

$$\langle \mathcal{A}\phi, \psi \rangle = \langle \mathcal{A}\psi, \phi \rangle, \quad \forall \phi, \psi \in H_0^1(\Omega).$$

The *bilinear* functional

$$\{\phi, \psi\} \mapsto \langle \mathcal{A}\phi, \psi \rangle : H_0^1(\Omega) \times H_0^1(\Omega) \to \mathbb{R},$$

will be denoted by $a(\cdot, \cdot)$ and is symmetric if and only if operator $\mathcal{A}$ is self-adjoint.

In order to fix ideas and to make things as simple as possible, we add to (1.1) the following *boundary condition*, of *Dirichlet* type:

$$y = 0 \quad \text{on } \Sigma \; (= \Gamma \times (0, T)). \tag{1.8}$$

The *initial condition* is

$$y(0) = y_0, \tag{1.9}$$

where y_0 is given in $L^2(\Omega)$. We shall assume that

$$v \in L^2(\mathcal{O} \times (0, T)). \tag{1.10}$$

We emphasize that the *choice* associated with (1.10) is by no means compulsory; we shall return to this issue. Actually, we began with (1.10) since it is the *simplest* possible choice, at least from a theoretical point of view. Indeed, it is a well-known fact (see, for example, J.L. Lions, 1961, Lions and Magenes, 1968) that system (1.1), (1.8), and (1.9) has a *unique solution* (denoted sometimes as $t \mapsto y(t; v)$ with $y(t; v) = x \mapsto y(x, t; v)$), which has the following properties:

$$y \in L^2(0, T; H_0^1(\Omega)), \quad \frac{\partial y}{\partial t} \in L^2(0, T; H^{-1}(\Omega)), \tag{1.11}$$

$$y \text{ is continuous from } [0, T] \text{ into } L^2(\Omega). \tag{1.12}$$

We are going to study the (*approximate*) *controllability* properties of problem (1.1), (1.8), and (1.9).

1.2 Approximate controllability

As a preliminary remark, we note that *exact* controllability is going to be difficult (unless $y_T = 0$; see Remark 1.7). Indeed, if we suppose that the coefficients of operator $\mathcal{A}$ are *smooth* (respectively, *real analytic*) then the solution y is, at time $T > 0$, *smooth outside* $\mathcal{O}$ (respectively, *real analytic outside* $\mathcal{O}$). Therefore, if the *target function* y_T is given in $L^2(\Omega)$ – which is a natural choice considering (1.12) – the *exact controllability condition*

$$y(T) = y_T,$$

will be, in general, *impossible*. This will become more precise below. For the time being, we start with the *approximate controllability*. In that direction, a key result is given by the following.

Proposition 1.3 *When* v *spans space* $L^2(\mathcal{O} \times (0, T))$*, then* $y(T; v)$ *spans an affine subspace which is dense in* $L^2(\Omega)$*.*

Proof.

(i) Let Y_0 be the solution of problem (1.1), (1.8), and (1.9) when $v = 0$. Then, the function $y(T; v) - Y_0$ describes a subspace of $L^2(\Omega)$ and we have to show the *density* of this subspace. It amounts to proving the above density result assuming that $y_0 = 0$.

(ii) We then apply the *Hahn–Banach theorem*, as in J.L. Lions (1968) (implying that the present proof is *not constructive*):

Let us consider therefore an element $g \in L^2(\Omega)$ such that

$$(y(T; v), g)_{L^2(\Omega)} = 0, \quad \forall v \in L^2(\mathcal{O} \times (0, T)). \tag{1.13}$$

Next, we introduce ψ as the solution of

$$-\frac{\partial \psi}{\partial t} + \mathcal{A}^* \psi = 0 \quad \text{in } Q, \tag{1.14}$$

where operator $\mathcal{A}^*$ is the adjoint of $\mathcal{A}$ and where ψ satisfies also

$$\psi = 0 \quad \text{on } \Sigma, \tag{1.15}$$

$$\psi(x, T) = g(x). \tag{1.16}$$

Then, multiplying both sides of (1.14) by $y(v)$ and applying the *Green's formula*, we obtain

$$(y(T; v), g)_{L^2(\Omega)} = \iint_{\mathcal{O} \times (0,T)} \psi \, v \, dx \, dt. \tag{1.17}$$

It follows from (1.17) that (1.13) is equivalent to

$$\psi = 0 \quad \text{in } \mathcal{O} \times (0, T). \tag{1.18}$$

It follows then from the (celebrated) *Mizohata uniqueness theorem* (Mizohata, 1958) that

$$\psi \equiv 0 \quad \text{in } Q; \tag{1.19}$$

combining (1.16) and (1.19) yields $g = 0$, which proves the proposition. □

Remark 1.4 Mizohata's theorem supposes that the coefficients of operator $\mathcal{A}$ are "sufficiently smooth" (see Saut and Scheurer, 1987 for generalizations).

Remark 1.5 Above, we have been assuming that $v \in L^2(\mathcal{O} \times (0, T))$; actually, the above *density result* holds true if v spans a *dense subspace* of $L^2(\mathcal{O} \times (0, T))$. Examples of such dense subspaces are $H^1(\mathcal{O} \times (0, T))$, $H_0^1(\mathcal{O} \times (0, T))$, and $\mathcal{D}(\mathcal{O} \times (0, T))$, the last space being the space of those C^∞ functions which have a compact support in $\mathcal{O} \times (0, T)$; the above list is not limitative. These many possible choices for the control space give a lot of flexibility to the formulation that follows (in Section 1.3).

Remark 1.6 Suppose that one would like to drive the system at time T close to a target function y_T *"containing" some singularities*. To fix ideas (but, here too, there is a lot of flexibility) we suppose that

$$y_T \in H^{-1}(\Omega). \tag{1.20}$$

If (1.20) holds, it may be sensible to admit fairly *general controls*, such as

$$v \in L^2(0, T; H^{-1}(\mathcal{O})), \tag{1.21}$$

or ever more general ones. We shall not pursue along these lines here.

Remark 1.7 Consider the *parabolic problem* (1.1), (1.8), and (1.9) when $v = 0$ in (1.1) and $T = +\infty$. Suppose also that operator $\mathcal{A}$ is as in Section 1.1, Example 1.2. Multiplying both sides of equation (1.1) by y we obtain

$$\frac{1}{2}\frac{\mathrm{d}}{\mathrm{d}t}\left(\|y(t)\|_{0,\Omega}^2\right) + \alpha \int_\Omega |\nabla y(t)|^2 \,\mathrm{d}x \le 0 \quad \text{for } t \in (0, +\infty), \tag{1.22}$$

where $\|\cdot\|_{0,\Omega} = \|\cdot\|_{L^2(\Omega)}$. Since the domain Ω is *bounded*, the following *Poincaré inequality* holds:

$$\lambda_0 \|z\|_{0,\Omega}^2 \le \int_\Omega |\nabla z|^2 \,\mathrm{d}x, \quad \forall z \in H_0^1(\Omega), \tag{1.23}$$

λ_0 (> 0) being the smallest eigenvalue of operator $-\Delta$ over the space $H_0^1(\Omega)$, that is, the smallest of those real numbers λ such that

$$-\Delta w = \lambda w \quad \text{in } \Omega, \quad w \in H_0^1(\Omega) \setminus \{0\}.$$

Combining relations (1.9), (1.22), and (1.23) we easily obtain

$$\|y(t)\|_{0,\Omega} \le e^{-\alpha\lambda_0 t} \|y_0\|_{0,\Omega} \quad \forall t \ge 0. \tag{1.24}$$

Left *uncontrolled*, the distributed system under consideration will go from y_0 to 0, the speed of convergence being controlled by the "size" of the product $\alpha\lambda_0$. Under these circumstances, a question which arises naturally is the following one: since the "natural tendency" of the system, if perturbed from $y = 0$, is to return to the above state at "exponential speed" (see (1.24)), is it possible to find a control v such that the target function $y_T = 0$ can be reached in finite time? If this is possible, the system has the so-called *null-controllability* property. Indeed, it is proved in Lebeau and Robbiano (1995) (see also Lebeau and Zuazua, 1998; Zuazua, 2002) that such property holds for any $T \in (0, +\infty)$ and $y_0 \in L^2(\Omega)$, if $\mathcal{A} = -\sum_{i=1}^{d} \frac{\partial}{\partial x_i} \sum_{j=1}^{d} a_{ij} \frac{\partial}{\partial x_j}$, the a_{ij}s verifying relation (1.3) and being sufficiently smooth (analytical, for example). More precisely, if the above assumptions on the a_{ij} hold, then, $\forall T$, $0 < T < +\infty$, and $y_0 \in L^2(\Omega)$, there exists $v \in L^2(\mathcal{O} \times (0, T))$, such that the corresponding solution of system (1.1), (1.8), and (1.9) verifies $y(T) = 0$. An "interesting" consequence of the result of Lebeau and Robbiano is that it implies the *exact controllability* property of the *backward heat equation*, if we assume that the initial state is 0. Indeed, if one makes the change of variable $T - t \to t$, it follows from the above reference that, $\forall y_T \in L^2(\Omega)$ there exists $v \in L^2(\mathcal{O} \times (0, T))$ such that,

$$\frac{\partial y}{\partial t} + \Delta y = v\chi_{\mathcal{O}\times(0,T)} \quad \text{in } Q, \tag{1.25}$$

$$y(0) = 0, \tag{1.26}$$

$$y = 0 \quad \text{on } \Gamma \times (0, T), \tag{1.27}$$

$$y(T) = y_T. \tag{1.28}$$

The backward heat equation is a well-known *unstable* model, for generic data. The above exact controllability result suggests that *unstable systems may be easier to control than stable ones*, a bold statement indeed (which appears to be true in many instances as we shall see in Chapter 4 when we will discuss the controllability of the *Kuramoto–Shivashinsky* (K.S.) equation, a model for *chaotic* behavior, among other things).

1.3 Formulation of the approximate controllability problem

As we have seen in Section 1.2, we do not restrict the generality by assuming that $y_0 = 0$ (it amounts to replacing y_T by $y_T - Y_0(T)$).

Let B be the unit ball of $L^2(\Omega)$; we want

$$y(T; v) \in y_T + \beta B, \tag{1.29}$$

with $\beta > 0$ and arbitrary small. According to Proposition 1.3 there are controls vs (actually, infinitely such vs) such that (1.29) holds true. Among all these vs, we want to find those which are solutions to the following minimization problem:

$$\inf_{v} \frac{1}{2} \iint_{\mathcal{O}\times(0,T)} v^2 \, dx \, dt, \quad v \in L^2(\mathcal{O} \times (0, T)), \quad y(T; v) \in y_T + \beta B. \tag{1.30}$$

Actually, problem (1.30) *has a unique solution. We want to construct numerical approximation schemes to find it.*

Before we proceed, a few remarks are now in order.

Remark 1.8 All that is stated above is true with

$$\begin{aligned} T > 0 &\quad \text{arbitrary small ,} \\ \mathcal{O} \subset \Omega &\quad \text{arbitrary “small,”} \\ \beta > 0 &\quad \text{also arbitrary small .} \end{aligned}$$

Letting $\beta \to 0$ *will be*, in general, *impossible*. This will be made explicit below.

Remark 1.9 Choices other than (1.30) are possible. The “obvious” candidates are

$$\inf_v \|v\|_{L^1(\mathcal{O}\times(0,T))}, \quad v \in L^1(\mathcal{O} \times (0,T)), \quad y(T;v) \in y_T + \beta B, \tag{1.31}$$

and

$$\inf_v \|v\|_{L^\infty(\mathcal{O}\times(0,T))}, \quad v \in L^\infty(\mathcal{O} \times (0,T)), \quad y(T;v) \in y_T + \beta B. \tag{1.32}$$

Other – more subtle – choices may be of interest. We shall return to this below.

1.4 Dual problem

We are going to apply the *Duality Theory of Convex Analysis* to problem (1.30). We define thus the following functionals and operator:

$$F_1(v) = \frac{1}{2} \iint_{\mathcal{O}\times(0,T)} v^2 \, dx \, dt, \tag{1.33}$$

$$F_2(g) = \begin{cases} 0 & \text{for } g \in y_T + \beta B, \\ +\infty & \text{otherwise} \end{cases} \tag{1.34}$$

(F_2 is a “proper” lower semicontinuous convex functional), and

$$Lv = y(T;v), \tag{1.35}$$

so that

$$L \in \mathcal{L}\left(L^2(\mathcal{O} \times (0,T)), L^2(\Omega)\right). \tag{1.36}$$

Then problem (1.30), where the infimium is taken over all vs satisfying (1.29), is *equivalent* to the following minimization problem:

$$\inf_{v\in L^2(\mathcal{O}\times(0,T))} \left(F_1(v) + F_2(Lv)\right). \tag{1.37}$$

We can apply now the *duality theory* of W. Fenchel and T.R. Rockafellar (cf. Ekeland and Temam, 1974). It gives

$$\inf_{v\in L^2(\mathcal{O}\times(0,T))}\Big(F_1(v)+F_2(Lv)\Big) = -\inf_{g\in L^2(\Omega)}\Big(F_1^*(L^*g)+F_2^*(-g)\Big), \tag{1.38}$$

where F_i^* is the conjugate function of F_i and L^* is the adjoint of operator L. We have then

$$F_1^*(v) = \sup_{\hat{v}\in L^2(\mathcal{O}\times(0,T))}\Big((v,\hat{v})-F_1(\hat{v})\Big) = F_1(v),$$

$$F_2^*(g) = \sup_{\hat{g}\in y_T+\beta B}(\hat{g},g) = (g,y_T)+\beta\,\|g\|,$$

where $\|g\| = \left(\int_\Omega |g|^2\,\mathrm{d}x\right)^{\frac{1}{2}}$ and where $(g,y_T) = \int_\Omega g y_T\,\mathrm{d}x$. Now, we compute L^*: g being given in $L^2(\Omega)$, we define ψ as the solution to system (1.14)–(1.16). Then, one verifies easily (actually, one uses (1.17)) that

$$L^*f = \psi\,\chi_{\mathcal{O}\times(0,T)}, \tag{1.39}$$

with $\chi_{\mathcal{O}\times(0,T)}$ being the characteristic function of $\mathcal{O}\times(0,T)$.

It follows then from (1.38) that

$$\inf_{v\in L^2(\mathcal{O}\times(0,T))}\Big(F_1(v)+F_2(Lv)\Big)$$
$$= -\inf_{g\in L^2(\Omega)}\left(\frac{1}{2}\iint_{\mathcal{O}\times(0,T)}|\psi_g|^2\,\mathrm{d}x\,\mathrm{d}t-(g,y_T)+\beta\,\|g\|\right), \tag{1.40}$$

where ψ_g is the solution to

$$-\frac{\partial\psi_g}{\partial t}+A^*\psi_g = 0 \text{ in } Q,\quad \psi_g(T)=g,\quad \psi_g=0 \text{ on } \Sigma. \tag{1.41}$$

Minimizing the functional in the right-hand side of (1.40), *with the state function verifying* (1.41) *is the dual problem.*

Remark 1.10 The *dual problem* (1.40), (1.41) has a unique solution. Let us denote by f this solution. Then, the solution u to problem (1.30) is given by

$$u = \psi\,\chi_{\mathcal{O}\times(0,T)}, \tag{1.42}$$

where ψ is the solution to problem (1.41) corresponding to f (that is, $\psi = \psi_f$).

Remark 1.11 We want to give now constructive algorithms for solving the dual problem, hence for the solution to the primal problem (using relation (1.42)).

Remark 1.12 As it is classical in this kind of question, relation (1.40) leads to lower and upper bounds, hence to some error estimates.

1.5 Direct solution to the dual problem

The function g being given in $L^2(\Omega)$, let us set

$$[g] = \|\psi_g\|_{L^2(\mathcal{O}\times(0,T))} \,. \tag{1.43}$$

We observe that $[g]$ is a norm on $L^2(\Omega)$. Indeed, if $[g] = 0$, then $\psi_g = 0$ in $\Omega \times (0, T)$, hence (according to the proof of Proposition 1.3) $g = 0$ follows. Let us now introduce a *variational inequality* expressing that the solution f of the dual problem (1.40), (1.41) realizes the minimum in the right-hand side of (1.40). This variational inequality reads as follows:

$$\begin{cases} f \in L^2(\Omega), \quad \forall g \in L^2(\Omega), \\ \displaystyle\iint_{\mathcal{O}\times(0,T)} \psi(\psi_g - \psi)\,dx\,dt - (y_T, g - f) + \beta\,\|g\| - \beta\,\|f\| \geq 0, \end{cases} \tag{1.44}$$

where, in (1.44), the function ψ_g is the solution to problem (1.41). Using relation (1.43), problem (1.44) can be reformulated as

$$\begin{cases} f \in L^2(\Omega), \\ [f, g - f] - (y_T, g - f) + \beta\,\|g\| - \beta\,\|f\| \geq 0, \quad \forall g \in L^2(\Omega). \end{cases} \tag{1.45}$$

Let us introduce now the "adjoint" state function y defined by

$$\frac{\partial y}{\partial t} + Ay = \psi\chi_{\mathcal{O}\times(0,T)} \text{ in } Q, \quad y(0) = 0, \quad y = 0 \text{ on } \Sigma. \tag{1.46}$$

Multiplying the first equation in (1.46) by $\psi_g - \psi$ and integrating over Q gives

$$\iint_{\mathcal{O}\times(0,T)} \psi(\psi_g - \psi)\,dx\,dt = (y(T), g - f). \tag{1.47}$$

Let us define now the *linear operator* Λ by

$$\Lambda g = y_g(T), \tag{1.48}$$

where, g being given, one computes ψ_g by (1.41) and then y_g by

$$\frac{\partial y_g}{\partial t} + Ay_g = \psi_g\chi_{\mathcal{O}\times(0,T)} \text{ in } Q, \quad y_g(0) = 0, \quad y_g = 0 \text{ on } \Sigma. \tag{1.49}$$

It follows then from relations (1.41), (1.44), and (1.46)–(1.49) that the *dual problem* (1.44), (1.45) reads also as follows:

$$\begin{cases} f \in L^2(\Omega), \\ (\Lambda f, g - f) - (y_T, g - f) + \beta\,\|g\| - \beta\,\|f\| \geq 0, \quad \forall g \in L^2(\Omega). \end{cases} \tag{1.50}$$

Remark 1.13 The equivalence between problems (1.44), (1.45), and (1.50) relies on the *following* relation:

$$[g, g'] = (\Lambda g, g'), \quad \forall g, g' \in L^2(\Omega). \tag{1.51}$$

Remark 1.14 The operator Λ satisfies

$$\Lambda \in \mathcal{L}(L^2(\Omega), L^2(\Omega)), \quad \Lambda = \Lambda^*, \quad \Lambda \geq 0. \tag{1.52}$$

It follows from (1.52) that the (unique) solution to problem (1.50) is also the solution to the following *minimization problem*:

$$\inf_{g \in L^2(\Omega)} \left(\frac{1}{2}(\Lambda g, g) - (y_T, g) + \beta \|g\| \right). \tag{1.53}$$

The above results can be summarized by the following.

Theorem 1.15

(i) *We have the identity*

$$\inf_{\substack{v \\ y(T;v) \in y_T + \beta B}} \frac{1}{2} \iint_{\mathcal{O} \times (0,T)} v^2 \, dx \, dt$$
$$= - \inf_{g \in L^2(\Omega)} \left(\frac{1}{2} \iint_{\mathcal{O} \times (0,T)} \psi_g^2 \, dx \, dt - (y_T, g) + \beta \|g\| \right), \tag{1.54}$$

where ψ_g is given by (1.41).

(ii) *The unique solution f of the dual problem is solution of the minimization problem* (1.53), *where the operator Λ is defined by relations* (1.41), (1.48), *and* (1.49).

(iii) *The unique solution u of (the primal) problem* (1.30) *is given by*

$$u = \psi \chi_{\mathcal{O} \times (0,T)}, \tag{1.55}$$

with ψ obtained from f via the solution of

$$-\frac{\partial \psi}{\partial t} + A^* \psi = 0 \text{ in } Q, \quad \psi(T) = f, \quad \psi = 0 \text{ on } \Sigma. \tag{1.56}$$

We have then $\Lambda f = y(T)$, where,

$$\frac{\partial y}{\partial t} + Ay = \psi \chi_{\mathcal{O} \times (0,T)} \text{ in } Q, \quad y(0) = 0, \quad y = 0 \text{ on } \Sigma. \tag{1.57}$$

Applications As a corollary – *that we have to make more precise* – one obtains the general principle of a solution method, namely,

(i) "Guess" the solution f of problem (1.53).
(ii) Solve (1.56) to compute the corresponding value of ψ.

(iii) Use an iterative method to compute the inf in g, using (1.53) or the problem in the right-hand side of (1.54).

This will be the task of Section 1.8. Before that, several remarks are in order.

Remark 1.16 The optimal control u – with respect to the *choice* of

$$\frac{1}{2}\iint_{\mathcal{O}\times(0,T)} v^2\,dx\,dt$$

as the cost function to minimize – is given by (1.55), where ψ is the solution of the *parabolic equation* (1.56). Therefore, ψ is *smooth* (indeed, the smoother the coefficients of operator $\mathcal{A}$, the smoother will be ψ). *In other words, the control u is a smooth function of x and t.*

This observation *excludes* the possibility of finding an optimal control of the "bang-bang" type. Of course, trying to find an optimal control satisfying some kind of bang-bang principle is by no mean compulsory, but knowing in advance that such a property takes place may be helpful. In that direction, the first idea coming to mind is to replace

$$\|v\|_{L^2(\mathcal{O}\times(0,T))} \text{ by } \|v\|_{L^\infty(\mathcal{O}\times(0,T))};$$

this possibility will be investigated in Section 1.7.

Remark 1.17 (Further comments on exact controllability.) *Exact controllability* corresponds to $\beta = 0$ in problem (1.53), or, equivalently, to

$$\inf_g\left(\frac{1}{2}[g]^2 - (y_T, g)\right), \quad g \in L^2(\Omega). \tag{1.58}$$

Let us denote by $\widehat{L^2(\Omega)}$ the completion of $L^2(\Omega)$ for the norm $[\cdot]$. Due to the smoothness properties of the solutions to parabolic equations such as (1.56) and (1.57), the space $\widehat{L^2(\Omega)}$ will contain (except for the case, without practical interest, where $\mathcal{O} = \Omega$) very singular distributions and even distributions of infinite order (outside $\mathcal{O}$), that is, ultra distributions. This implies that problem (1.58) has a unique solution f_0 if and only if $y_T \in (\widehat{L^2(\Omega)})'$, the dual of space $\widehat{L^2(\Omega)}$. We also have the following *convergence result*: let f_β be the unique solution to problem (1.50), (1.53), then $\lim_{\beta\to 0} f_\beta = f_0$ in $L^2(\Omega)$ *if and only if* $y_T \in (\widehat{L^2(\Omega)})'$.

Remark 1.18 Another way of expressing the above results is to observe that Λ is an *isomorphism* from $\widehat{L^2(\Omega)}$ onto its dual space. This is closely related to the *Hilbert Uniqueness Method* (HUM) as discussed in J.L. Lions (1988a,b).

1.6 Penalty arguments

For those problems involving "many constraints" of different nature, *penalty* arguments can be used in many ways. In the present section we are going to "penalize" the constraint

$$y(T; v) \in y_T + \beta B. \tag{1.59}$$

This also can be done in many ways!

One possibility is to introduce a *smooth* functional over $L^2(\Omega)$ which vanishes on the ball $y_T + \beta B$ and is strictly positive outside this ball. Let $\mu(\cdot)$ be such a functional; then one can consider

$$\inf_{v \in L^2(\mathcal{O}\times(0,T))} \left(\frac{1}{2} \iint_{\mathcal{O}\times(0,T)} v^2 \, dx \, dt + k\mu(y(T; v)) \right), \tag{1.60}$$

with $k > 0$ "large." Another possibility is the following: one introduces

$$J_k(v) = \frac{1}{2} \iint_{\mathcal{O}\times(0,T)} v^2 \, dx \, dt + \frac{k}{2} \|y(T; v) - y_T\|^2, \tag{1.61}$$

where $k > 0$ is "large," and where $\|\cdot\|$ denotes the $L^2(\Omega)$-norm. One considers then the following problem:

$$\inf_{v \in L^2(\mathcal{O}\times(0,T))} J_k(v). \tag{1.62}$$

Problem (1.62) *has a unique solution*, denoted by u_k. Let us verify the following result:

There exists k large enough such that the solution u_k *of problem* (1.62) *verifies* $\|y(T; u_k) - y_T\| \le \beta$, that is,

$$y(T; u_k) \in y_T + \beta B. \tag{1.63}$$

Before proving (1.63), let us make the following remarks:

Remark 1.19 It follows from (1.63) that u_k is for k large enough a control such that $y(T; u_k) \in y_T + \beta B$. Of course, it has no reason to coincide with the solution u_β of problem

$$\inf_v \frac{1}{2} \iint_{\mathcal{O}\times(0,T)} v^2 \, dx \, dt, \quad v \in L^2(\mathcal{O} \times (0, T)), \quad y(T; v) \in y_T + \beta B.$$

Remark 1.20 The proof which follows is *not constructive*, therefore, it does not give a "constructive choice" for k. This is a difficulty since β "disappears" in problem (1.62). *We shall make below a constructive proposal for the choice of k.*

Remark 1.21 Of course, k being given, the *optimality system* for problem (1.62) is quite classical. It follows from, for example, J. L. Lions (1968, 1971), that the optimality system associated with problem (1.62) reads as follows:

$$\begin{cases} \dfrac{\partial y}{\partial t} + Ay = \psi \chi_{\mathcal{O}\times(0,T)} \text{ in } Q, \quad y(0) = 0, \quad y = 0 \text{ on } \Sigma, \\ -\dfrac{\partial \psi}{\partial t} + A^*\psi = 0 \text{ in } Q, \quad \psi(T) = k(y_T - y(T)), \quad \psi = 0 \text{ on } \Sigma. \end{cases} \tag{1.64}$$

The optimal control u_k is given by $\psi\chi_{\mathcal{O}\times(0,T)}$ where ψ is obtained from the solution of the *two-point boundary value problem* (1.64) (two points with respect to the time variable, the two points being $t = 0$ and $t = T$). It is worth noticing that if one denotes the function $\psi(T)$ by f, then f is solution of the functional equation

$$\left(k^{-1}\mathbf{I} + \Lambda\right)f = y_T, \tag{1.65}$$

where the operator Λ is still defined by relations (1.41), (1.48), and (1.49) (see Section 1.5).

Proof of (1.63). Given $\varepsilon > 0$, there exists a control w such that

$$\|y(T;w) - y_T\| \leq \varepsilon. \tag{1.66}$$

This follows from the *approximate controllability* result (Proposition 1.3 in Section 1.2) and it is not constructive. We have then

$$J_k(u_k) \leq \frac{1}{2}\iint_{\mathcal{O}\times(0,T)} w^2\,\mathrm{d}x\,\mathrm{d}t + \frac{k\varepsilon^2}{2}, \tag{1.67}$$

so that

$$\|y(T;u_k) - y_T\|^2 \leq \frac{1}{k}\iint_{\mathcal{O}\times(0,T)} w^2\,\mathrm{d}x\,\mathrm{d}t + \varepsilon^2. \tag{1.68}$$

We choose $\varepsilon = \frac{\beta}{\sqrt{2}}$, then w is chosen so that (1.66) holds and we choose k so that

$$\frac{1}{k}\iint_{\mathcal{O}\times(0,T)} w^2\,\mathrm{d}x\,\mathrm{d}t \leq \frac{1}{2}\beta^2.$$

Then, relation (1.68) implies (1.63). □

Remark 1.22 In general (that is, for y_T generically given in $L^2(\Omega)$) the above process does not converge as $k \to +\infty$ (otherwise it would give exact controllability at the limit!).

Remark 1.23 It remains to solve the optimality system (1.64) *if we have a way to choose k*.

This is what we propose now:

It follows from the above sections that the dual problem of (1.30) can be formulated by (1.50), namely,

$$\begin{cases} f^* \in L^2(\Omega), \quad \forall g \in L^2(\Omega), \\ (\Lambda f^*, g - f^*) - (y_T, g - f^*) + \beta\,\|g\| - \beta\,\|f^*\| \geq 0, \end{cases} \tag{1.69}$$

(where the solution has been denoted by f^*, instead of f). Similarly, the dual of the penalized problem (1.62) reads as follows:

$$(k^{-1}\mathbf{I} + \Lambda)f = y_T. \tag{1.70}$$

Multiplying both sides of (1.70) by f we obtain

$$(\Lambda f, f) + k^{-1} \|f\|^2 = (y_T, f). \tag{1.71}$$

Taking now $g = 0$ and $g = 2f^*$ in (1.69) yields

$$(\Lambda f^*, f^*) + \beta \|f^*\| = (y_T, f^*). \tag{1.72}$$

Suppose that $f = f^*$; it follows then from (1.71) and (1.72) that

$$k^{-1} \|f\|^2 = \beta \|f\|,$$

that is, if $f \neq 0$,

$$k = \frac{1}{\beta} \|f\|. \tag{1.73}$$

We propose thus the following rule:

After a few iterations, where k is *a priori* given, we take k variable with n and define it by

$$k_n = \frac{1}{\beta} \|f^n\|. \tag{1.74}$$

Remark 1.24 As shown above, the dual of the penalized problem (1.62) can be written as $(k^{-1}\mathbf{I} + \Lambda)f = y_T$; actually, problem (1.69), the dual of (1.30), enjoys the following alternative formulation:

$$y_T \in \beta \partial j(f^*) + \Lambda f^*; \tag{1.75}$$

in (1.75), $\partial j(\cdot)$ denotes the *subgradient* (see, for example, Ekeland and Temam, 1974 for this notion) of the *convex* functional $j(\cdot)$ defined by

$$j(g) = \|g\|_{L^2(\Omega)}, \quad \forall g \in L^2(\Omega). \tag{1.76}$$

Intuitively, problem (1.70) being linear is easier to solve than (1.75) which is nonlinear, nondifferentiable, and so on. In fact, we shall see in Section 1.8 that if one has a method for solving problem (1.70), it can be used in a very simple way to solve problem (1.75).

1.7 L^∞ cost functions and bang-bang controls

We consider the same "model" problem as before, namely,

$$\frac{\partial y}{\partial t} + Ay = v\chi_{\mathcal{O}\times(0,T)} \text{ in } Q, \quad y(0) = 0, \quad y = 0 \text{ on } \Sigma. \tag{1.77}$$

Given $T > 0$ and $y_T \in L^2(\Omega)$, we consider those controls vs such that

$$y(T) \in y_T + \beta B, \tag{1.78}$$

where, in (1.78), β is a positive number and B is the unit ball of $L^2(\Omega)$. Next, we consider the following *control problem*:

$$\inf_v \|v\|_{L^\infty(\mathcal{O}\times(0,T))}, \tag{1.79}$$

with v subjected to (1.77) and (1.78). A few remarks are in order.

Remark 1.25 This remark is purely technical. The space described by $y(T; v)$ is *dense* in $L^2(\Omega)$ when v spans the space of the C^∞ functions with compact support in $\mathcal{O} \times (0, T)$, so that the infimium in (1.79) is *always* a finite number, *no matter how small* β (> 0) *is*.

Remark 1.26 The choice of the L^∞- norm in (1.79) is less convenient than the choice of the L^2-norm, but it is not an unreasonable choice. It leads to new difficulties, mostly due to the *nondifferentiability* of the L^∞-norm (and of *any* power of it). Below, we discuss what to do in order to handle this type of cost function, which leads to *bang-bang* type results.

Remark 1.27 Of course, instead of (1.79), one can consider, more generally,

$$\inf_v \|v\|_{L^s(\mathcal{O}\times(0,T))}, \tag{1.80}$$

with v still subjected to (1.77), (1.78), and s chosen arbitrarily in $[1, +\infty]$, that is,

$$1 \le s \le +\infty. \tag{1.81}$$

Indeed, if $s \in (1, +\infty)$ it is more convenient to use $v \to s^{-1} \|v\|^s_{L^s(\mathcal{O}\times(0,T))}$ as cost function, since it has better *differentiability properties* and does not change the solution of problem (1.80).

Let us consider the case $s = 1$; then for *any* v in $L^1(\mathcal{O} \times (0, T))$ the function $y(T; v)$ belongs to $L^2(\Omega)$ if and only if $d \le 2$ (see, for example, Ladyzenskaya, Solonnikov, and Ural'ceva, 1968 for this result). Actually, this does not modify the statement of problem (1.80) (with $s = 1$), since if $d > 2$, we can always restrict ourselves to those controls v in $L^1(\mathcal{O} \times (0, T))$, such that $y(T; v) \in L^2(\Omega)$.

Remark 1.28 There is still another variant that we shall not consider in this book, namely to replace in (1.78) the unit ball B of $L^2(\Omega)$ by the unit ball of $L^r(\Omega)$. We refer to Fabre, Puel, and Zuazua (1993) for a discussion of this case.

Remark 1.29 For technical reasons (whose explanation will appear later on) we are going to consider the problem in the following form:

$$\inf_v \frac{1}{2} \|v\|^2_{L^\infty(\mathcal{O}\times(0,T))}, \tag{1.82}$$

or

$$\inf_v \frac{1}{2} \|v\|^2_{L^s(\mathcal{O}\times(0,T))}, \tag{1.83}$$

with v subjected to (1.77) and (1.78).

Synopsis In the following, we propose an *approximation method* for problem (1.82), which leads to: (i) numerical methods and (ii) connections with results from Fabre, Puel, and Zuazua (1993). The results in the above publication have been found by a *duality* approach, which leads – among other things – to some very interesting formulae; we will present these formulae.

Approximation by penalty and regularization (I) We begin by considering the following problem:

$$\inf_{v} J_k^s(v), \tag{1.84}$$

where, in (1.84), the cost function $J_k^s(\cdot)$ is defined by

$$J_k^s(v) = \frac{1}{2}\, \|v\|_{L^s}^2 + \frac{1}{2} k\, \|y(T; v) - y_T\|_{L^2(\Omega)}^2; \tag{1.85}$$

in (1.85), L^s stands for $L^s(\mathcal{O} \times (0, T))$ and $y(\cdot; v)$ is the solution of (1.77). The idea here is to have $k\ (> 0)$ *large* to "force" (*penalty*) the final condition $y(T; v) = y_T$, and to have s large, as an approximation of $s = +\infty$ (*regularization*). Problem (1.84) has a *unique* solution and we are going to write the corresponding *optimality conditions*. One can easily verify that,

$$\begin{cases} \forall v, w \in L^s, \\ \dfrac{\mathrm{d}}{\mathrm{d}\lambda} \left(\dfrac{1}{2}\, \|v + \lambda w\|_{L^s}^2 \right) \Bigg|_{\lambda=0} = \|v\|_{L^s}^{2-s} \displaystyle\iint_{\mathcal{O}\times(0,T)} v\, |v|^{s-2}\, w \,\mathrm{d}x\,\mathrm{d}t. \end{cases} \tag{1.86}$$

The *quadratic* part of $J_k^s(\cdot)$ gives no problem and one verifies easily that if $\nabla J_k^s(\cdot)$ denotes the *derivative* (gradient) of $J_k^s(\cdot)$, then

$$\nabla J_k^s(v) \in L^{s'}, \quad \forall v \in L^s, \text{ with } s' = \frac{s}{s-1}, \tag{1.87}$$

and, from (1.86),

$$\begin{aligned} \iint_{\mathcal{O}\times(0,T)} \nabla J_k^s(v) v w \,\mathrm{d}x\,\mathrm{d}t = \|v\|_{L^s}^{2-s} &\iint_{\mathcal{O}\times(0,T)} v\, |v|^{s-2}\, w \,\mathrm{d}x\,\mathrm{d}t \\ &- \iint_{\mathcal{O}\times(0,T)} p w \,\mathrm{d}x\,\mathrm{d}t, \quad \forall v, w \in L^s, \end{aligned} \tag{1.88}$$

where, in (1.88), p is the solution of the *adjoint state equation*

$$-\frac{\partial p}{\partial t} + \mathcal{A}^* p = 0 \text{ in } Q, \quad p(T) = k(y_T - y(T; v)), \quad p = 0 \text{ on } \Sigma, \tag{1.89}$$

with, in (1.89), $y(T; v)$ obtained from v by the solution of problem (1.77); in (1.87), the exponent s' is the conjugate of s since $\frac{1}{s} + \frac{1}{s'} = 1$.

Let us denote by u the solution of problem (1.84); this solution verifies $\nabla J_k^s(u) = 0$, which implies (from (1.88)) that

$$\|u\|_{L^s}^{2-s}\, u\, |u|^{s-2} = p\chi_{\mathcal{O}\times(0,T)}, \tag{1.90}$$

where, in (1.90), we still denote by p the particular solution of the adjoint system (1.89) corresponding to $v = u$. Relation (1.90) is equivalent to

$$u = \|p\|_{L^{s'}}^{2-s'} p\,|p|^{s'-2} \chi_{\mathcal{O}\times(0,T)}. \tag{1.91}$$

We have therefore obtained the following *optimality system* for problem (1.84):

$$\begin{cases} \dfrac{\partial y}{\partial t} + \mathcal{A}y = \|p\|_{L^{s'}}^{2-s'} p\,|p|^{s'-2} \chi_{\mathcal{O}\times(0,T)} \text{ in } Q, \\ \quad y(0) = 0, \quad y = 0 \text{ on } \Sigma, \\ -\dfrac{\partial p}{\partial t} + \mathcal{A}^* p = 0 \text{ in } Q, \\ \quad p(T) = k(y_T - y(T)), \quad p = 0 \text{ on } \Sigma. \end{cases} \tag{1.92}$$

The above result holds for any fixed arbitrarily large s; the same observation applies to k. Let us summarize:

The optimality system (1.92) *has a unique solution and the optimal control u is given by relation* (1.91).

Approximation by penalty and regularization (II) Suppose now that $s \to +\infty$, that is, $s' \to 1$ in (1.91) and (1.92), the parameter k being fixed. We make the assumption (actually this is a *conjecture*; see Fabre, Puel, and Zuazua, 1993 for a discussion of this issue) that

$$p \neq 0 \quad \text{a.e. in } Q, \tag{1.93}$$

(unless $p \equiv 0$). Then, the limit of (1.92) is given by

$$\begin{cases} \dfrac{\partial y}{\partial t} + \mathcal{A}y = \|p\|_{L^1} \operatorname{sign}(p) \chi_{\mathcal{O}\times(0,T)} \text{ in } Q, \\ \quad y(0) = 0, \quad y = 0 \text{ on } \Sigma, \\ -\dfrac{\partial p}{\partial t} + \mathcal{A}^* p = 0 \text{ in } Q, \\ \quad p(T) = k(y_T - y(T)), \quad p = 0 \text{ on } \Sigma. \end{cases} \tag{1.94}$$

Remark 1.30 We observe that (1.94) has been obtained by taking the limit in (1.92) as $s \to +\infty$. This convergence result is not difficult to prove if we suppose that (1.93) holds; see the above reference for further details and results.

Remark 1.31 It follows from (1.94) (or (1.91)) that the *optimal control u* is given by

$$u = \|p\|_{L^1} \operatorname{sign}(p) \chi_{\mathcal{O}\times(0,T)}, \tag{1.95}$$

which is definitely a *bang-bang* result.

Remark 1.32 What has been discussed above is simple, thanks to the choice of (1.82) as control problem, which leads, in turn, to the *regularized* and *regularized–penalized* problems (1.83) and (1.84). This approach and the corresponding results are closely related to those in Fabre, Puel, and Zuazua (1993); in fact, these authors start from the *dual formulation* that is discussed below.

Dual formation (I) We can use *duality* as in the L^2 case. We obtain then the following duality relation:

$$\inf_{v\in L^s}\left(\frac{1}{2}\|v\|_{L^s}^2+\frac{1}{2}k\,\|y(T;v)-y_T\|_{L^2(\Omega)}^2\right)$$
$$=-\inf_g\left(\frac{1}{2}\,\|\psi_g\|_{L^{s'}}^2+\frac{1}{2k}\,\|g\|_{L^2(\Omega)}^2-(y_T,g)_{L^2(\Omega)}\right),\quad(1.96)$$

where, in (1.96), ψ_g is obtained from g via the solution of

$$-\frac{\partial\psi_g}{\partial t}+\mathcal{A}^*\psi_g=0\text{ in }Q,\quad\psi_g(T)=g,\quad\psi_g=0\text{ on }\Sigma.\qquad(1.97)$$

As already mentioned (see Remark 1.32), in Fabre, Puel, and Zuazua (1993) the authors start from the formulation (1.96), (1.97), *directly* with $s'=1$; this has to be understood in the following manner: one considers

$$\inf_{g\in L^2(\Omega)}\left(\frac{1}{2}\,\|\psi_g\|_{L^1}^2+\frac{1}{2k}\,\|g\|_{L^2(\Omega)}^2-(y_T,g)_{L^2(\Omega)}\right),\qquad(1.98)$$

as the *primal problem* (with ψ_g still defined by (1.97)); then, the *dual problem* is

$$\inf_{v\in L^\infty}\left(\frac{1}{2}\,\|v\|_{L^\infty}^2+\frac{1}{2}k\,\|y(T;v)-y_T\|_{L^(\Omega)}^2\right).\qquad(1.99)$$

Dual formulation (II) Our goal, now, is to achieve (1.78), namely,

$$y(T;v)\in y_T+\beta B.$$

Using the *penalized* formulation one obtains $y(T;v)$ "close" to y_T. In order to have $y(T;v)$ satisfying (1.78) one has to choose k in a suitable fashion; this can be done as follows:

Observe first that, from (1.96), the *dual* problem of (1.84) is given by

$$\inf_g\left(\frac{1}{2}\,\|\psi_g\|_{L^{s'}}^2+\frac{1}{2k}\,\|g\|_{L^2(\Omega)}^2-(y_T,g)_{L^2(\Omega)}\right).\qquad(1.100)$$

Next, let us denote by f the solution of problem (1.100); it satisfies (with obvious notation) the following *variational equation* in $L^2(\Omega)$:

$$\begin{cases}f\in L^2(\Omega),\ \forall g\in L^2(\Omega)\text{ we have}\\ \|\psi\|_{L^{s'}}^{2-s'}\displaystyle\iint_{\mathcal{O}\times(0,T)}|\psi|^{s'-2}\,\psi\psi_g\,dx\,dt+\frac{1}{k}\int_\Omega fg\,dx=\int_\Omega y_Tg\,dx,\end{cases}\qquad(1.101)$$

which implies in turn that

$$\|\psi\|^2_{L^{s'}} + \frac{1}{k}\,\|f\|^2_{L^2(\Omega)} = (y_T, f)_{L^2(\Omega)}. \tag{1.102}$$

Consider now the control problem

$$\inf_v \frac{1}{2}\,\|v\|_{L^s}, \quad v \text{ verifies (1.77) and (1.78);} \tag{1.103}$$

its *dual problem* is given by

$$\inf_{g\in L^2(\Omega)} \left(\frac{1}{2}\,\|\psi_g\|^2_{L^{s'}} + \beta\,\|g\|_{L^2(\Omega)} - (y_T, g)_{L^2(\Omega)} \right). \tag{1.104}$$

Denote by f^* the solution to problem (1.104); f^* satisfies the following *variational inequality* in $L^2(\Omega)$:

$$\begin{cases} f^* \in L^2(\Omega); \quad \forall g \in L^2(\Omega) \text{ we have (with obvious notation)} \\ \|\psi^*\|^{2-s'}_{L^{s'}} \displaystyle\iint_{\mathcal{O}\times(0,T)} |\psi^*|^{s'-2}\,\psi^*(\psi_g - \psi^*)\,\mathrm{d}x\,\mathrm{d}t \\ \qquad + \beta\,\|g\|_{L^2(\Omega)} - \beta\,\|f^*\|_{L^2(\Omega)} \geq \displaystyle\int_\Omega y_T(g - f^*)\,\mathrm{d}x. \end{cases} \tag{1.105}$$

Taking successively $g = 0$ and $g = 2f^*$ in (1.105), we obtain

$$\|\psi^*\|^2_{L^{s'}} + \beta\,\|f^*\|_{L^2(\Omega)} = (y_T, f^*)_{L^2(\Omega)}. \tag{1.106}$$

The positive number β being given, we look for k such that $f = f^*$, which implies in turn that $\psi = \psi^*$ and, therefore, that the *primal problems* (1.84) and (1.103) have the same solution, namely, $u\ (= \|\psi\|^{2-s'}_{L^{s'}}\ |\psi|^{s'-2}\,\psi\chi_{\mathcal{O}\times(0,T)})$. Suppose that $f = f^*$, it follows then from (1.102) and (1.106) that

$$\frac{1}{k}\,\|f\|^2_{L^2(\Omega)} = \beta\,\|f\|_{L^2(\Omega)}.$$

If $f \neq 0$ we thus have

$$k = \frac{1}{\beta}\,\|f\|_{L^2(\Omega)}. \tag{1.107}$$

Relation (1.107) suggests the following approach for solving problem (1.104) using solution methods for problem (1.100):

Suppose that we have an *iterative* procedure producing $f^1, f^2, \ldots, f^n, \ldots$; we shall use a constant parameter k for several iterations and then a *variable* one defined by

$$k^n = \frac{1}{\beta}\,\|f^n\|_{L^2(\Omega)}. \tag{1.108}$$

We shall conclude this section with the following remark.

Remark 1.33 A control problem closely related to those discussed above is the one defined by

$$\inf_{v \in C_f} \frac{1}{2} \|y(T;v) - y_T\|^2_{L^2(\Omega)}, \tag{1.109}$$

where, in (1.109), C_f is the *closed convex set* of $L^\infty(\mathcal{O} \times (0,T))$ defined by

$$C_f = \{ v \mid v \in L^\infty(\mathcal{O} \times (0,T)), |v(x,t)| \leq C \text{ a.e. in } \mathcal{O} \times (0,T) \}. \tag{1.110}$$

Actually, problem (1.109) is fairly easy to solve if we have solution methods for problem (1.84) with $s = 2$; such methods will be discussed in the following section, together with applications to the solution of problems such as (1.109).

1.8 Numerical methods

1.8.1 Generalities. Synopsis

In this section, we shall address the *numerical solution of the approximate controllability* problems discussed in the preceding sections (the notation of which we shall keep); we shall start our discussion with the solution of the following two fundamental control problems:

First control problem This problem is defined by

$$\inf_{v \in \mathcal{U}_f} \frac{1}{2} \iint_{\mathcal{O}\times(0,T)} v^2 \, dx \, dt, \tag{1.111}$$

with the *control space* $\mathcal{U}_f$ defined by

$$\mathcal{U}_f = \{ v \mid v \in L^2(\mathcal{O} \times (0,T)), \ y(T) \in y_T + \beta B \}, \tag{1.112}$$

where, in (1.112), the *target function* y_T is given in $L^2(\Omega)$, B is the unit ball of $L^2(\Omega)$, β is a positive number and where the state function y is the solution to the following *parabolic problem*:

$$\frac{\partial y}{\partial t} + \mathcal{A} y = v \chi_{\mathcal{O}\times(0,T)} \quad \text{in } Q, \tag{1.113}$$

$$y(0) = y_0 \ (\in L^2(\Omega)), \tag{1.114}$$

$$y = 0 \quad \text{on } \Sigma. \tag{1.115}$$

The control problem (1.111) *has a unique solution.*

Second control problem It is the variant of problem (1.111) defined by

$$\inf_{v \in L^2(\mathcal{O}\times(0,T))} \left(\frac{1}{2} \iint_{\mathcal{O}\times(0,T)} v^2 \, dx \, dt + \frac{k}{2} \|y(T) - y_T\|^2_{L^2(\Omega)} \right), \tag{1.116}$$

where, in (1.116), k is a positive parameter and the function y is still defined by (1.113)–(1.115).

The control problem (1.116) *has a unique solution.*

The solution of the control problems (1.111) and (1.116) can be achieved by methods acting *directly* on the control v; these methods have the advantage of being easy to generalize (in principle) to control problems for *nonlinear* state equations as shown in some of the following sections. In the particular case of problems (1.111) and (1.116) where the state equation (namely, (1.113)–(1.115)) is *linear* and the cost functions *quadratic*, instead of solving (1.111) and (1.116) directly, we can solve equivalent problems obtained by applying, for example, *Convex Duality Theory*, as already shown in Sections 1.5 and 1.6. In fact, these dual problems can be viewed as *identification problems* for the *final data* of a *backward* (in time) *adjoint equation*, in the spirit of the *Reverse Hilbert Uniqueness Method* (RHUM) discussed in J.L. Lions (1988b); from our point of view, these dual problems are better suited to numerical calculations than the original ones (for a discussion concerning the exact and approximate *boundary controllability* of the *heat equation*, which includes numerical methods, see, for example, Glowinski, 1992b; Carthel, Glowinski, and Lions, 1994).

It follows from Section 1.5 (respectively, Section 1.6) that a dual problem to (1.111) (respectively, (1.116)) is defined by the following *variational inequality*:

$$\begin{cases} f \in L^2(\Omega); \quad \forall g \in L^2(\Omega) \text{ we have} \\ (\Lambda f, g - f)_{L^2(\Omega)} + \beta \, \|g\|_{L^2(\Omega)} - \beta \, \|f\|_{L^2(\Omega)} \\ \qquad \geq (y_T - Y_0(T), g - f)_{L^2(\Omega)}, \end{cases} \tag{1.117}$$

(respectively, by the following *linear* equation:

$$\left(k^{-1}\mathbf{I} + \Lambda\right) f = y_T - Y_0(T)), \tag{1.118}$$

where, in (1.117) and (1.118), the operator Λ is the one defined in Section 1.5 and where the function Y_0 is defined by

$$\frac{\partial Y_0}{\partial t} + \mathcal{A} Y_0 = 0 \quad \text{in } Q, \tag{1.119}$$

$$Y_0(0) = y_0, \tag{1.120}$$

$$Y_0 = 0 \quad \text{on } \Sigma. \tag{1.121}$$

In the following subsections, we shall discuss the numerical solution of problem (1.118) by methods combining *conjugate gradient algorithms* with *finite difference* and *finite elements' discretizations*. We shall then apply the resulting methodology to the solution of the *nonlinear* problem (1.117).

1.8.2 Conjugate gradient solution of problem (1.118)

From now on we shall denote by $(\cdot,\cdot)$ and $\|\cdot\|$ the canonical $L^2(\Omega)$-scalar product and $L^2(\Omega)$-norm, respectively. The various approximations of problem (1.118) can be solved by iterative methods closely related to the algorithm discussed in this subsection. Writing problem (1.118) in *variational form* we obtain

$$\begin{cases} f \in L^2(\Omega), \\ k^{-1}(f,g) + (\Lambda f, g) = (y_T - Y_0(T), g), \quad \forall g \in L^2(\Omega). \end{cases} \tag{1.122}$$

From the *symmetry*, *continuity*, and *positive definiteness* of the bilinear functional $\{g,g'\} \to (\Lambda g, g')$, the variational problem (1.122) is a particular case of the following general problem

$$\begin{cases} u \in V, \\ a(u,v) = L(v), \quad \forall v \in V, \end{cases} \tag{1.123}$$

where:

(i) V is a real *Hilbert space* for the scalar product $(\cdot,\cdot)$ and the corresponding norm $\|\cdot\|$.
(ii) $a : V \times V \to \mathbb{R}$ is *bilinear*, *continuous*, *symmetric*, and V*-elliptic* (that is, there exists $\alpha > 0$ such that

$$a(v,v) \geq \alpha \|v\|^2, \quad \forall v \in V).$$

(iii) $L : V \to \mathbb{R}$ is *linear* and *continuous*.

If properties (i)–(iii) hold, then problem (1.123) has a *unique* solution (for this result that goes back to Hilbert, see, for example, J.L. Lions, 1968; Ekeland and Temam, 1974; Glowinski, 1984).

Problem (1.123) can be solved by the following *conjugate gradient algorithm*:

Algorithm 1.34 *(Conjugate gradient solution of problem (1.123))*

Step 0: Initialization

$$u^0 \textit{ is given in } V; \tag{1.124}$$

solve

$$g^0 \in V, \quad (g^0, v) = a(u^0, v) - L(v), \quad \forall v \in V. \tag{1.125}$$

If $\|g^0\| / \|u^0\| \leq \varepsilon$ *take* $u = u^0$*; otherwise, set*

$$w^0 = g^0. \tag{1.126}$$

For $n \geq 0$, u^n, g^n, *and* w^n *being known with* g^n *and* w^n *both different from* 0, *compute* u^{n+1}, g^{n+1}, *and, if necessary,* w^{n+1} *as follows:*

Step 1: Steepest descent

Compute

$$\rho_n = \|g^n\|^2 / a(w^n, w^n), \tag{1.127}$$

and take

$$u^{n+1} = u^n - \rho_n w^n. \tag{1.128}$$

Solve

$$g^{n+1} \in V, \quad (g^{n+1}, v) = (g^n, v) - \rho_n a(w^n, v), \quad \forall v \in V. \tag{1.129}$$

Step 2: Convergence testing and construction of the new descent direction
If $\|g^{n+1}\| / \|g^0\| \leq \varepsilon$ *take* $u = u^{n+1}$*; otherwise compute*

$$\gamma_n = \left\|g^{n+1}\right\|^2 \Big/ \|g^n\|^2, \tag{1.130}$$

$$w^{n+1} = g^{n+1} + \gamma_n w^n. \tag{1.131}$$

Do $n = n + 1$ *and return to* (1.127).

Despite its apparent simplicity Algorithm 1.34 is one of the most powerful tools of Scientific Computing; it is currently used to solve very complicated problems from Science and Engineering which may involve many millions of unknowns. Large-scale applications of the above algorithm will be found in several parts of this book (application to the solution of problems from Fluid Dynamics can be found in, for example, Glowinski, 2003). The popularity of the conjugate gradient methodology is clearly related to its *good convergence properties*. Concerning these *convergence properties* it is shown in the above publication that:

(i) Suppose that $\varepsilon = 0$ in (1.124)–(1.131); then

$$\lim_{n \to +\infty} \left\|u^n - u\right\| = 0, \quad \forall u_0 \in V, \tag{1.132}$$

u being the solution of the variational problem (1.123).

(ii) Suppose that $\varepsilon = 0$ in (1.124)–(1.131), that the space V is *finite dimensional* and that *round-off errors* do not take place; Algorithm 1.34 enjoys then the *finite termination property*, namely, there exists $n_0 \leq N$ such that $u_{n_0} = u$ (with N the dimension of V).

Actually, we can "quantify" the convergence result (1.132) since it follows from, for example, Daniel (1970) that

$$\left\|u^n - u\right\|_a \leq 2 \left\|u^0 - u\right\|_a \left(\frac{\sqrt{\nu_a} - 1}{\sqrt{\nu_a} + 1}\right)^n, \tag{1.133}$$

where, in (1.133), the norm $\|\cdot\|_a$ is defined by $\|v\|_a = \sqrt{a(v, v)}, \forall v \in V$, and where ν_a denotes the *condition number* of the bilinear functional $a(\cdot, \cdot)$; this condition number is defined by $\nu_a = \|A\| \left\|A^{-1}\right\|$, where A is the unique operator in $\mathcal{L}(V, V)$ defined by

$$a(v, w) = (Av, w), \quad \forall v, w \in V.$$

For more information on the convergence and implementation of Algorithm 1.34 (including the choice of ε) see, for example, Glowinski (2003, Chapter III) and the references therein.

Application to the solution of problem (1.118) Before applying Algorithm 1.34 to the solution of problem (1.118), let us recall the definition of operator Λ; it follows from Section 1.5 that operator Λ is defined by

$$\Lambda g = \phi(T), \tag{1.134}$$

where the function ϕ is obtained from g as follows:

Solve the *backward* equation

$$-\frac{\partial \psi}{\partial t} + \mathcal{A}^* \psi = 0 \text{ in } Q, \quad \psi = 0 \text{ on } \Sigma, \quad \psi(T) = g, \tag{1.135}$$

and then the *forward* equation

$$\frac{\partial \phi}{\partial t} + \mathcal{A}\phi = \psi \chi_{\mathcal{O}\times(0,T)} \text{ in } Q, \quad \phi = 0 \text{ on } \Sigma, \quad \phi(0) = 0. \tag{1.136}$$

Applying now Algorithm 1.34 to the solution of problem (1.118) we obtain the following iterative method:

Algorithm 1.35 *(Conjugate gradient solution of problem (1.118))*

$$f^0 \textit{ is given in } L^2(\Omega); \tag{1.137}$$

solve first

$$-\frac{\partial p^0}{\partial t} + \mathcal{A}^* p^0 = 0 \textit{ in } Q, \quad p^0 = 0 \textit{ on } \Sigma, \quad p^0(T) = f^0, \tag{1.138}$$

and set

$$u^0 = p^0 \chi_{\mathcal{O}\times(0,T)}. \tag{1.139}$$

Solve now

$$\frac{\partial y^0}{\partial t} + \mathcal{A} y^0 = u^0 \textit{ in } Q, \quad y^0 = 0 \textit{ on } \Sigma, \quad y^0(0) = y_0, \tag{1.140}$$

and compute

$$g^0 = k^{-1} f^0 + y^0(T) - y_T. \tag{1.141}$$

If $\|g^0\|_{L^2(\Omega)} \big/ \|f^0\|_{L^2(\Omega)} \leq \varepsilon$ *take* $f = f^0$ *and* $u = u^0$*; otherwise, set*

$$w^0 = g^0. \tag{1.142}$$

Then, for $n \geq 0$*, assuming that* f^n*,* g^n*, and* w^n *are known, compute* f^{n+1}*,* g^{n+1}*, and* w^{n+1} *as follows:*

Solve

$$-\frac{\partial \bar{p}^n}{\partial t} + \mathcal{A}^* \bar{p}^n = 0 \textit{ in } Q, \quad \bar{p}^n = 0 \textit{ on } \Sigma, \quad \bar{p}^n(T) = w^n, \tag{1.143}$$

and set

$$\bar{u}^n = \bar{p}^n \chi_{\mathcal{O}\times(0,T)}. \tag{1.144}$$

Solve now

$$\frac{\partial \bar{y}^n}{\partial t} + \mathcal{A}\bar{y}^n = \bar{u}^n \text{ in } Q, \quad \bar{y}^n = 0 \text{ on } \Sigma, \quad \bar{y}^n(0) = 0, \tag{1.145}$$

and compute

$$\bar{g}^n = k^{-1}w^n + \bar{y}^n(T), \tag{1.146}$$

$$\rho_n = \|g^n\|^2_{L^2(\Omega)} \Big/ \int_\Omega \bar{g}^n w^n \, dx, \tag{1.147}$$

and then

$$f^{n+1} = f^n - \rho_n w^n, \tag{1.148}$$

$$g^{n+1} = g^n - \rho_n \bar{g}^n. \tag{1.149}$$

If $\|g^{n+1}\|_{L^2(\Omega)} \big/ \|g^0\|_{L^2(\Omega)} \leq \varepsilon$, *take* $f = f^{n+1}$, *and solve* (1.135) *with* $g = f$ *to obtain* $u = \psi\chi_{\mathcal{O}\times(0,T)}$ *as the solution of problem* (1.116)*; if the above stopping test is not satisfied, compute*

$$\gamma_n = \left\|g^{n+1}\right\|^2_{L^(\Omega)} \Big/ \|g^n\|^2_{L^2(\Omega)}, \tag{1.150}$$

and then

$$w^{n+1} = g^{n+1} + \gamma_n w^n. \tag{1.151}$$

Do $n = n + 1$ *and go to* (1.143).

Remark 1.36 It is fairly easy to show that

$$\left\|k^{-1}\mathbf{I} + \Lambda\right\| = k^{-1} + \|\Lambda\|, \qquad \left\|\left(k^{-1}\mathbf{I} + \Lambda\right)^{-1}\right\| = k,$$

implying that the *condition number* of the bilinear functional in the left-hand side of (1.122) is equal to $\|\Lambda\| \, k + 1$. It follows from this result, and from (1.133), that one can expect that, *generically* speaking, the number of iterations sufficient to achieve convergence will vary like $\sqrt{k} \ln\left(\frac{k^{3/4}}{\sqrt{\varepsilon}}\right)$ for *large* values of k.

1.8.3 Time discretization of problem (1.118)

The crucial point here is to approximate properly the operator Λ defined by (1.134)–(1.136) in Section 1.8.2. Assuming that T is *bounded* and that the operator $\mathcal{A}$ is independent of t, we introduce a *time discretization step* defined by $\Delta t = T/N$, where N is a *positive integer*. Using an *implicit Euler* time discretization scheme, we approximate (1.135) by

$$\psi^{N+1} = g, \quad g \in L^2(\Omega); \tag{1.152a}$$

then, assuming that ψ^{n+1} is known, solve the following Dirichlet problem for $n = N$, $N-1, \dots, 1$,

$$-\frac{\psi^{n+1} - \psi^n}{\Delta t} + \mathcal{A}^* \psi^n = 0 \text{ in } \Omega, \quad \psi^n = 0 \text{ on } \Gamma, \tag{1.152b}$$

where $\psi^n \approx \psi(n\Delta t)$ $(\psi(n\Delta t) : x \to \psi(x, n\Delta t))$. Next, using similar notation, we approximate (1.136) *by*

$$\phi^0 = 0; \tag{1.153a}$$

then, assuming that ϕ^{n-1} is known, we solve the following Dirichlet problem for $n = 1, \dots, N$,

$$\frac{\phi^n - \phi^{n-1}}{\Delta t} + \mathcal{A}\phi^n = \psi^n \chi_{\mathcal{O}} \text{ in } \Omega, \quad \phi^n = 0 \text{ on } \Gamma. \tag{1.153b}$$

Finally, we approximate Λ by $\Lambda^{\Delta t}$ defined by

$$\Lambda^{\Delta t} g = \phi^N. \tag{1.154}$$

From the *ellipticity properties* of operators $\mathcal{A}$ and $\mathcal{A}^*$ (see Section 1.1), the *Dirichlet* problems (1.152b) and (1.153b) have a unique solution; we have, furthermore, the following.

Theorem 1.37 *Operator $\Lambda^{\Delta t}$ is symmetric and positive semidefinite from $L^2(\Omega)$ into $L^2(\Omega)$.*

Proof. Consider a pair $\{g, \tilde{g}\} \in L^2(\Omega) \times L^2(\Omega)$. We have then (with obvious notation)

$$(\Lambda^{\Delta t} g, \tilde{g})_{L^2(\Omega)} = \int_\Omega \phi^N \tilde{\psi}^{N+1} \, dx. \tag{1.155}$$

We also have, since $\phi^0 = 0$,

$$\Delta t \sum_{n=1}^{N} \left(\phi^n \frac{\tilde{\psi}^{n+1} - \tilde{\psi}^n}{\Delta t} + \tilde{\psi}^n \frac{\phi^n - \phi^{n-1}}{\Delta t} \right) = \phi^N \tilde{\psi}^{N+1}. \tag{1.156}$$

Integrating (1.156) over Ω and taking (1.152b) and (1.153b) into account we obtain from (1.155)

$$\begin{aligned}
(\Lambda^{\Delta t} g, \tilde{g})_{L^2(\Omega)} &= \int_\Omega \phi^N \tilde{\psi}^{N+1} \, dx \\
&= \Delta t \sum_{n=1}^{N} \int_\Omega \left(\phi^n \mathcal{A}^* \tilde{\psi}^n - \tilde{\psi}^n \mathcal{A} \phi^n \right) dx + \Delta t \sum_{n=1}^{N} \int_{\mathcal{O}} \psi^n \tilde{\psi}^n \, dx \\
&= \Delta t \sum_{n=1}^{N} \int_{\mathcal{O}} \psi^n \tilde{\psi}^n \, dx, \quad \forall g, \tilde{g} \in L^2(\Omega),
\end{aligned} \tag{1.157}$$

which completes the proof of the theorem. □

Next, we compute the discrete analogue of Y_0 via:

$$Y_0^0 = y_0, \tag{1.158a}$$

and for $n = 1, \ldots, N$, assuming that Y_0^{n-1} is known, solve the following (well-posed) elliptic problem:

$$\frac{Y_0^n - Y_0^{n-1}}{\Delta t} + \mathcal{A}Y_0^n = 0 \text{ in } \Omega, \quad Y_0^n = 0 \text{ on } \Gamma. \tag{1.158b}$$

Finally, we approximate problem (1.118) by

$$\begin{cases} f^{\Delta t} \in L^2(\Omega), \\ \left(k^{-1}f^{\Delta t} + \Lambda^{\Delta t} f^{\Delta t}, g\right)_{L^2(\Omega)} = \left(y_T - Y_0^N, g\right)_{L^2(\Omega)}, \quad \forall g \in L^2(\Omega). \end{cases} \tag{1.159}$$

Problem (1.159) can be solved by a *time-discrete analogue* of Algorithm 1.35.

Remark 1.38 The *Euler's schemes* which have been used to time-discretize problem (1.118) are *first-order accurate*, only; for some applications this may require very small time steps Δt in order to obtain an acceptable level of accuracy. A simple way to improve this situation is to use second-order schemes like those described in Section 1.8.5 (variants of these schemes have been successfully used to solve *boundary controllability* problems for the *heat equation* in Carthel, Glowinski, and Lions, 1994; see also Chapter 2).

1.8.4 Full discretization of problem (1.118)

We suppose from now on – and for simplicity – that Ω and $\mathcal{O}$ are *polygonal* domains of $\mathbb{R}^2$ (if Ω and/or $\mathcal{O}$ are nonpolygonal we shall approximate them by polygonal domains). We introduce then a first *finite element triangulation* $\mathcal{T}_h$ of Ω (h: largest length of the edges of the triangles of $\mathcal{T}_h$) as in, for example, Dean, Glowinski, and Li (1989), Glowinski, Li, and Lions (1990) and Glowinski (1992a); we suppose that both $\bar{\Omega}$ and $\bar{\mathcal{O}}$ are unions of triangles of $\mathcal{T}_h$. Next, we approximate $H^1(\Omega)$, $L^2(\Omega)$, and $H_0^1(\Omega)$ by the following *finite-dimensional* spaces (with P_1 the space of the polynomials in two variables of degree ≤ 1)

$$V_h = \{ z_h \,|\, z_h \in C^0(\bar{\Omega}),\, z_h|_T \in P_1,\, \forall T \in \mathcal{T}_h \}, \tag{1.160}$$

and

$$V_{0h} = \{ z_h \,|\, z_h \in V_h,\, z_h = 0 \text{ on } \Gamma \} \,(= V_h \cap H_0^1(\Omega)), \tag{1.161}$$

respectively. We introduce now a second finite element triangulation $\mathcal{T}_H$ of Ω (we may take $\mathcal{T}_h = \mathcal{T}_H$, but $\mathcal{T}_H$ *coarser* than $\mathcal{T}_h$ is a feasible alternative) and we associate with $\mathcal{T}_H$ the following two finite-dimensional spaces:

$$E_H = \{g \mid g \in C^0(\bar{\Omega}),\ g|_T \in P_1,\ \forall T \in \mathcal{T}_H\}, \tag{1.162}$$

$$E_{0H} = \{g \mid g \in E_H,\ g = 0 \text{ on } \Gamma\} \ (= E_H \cap H_0^1(\Omega)). \tag{1.163}$$

Since the *closure* of $H_0^1(\Omega)$ in $L^2(\Omega)$ is $L^2(\Omega)$ we can use either V_h or V_{0h} (respectively, E_H or E_{0H}) to approximate $L^2(\Omega)$.

At this stage, it is convenient to (re)introduce $a : H_0^1(\Omega) \times H_0^1(\Omega) \to \mathbb{R}$, namely, the *bilinear* functional associated with the elliptic operator $\mathcal{A}$; it is defined by

$$a(y,z) = \langle \mathcal{A}y, z\rangle, \quad \forall y,z \in H_0^1(\Omega), \tag{1.164}$$

where, in (1.164), $\langle\cdot\rangle$ denotes the duality pairing between $H^{-1}(\Omega)$ and $H_0^1(\Omega)$. Similarly, we have

$$a(z,y) = \langle \mathcal{A}^* y, z\rangle, \quad \forall y,z \in H_0^1(\Omega). \tag{1.165}$$

From the properties of operator $\mathcal{A}$ (see Section 1.1), the above bilinear functional is *continuous* over $H_0^1(\Omega) \times H_0^1(\Omega)$ and $H_0^1(\Omega)$-*elliptic.*

We approximate problem (1.118) by

$$\begin{cases} f_{hH}^{\Delta t} \in E_{0H}; \quad \forall g \in E_{0H} \text{ we have} \\ \displaystyle\int_\Omega \left(k^{-1} f_{hH}^{\Delta t} + \Lambda_{hH}^{\Delta t} f_{hH}^{\Delta t}\right) g \, dx = \int_\Omega \left(y_T - Y_{0h}^N\right) g \, dx, \end{cases} \tag{1.166}$$

where, in (1.166), Y_{0h}^N is obtained from the *full discretization* of problem (1.119)–(1.121), namely,

$$Y_{0h}^0 = y_{0h} \quad \textit{with } y_{0h}\ (\in V_h) \textit{ an approximation of } y_0. \tag{1.167a}$$

For $n = 1, \ldots, N$, assuming that Y_{0h}^{n-1} is known, compute Y_{0h}^n via the solution of the following (approximate and well-posed) elliptic problem:

$$\begin{cases} Y_{0h}^n \in V_{0h}, \\ \displaystyle\int_\Omega \frac{Y_{0h}^n - Y_{0h}^{n-1}}{\Delta t} z \, dx + a(Y_{0h}^n, z) = 0, \quad \forall z \in V_{0h}. \end{cases} \tag{1.167b}$$

The operator $\Lambda_{hH}^{\Delta t}$ is defined by

$$\Lambda_{hH}^{\Delta t} g = \phi_h^N, \quad \forall g \in E_{0H}, \tag{1.168}$$

where in order to compute ϕ_h^n we solve sequentially the following two discrete parabolic problems:

First problem

$$\psi_h^{N+1} = g; \tag{1.169a}$$

for $n = N, N-1, \ldots, 1$, *we compute* ψ_h^n *from* ψ_h^{n+1} *via the solution of the following discrete Dirichlet problem:*

$$\begin{cases} \psi_h^n \in V_{0h}, \\ \displaystyle\int_\Omega \frac{\psi_h^n - \psi_h^{n+1}}{\Delta t} z \, dx + a(z, \psi_h^n) = 0, \quad \forall z \in V_{0h}. \end{cases} \tag{1.169b}$$

Second problem

$$\phi_h^0 = 0; \tag{1.170a}$$

for $n = 1, \ldots, N$, *we compute* ϕ_h^n *from* ϕ_h^{n-1} *via the solution of the following discrete Dirichlet problem:*

$$\begin{cases} \phi_h^n \in V_{0h}, \\ \displaystyle\int_\Omega \frac{\phi_h^n - \phi_h^{n-1}}{\Delta t} z \, dx + a(\phi_h^n, z) = \int_{\mathcal{O}} \psi_h^n z \, dx, \quad \forall z \in V_{0h}. \end{cases} \tag{1.170b}$$

The discrete elliptic problems (1.169b) and (1.170b) have a unique solution (this follows from the properties of the bilinear functional $a(\cdot, \cdot)$). Concerning the properties of operator $\Lambda_{hH}^{\Delta t}$ we can prove the following *fully discrete* analogue of relation (1.157):

$$\int_\Omega (\Lambda_{hH}^{\Delta t} g)\tilde{g} \, dx = \Delta t \sum_{n=1}^{N} \int_{\mathcal{O}} \psi_h^n \tilde{\psi}_h^n \, dx, \quad \forall g, \tilde{g} \in E_{0H}, \tag{1.171}$$

which shows that operator $\Lambda_{hH}^{\Delta t}$ is *symmetric* and *positive semidefinite*, implying in turn that the problem (1.166) has a *unique solution* and can be solved by a *conjugate gradient algorithm* (described in the following section).

Remark 1.39 We can apply the *trapezoidal rule* to evaluate the various $L^2(\Omega)$-scalar products taking place in (1.166), (1.167), (1.169), and (1.170) (see Glowinski, Li, and Lions, 1990; Glowinski, 1992a; for more details on the use of *numerical integration* in the context of control problems).

Remark 1.40 Instead of E_{0H} we can take the space E_H to approximate problem (1.118); the corresponding approximate problem is still well posed and can be solved by a conjugate gradient algorithm.

1.8.5 Iterative solution of problem (1.166)

From the properties of operator $\Lambda_{hH}^{\Delta t}$ shown in the preceding section, the bilinear functional in (1.166) is *symmetric* and *positive definite* (in fact *uniformly* with respect to h, H, and Δt). Thus problem (1.166) can be solved by a *conjugate gradient algorithm*, which is a discrete analogue of Algorithm 1.35.

Description of the algorithm For simplicity, we shall drop the subscripts h, H, and superscript Δt from $f_{hH}^{\Delta t}$.

Algorithm 1.41 *(Conjugate gradient solution of problem (1.166))*

Initialization

$$f_0 \text{ is given in } E_{0H}; \tag{1.172}$$

assuming that p_0^{n+1} is known, solve the following (well-posed) discrete Dirichlet problem for $n = N, \ldots, 1$:

$$\int_\Omega \frac{p_0^n - p_0^{n+1}}{\Delta t} z \, dx + a(z, p_0^n) = 0, \quad \forall z \in V_{0h}; \quad p_0^n \in V_{0h}, \tag{1.173a}$$

with

$$p_0^{N+1} = f_0, \tag{1.173b}$$

and set

$$u_0^n = p_0^n|_{\mathcal{O}}. \tag{1.174}$$

Assuming that y_0^{n-1} is known, solve the following (well-posed) discrete Dirichlet problem for $n = 1, \ldots, N$:

$$\begin{cases} y_0^n \in V_{0h}, \\ \displaystyle\int_\Omega \frac{y_0^n - y_0^{n-1}}{\Delta t} z \, dx + a(y_0^n, z) = \int_{\mathcal{O}} u_0^n z \, dx, \quad \forall z \in V_{0h}, \end{cases} \tag{1.175a}$$

with

$$y_0^0 = y_{0h}. \tag{1.175b}$$

Finally, solve the following variational problem:

$$\begin{cases} g_0 \in E_{0H}, \\ \displaystyle\int_\Omega g_0 g \, dx = \int_\Omega \left(k^{-1} f_0 + y_0^N - y_T\right) g \, dx, \quad \forall g \in E_{0H}, \end{cases} \tag{1.176}$$

and set

$$w_0 = g_0. \tag{1.177}$$

Then for $m \geq 0$, assuming that f_m, g_m, w_m are known, compute f_{m+1}, g_{m+1}, w_{m+1} as follows:

Steepest descent *Assuming that $\bar{p}_m^{n+1}$ is known, solve for $n = N, \ldots, 1$, the following (well-posed) problem:*

$$\begin{cases} \bar{p}_m^n \in V_{0h}, \\ \displaystyle\int_\Omega \frac{\bar{p}_m^n - \bar{p}_m^{n+1}}{\Delta t} z \, dx + a(z, \bar{p}_m^n) = 0, \quad \forall z \in V_{0h}, \end{cases} \tag{1.178a}$$

with

$$\bar{p}_m^{N+1} = w_m, \tag{1.178b}$$

and set

$$\bar{u}_m^n = \bar{p}_m^n|_{\mathcal{O}}. \tag{1.179}$$

Assuming that $\bar{y}_m^{n-1}$ *is known, solve for* $n = 1, \ldots, N$*, the following (well-posed) problem:*

$$\begin{cases} \bar{y}_m^n \in V_{0h}, \\ \displaystyle\int_\Omega \frac{\bar{y}_m^n - \bar{y}_m^{n-1}}{\Delta t} z \,\mathrm{d}x + a(\bar{y}_m^n, z) = \int_{\mathcal{O}} \bar{u}_m^n z \,\mathrm{d}x, \quad \forall z \in V_{0h}, \end{cases} \tag{1.180a}$$

with

$$\bar{y}_m^0 = 0. \tag{1.180b}$$

Next, solve

$$\begin{cases} \bar{g}_m \in E_{0H}, \\ \displaystyle\int_\Omega \bar{g}_m g \,\mathrm{d}x = \int_\Omega \left(k^{-1} w_m + \bar{y}_m^N\right) g \,\mathrm{d}x, \quad \forall g \in E_{0H}, \end{cases} \tag{1.181}$$

compute

$$\rho_m = \int_\Omega |g_m|^2 \,\mathrm{d}x \Big/ \int_\Omega \bar{g}_m w_m \,\mathrm{d}x, \tag{1.182}$$

and then

$$f_{m+1} = f_m - \rho_m w_m. \tag{1.183}$$

Testing the convergence and construction of the new descent direction *Compute*

$$g_{m+1} = g_m - \rho_m \bar{g}_m. \tag{1.184}$$

If $\|g_{m+1}\|_{L^2(\Omega)} / \|g_0\|_{L^2(\Omega)} \le \varepsilon$*, take* $f = f_{m+1}$ *and solve* (1.169) *(with* $g = f$*) to obtain* $u^n = \psi_{\mathcal{O}}^n$*, for* $n = 1, \ldots, N$*; if the above stopping test is not verified compute*

$$\gamma_m = \|g_{m+1}\|_{L^2(\Omega)}^2 \Big/ \|g_m\|_{L^2(\Omega)}^2, \tag{1.185}$$

and then

$$w_{m+1} = g_{m+1} + \gamma_m w_m. \tag{1.186}$$

Do $m = m + 1$ *and go to* (1.178a).

Remark 1.42 The computer implementation of Algorithm 1.41 requires the solution of the *discrete Dirichlet problems* (1.173a), (1.175a), and (1.178a), (1.180a); to solve these (linear) problems we can use either *direct methods* (such as *Cholesky's* if the bilinear functional $a(\cdot,\cdot)$ is *symmetric*) or *iterative methods* (such as *conjugate gradient, relaxation, multigrid*, etc.). To initialize the iterative methods we shall use the solution of the corresponding problem at the previous time step.

A variant of Algorithm 1.41 has been employed in Carthel, Glowinski, and Lions (1994), to solve exact and approximate *boundary controllability* problems for the *heat equation*; see also Chapter 2, Section 2.5 of this volume.

1.8.6 On the use of second-order accurate time-discretization schemes for the solution of problem (1.116)

We complete, now, Remark 1.38 and follow closely Carthel, Glowinski, and Lions (1994, Section 4.6).

1.8.6.1 Generalities

The numerical methods described in Sections 1.8.3–1.8.5 rely on a *first-order accurate* time-discretization scheme (namely, the *backward Euler* scheme). In order to decrease the computational cost for a *given accuracy* (or increase the accuracy for the same computational cost), it makes sense to use higher-order time discretization schemes. A natural choice in that direction seems to be the *Crank–Nicolson* scheme (see, for example, Raviart and Thomas (1983, Chapter 7) since it is a *one-step, second-order accurate* time-discretization scheme, which is, in addition, no more complicated to implement in practice than the backward Euler scheme. Unfortunately, it is well known that the Crank–Nicolson scheme is not well suited (unless one takes Δt of the order of h^2) to simulate *fast transient phenomena* and/or to carry out numerical integration on *long time intervals* $[0, T]$. From these drawbacks a more natural choice is the *two-step implicit* scheme described below; this scheme is *second-order accurate*, has much better properties than Crank–Nicolson concerning fast transients and long time intervals, and is no more complicated to implement in practice than the backward Euler scheme (for a discussion of multistep schemes applied to the time discretization of parabolic problems, see, for example, Thomee, 1990, Section 6).

1.8.6.2 A second-order accurate time approximation of problem (1.116)

In order to solve the control problem (1.116) via the solution of the functional equation (1.118), the crucial point is – again – to properly approximate the operator Λ and the function Y_0 defined in Section 1.8.1.

Approximation of operator Λ Focusing on the time discretization, we approximate Λ by $\Lambda^{\Delta t}$ defined as follows (we use the notation of Section 1.8.3):
Let us consider $g \in L^2(\Omega)$, then

$$\Lambda^{\Delta t} g = 2\phi^{N-1} - \phi^{N-2}, \tag{1.187}$$

where to obtain ϕ^{N-1} and ϕ^{N-2} we solve first, for $n = N-1, \ldots, 1$, the following (well-posed) Dirichlet problem:

$$\frac{\frac{3}{2}\psi^n - 2\psi^{n+1} + \frac{1}{2}\psi^{n+2}}{\Delta t} + \mathcal{A}^*\psi^n = 0 \text{ in } \Omega, \quad \psi^n = 0 \text{ on } \Gamma, \tag{1.188}$$

with

$$\psi^N = 2g, \quad \psi^{N+1} = 4g, \tag{1.189}$$

then, with $\phi^0 = 0$,

$$\frac{\phi^1 - \phi^0}{\Delta t} + \left(\frac{2}{3}\mathcal{A}\phi^1 + \frac{1}{3}\mathcal{A}\phi^0\right) = \frac{2}{3}\psi^1\chi_{\mathcal{O}} \text{ in } \Omega, \quad \phi^1 = 0 \text{ on } \Gamma, \tag{1.190}$$

and finally, for $n = 2, \ldots, N-1$,

$$\frac{\frac{3}{2}\phi^n - 2\phi^{n-1} + \frac{1}{2}\phi^{n-2}}{\Delta t} + \mathcal{A}\phi^n = \psi^n\chi_{\mathcal{O}} \text{ in } \Omega, \quad \phi^n = 0 \text{ on } \Gamma. \tag{1.191}$$

It can be shown that

$$\int_\Omega (\Lambda^{\Delta t} g)\,\tilde{g}\,\mathrm{d}x = \Delta t \sum_{n=1}^{N-1} \int_{\mathcal{O}} \psi^n \tilde{\psi}^n\,\mathrm{d}x, \quad \forall g, \tilde{g} \in L^2(\Omega),$$

that is, Theorem 1.37 still holds for this new operator $\Lambda^{\Delta t}$ (in fact, $\Lambda^{\Delta t}$ has been defined so that the above relation holds; see also Remark 1.43 below).

Approximation of Y_0 To compute the discrete analogue of Y_0, we take $Y_0^0 = y_0$ and solve the Dirichlet problem:

$$\frac{Y_0^1 - Y_0^0}{\Delta t} + \left(\frac{2}{3}\mathcal{A}Y_0^1 + \frac{1}{3}\mathcal{A}Y_0^0\right) = 0 \text{ in } \Omega, \quad Y_0^1 = 0 \text{ on } \Gamma, \tag{1.192a}$$

and then for $n = 2, \ldots, N-1$,

$$\frac{\frac{3}{2}Y_0^n - 2Y_0^{n-1} + \frac{1}{2}Y_0^{n-2}}{\Delta t} + \mathcal{A}Y_0^n = 0 \text{ in } \Omega, \quad Y_0^n = 0 \text{ on } \Gamma. \tag{1.192b}$$

Approximation of problem (1.118) We approximate problem (1.118) by

$$\begin{cases} f^{\Delta t} \in L^2(\Omega); \quad \forall g \in L^2(\Omega), \text{ we have} \\ \left(k^{-1}f^{\Delta t} + \Lambda^{\Delta t}f^{\Delta t}, g\right)_{L^2(\Omega)} = \left(y_T - 2Y_0^{N-1} + Y_0^{N-2}, g\right)_{L^2(\Omega)}. \end{cases} \tag{1.193}$$

Problem (1.193) can be solved by a discrete variant of Algorithm 1.35. Also, the finite element discretization discussed in Section 1.8.4 can be applied easily to problem (1.193) and the resulting fully discrete problem can be solved by a variant of Algorithm 1.41.

Remark 1.43 The definition of $\Lambda^{\Delta t}$ via relations (1.187)–(1.191) may look somewhat artificial; in fact, it can be shown that the control obtained via the solution of (1.193) is the *unique* solution of the following (time-discrete) control problem:

$$\min_{\{v^n\}_{n=1}^{N-1} \in (L^2(\mathcal{O}))^{N-1}} J^{\Delta t}(v^1, \ldots, v^{N-1}), \tag{1.194}$$

where, in (1.194), we have

$$
\begin{aligned}
J^{\Delta t}(v^1,\ldots,v^{N-1}) = \frac{\Delta t}{2}\sum_{n=1}^{N-1}\int_{\mathcal{O}} |v^n|^2\,dx \\
+\frac{k}{2}\left\|2y^{N-1}-y^{N-2}-y_T\right\|^2_{L^2(\Omega)},
\end{aligned} \tag{1.195}
$$

and where y^{N-1} and y^{N-2} are obtained from $\{v^n\}_{n=1}^{N-1}$ via the solution of the following parabolic problem:

$$
y^0 = y_0, \tag{1.196}
$$

$$
\frac{y^1-y^0}{\Delta t}+\mathcal{A}\left(\frac{2}{3}y^1+\frac{1}{3}y^0\right)=\frac{2}{3}v^1\chi_{\mathcal{O}} \text{ in } \Omega, \quad y^1=0 \text{ on } \Gamma, \tag{1.197}
$$

and for $n = 2,\ldots,N-1$,

$$
\frac{\frac{3}{2}y^n-2y^{n-1}+\frac{1}{2}y^{n-2}}{\Delta t}+\mathcal{A}y^n = v^n\chi_{\mathcal{O}} \text{ in } \Omega, \quad y^n=0 \text{ on } \Gamma. \tag{1.198}
$$

In principle, $2y^{N-1}-y^{N-2}$ is an $O(|\Delta t|^2)$ accurate approximation of $y(T)$ obtained by *extrapolation*.

Remark 1.44 A variant of the previously mentioned second-order accurate time-discretization scheme has been successfully applied in Carthel, Glowinski, and Lions (1994, Section 7) to the solution of exact and approximate boundary controllability problems for the *heat equation*; see also Section 2.5 of this volume.

1.8.7 Convergence of the approximate solutions of problem (1.116), (1.118)

In this section, we are going to discuss the *convergence* of the solution of the *fully discrete* problem (1.166) – and of the corresponding approximate solution of problem (1.116), (1.118) – as $\{\Delta t, h, H\}\to \mathbf{0}$.

Problem (1.166) has been defined in Section 1.8.4 (whose notation is kept) by

$$
\begin{cases}
f_{hH}^{\Delta t}\in E_{0H}; \quad \forall g\in E_{0H} \text{ we have} \\
\displaystyle\int_\Omega \left(k^{-1}f_{hH}^{\Delta t}+\Lambda_{hH}^{\Delta t}f_{hH}^{\Delta t}\right)g\,dx = \int_\Omega \left(y_T-Y_{0h}^n\right)g\,dx.
\end{cases} \tag{1.199}
$$

Concerning the convergence of $\{f_{hH}^{\Delta t}\}_{\{\Delta t,h,H\}}$ as $\{\Delta t,h,H\}\to\mathbf{0}$, we have the following:

Theorem 1.45 *We suppose that*

$$
\lim_{h\to 0}\|y_{0h}-y_0\|_{L^2(\Omega)}=0, \tag{1.200}
$$

and that

$$\text{the angles of } \mathcal{T}_h \text{ are bounded away from } 0 \tag{1.201}$$

(that is, there exists $\theta_0 > 0$, such that $\theta \geq \theta_0$, $\forall \theta$ angle of $\mathcal{T}_h$, $\forall h$). Then

$$\lim_{\{\Delta t, h, H\} \to \mathbf{0}} \|f_{hH}^{\Delta t} - f\|_{L^2(\Omega)} = 0, \tag{1.202}$$

and

$$\lim_{\{\Delta t, h, H\} \to \mathbf{0}} \left\| \psi_{hH}^{\Delta t} \chi_{\mathcal{O}} - u \right\|_{L^2(\mathcal{O} \times (0,T))} = 0, \tag{1.203}$$

where, in (1.202), *f and $f_{hH}^{\Delta t}$ are the solution of problems* (1.118) *and* (1.166), (1.199), *respectively, and where, in* (1.203), *u is the solution of the control problem* (1.116) *and $\psi_{hH}^{\Delta t} \chi_{\mathcal{O}}$ the discrete control corresponding to $f_{hH}^{\Delta t}$ via* (1.169), *with $\psi_h^{N+1} = f_{hH}^{\Delta t}$ in* (1.169a).

Proof. To simplify the presentation, we split the proof in several steps.

(i) *Estimates.* Taking $g = f_{hH}^{\Delta t}$ in (1.199), we obtain, since the operator $\Lambda_{hH}^{\Delta t}$ is *positive semidefinite* (see Section 1.8.4, relation (1.171)), that

$$\|f_{hH}^{\Delta t}\|_{L^2(\Omega)} \leq k \left\| y_T - Y_{0h}^N \right\|_{L^2(\Omega)}, \quad \forall \{\Delta t, h, H\}. \tag{1.204}$$

It follows then from standard results on the *finite element approximation of parabolic problems* (see, for example, Raviart and Thomas, 1983, Chapter 7, Section 8; and Fujita and Suzuki, 1991, Chapter 2, Section 8) that properties (1.200) and (1.201) imply that

$$\lim_{\{\Delta t, h\} \to \mathbf{0}} \left\| Y_{0h}^N - Y_0(T) \right\|_{L^2(\Omega)} = 0, \tag{1.205}$$

where the function Y_0 is *the* solution of the parabolic problem (1.119)–(1.121). It follows from (1.205) that the family $\{Y_{0h}^N\}_{\{\Delta t, h\}}$ is *bounded* in $L^2(\Omega)$, which implies in turn that the right-hand side of (1.204), and therefore $\{\|f_{hH}^{\Delta t}\|_{L^2(\Omega)}\}_{\{\Delta t, h, H\}}$ are *bounded*. Since the family $\{f_{hH}^{\Delta t}\}_{\{\Delta t, h, H\}}$ is bounded in $L^2(\Omega)$ we can extract from it a subsequence – still denoted by $\{f_{hH}^{\Delta t}\}_{\{\Delta t, h, H\}}$ – such that

$$\lim_{\{\Delta t, h, H\} \to 0} f_{hH}^{\Delta t} = f^* \text{ weakly in } L^2(\Omega) . \tag{1.206}$$

(ii) *Weak convergence.* To show that $f^* = f$, it is convenient to introduce Π_H, the $L^2(\Omega)$-projection operator from $L^2(\Omega)$ into E_{0H}; we have then

$$\lim_{H \to 0} \|\Pi_H g - g\|_{L^2(\Omega)} = 0, \quad \forall g \in L^2(\Omega). \tag{1.207}$$

In Lemma 1.46, hereafter, we shall prove that

$$\lim_{\{\Delta t, h, H\} \to \mathbf{0}} \left\| \Lambda_{hH}^{\Delta t} \Pi_H g - \Lambda g \right\|_{L^2(\Omega)} = 0, \quad \forall g \in L^2(\Omega). \tag{1.208}$$

It follows then from (1.199), (1.204)–(1.208), and from the *symmetry* of operators Λ and $\Lambda_{hH}^{\Delta t}$ that, $\forall g \in L^2(\Omega)$,

$$\lim_{\{\Delta t,h,H\}\to\mathbf{0}} \int_\Omega \left(k^{-1} f_{hH}^{\Delta t} + \Lambda_{hH}^{\Delta t} f_{hH}^{\Delta t}\right) \Pi_H g \, dx$$

$$= \lim_{\{\Delta t,h,H\}\to\mathbf{0}} \left(k^{-1} \int_\Omega f_{hH}^{\Delta t}(\Pi_H g) \, dx + \int_\Omega (\Lambda_{hH}^{\Delta t} \Pi_H g) f_{hH}^{\Delta t} \, dx\right)$$

$$= \int_\Omega \left(k^{-1} f^* + \Lambda f^*\right) g \, dx$$

$$= \lim_{\{\Delta t,h,H\}\to\mathbf{0}} \int_\Omega \left(y_T - Y_{0h}^N\right) \Pi_H g \, dx$$

$$= \int_\Omega (y_T - Y_0(T)) \, g \, dx.$$

Assuming that (1.208) holds, we have proved thus that f^* is a solution of problem (1.118); since the above problem has a *unique* solution we have $f^* = f$ and also the convergence to f of the *whole* family $\{f_{hH}^{\Delta t}\}_{\{\Delta t,h,H\}}$ as $\{\Delta t, h, H\} \to \mathbf{0}$.

(iii) *Strong convergence*. Let us introduce $\bar{f}_{hH}^{\Delta t} = f_{hH}^{\Delta t} - \Pi_H f$; we clearly have

$$\lim_{\{\Delta t,h,H\}\to\mathbf{0}} \bar{f}_{hH}^{\Delta t} = 0 \text{ weakly in } L^2(\Omega). \tag{1.209}$$

We also have, for any triple $\{\Delta t, h, H\}$,

$$k^{-1} \|\bar{f}_{hH}^{\Delta t}\|_{L^2(\Omega)}^2 \le \int_\Omega \left(k^{-1} \bar{f}_{hH}^{\Delta t} + \Lambda_{hH}^{\Delta t} \bar{f}_{hH}^{\Delta t}\right) \bar{f}_{hH}^{\Delta t} \, dx. \tag{1.210}$$

Concerning the right-hand side of (1.210), we have, from (1.199),

$$\int_\Omega \left(k^{-1} \bar{f}_{hH}^{\Delta t} + \Lambda_{hH}^{\Delta t} \bar{f}_{hH}^{\Delta t}\right) \bar{f}_{hH}^{\Delta t} \, dx$$

$$= \int_\Omega \left(k^{-1} f_{hH}^{\Delta t} + \Lambda_{hH}^{\Delta t} f_{hH}^{\Delta t}\right) \bar{f}_{hH}^{\Delta t} \, dx - \int_\Omega \left(k^{-1} \Pi_H f + \Lambda_{hH}^{\Delta t} \Pi_H f\right) \bar{f}_{hH}^{\Delta t} \, dx$$

$$= \int_\Omega \left(y_T - Y_{0h}^N\right) \bar{f}_{hH}^{\Delta t} \, dx - \int_\Omega \left(k^{-1} \Pi_H f + \Lambda_{hH}^{\Delta t} \Pi_H f\right) \bar{f}_{hH}^{\Delta t} \, dx.$$

Taking the limit in the above relations, and in (1.210), as $\{\Delta t, h, H\} \to \mathbf{0}$, we obtain from (1.204)–(1.209) that

$$0 \le \liminf_{\{\Delta t,h,H\}\to\mathbf{0}} \|\bar{f}_{hH}^{\Delta t}\|_{L^2(\Omega)}^2 \le \limsup_{\{\Delta t,h,H\}\to\mathbf{0}} \|\bar{f}_{hH}^{\Delta t}\|_{L^2(\Omega)}^2 \le 0; \tag{1.211}$$

we have thus proved that

$$\lim_{\{\Delta t,h,H\}\to\mathbf{0}} \|\bar{f}_{hH}^{\Delta t}\|_{L^2(\Omega)} = 0,$$

which combined with (1.207) (with $g = f$) implies in turn the convergence property (1.202).

(iv) *Convergence of the discrete control.* The solution u of the control problem (1.116) satisfies $u = \psi\chi_{\mathcal{O}\times(0,T)}$, where ψ is the solution of the parabolic problem (1.135) when $\psi(T) = f$, f being the solution of problem (1.118). Similarly, we associate to the solution $f_{hH}^{\Delta t}$ of problem (1.166), (1.199) the solution $\{\psi_{hH}^n\}_{n=1}^N$ of problem (1.169) corresponding to $\psi_h^{N+1} = f_{hH}^{\Delta t}$ in (1.169a) or, equivalently, the piecewise constant function $\psi_{hH}^{\Delta t}$ of t defined by

$$\psi_{hH}^{\Delta t} = \sum_{n=1}^{N} \psi_{hH}^n I_n, \tag{1.212}$$

where I_n is the *characteristic function* of the interval $(0,T) \cap \big((n-\frac{1}{2})\Delta t, (n+\frac{1}{2})\Delta t\big)$. Since $\lim_{\{\Delta t,h,H\}\to\mathbf{0}} \|f_{hH}^{\Delta t} - f\|_{L^2(\Omega)} = 0$, it follows from Raviart and Thomas (1983), and from Lemma 1.46, that

$$\lim_{\{\Delta t,h,H\}\to\mathbf{0}} \|\psi_{hH}^{\Delta t} - \psi\|_{L^2(Q)} = 0,$$

which implies in turn that

$$\lim_{\{\Delta t,h,H\}\to\mathbf{0}} \|\psi_{hH}^{\Delta t}\chi_{\mathcal{O}\times(0,T)} - u\|_{L^2(\mathcal{O}\times(0,T))} = 0,$$

that is, relation (1.203) holds. □

The proof of Theorem 1.45 will be complete once we have proved the following:

Lemma 1.46 *Suppose that the angle condition* (1.201) *holds and consider a family* $\{g_H\}_H$ *such that* $g_H \in E_{0H}$, $\forall H$, *and*

$$\lim_{H\to 0} \|g_H - g\|_{L^2(\Omega)} = 0. \tag{1.213}$$

If with g_H *we associate* $\Lambda_{hH}^{\Delta t} g_H$, $\{\psi_{ghH}^n\}_{n=1}^N$, $\{\phi_{ghH}^n\}_{n=1}^N$ *via* (1.168)–(1.170), *we then have*

$$\lim_{\{\Delta t,h,H\}\to\mathbf{0}} \left\|\psi_{ghH}^{\Delta t} - \psi_g\right\|_{L^2(Q)} = 0, \tag{1.214}$$

$$\lim_{\{\Delta t,h,H\}\to\mathbf{0}} \left\|\Lambda_{hH}^{\Delta t} g_H - \Lambda g\right\|_{L^2(\Omega)} = 0, \tag{1.215}$$

where, in (1.214), $\psi_{ghH}^{\Delta t}$ *is defined from* $\{\psi_{ghH}^n\}_{n=1}^N$ *by* $\psi_{ghH}^{\Delta t} = \sum_{n=1}^N \psi_{ghH}^n I_n$ *and where* ψ_g *is the solution of*

$$-\frac{\partial \psi_g}{\partial t} + \mathcal{A}^*\psi_g = 0 \text{ in } Q, \quad \psi_g = 0 \text{ on } \Sigma, \quad \psi_g(T) = g. \tag{1.216}$$

Proof.

(i) *Proof of* (1.214): For convenience, extend $\{\psi_{ghH}^n\}_{n=1}^N$ to $n = 0$ by solving problem (1.169b) for $n = 0$, and still denote by $\psi_{ghH}^{\Delta t}$ the function $\sum_{n=0}^N \psi_{ghH}^n I_n$;

it follows then from Raviart and Thomas (1983) that

$$\lim_{\{\Delta t,h,H\}\to\mathbf{0}} \left\|\psi_{ghH}^{\Delta t} - \psi_g\right\|_{L^\infty(0,T;L^2(\Omega))} = 0, \tag{1.217}$$

if we can show that the convergence property

$$\lim_{\{\Delta t,h,H\}\to\mathbf{0}} \left\|\psi_{ghH}^{N} - g\right\|_{L^2(\Omega)} = 0, \tag{1.218}$$

holds. To show (1.218), observe first that ψ_{ghH}^N is the unique solution of the discrete elliptic problem

$$\begin{cases} \psi_{ghH}^N \in V_{0h}, \\ \displaystyle\int_\Omega \psi_{ghH}^N v_h \, dx + \Delta t \, a(v_h, \psi_{ghH}^N) = \int_\Omega g_H v_h \, dx, \quad \forall v_h \in V_{0h}. \end{cases} \tag{1.219}$$

Taking $v_h = \psi_{ghH}^N$ in (1.219), we obtain from, the $H_0^1(\Omega)$-ellipticity of $a(\cdot,\cdot)$ (see Section 1.1) that

$$\left\|\psi_{ghH}^N\right\|_{L^2(\Omega)} \leq C, \quad \forall\{\Delta t, h, H\}, \tag{1.220}$$

$$(\Delta t)^{1/2} \left\|\psi_{ghH}^N\right\|_{H_0^1(\Omega)} \leq C, \quad \forall\{\Delta t, h, H\}, \tag{1.221}$$

where, in (1.220), (1.221), (and in the following), C denotes various quantities independent of Δt, h, and H. Since, from (1.220), the family $\{\psi_{ghH}^N\}_{\{\Delta t,h,H\}}$ is bounded in $L^2(\Omega)$, we can extract a subsequence – still denoted by $\{\psi_{ghH}^N\}_{\{\Delta t,h,H\}}$ – such that

$$\lim_{\{\Delta t,h,H\}\to\mathbf{0}} \psi_{ghH}^N = g^* \text{ weakly in } L^2(\Omega)\,. \tag{1.222}$$

Consider, next, $v \in \mathcal{D}(\Omega)$ and denote by $r_h v$ the *linear interpolate* of v on $\mathcal{T}_h$; since the *angle condition* (1.201) holds, it follows from, for example, Ciarlet (1978, 1991), Raviart and Thomas (1983), Glowinski (1984, Appendix 1), that

$$\lim_{h\to 0} \|r_h v - v\|_{H_0^1(\Omega)} = 0, \quad \forall v \in \mathcal{D}(\Omega). \tag{1.223}$$

Take now $v_h = r_h v$ in (1.219); it follows then from (1.221), (1.223), and from the continuity of $a(\cdot,\cdot)$ over $H_0^1(\Omega) \times H_0^1(\Omega)$, that, $\forall v \in \mathcal{D}(\Omega)$,

$$\left|\int_\Omega \psi_{ghH}^N r_h v \, dx - \int_\Omega g_H r_h v \, dx\right| \leq C \|v\|_{H_0^1(\Omega)} |\Delta t|^{\frac{1}{2}}\,. \tag{1.224}$$

Taking the limit in (1.224), as $\{\Delta t, h, H\} \to \mathbf{0}$, it follows then from (1.213), (1.222), and (1.223) that

$$\int_\Omega g^* v \, dx = \int_\Omega g v \, dx, \quad \forall v \in \mathcal{D}(\Omega). \tag{1.225}$$

Since $\mathcal{D}(\Omega)$ is *dense* in $L^2(\Omega)$, it follows from (1.225) that $g^* = g$, and also that the whole family $\{\psi_{ghH}^N\}_{\{\Delta t,h,H\}}$ converges *weakly* to g. To prove the *strong convergence*, observe that

$$\begin{aligned}
\int_\Omega \left|\psi_{ghH}^N - g\right|^2 dx &= \int_\Omega |g|^2 \, dx - 2\int_\Omega \psi_{ghH}^N g \, dx + \int_\Omega \left|\psi_{ghH}^N\right|^2 dx \\
&\le \int_\Omega |g|^2 \, dx - 2\int_\Omega \psi_{ghH}^N g \, dx + \int_\Omega \left|\psi_{ghH}^N\right|^2 dx + \Delta t \, a(\psi_{ghH}^N, \psi_{ghH}^N) \\
&= \int_\Omega |g|^2 \, dx - 2\int_\Omega \psi_{ghH}^N g \, dx + \int_\Omega g_H \psi_{ghH}^N \, dx.
\end{aligned} \tag{1.226}$$

It follows then from (1.213), (1.226), and from the weak convergence of $\{\psi_{ghH}^N\}_{\{\Delta t,h,H\}}$ to g in $L^2(\Omega)$ that the convergence property (1.218) holds; it implies (1.217) and therefore (1.214).

(ii) *Proof of* (1.215): With the solution ψ_g of (1.216) we associate ϕ_g solution of

$$\frac{\partial \phi_g}{\partial t} + \mathcal{A}\phi_g = \psi_g|_{\mathcal{O}\times(0,T)} \text{ in } Q, \quad \phi_g = 0 \text{ on } \Sigma, \quad \phi_g(0) = 0. \tag{1.227}$$

We have then

$$\Lambda g = \phi_g(T). \tag{1.228}$$

Similarly, we associate $\{\psi_{ghH}^n\}_{n=0}^N$ with $\{\phi_{ghH}^n\}_{n=0}^N$ defined by

$$\phi_{ghH}^0 = 0, \tag{1.229a}$$

and, for $n = 1, \ldots, N$, by the solution of the following discrete elliptic problem:

$$\begin{cases}
\phi_{ghH}^n \in V_{0h}, \quad \forall v_h \in V_{0h}, \\
\displaystyle\int_\Omega \frac{\phi_{ghH}^n - \phi_{ghH}^{n-1}}{\Delta t} v_h \, dx + a(\phi_{ghH}^n, v_h) = \int_{\mathcal{O}} \psi_{ghH}^n v_h \, dx.
\end{cases} \tag{1.229b}$$

We have

$$\Lambda_{hH}^{\Delta t} g_H = \phi_{ghH}^N. \tag{1.230}$$

In order to prove (1.215) it is quite convenient to associate with ψ_g the family $\{\theta_{ghH}^n\}_{n=0}^N$ defined by

$$\theta_{gh}^0 = 0, \tag{1.231a}$$

and, for $n = 1, \ldots, N$, by the solution of the following discrete elliptic problem:

$$\begin{cases} \theta_{gh}^n \in V_{0h}, \quad \forall v_h \in V_{0h}, \\ \displaystyle\int_\Omega \frac{\theta_{gh}^n - \theta_{gh}^{n-1}}{\Delta t} v_h \,\mathrm{d}x + a(\theta_{gh}^n, v_h) = \int_{\mathcal{O}} \psi_g(n\Delta t) v_h \,\mathrm{d}x. \end{cases} \tag{1.231b}$$

Let us define $\phi_{ghH}^{\Delta t}$ and $\theta_{gh}^{\Delta t}$ by

$$\phi_{ghH}^{\Delta t} = \sum_{n=1}^{N} \phi_{ghH}^n I_n, \tag{1.232}$$

$$\theta_{ghH}^{\Delta t} = \sum_{n=1}^{N} \theta_{ghH}^n I_n, \tag{1.233}$$

respectively. Since

$$\psi_g \in C^0([0, T]; L^2(\Omega)),$$

it follows from Raviart and Thomas (1983), that

$$\lim_{\{\Delta t, h, H\} \to \mathbf{0}} \theta_{ghH}^{\Delta t} = \phi_g \quad \text{strongly in } L^2(0, T; H_0^1(\Omega)), \tag{1.234}$$

$$\lim_{\{\Delta t, h, H\} \to \mathbf{0}} \max_{1 \le n \le N} \left\| \theta_{ghH}^n - \phi_g(n\Delta t) \right\|_{L^2(\Omega)} = 0. \tag{1.235}$$

Actually, similar convergence results hold for $\phi_{ghH}^{\Delta t}$. To show them, denote by $\bar{\phi}_{ghH}^{\Delta t}$ the difference $\phi_{ghH}^{\Delta t} - \theta_{ghH}^{\Delta t}$; we clearly have

$$\bar{\phi}_{ghH}^0 = 0, \tag{1.236a}$$

and for $n = 1, \ldots, N$

$$\begin{cases} \bar{\phi}_{ghH}^n \in V_{0h}, \\ \displaystyle\int_\Omega \frac{\bar{\phi}_{ghH}^n - \bar{\phi}_{ghH}^{n-1}}{\Delta t} v_h \,\mathrm{d}x + a(\bar{\phi}_{ghH}^n, v_h) \\ \displaystyle\qquad = \int_{\mathcal{O}} \left(\psi_{ghH}^n - \psi_g(n\Delta t) \right) v_h \,\mathrm{d}x, \quad \forall v_h \in V_{0h}. \end{cases} \tag{1.236b}$$

Take $v_h = \bar{\phi}_{ghH}^n$ in (1.236b) and remember that $a(v, v) \ge \alpha \|v\|^2$, $\forall v \in H_0^1(\Omega)$, with $\alpha > 0$ (see Section 1.1); we have then from the Schwarz inequality in $L^2(\Omega)$ and from the relation

$$2\eta\xi \le c\eta^2 + c^{-1}\xi^2, \quad \forall \eta, \xi \in \mathbb{R}, \forall c > 0,$$

that

$$\frac{1}{2\Delta t}\left(\left\|\bar{\phi}^{n}_{ghH}\right\|^2_{L^2(\Omega)} - \left\|\bar{\phi}^{n-1}_{ghH}\right\|^2_{L^2(\Omega)}\right) + \alpha\left\|\bar{\phi}^{n}_{ghH}\right\|^2_{H_0^1(\Omega)} \leq \left\|\psi^{n}_{ghH} - \psi_g(n\Delta t)\right\|_{L^2(\Omega)}\left\|\bar{\phi}^{n}_{ghH}\right\|_{L^2(\Omega)},$$

which implies, in turn, since the injection from $H_0^1(\Omega)$ into $L^2(\Omega)$ is continuous, that, $\forall n \geq 1$, $\forall c > 0$, we have

$$\frac{1}{2\Delta t}\left(\left\|\bar{\phi}^{n}_{ghH}\right\|^2_{L^2(\Omega)} - \left\|\bar{\phi}^{n-1}_{ghH}\right\|^2_{L^2(\Omega)}\right) + \gamma\left\|\bar{\phi}^{n}_{ghH}\right\|^2_{L^2(\Omega)} \leq \frac{1}{2}\left(c^{-1}\left\|\psi^{n}_{ghH} - \psi_g(n\Delta t)\right\|^2_{L^2(\Omega)} + c\left\|\bar{\phi}^{n}_{ghH}\right\|^2_{L^2(\Omega)}\right), \tag{1.237}$$

where, in (1.237), γ is a *positive constant*. Taking $c = 2\gamma$ in (1.237), we obtain, $\forall n = 1, \ldots, N$,

$$\left\|\bar{\phi}^{n}_{ghH}\right\|^2_{L^2(\Omega)} - \left\|\bar{\phi}^{n-1}_{ghH}\right\|^2_{L^2(\Omega)} \leq \frac{\Delta t}{2\gamma}\left\|\psi^{n}_{ghH} - \psi_g(n\Delta t)\right\|^2_{L^2(\Omega)},$$

which implies, by summation from $n = 1$ to $n = N$, that

$$\left\|\bar{\phi}^{N}_{ghH}\right\|^2_{L^2(\Omega)} \leq \frac{\Delta t}{2\gamma}\sum_{n=1}^{N}\left\|\psi^{n}_{ghH} - \psi_g(n\Delta t)\right\|^2_{L^2(\Omega)}. \tag{1.238}$$

It follows then from (1.214) and (1.238) that

$$\lim_{\{\Delta t,h,H\}\to\mathbf{0}}\left\|\phi^{N}_{ghH} - \theta^{N}_{ghH}\right\|_{L^2(\Omega)} = 0,$$

which combined with (1.235) implies that

$$\lim_{\{\Delta t,h,H\}\to\mathbf{0}}\left\|\phi^{N}_{ghH} - \phi_g(T)\right\|_{L^2(\Omega)} = 0. \tag{1.239}$$

Finally, relations (1.228), (1.230), and (1.239) imply the convergence result (1.215). □

1.8.8 Solution methods for problem (1.117)

In this section, we discuss the solution of the *variational inequality* problem (1.117). Using a *duality* argument, we can prove that the above problem is equivalent to the control problem (1.111). We observe also that (1.117) can be written as the following nonlinear (multivalued) equation in $L^2(\Omega)$:

$$y_T - Y_0(T) \in \Lambda f + \beta\partial j(f), \tag{1.240}$$

where, in (1.240), $\partial j(f)$ denotes the *subgradient* (see, for example, Ekeland and Temam, 1974 for this notion) at f of the *convex functional* $j(\cdot)$ defined by

$$j(g) = \|g\|_{L^2(\Omega)}, \quad \forall g \in L^2(\Omega).$$

Equation (1.240) strongly suggests the use of *operator-splitting methods* like those discussed in, for example, P.L. Lions and B. Mercier (1979), Fortin and Glowinski (1983), Glowinski and Le Tallec (1989), Marchuk (1990), and Glowinski (2003). A simple way to derive such methods is to associate with (1.240) a *time-dependent* equation (for a *pseudotime* τ) such as

$$\begin{cases} \dfrac{df}{d\tau} + \Lambda f + \beta \partial j(f) \ni y_T - Y_0(T) \quad \text{on } (0, +\infty), \\ f(0) = f_0 \ (\in L^2(\Omega)). \end{cases} \tag{1.241}$$

Next, we use time discretization by operator-splitting to solve the initial value problem (1.241) in order to "capture" its steady-state solution, namely, the solution of (1.240). A natural choice for the solution of (1.241) is the *Peaceman–Rachford scheme* (cf. Peaceman and Rachford, 1955), which for the present problem provides

$$f^0 = f_0; \tag{1.242}$$

then, for $k \geq 0$, compute $f^{k+\frac{1}{2}}$ and f^{k+1}, from f^k, by solving

$$\frac{f^{k+\frac{1}{2}} - f^k}{\Delta\tau/2} + \beta \partial j(f^{k+\frac{1}{2}}) + \Lambda f^k \ni y_T - Y_0(T), \tag{1.243}$$

and

$$\frac{f^{k+1} - f^{k+\frac{1}{2}}}{\Delta\tau/2} + \beta \partial j(f^{k+\frac{1}{2}}) + \Lambda f^{k+1} \ni y_T - Y_0(T), \tag{1.244}$$

where $\Delta\tau$ (> 0) is a (pseudo) time-discretization step. The *convergence* of $\{f^k\}_{k \geq 0}$ to the solution f of (1.117), (1.240) is a direct consequence of, for example, P.L. Lions and Mercier (1979), Gabay (1982, 1983), and Glowinski and Le Tallec (1989); the convergence results proved in the above references apply to the present problem since operator Λ (respectively, functional $j(\cdot)$) is *linear*, *continuous*, and *positive definite* (respectively, *convex* and *continuous*) over $L^2(\Omega)$. Among the many possible variants of the above algorithm, let us mention the one obtained by applying to problem (1.241) the θ-scheme discussed in, for example, Glowinski and Le Tallec (1989) and Glowinski (2003); we obtain then (with $0 < \theta \leq 1/3$; see the above two references) the following algorithm:

$$f^0 = f_0; \tag{1.245}$$

then, for $k \geq 0$, compute $f^{k+\theta}$, $f^{k+1-\theta}$, and f^{k+1}, from f^k, by solving (with $\theta' = 1 - 2\theta$)

$$\frac{f^{k+\theta} - f^k}{\theta \Delta\tau} + \beta \partial j(f^{k+\theta}) + \Lambda f^k \ni y_T - Y_0(T), \tag{1.246}$$

$$\frac{f^{k+1-\theta} - f^{k+\theta}}{\theta' \Delta\tau} + \beta \partial j(f^{k+\theta}) + \Lambda f^{k+1-\theta} \ni y_T - Y_0(T), \tag{1.247}$$

$$\frac{f^{k+1} - f^{k+1-\theta}}{\theta \Delta\tau} + \beta j(f^{k+1}) + \Lambda f^{k+1-\theta} \ni y_T - Y_0(T). \tag{1.248}$$

In practice, it may pay to use a *variable* $\Delta\tau$. Concerning now the solution of the various subproblems in the above two algorithms, we will make the following observations:

(i) Assuming that we know how to solve problems (1.243) and (1.246), (1.248), the functions f^{k+1} and $f^{k+1-\theta}$ are obtained via the solution of linear problems similar to the one whose solution has been discussed in Sections 1.8.2–1.8.7; in particular, we can use the conjugate gradient algorithm (1.172)–(1.186) (that is, Algorithm 1.41) to solve finite element approximations of problems (1.244) and (1.247).

(ii) Problems (1.243) and (1.246), (1.248) are fairly easy to solve. Consider, for example, problem (1.243); it is clearly equivalent to the following minimization problem:

$$\begin{cases} f^{k+\frac{1}{2}} \in L^2(\Omega), \\ J_k(f^{k+\frac{1}{2}}) \leq J_k(v), \quad \forall v \in L^2(\Omega), \end{cases} \tag{1.249}$$

with

$$\begin{aligned} J_k(v) = \frac{1}{2}\int_\Omega |v|^2 \, dx + \frac{1}{2}\beta \Delta\tau \, \|v\|_{L^2(\Omega)} - \int_\Omega f^k v \, dx \\ - \frac{1}{2}\Delta\tau \int_\Omega \left(y_T - Y_0(T) - \Lambda f^k\right) v \, dx, \quad \forall v \in L^2(\Omega). \end{aligned} \tag{1.250}$$

To solve problem (1.249), we define $f_*^{k+\frac{1}{2}}$ as

$$f_*^{k+\frac{1}{2}} = f^k + \frac{1}{2}\Delta\tau \left(y_T - Y_0(T) - \Lambda f^k\right), \tag{1.251}$$

and observe that the solution of problem (1.249) is clearly of the form

$$f^{k+\frac{1}{2}} = \lambda^{k+\frac{1}{2}} f_*^{k+\frac{1}{2}} \quad \text{with } \lambda^{k+\frac{1}{2}} \geq 0. \tag{1.252}$$

To obtain $\lambda^{k+\frac{1}{2}}$ we minimize with respect to λ the polynomial

$$\|f_*^{k+\frac{1}{2}}\|^2_{L^2(\Omega)} \left(\frac{1}{2}\lambda^2 - \lambda\right) + \frac{1}{2}\beta \Delta\tau \|f_*^{k+\frac{1}{2}}\|_{L^2(\Omega)} \lambda \left(= J_k\left(\lambda f_*^{k+\frac{1}{2}}\right)\right).$$

We obtain then (since $\lambda^{k+\frac{1}{2}} \geq 0$),

$$\lambda^{k+\frac{1}{2}} = \begin{cases} 1 - \dfrac{\beta \Delta\tau}{2\left\|f_*^{k+\frac{1}{2}}\right\|_{L^2(\Omega)}} & \text{if } \left\|f_*^{k+\frac{1}{2}}\right\|_{L^2(\Omega)} \geq \frac{1}{2}\beta\Delta\tau, \\ 0 & \text{if } \left\|f_*^{k+\frac{1}{2}}\right\|_{L^2(\Omega)} < \frac{1}{2}\beta\Delta\tau. \end{cases} \tag{1.253}$$

The same method applies to the solution of problems (1.246) and (1.248).

Remark 1.47 Concerning the calculation of f^{k+1} we shall use equation (1.243) to rewrite (1.244) as

$$\frac{f^{k+1} - 2f^{k+\frac{1}{2}} + f^k}{\frac{1}{2}\Delta\tau} + \Lambda f^{k+1} = \Lambda f^k, \tag{1.254}$$

which is better suited for practical computations. A similar observation holds for the computation of $f^{k+1-\theta}$ in (1.247).

Remark 1.48 Another variant of algorithm (1.242)–(1.244) is obtained by applying the well-known *Marchuk–Yanenko* operator-splitting scheme to the solution of the initial value problem (1.241) (the Marchuk–Yanenko scheme is discussed in, for example, Marchuk, 1990; Glowinski, 2003); the resulting scheme reads as follows:

$$f^0 = f_0, \tag{1.255}$$

then, for $k \geq 0$, compute $f^{k+\frac{1}{2}}$ and f^{k+1} from f^k via the solution of

$$\frac{f^{k+\frac{1}{2}} - f^k}{\Delta\tau} + \Lambda f^{k+\frac{1}{2}} = y_T - Y_0(T), \tag{1.256}$$

$$\frac{f^{k+1} - f^{k+\frac{1}{2}}}{\Delta\tau} + \beta\partial j(f^{k+1}) \ni 0. \tag{1.257}$$

The various comments concerning the solution of the problems (1.243) and (1.244) still apply to the solution of the problems (1.256) and (1.257).

1.8.9 Splitting methods for nonquadratic cost functions and control constrained problems

1.8.9.1 Generalities

In Section 1.7, we have considered control problems such as (or closely related to)

$$\min_{v \in L^s(\mathcal{O}\times(0,T))} \left(\frac{1}{s} \iint_{\mathcal{O}\times(0,T)} \|v\|^s \, dx\, dt + \frac{1}{2} k \, \|y(T) - y_T\|^2_{L^2(\Omega)} \right), \tag{1.258}$$

where, in (1.258), $s \in [1, +\infty)$, $k > 0$, and where y is defined by (1.113)–(1.115); the case $s = 2$ has been treated in Sections 1.8.2–1.8.7. Solving the problem (1.258) for large values of s provides solution close to those obtained with cost functions containing terms such as $\|v\|_{L^\infty(\mathcal{O}\times(0,T))}$. Another control problem of interest is defined by

$$\min_{v\in\mathcal{C}_f} \|y(T) - y_T\|_{L^2(\Omega)}, \tag{1.259}$$

with

$$\mathcal{C}_f = \{\, v \mid v \in L^\infty(\mathcal{O} \times (0,T)),\ |v(x,t)| \leq C \text{ a.e. in } \mathcal{O} \times (0,T) \,\}$$

and y still defined by the relations (1.113)–(1.115). The *convex* set $\mathcal{C}_f$ is clearly *closed* in $L^2(\mathcal{O} \times (0,T))$; we shall denote by $I_{\mathcal{C}_f}$ its *indicator functional* in $L^2(\mathcal{O} \times (0,T))$ (namely, $I_{\mathcal{C}_f}(v) = 0$ if $v \in \mathcal{C}_f$, $I_{\mathcal{C}_f}(v) = +\infty$ if $v \in L^2(\mathcal{O} \times (0,T)) \setminus \mathcal{C}_f$). Problem (1.259) is clearly equivalent to

$$\min_{v\in L^2(\mathcal{O}\times(0,T))} \left(I_{\mathcal{C}_f}(v) + \frac{1}{2}\|y(T) - y_T\|^2_{L^2(\Omega)} \right). \tag{1.260}$$

In the following subsections we shall show that the problems (1.258) and (1.259), (1.260) are fairly easy to solve if one has a solver for the problem (1.116) (that is, for the problem (1.258) when $s = 2$).

1.8.9.2 Solution of problem (1.258)

Suppose for the time being that $s > 1$ and let us denote by $J(\cdot)$ the *strictly convex* functional defined by

$$J(v) = \frac{1}{s}\iint_{\mathcal{O}\times(0,T)} |v|^s \,\mathrm{d}x\,\mathrm{d}t + \frac{1}{2}k\,\|y(T) - y_T\|^2_{L^2(\Omega)}, \tag{1.261}$$

where, in (1.261), y is obtained from v via (1.113)–(1.115). Next, we define $J_1(\cdot)$ and $J_2(\cdot)$ by

$$J_1(v) = \frac{1}{s}\iint_{\mathcal{O}\times(0,T)} |v|^s \,\mathrm{d}x\,\mathrm{d}t \tag{1.262}$$

and

$$J_2(v) = \frac{1}{2}k\,\|y(T) - y_T\|^2_{L^2(\Omega)}, \tag{1.263}$$

where y is obtained from v via (1.113)–(1.115), respectively. Both functionals are clearly differentiable in $L^s(\mathcal{O} \times (0,T))$ and we have

$$\langle J_1'(v), w\rangle = \iint_{\mathcal{O}\times(0,T)} |v|^{s-2}\, vw \,\mathrm{d}x\,\mathrm{d}t, \quad \forall v, w \in L^s(\mathcal{O} \times (0,T)), \tag{1.264}$$

$$\langle J_2'(v), w\rangle = -\iint_{\mathcal{O}\times(0,T)} pw \,\mathrm{d}x\,\mathrm{d}t, \quad \forall v, w \in L^s(\mathcal{O} \times (0,T)), \tag{1.265}$$

where, in (1.264) and (1.265), $\langle \cdot, \cdot \rangle$ denotes the *duality pairing* between $L^{s'}(\mathcal{O} \times (0,T))$ and $L^s(\mathcal{O} \times (0,T))$ (with $s' = \frac{s}{s-1}$, and where p is the solution of the *adjoint state equation*

$$-\frac{\partial p}{\partial t} + \mathcal{A}^* p = 0 \text{ in } Q, \quad p = 0 \text{ on } \Sigma, \quad p(T) = k\,(y_T - y(T)). \tag{1.266}$$

If u is *the* solution to the control problem (1.258), it is characterized by the relation $J'(u) = 0$, which takes here the following form:

$$J_1'(u) + J_2'(u) = 0. \tag{1.267}$$

In order to solve (1.258), via (1.267), we associate with (1.267) the following initial value problem in $L^s(\mathcal{O} \times (0,T))$ (for the pseudotime variable τ):

$$\begin{cases} \dfrac{du}{d\tau} + J_1'(u) + J_2'(u) = 0 \text{ on } (0, +\infty), \\ u(0) = u_0. \end{cases} \tag{1.268}$$

To obtain the *steady-state solution* of (1.268) (that is, the solution of (1.258), (1.267)) we integrate (1.268) from 0 to $+\infty$ by *operator-splitting*; if one uses the *Peaceman–Rachford scheme* (see Section 1.8.8), we obtain

$$u^0 = u_0, \tag{1.269}$$

and for $n \geq 0$, assuming that u^n is known

$$\frac{u^{n+\frac{1}{2}} - u^n}{\frac{1}{2}\Delta\tau} + J_1'(u^{n+\frac{1}{2}}) + J_2'(u^n) = 0, \tag{1.270}$$

$$\frac{u^{n+1} - u^{n+\frac{1}{2}}}{\frac{1}{2}\Delta\tau} + J_1'(u^{n+\frac{1}{2}}) + J_2'(u^{n+1}) = 0. \tag{1.271}$$

Equation (1.270) can also be written

$$\frac{u^{n+\frac{1}{2}} - u^n}{\frac{1}{2}\Delta\tau} + \left|u^{n+\frac{1}{2}}\right|^{s-2} u^{n+\frac{1}{2}} - p^n \chi_{\mathcal{O}\times(0,T)} = 0, \tag{1.272}$$

where p^n is obtained from u^n via

$$\frac{\partial y^n}{\partial t} + \mathcal{A} y^n = u^n \chi_{\mathcal{O}\times(0,T)} \text{ in } Q, \quad y^n = 0 \text{ on } \Sigma, \quad y^n(0) = 0, \tag{1.273}$$

$$-\frac{\partial p^n}{\partial t} + \mathcal{A}^* p^n = 0 \text{ in } Q,\ p^n = 0 \text{ on } \Sigma,\ p^n(T) = k\left(y_T - y^n(T)\right). \tag{1.274}$$

We obtain thus $u^{n+\frac{1}{2}}$ from u^n by solving the *nonlinear problem*

$$u^{n+\frac{1}{2}} + \frac{1}{2}\Delta\tau \left|u^{n+\frac{1}{2}}\right|^{s-2} u^{n+\frac{1}{2}} = u^n + \frac{1}{2}\Delta\tau p^n \chi_{\mathcal{O}\times(0,T)}. \tag{1.275}$$

Problem (1.275) can be solved *pointwise* in $\mathcal{O} \times (0, T)$; at almost every point of $\mathcal{O} \times (0, T)$ (in practice at the nodes of a *finite difference* or *finite element grid*) we shall have to solve a *one variable* equation of the following form:

$$\xi + \frac{1}{2}\Delta\tau\, |\xi|^{s-2}\, \xi = b, \tag{1.276}$$

which has, $\forall b \in \mathbb{R}$, a *unique* solution. Problem (1.271) is equivalent to the following *minimization problem*:

$$\min_{v \in L^2(\mathcal{O}\times(0,T))} j_n(v), \tag{1.277}$$

where the functional $j_n(\cdot)$ is defined by

$$\begin{aligned} j_n(v) = &\frac{1}{2}\iint_{\mathcal{O}\times(0,T)} v^2 \,\mathrm{d}x\,\mathrm{d}t \\ &- \iint_{\mathcal{O}\times(0,T)} \left(u^{n+\frac{1}{2}} + \frac{1}{2}\Delta\tau \left| u^{n+\frac{1}{2}} \right|^{s-2} u^{n+\frac{1}{2}} \right) v \,\mathrm{d}x\,\mathrm{d}t \\ &+ \frac{1}{4} k \Delta\tau\, \|y(T) - y_T\|^2_{L^2(\Omega)}, \end{aligned} \tag{1.278}$$

with y obtained from v via (1.113)–(1.115). Problem (1.277) is a simple variant of problem (1.116); it can therefore be solved by the numerical methods described in Sections 1.8.2–1.8.7.

Remark 1.49 From a formal point of view the above method still applies if $s = 1$. In such a case we shall replace (1.275) by the following minimization problem:

$$\begin{aligned} \min_{v \in L^2(\mathcal{O}\times(0,T))} &\left(\frac{1}{2}\iint_{\mathcal{O}\times(0,T)} v^2 \,\mathrm{d}x\,\mathrm{d}t \right. \\ &\left. + \frac{1}{2}\Delta\tau \iint_{\mathcal{O}\times(0,T)} |v| \,\mathrm{d}x\,\mathrm{d}t - \iint_{\mathcal{O}\times(0,T)} \left(u^n + \frac{1}{2}\Delta\tau p^n \right) v \,\mathrm{d}x\,\mathrm{d}t \right), \end{aligned} \tag{1.279}$$

whose solution $u^{n+\frac{1}{2}}$ is given (in *closed form*) by, $\forall \{x, t\} \in \mathcal{O} \times (0, T)$,

$$u^{n+\frac{1}{2}}(x,t) = \begin{cases} 0, \quad \text{if } \left| \left(u^n + \frac{1}{2}\Delta\tau p^n \right)(x,t) \right| \le \frac{1}{2}\Delta\tau, \\[2ex] \left(u^n + \frac{1}{2}\Delta\tau p^n \right)(x,t) \\ \quad - \frac{1}{2}\Delta\tau \operatorname{sign}\left(u^n + \frac{1}{2}\Delta\tau p^n \right)(x,t), \\ \quad \text{if } \left| \left(u^n + \frac{1}{2}\Delta\tau p^n \right)(x,t) \right| > \frac{1}{2}\Delta\tau. \end{cases} \tag{1.280}$$

Concerning now the calculation of u^{n+1}, we observe that this function is the solution of

$$\frac{u^{n+1}-2u^{n+\frac{1}{2}}+u^n}{\frac{1}{2}\Delta\tau}+J_2'(u^{n+1})=J_2'(u^n),$$

which is equivalent to the minimization problem

$$\min_{v\in L^2(\mathcal{O}\times(0,T))}\left(\frac{1}{2}\iint_{\mathcal{O}\times(0,T)} v^2\,\mathrm{d}x\,\mathrm{d}t+\frac{1}{4}k\Delta\tau\,\|y(T)-y_T\|^2_{L^2(\Omega)}\right.$$
$$\left.-\iint_{\mathcal{O}\times(0,T)}\left(2u^{n+\frac{1}{2}}-u^n-\frac{1}{2}\Delta\tau p^n\right)v\,\mathrm{d}x\,\mathrm{d}t\right), \tag{1.281}$$

where y is a function of v via the solution of (1.113)–(1.115). Problem (1.281) is also a variant of problem (1.116).

Remark 1.50 Equation (1.268) and algorithm (1.269)–(1.271) are largely formal if $1 \leq s < 2$; however, they make full sense for the discrete analogues of problem (1.258) obtained by finite difference and finite element approximations close to those discussed in Sections 1.8.2–1.8.7.

1.8.9.3 Solution of problem (1.259), (1.260)

We follow the approach taken in Section 1.8.9.2; we introduce therefore J_1 and J_2 defined by

$$J_1(v)=I_{\mathcal{C}_f}(v), \tag{1.282}$$

and

$$J_2(v)=\frac{1}{2}\,\|y(T)-y_T\|^2_{L^2(\Omega)}, \tag{1.283}$$

respectively, with y obtained from v via the solution of (1.113)–(1.115). The solution u of problem (1.259), (1.260) is *characterized* therefore by

$$0\in\partial J_1(u)+J_2'(u), \tag{1.284}$$

where, in (1.284), ∂J_1 is the subgradient of J_1, and where J_2' is defined by (1.265), (1.266) with $k=1$.

We associate with (1.284) the following *initial value problem* in $L^2(\mathcal{O}\times(0,T))$:

$$\begin{cases}\dfrac{\mathrm{d}u}{\mathrm{d}\tau}+\partial J_1(u)+J_2'(u)\ni 0,\\ u(0)=u_0\ (\in\mathcal{C}_f).\end{cases} \tag{1.285}$$

Applying as in Section 1.8.9.2 the *Peaceman–Rachford scheme* to the solution of (1.285), we obtain

$$u^0=u_0, \tag{1.286}$$

and for $n \geq 0$, assuming that u^n is known

$$\frac{u^{n+\frac{1}{2}} - u^n}{\frac{1}{2}\Delta\tau} + \partial J_1(u^{n+\frac{1}{2}}) + J_2'(u^n) \ni 0, \tag{1.287}$$

$$\frac{u^{n+1} - u^{n+\frac{1}{2}}}{\frac{1}{2}\Delta\tau} + \partial J_1(u^{n+\frac{1}{2}}) + J_2'(u^{n+1}) \ni 0. \tag{1.288}$$

"Equation" (1.287) is equivalent to the minimization problem

$$\min_{v \in \mathcal{C}_f} \left(\frac{1}{2} \iint_{\mathcal{O}\times(0,T)} v^2 \, dx \, dt - \iint_{\mathcal{O}\times(0,T)} \left(u^n + \frac{1}{2}\Delta\tau p^n \right) v \, dx \, dt \right), \tag{1.289}$$

where, in (1.289), p^n is obtained from u^n via (1.273), (1.274) with $k = 1$; we have then

$$u^{n+\frac{1}{2}}(x,t) = \min\left(C, \max\left(-C, \left(u^n + \frac{1}{2}\Delta\tau p^n \right)(x,t) \right) \right),$$

$$\text{a.e. on } \mathcal{O} \times (0,T). \tag{1.290}$$

Summing (1.287) and (1.288) implies that

$$\frac{u^{n+1} - 2u^{n+\frac{1}{2}} + u^n}{\frac{1}{2}\Delta\tau} + J_2'(u^{n+1}) = J_2'(u^n),$$

which is equivalent to the following minimization problem:

$$\min_{v \in L^2(\mathcal{O}\times(0,T))} \left(\frac{1}{2} \iint_{\mathcal{O}\times(0,T)} v^2 \, dx \, dt + \frac{1}{4}\Delta\tau \, \|y(T) - y_T\|^2_{L^2(\Omega)} \right.$$
$$\left. - \iint_{\mathcal{O}\times(0,T)} \left(2u^{n+\frac{1}{2}} - u^n - \frac{1}{2}\Delta\tau p^n \right) v \, dx \, dt \right), \tag{1.291}$$

where y is a function of v via the solution of (1.113)–(1.115). Problem (1.291) is also a variant of problem (1.116).

1.9 Relaxation of controllability

1.9.1 Generalities

Let $\mathcal{H}$ be a *Hilbert space* and let C be a *linear operator* such that

$$C \in \mathcal{L}(L^2(\Omega), \mathcal{H}), \tag{1.292}$$

and

$$\text{the range of } C \text{ is } \textit{dense} \text{ in } \mathcal{H}. \tag{1.293}$$

We consider again the *state equation*

$$\frac{\partial y}{\partial t} + \mathcal{A}y = v\chi_{\mathcal{O}\times(0,T)} \text{ in } Q, \quad y(0) = 0, \quad y = 0 \text{ on } \Sigma, \tag{1.294}$$

and we look now for the solution of

$$\inf_{v} \frac{1}{2} \iint_{\mathcal{O}\times(0,T)} v^2 \,\mathrm{d}x\,\mathrm{d}t, \tag{1.295}$$

for all vs such that

$$Cy(T; v) \in h_T + \beta B_H, \tag{1.296}$$

where h_T is given in $\mathcal{H}$ and $B_{\mathcal{H}}$ denotes the *unit ball* of $\mathcal{H}$.

Remark 1.51 If $\mathcal{H} = L^2(\Omega)$, $C =$ identity, and $h_T = y_T$, then problem (1.295), (1.296) is exactly a controllability problem which has been discussed before.

1.9.2 Examples of operator C

Example 1.52 Let ω be an open set in Ω and χ_ω be its *characteristics* function. Then,

$$Cy = y\chi_\omega \tag{1.297}$$

corrcsponds to

$$\mathcal{H} = L^2(\omega). \tag{1.298}$$

Here, we want to reach (or to get close) to a given state *on the subset* ω.

Example 1.53 Let $\theta_1, \ldots, \theta_N$ be N given *linearly independent* elements of $L^2(\Omega)$. Then,

$$Cy = \{(y, \theta_i)_{L^2(\Omega)}\}_{i=1}^N \tag{1.299}$$

corresponds to $\mathcal{H} = \mathbb{R}^N$.

The same considerations as in previous sections apply. In particular, we will write down explicitly a dual formulation of problems (1.295) and (1.296) in the particular cases of Examples 1.52 and 1.53.

1.9.3 Dual formulation in the case of Example 1.52

Let g be given in $L^2(\omega)$. We introduce ψ_g defined by

$$-\frac{\partial \psi_g}{\partial t} + \mathcal{A}^*\psi_g = 0 \text{ in } Q, \quad \psi_g(T) = g\chi_\omega, \quad \psi_g = 0 \text{ on } \Sigma. \tag{1.300}$$

A *dual problem*, for the problem of Example 1.52, is then

$$\min_{g\in L^2(\omega)} \left(\frac{1}{2} \iint_{\mathcal{O}\times(0,T)} \psi_g^2 \,\mathrm{d}x\,\mathrm{d}t - (g, h_T)_{L^2(\omega)} + \beta \,\|g\|_{L^2(\omega)} \right). \tag{1.301}$$

If f is the (unique) solution of the dual problem (1.301) the solution u of the corresponding control problem (1.295) is given by $u = \psi \chi_{\mathcal{O}\times(0,T)}$, where ψ is the solution of (1.300) corresponding to $g = f$.

1.9.4 Dual formulation in the case of Example 1.53

Let $\mathbf{g} = \{g_i\}_{i=1}^N$ be given in $\mathbb{R}^N$. We define ψ_g by

$$-\frac{\partial \psi_g}{\partial t} + \mathcal{A}^* \psi_g = 0 \text{ in } Q, \quad \psi_g(T) = \sum_{i=1}^N g_i \theta_i, \quad \psi_g = 0 \text{ on } \Sigma. \tag{1.302}$$

The dual problem is then analogous to (1.301) but with $\mathbb{R}^N$ playing the role of $L^2(\omega)$:

$$\min_{\mathbf{g}\in\mathbb{R}^N} \left(\frac{1}{2} \iint_{\mathcal{O}\times(0,T)} \psi_g^2 \, \mathrm{d}x \, \mathrm{d}t - (\mathbf{g}, \mathbf{h}_T)_{\mathbb{R}^N} + \beta \, \|\mathbf{g}\|_{\mathbb{R}^N} \right). \tag{1.303}$$

If $\mathbf{f}$ is the unique solution of problem (1.303) the solution u of the corresponding control problem (1.295) is given by $u = \psi \chi_{\mathcal{O}\times(0,T)}$, where ψ is the solution of (1.302) corresponding to $\mathbf{g} = \mathbf{f}$.

1.9.5 Further comments

Remark 1.54 We can also consider *time averages* as shown in J.L. Lions (1993).

Concerning now the numerical solution of problem (1.295), it can be achieved by methods directly inspired by those discussed in Section 1.8. It is, in particular, quite convenient to introduce an operator $\Lambda \in \mathcal{L}(\mathcal{H}, \mathcal{H}')$ which will play for problem (1.295) the role played for problems (1.111) and (1.116) by the operator Λ defined in Section 1.5 (see also Section 1.8.2). Considering, first, Example 1.52, the dual problem (1.301) can also be written as

$$\begin{cases} f \in L^2(\omega), \quad \forall g \in L^2(\omega), \\ (\Lambda f, g - f)_{L^2(\omega)} + \beta \, \|g\|_{L^2(\omega)} - \beta \, \|f\|_{L^2(\omega)} \geq (h_T, g - f)_{L^2(\omega)}, \end{cases} \tag{1.304}$$

where, in (1.304), operator Λ is defined as follows:

$$\Lambda g = \varphi_g(T) \chi_\omega, \quad \forall g \in L^2(\omega), \tag{1.305}$$

with $\varphi_g(T)$ obtained from g via (1.300) and

$$\frac{\partial \varphi_g}{\partial t} + \mathcal{A} \varphi_g = \psi_g \chi_{\mathcal{O}\times(0,T)} \text{ in } Q, \quad \varphi_g(0) = 0, \quad \varphi_g = 0 \text{ on } \Sigma. \tag{1.306}$$

Operator $\Lambda \in \mathcal{L}\left(L^2(\omega), L^2(\omega)\right)$ and is *symmetric* and *positive definite* over $L^2(\omega)$. The numerical methods discussed in Section 1.8.8 can be easily modified in order to accommodate problem (1.304).

Consider, now, Example 1.53; the dual problem (1.303) can be written as

$$\begin{cases} \mathbf{f} \in \mathbb{R}^N, \quad \forall \mathbf{g} \in \mathbb{R}^N, \\ (\mathbf{\Lambda f}, \mathbf{g} - \mathbf{f})_{\mathbb{R}^N} + \beta \, \|\mathbf{g}\|_{\mathbb{R}^N} - \beta \, \|\mathbf{f}\|_{\mathbb{R}^N} \geq (\mathbf{h}_T, \mathbf{g} - \mathbf{f})_{\mathbb{R}^N}, \end{cases} \tag{1.307}$$

where, in (1.307), $\mathbf{\Lambda}$ is the $m \times m$ *symmetric* and *positive definite matrix* defined by

$$\mathbf{\Lambda} = (\lambda_{ij})_{1 \leq i,j \leq m}, \quad \lambda_{ij} = \int_\Omega \phi_i(T)\theta_j \, dx, \tag{1.308}$$

with, in (1.308), ϕ_i defined from θ_i by

$$-\frac{\partial \psi_i}{\partial t} + \mathcal{A}^* \psi_i = 0 \text{ in } Q, \quad \psi_i(T) = \theta_i, \quad \psi_i = 0 \text{ on } \Sigma, \tag{1.309}$$

$$\frac{\partial \phi_i}{\partial t} + \mathcal{A}\phi_i = \psi_i \chi_{\mathcal{O} \times (0,T)} \text{ in } Q, \quad \phi_i(0) = 0, \quad \phi_i = 0 \text{ on } \Sigma. \tag{1.310}$$

Remark 1.55 Problem has clearly the "flavor" of a *Galerkin method* (like those discussed in Section 1.8 to solve problems (1.117) and (1.118)).

To conclude this section we shall discuss a solution method for problem (1.307); this method is applicable if N is not too large, since it relies on the *explicit* construction of matrix $\mathbf{\Lambda}$. Our solution method is based on the fact that, according to, for example, Glowinski, Lions, and Trémolières (1976, Chapter 2) and (1981, Chapter 2 and Appendix 2), problem (1.307) is *equivalent* to the following *nonlinear system*:

$$\begin{cases} \mathbf{\Lambda f} + \beta \mathbf{p} = \mathbf{h}_T, \\ (\mathbf{p}, \mathbf{f})_{\mathbb{R}^N} = \|\mathbf{f}\|_{\mathbb{R}^N}, \quad \|\mathbf{p}\|_{\mathbb{R}^N} \leq 1, \end{cases} \tag{1.311}$$

which has a *unique* solution since $\beta > 0$. System (1.311) is in turn *equivalent* to

$$\begin{cases} \mathbf{\Lambda f} + \beta \mathbf{p} = \mathbf{h}_T, \\ \mathbf{p} = P_B(\mathbf{p} + \rho \mathbf{f}), \quad \forall \rho > 0, \end{cases} \tag{1.312}$$

where, in (1.312), $P_B : \mathbb{R}^N \to \mathbb{R}^N$ is the *orthogonal projector* from $\mathbb{R}^N$ onto the *closed unit ball* B of $\mathbb{R}^N$; we clearly have, $\forall \mathbf{g} \in \mathbb{R}^N$,

$$P_B(\mathbf{g}) = \begin{cases} \mathbf{g} & \text{if } \mathbf{g} \in B, \\ \mathbf{g} / \|\mathbf{g}\|_{\mathbb{R}^N} & \text{if } \mathbf{g} \notin B. \end{cases}$$

Relations (1.312) suggest the following *iterative method* (of the *fixed point* type):

$$\mathbf{p}^0 \text{ is given in } B \text{ (we can take, for example, } \mathbf{p}^0 = \mathbf{0}); \tag{1.313}$$

then for $n \geq 0$, assuming that $\mathbf{p}^n$ is known, we compute $\mathbf{f}^n$, and then $\mathbf{p}^{n+1}$, by

$$\mathbf{\Lambda f}^n = \mathbf{h}_T - \beta \mathbf{p}^n, \tag{1.314}$$

$$\mathbf{p}^{n+1} = P_B\left(\mathbf{p}^n + \rho \mathbf{f}^n\right). \tag{1.315}$$

Concerning the convergence of algorithm (1.313)–(1.315) we have the following.

Proposition 1.56 *Suppose that*

$$0 < \rho < 2\frac{\mu_1}{\beta}, \tag{1.316}$$

where μ_1 is the smallest eigenvalue of matrix $\mathbf{\Lambda}$. Then, $\forall \mathbf{p}^0 \in B$, we have

$$\lim_{n \to +\infty} \left\{\mathbf{f}^n, \mathbf{p}^n\right\} = \{\mathbf{f}, \mathbf{p}\}, \tag{1.317}$$

where $\{\mathbf{f}, \mathbf{p}\}$ is the solution of (1.311).

Proof. The convergence result (1.317) is a direct consequence of Glowinski, Lions and Trémolières (1976, Chapter 2) and (1981, Chapter 2 and Appendix 2) (see also Ciarlet, 1989, Chapter 9); however, we shall prove it here for the sake of completeness. Introduce therefore $\bar{\mathbf{f}}^n = \mathbf{f}^n - \mathbf{f}$ and $\bar{\mathbf{p}}^n = \mathbf{p}^n - \mathbf{p}$; we have

$$\mathbf{\Lambda}\bar{\mathbf{f}}^n = -\beta \bar{\mathbf{p}}^n. \tag{1.318}$$

We also have (since the operator P_B is a *contraction*)

$$\left\|\bar{\mathbf{p}}^{n+1}\right\|_{\mathbb{R}^N} \leq \left\|\bar{\mathbf{p}}^n + \rho \bar{\mathbf{f}}^n\right\|_{\mathbb{R}^N},$$

which implies in turn that

$$\begin{aligned}\left\|\bar{\mathbf{p}}^{n+1}\right\|^2_{\mathbb{R}^N} &\leq \left\|\bar{\mathbf{p}}^n\right\|^2_{\mathbb{R}^N} + 2\rho\left(\bar{\mathbf{p}}^n, \bar{\mathbf{f}}^n\right)_{\mathbb{R}^N} + \rho^2\left\|\bar{\mathbf{f}}^n\right\|^2_{\mathbb{R}^N} \\ &= \left\|\bar{\mathbf{p}}^n\right\|^2_{\mathbb{R}^N} - \frac{2\rho}{\beta}\left(\mathbf{\Lambda}\bar{\mathbf{f}}^n, \bar{\mathbf{f}}^n\right)_{\mathbb{R}^N} + \rho^2\left\|\bar{\mathbf{f}}^n\right\|^2_{\mathbb{R}^N}.\end{aligned} \tag{1.319}$$

It follows from (1.319) that

$$\begin{aligned}\left\|\bar{\mathbf{p}}^n\right\|^2_{\mathbb{R}^N} - \left\|\bar{\mathbf{p}}^{n+1}\right\|^2_{\mathbb{R}^N} &\geq \frac{2\rho}{\beta}\left(\mathbf{\Lambda}\bar{\mathbf{f}}^n, \bar{\mathbf{f}}^n\right)_{\mathbb{R}^N} - \rho^2\left\|\bar{\mathbf{f}}^n\right\|^2_{\mathbb{R}^N} \\ &\geq \rho\left(\frac{2\mu_1}{\beta} - \rho\right)\left\|\bar{\mathbf{f}}^n\right\|^2_{\mathbb{R}^N},\end{aligned} \tag{1.320}$$

where μ_1 (> 0) is the *smallest eigenvalue* of matrix $\mathbf{\Lambda}$. Suppose that (1.316) holds; it follows then from (1.320) that the sequence $\{\|\bar{\mathbf{p}}^n\|^2_{\mathbb{R}^N}\}_{n\geq 0}$ is *decreasing*; since it has 0 as a lower bound it converges to some (≥ 0) limit, implying that

$$\lim_{n \to +\infty}\left(\left\|\bar{\mathbf{p}}^n\right\|^2_{\mathbb{R}^N} - \left\|\bar{\mathbf{p}}^{n+1}\right\|^2_{\mathbb{R}^N}\right) = 0. \tag{1.321}$$

Combining (1.316), (1.320), and (1.321) we obtain that $\lim_{n\to+\infty} \|\bar{\mathbf{f}}^n\|_{\mathbb{R}^N} = 0$; we have thus shown that $\lim_{n\to+\infty} \mathbf{f}^n = \mathbf{f}$. The convergence of $\{\mathbf{p}^n\}_{n\geq 0}$ to $\mathbf{p}$ follows from the convergence of $\{\mathbf{f}^n\}_{n\geq 0}$ and from (1.314) (or (1.318)). □

Remark 1.57 If N is not too large, so that matrix $\mathbf{\Lambda}$ can be constructed (via relations (1.308)–(1.310)) at a reasonable cost, we shall use the *Cholesky factorization method* (see, for example, Ciarlet, 1989, Chapter 4) to solve the various systems (1.314). If N is very large, we can expect $\mathbf{\Lambda}$ to be *ill-conditioned* and expensive to construct and factorize; therefore, instead of using algorithm (1.313)–(1.315) we suggest solving problem (1.307) by simple variants of the methods used in Section 1.8.8 to solve problem (1.117).

1.10 Pointwise control

1.10.1 Generalities

A rather natural question in the present framework is to consider situations where in (1.1) the open set $\mathcal{O}$ is replaced by a "small" set, in particular a set of measure 0. One has then to consider functions which are not in $L^2(\Omega)$ (for a given t). Many situations can be considered. We confine ourselves here with the case where $\mathcal{O}$ *is reduced to a point*, namely,

$$\mathcal{O} = \{b\}, \quad b \in \Omega. \tag{1.322}$$

Then, if $\delta(x-b)$ denotes the *Dirac measure* at b, the *state function* y is given by

$$\frac{\partial y}{\partial t} + \mathcal{A}y = v(t)\delta(x-b) \text{ in } Q, \quad y(0) = 0, \quad y = 0 \text{ on } \Sigma. \tag{1.323}$$

In (1.323) the control v is now a function of t only. We shall assume that

$$v \in L^2(0,T). \tag{1.324}$$

Problem (1.323) has a *unique solution*, which is defined by *transposition*, as in Lions and Magenes (1968). It follows from this reference that, if $d \leq 3$ one has

$$y \in L^2(Q), \quad \frac{\partial y}{\partial t} \in L^2(0,T;H^{-2}(\Omega)), \tag{1.325}$$

so that

$$t \to y(t;v) \text{ is continuous from } [0,T] \text{ into } H^{-1}(\Omega). \tag{1.326}$$

When the control v spans the space $L^2(0,T)$, $y(T;v)$ spans a subspace of $H^{-1}(\Omega)$. Let us look for the orthogonal of (the closure of) this subspace. Let f be given in $H_0^1(\Omega)$ such that

$$\langle y(T;v), f\rangle = 0, \quad \forall v \in L^2(0,T), \tag{1.327}$$

where $\langle\cdot,\cdot\rangle$ denotes the duality pairing between $H^{-1}(\Omega)$ and $H_0^1(\Omega)$. Let ψ be the solution of

$$-\frac{\partial \psi}{\partial t} + \mathcal{A}^*\psi = 0 \text{ in } Q, \quad \psi(T) = f, \quad \psi = 0 \text{ on } \Sigma. \tag{1.328}$$

Then

$$\langle y(T;v), f\rangle = \int_0^T \psi(b,t)v(t)\,dt. \tag{1.329}$$

Remark 1.58 If $d = 1$, the "function" $\{x,t\} \to v(t)\delta(x-b)$ belongs to

$$L^2(0,T;H^{-1}(\Omega));$$

this property implies in turn that

$$y \in L^2(0,T;H_0^1(\Omega)) \cap C^0([0,T];L^2(\Omega)), \quad \frac{\partial y}{\partial t} \in L^2(0,T;H^{-1}(\Omega)).$$

Remark 1.59 Since $f \in H_0^1(\Omega)$ the solution ψ of problem (1.328) satisfies

$$\psi \in L^2(0,T;H^2(\Omega) \cap H_0^1(\Omega)),$$

so that, if $d \leq 3$, $\psi(b,t)$ makes sense with the function $t \to \psi(b,t)$ belonging to $L^2(0,T)$ (since the injection from $H^2(\Omega)$ into $C^0(\bar{\Omega})$ is continuous), implying the validity of (1.329).

It follows then from relations (1.327) and (1.329) that f belongs to the orthogonal of the space $\{y(T;v)\}_{v \in L^2(0,T)}$ if and only if

$$\psi(b,t) = 0. \tag{1.330}$$

The above results can be summarized as follows:

$$y(T;v) \text{ spans a dense subset of } H^{-1}(\Omega) \text{ when } v \text{ describes } L^2(0,T)$$
$$\text{if and only if } b \text{ is such that (1.330) implies } \psi = 0. \tag{1.331}$$

This is a condition on b that we will further discuss in the next section.

1.10.2 On the concept of strategic point. Formulation of a control problem

We assume that

$$\mathcal{A} = \mathcal{A}^*, \quad \mathcal{A} \text{ independent of } t. \tag{1.332}$$

Taking advantage of the fact that Ω is *bounded*, we introduce the *eigenfunctions* and *eigenvalues* of $\mathcal{A}$, that is,

$$\mathcal{A}w_j = \lambda_j w_j, \quad w_j = 0 \text{ on } \Gamma, \quad w_j \neq 0, \tag{1.333}$$

and impose $\|w_j\|_{L^2(\Omega)} = 1$. Then (assuming in order to simplify the presentation that the spectrum of $\mathcal{A}$ is *simple*)

$$\psi(t) = \sum_{j=1}^{+\infty} (f, w_j)\, w_j \exp\left(-\lambda_j(T-t)\right), \tag{1.334}$$

where, in (1.334), $(\cdot,\cdot)$ denotes the scalar product of $L^2(\Omega)$. We shall say that b is a *strategic point* in Ω if

$$w_j(b) \neq 0, \quad \forall j = 1, \dots, +\infty. \tag{1.335}$$

Then, (1.330) implies $(f, w_j) = 0$, $\forall j = 1, \dots, +\infty$, that is, $f = 0$. In this case, (1.331) *is true if and only if b is a strategic point.*

We suppose from now on that (1.331) holds true. We are then looking for the solution of the following control problem:

$$\inf_{v \in \mathcal{U}_f} \frac{1}{2} \int_0^T v^2 \, dt, \tag{1.336}$$

with

$$\mathcal{U}_f = \{ v \mid v \in L^2(0,T), \, y(T; v) \in y_T + \beta B_{-1} \}, \tag{1.337}$$

where y_T is given in $H^{-1}(\Omega)$ and where B_{-1} denotes the unit ball of $H^{-1}(\Omega)$.

1.10.3 Duality results

The *dual problem* of (1.336) is as follows: One looks (with obvious notation) for the solution of

$$\inf_{g \in H_0^1(\Omega)} \left(\frac{1}{2} \int_0^T \left| \psi_g(b,t) \right|^2 \, dt - \langle y_T, g \rangle + \beta \, \|g\|_{H_0^1(\Omega)} \right), \tag{1.338}$$

where, in (1.338), ψ_g is obtained from g via the solution of

$$-\frac{\partial \psi_g}{\partial t} + \mathcal{A}^* \psi_g = 0 \text{ in } Q, \quad \psi_g(T) = g, \quad \psi_g = 0 \text{ on } \Sigma; \tag{1.339}$$

the minima in (1.336) and (1.338) are opposite. If f is the solution of problem (1.338) then the *optimal control u* (that is, the solution of the "primal" problem (1.336)) is given by

$$u(t) = \psi(b,t), \tag{1.340}$$

where ψ is the solution of (1.339) corresponding to $g = f$.

1.10.4 Iterative solution of the dual problem

From a practical point of view it is convenient to introduce the *linear operator* $\Lambda \in \mathcal{L}(H_0^1(\Omega), H^{-1}(\Omega))$ defined by

$$\Lambda g = \phi_g(T), \tag{1.341}$$

where, in (1.341), ϕ_g is obtained from g via (1.339) and

$$\frac{\partial \phi_g}{\partial t} + \mathcal{A} \phi_g = \psi_g(b,t) \delta(x-b) \text{ in } Q, \quad \phi_g(0) = 0, \quad \phi_g = 0 \text{ on } \Sigma. \tag{1.342}$$

We can easily show that (with obvious notation)

$$\langle \Lambda g_1, g_2 \rangle = \int_0^T \psi_1(b,t)\psi_2(b,t)\,dt, \quad \forall g_1, g_2 \in H_0^1(\Omega). \tag{1.343}$$

Relation (1.343) implies that operator Λ is *self-adjoint* and *positive semi-definite*; it is *positive definite* if b is *strategic*. Combining (1.338) and (1.343) we can rewrite (1.338) as follows:

$$\inf_{g \in H_0^1(\Omega)} \left(\frac{1}{2}\langle \Lambda g, g \rangle - \langle y_T, g \rangle + \beta \|g\|_{H_0^1(\Omega)} \right). \tag{1.344}$$

The minimization problem (1.344) is *equivalent* to the following *variational inequality*:

$$\begin{cases} f \in H_0^1(\Omega), \quad \forall g \in H_0^1(\Omega), \\ \langle \Lambda f, g - f \rangle + \beta \|g\|_{H_0^1(\Omega)} - \beta \|f\|_{H_0^1(\Omega)} \geq \langle y_T, g - f \rangle, \end{cases} \tag{1.345}$$

which can also be written as

$$y_T \in \Lambda f + \beta \partial j(f), \tag{1.346}$$

where $\partial j(\cdot)$ is the *subgradient* of the convex functional $j\colon H_0^1(\Omega) \to \mathbb{R}$ defined by

$$j(g) = \|g\|_{H_0^1(\Omega)}, \quad \forall g \in H_0^1(\Omega).$$

As seen previously (in Section 1.8.8, particularly) to solve problem (1.346), we can associate with it the following *initial value problem* in $H_0^1(\Omega)$ (with $\Delta = \nabla^2$ the *Laplace operator*):

$$\begin{cases} \dfrac{\partial(-\Delta f)}{\partial \tau} + \Lambda f + \beta \partial j(f) \ni y_T \quad \text{on } (0, +\infty), \\ f(0) = f_0, \end{cases} \tag{1.347}$$

and, then, integrate (1.347) from $\tau = 0$ to $\tau = +\infty$ in order to obtain the steady-state solution of (1.347), that is, the solution of problem (1.346). As in Section 1.8.8, the *Peaceman–Rachford scheme* is well suited to the solution of problem (1.347); we obtain

$$f^0 = f_0, \tag{1.348}$$

then for $n \geq 0$, f^n being known, we obtain $f^{n+\frac{1}{2}}$ and f^{n+1} from

$$\frac{-\Delta\left(f^{n+\frac{1}{2}} - f^n\right)}{\frac{1}{2}\Delta\tau} + \beta \partial j(f^{n+\frac{1}{2}}) + \Lambda f^n \ni y_T, \tag{1.349}$$

$$\frac{-\Delta\left(f^{n+1} - f^{n+\frac{1}{2}}\right)}{\frac{1}{2}\Delta\tau} + \beta \partial j(f^{n+\frac{1}{2}}) + \Lambda f^{n+1} \ni y_T. \tag{1.350}$$

Problem (1.349) is equivalent to the following *minimization problem*:

$$\min_{g \in H_0^1(\Omega)} \left(\frac{1}{2} \int_\Omega |\nabla g|^2 \, dx + \frac{\beta \Delta \tau}{2} \left(\int_\Omega |\nabla g|^2 \, dx \right)^{\frac{1}{2}} - \int_\Omega \nabla f^n \cdot \nabla g \, dx - \frac{\Delta \tau}{2} \langle y_T - \Lambda f^n, g \rangle \right). \tag{1.351}$$

Problem (1.351) has a *unique* solution $f^{n+\frac{1}{2}} \in H_0^1(\Omega)$; this solution is given by

$$f^{n+\frac{1}{2}} = \lambda^{n+\frac{1}{2}} f_*^{n+\frac{1}{2}}, \tag{1.352}$$

where, in (1.352),

(i) $f_*^{n+\frac{1}{2}}$ is *the* solution of the following *linear Dirichlet problem*:

$$\begin{cases} f_*^{n+\frac{1}{2}} \in H_0^1(\Omega), \quad \forall g \in H_0^1(\Omega), \\ \displaystyle\int_\Omega \nabla f_*^{n+\frac{1}{2}} \cdot \nabla g \, dx = \int_\Omega \nabla f^n \cdot \nabla g \, dx + \frac{1}{2} \Delta \tau \langle y_T - \Lambda f^n, g \rangle, \end{cases} \tag{1.353}$$

(that is, $f_*^{n+\frac{1}{2}}$ satisfies $-\Delta \left(f_*^{n+\frac{1}{2}} - f^n \right) = \frac{1}{2} \Delta \tau \, (y_T - \Lambda f^n)$ in Ω, $f_*^{n+\frac{1}{2}} = 0$ on Γ).

(ii) $\lambda^{n+\frac{1}{2}} \geq 0$ and is *the* minimizer over $\mathbb{R}_+$ of the *quadratic polynomial*

$$\lambda \to \left(\frac{1}{2} \lambda^2 - \lambda \right) \| f_*^{n+\frac{1}{2}} \|_{H_0^1(\Omega)}^2 + \frac{1}{2} \beta \Delta \tau \lambda \| f_*^{n+\frac{1}{2}} \|_{H_0^1(\Omega)};$$

we then have

$$\lambda^{n+\frac{1}{2}} = \begin{cases} 1 - \dfrac{\beta \Delta \tau}{2 \| f_*^{n+\frac{1}{2}} \|_{H_0^1(\Omega)}^2}, & \text{if } \| f_*^{n+\frac{1}{2}} \|_{H_0^1(\Omega)}^2 \geq \frac{1}{2} \beta \Delta \tau, \\ 0 & \text{if } \| f_*^{n+\frac{1}{2}} \|_{H_0^1(\Omega)}^2 < \frac{1}{2} \beta \Delta \tau. \end{cases} \tag{1.354}$$

Now, in order to compute f^{n+1} we observe that (1.349) and (1.350) imply

$$-\frac{\Delta \left(f^{n+1} - 2 f^{n+\frac{1}{2}} + f^n \right)}{\frac{1}{2} \Delta \tau} + \Lambda f^{n+1} = \Lambda f^n,$$

that is,

$$-\frac{2}{\Delta \tau} \Delta f^{n+1} + \Lambda f^{n+1} = \Lambda f^n - \frac{2}{\Delta \tau} \Delta \left(2 f^{n+\frac{1}{2}} - f^n \right). \tag{1.355}$$

Problem (1.355) is a particular case of the *"generalized" elliptic problem*

$$-r^{-1}\Delta f + \Lambda f = \text{rhs}, \tag{1.356}$$

where rhs $\in H^{-1}(\Omega)$, $r > 0$ (and the operator Λ is "*non local*"); the solution of problems like (1.356) will be addressed in the following section.

1.10.5 Solution of problem (1.356)

1.10.5.1 Generalities

From the properties of the operators $-\Delta$ and Λ (*ellipticity* and *symmetry*) problem (1.356) can be solved by a *conjugate gradient algorithm* (like the one discussed in Section 1.8.2) operating in $H_0^1(\Omega)$. We think, however, that it may be instructive to discuss first a class of control problems closely related to problem (1.336) for which the *dual problems* are in the same form as in (1.356). Let us consider therefore the following class of *approximate pointwise controllability problems*:

$$\min_{v \in L^2(0,T)} \left(\frac{1}{2} \int_0^T v^2 \, dt + \frac{k}{2} \|y(T) - y_T\|_{-1}^2 \right), \tag{1.357}$$

obtained by *penalization* of the final condition $y(T) = y_T$. In (1.357),

(i) The *penalty* parameter k is *positive.*
(ii) The function y is obtained from v via the solution of problem (1.323).
(iii) The *target* "function" y_T belongs to $H^{-1}(\Omega)$.
(iv) The $H^{-1}(\Omega)$-norm $\|\cdot\|_{-1}$ is defined, $\forall g \in H^{-1}(\Omega)$, by

$$\|g\|_{-1} = \|\varphi_g\|_{H_0^1(\Omega)} \left(= \left(\int_\Omega |\nabla \varphi_g|^2 \, dx \right)^{\frac{1}{2}} \right), \tag{1.358a}$$

with φ_g the solution in $H_0^1(\Omega)$ of the linear Dirichlet problem

$$-\Delta \varphi_g = g \text{ in } \Omega, \quad \varphi_g = 0 \text{ on } \Gamma. \tag{1.358b}$$

Problem (1.357) has a *unique* solution u which is characterized by the existence of p belonging to $L^2(0, T; H^2(\Omega) \cap H_0^1(\Omega)) \cap C^0([0, T]; H_0^1(\Omega))$ such that $\{u, y, p\}$ satisfies the following *optimality system*:

$$\frac{\partial y}{\partial t} + \mathcal{A}y = u\delta(x - b) \text{ in } Q, \quad y = 0 \text{ on } \Sigma, \quad y(0) = 0, \tag{1.359}$$

$$\begin{cases} -\dfrac{\partial p}{\partial t} + \mathcal{A}^* p = 0 \text{ in } Q, \quad p = 0 \text{ on } \Sigma, \\ p(T) \in H_0^1(\Omega) \text{ with } p(T) = k(-\Delta)^{-1}(y_T - y(T)), \end{cases} \tag{1.360}$$

$$u(t) = p(b, t). \tag{1.361}$$

Let us define $f \in H_0^1(\Omega)$ by $f = p(T)$; it follows then from (1.359)–(1.361) that f is solution of the (*dual*) problem

$$-k^{-1}\Delta f + \Lambda f = y_T. \tag{1.362}$$

Concerning the solution of problem (1.357) we have two options: namely, we can use either the *primal formulation* (1.357) or the *dual formulation* (1.362). Both approaches will be discussed in the following two paragraphs.

1.10.5.2 Direct solution of problem (1.357)

Solving the control problem (1.357) *directly* (that is, in $L^2(0,T)$) is worth considering for the following reasons:

(i) It can be generalized to pointwise control problems with *nonlinear* state equations.
(ii) The space $L^2(0,T)$ is a space of *one variable* functions, even for multidimensional domains Ω (that is, $\Omega \subset \mathbb{R}^d$ with $d \geq 2$).
(iii) The structure of the space $L^2(0,T)$ is quite simple, making the implementation of conjugate gradient algorithms operating in this space fairly easy.

Let us denote by $J(\cdot)$ the functional in (1.357); the solution u of problem (1.357) satisfies $J'(u) = 0$ where $J'(u)$ is the gradient at u of the functional $J(\cdot)$. Let us consider $v \in L^2(0,T)$; we can identify $J'(v)$ with an element of $L^2(0,T)$ and we have

$$\int_0^T J'(v)w\,\mathrm{d}t = \int_0^T \big(v(t) - p(b,t)\big)w(t)\,\mathrm{d}t, \quad \forall v, w \in L^2(0,T), \tag{1.363}$$

where, in (1.363), p is obtained from v via (1.323) and the corresponding *adjoint equation*, namely,

$$\begin{cases} -\dfrac{\partial p}{\partial t} + \mathcal{A}^* p = 0 \text{ in } Q, \quad p = 0 \text{ on } \Sigma, \\ p(T) \in H_0^1(\Omega) \text{ with } p(T) = k(-\Delta)^{-1}\left(y_T - y(T)\right). \end{cases} \tag{1.364}$$

Writing $J'(u) = 0$ in *variational form*, namely,

$$\begin{cases} u \in L^2(0,T), \\ \displaystyle\int_0^T J'(u)v\,\mathrm{d}t = 0, \quad \forall v \in L^2(0,T), \end{cases}$$

and taking into account the fact that operator $v \to J'(v)$ is *affine* with respect to v (with a linear part associated with an $L^2(0,T)$-elliptic operator), we observe that problem (1.357) is a particular case of problem (1.123) (see Section 1.8.2); it can be solved therefore by the *conjugate gradient algorithm* (1.124)–(1.131). In the particular case considered here this algorithm takes the following form.

Algorithm 1.60 *(Conjugate gradient solution of problem (1.357))*

$$u^0 \in L^2(0,T) \text{ is given;} \tag{1.365}$$

solve

$$\frac{\partial y^0}{\partial t} + \mathcal{A}y^0 = u^0\delta(x-b) \text{ in } Q, \quad y^0 = 0 \text{ on } \Sigma, \quad y^0(0) = 0, \tag{1.366}$$

then

$$\begin{cases} -\dfrac{\partial p^0}{\partial t} + \mathcal{A}^* p^0 = 0 \text{ in } Q, \quad p^0 = 0 \text{ on } \Sigma, \\ p^0(T) \in H_0^1(\Omega), \text{ with } p^0(T) = k(-\Delta)^{-1}\left(y_T - y^0(T)\right), \end{cases} \tag{1.367}$$

$$\begin{cases} g^0 \in L^2(0,T), \\ \displaystyle\int_0^T g^0 v \, dt = \int_0^T \left(u^0(t) - p^0(b,t)\right) v(t) \, dt, \quad \forall v \in L^2(0,T), \end{cases} \tag{1.368}$$

and set

$$w^0 = g^0. \tag{1.369}$$

Assuming that u^n, g^n, and w^n are known (with the last two different from 0), we obtain u^{n+1}, g^{n+1}, and (if necessary) w^{n+1} as follows:
Solve

$$\frac{\partial \bar{y}^n}{\partial t} + \mathcal{A}\bar{y}^n = w^n\delta(x-b) \text{ in } Q, \quad \bar{y}^n = 0 \text{ on } \Sigma, \quad \bar{y}^n(0) = 0, \tag{1.370}$$

and then

$$\begin{cases} -\dfrac{\partial \bar{p}^n}{\partial t} + \mathcal{A}^* \bar{p}^n = 0 \text{ in } Q, \quad \bar{p}^n = 0 \text{ on } \Sigma, \\ \bar{p}^n(T) \in H_0^1(\Omega) \text{ with } \bar{p}^n(T) = k(-\Delta)^{-1}(-\bar{y}^n(T)), \end{cases} \tag{1.371}$$

and

$$\begin{cases} \bar{g}^n \in L^2(0,T), \\ \displaystyle\int_0^T \bar{g}^n v \, dt = \int_0^T (w^n(t) - \bar{p}^n(b,t)) v(t) \, dt, \quad \forall v \in L^2(0,T). \end{cases} \tag{1.372}$$

Compute then

$$\rho_n = \int_0^T |g^n|^2 \, dt \bigg/ \int_0^T \bar{g}^n w^n \, dt, \tag{1.373}$$

and update u^n and g^n by

$$u^{n+1} = u^n - \rho_n w^n, \tag{1.374}$$

$$g^{n+1} = g^n - \rho_n \bar{g}^n, \tag{1.375}$$

respectively. If $\|g^{n+1}\|_{L^2(0,T)} \Big/ \|g^0\|_{L^2(0,T)} \leq \varepsilon$, *take* $u = u^{n+1}$; *otherwise, compute*

$$\gamma_n = \frac{\|g^{n+1}\|^2_{L^2(0,T)}}{\|g^n\|^2_{L^2(0,T)}}, \tag{1.376}$$

and update w^n *by*

$$w^{n+1} = g^{n+1} + \gamma_n w^n. \tag{1.377}$$

Do $n = n + 1$ *and go to* (1.370).

A finite element/finite difference implementation of the above algorithm will be briefly discussed in Section 1.10.6.

1.10.5.3 A duality method for the solution of problem (1.357)

Suppose that we can solve the *dual* problem (1.362), then from f $(= p(T))$ we can compute p, via (1.360), and obtain the control u via (1.361). Problem (1.362) is equivalent to

$$\begin{cases} f \in H_0^1(\Omega), \\ k^{-1} \displaystyle\int_\Omega \nabla f \cdot \nabla g \, dx + \langle \Lambda f, g \rangle = \langle y_T, g \rangle, \quad \forall g \in H_0^1(\Omega). \end{cases} \tag{1.378}$$

From the *symmetry*, *positivity*, and *continuity* of Λ (see Section 1.10.4) the *bilinear functional* in the left-hand side of (1.378) is *continuous*, *symmetric*, and $H_0^1(\Omega)$-*elliptic* (we have indeed

$$k^{-1} \int_\Omega |\nabla g|^2 \, dx + \langle \Lambda g, g \rangle \geq k^{-1} \|g\|^2_{H_0^1(\Omega)}, \quad \forall g \in H_0^1(\Omega));$$

problem (1.378) (and therefore (1.357)) can be solved by a conjugate gradient algorithm operating this time in $H_0^1(\Omega)$. This algorithm – which is closely related to algorithm (1.137)–(1.151) – reads as follows:

Algorithm 1.61 *(Conjugate gradient solution of problem (1.378))*

$$f^0 \in H_0^1(\Omega) \textit{ is given;} \tag{1.379}$$

solve

$$-\frac{\partial p^0}{\partial t} + A^* p^0 = 0 \textit{ in } Q, \quad p^0 = 0 \textit{ on } \Sigma, \quad p^0(T) = f^0, \tag{1.380}$$

then

$$\frac{\partial y^0}{\partial t} + \mathcal{A}y^0 = p^0(b,t)\delta(x-b) \text{ in } Q, \quad y^0 = 0 \text{ on } \Sigma, \quad y^0(0) = 0, \tag{1.381}$$

$$\begin{cases} g^0 \in H_0^1(\Omega), \quad \forall \phi \in H_0^1(\Omega), \\ \int_\Omega \nabla g^0 \cdot \nabla \phi \, \mathrm{d}x = k^{-1} \int_\Omega \nabla f^0 \cdot \nabla \phi \, \mathrm{d}x + \langle y^0(T) - y_T, \phi \rangle, \end{cases} \tag{1.382}$$

and set

$$w^0 = g^0. \tag{1.383}$$

Assuming that f^n, g^n and w^n are known (with the last two different from 0), we obtain f^{n+1}, g^{n+1}, and w^{n+1} as follows:
Solve

$$-\frac{\partial \bar{p}^n}{\partial t} + \mathcal{A}^*\bar{p}^n = 0 \text{ in } Q, \quad \bar{p}^n = 0 \text{ on } \Sigma, \quad \bar{p}^n(T) = w^n, \tag{1.384}$$

then

$$\frac{\partial \bar{y}^n}{\partial t} + \mathcal{A}\bar{y}^n = \bar{p}^n(b,t)\delta(x-b) \text{ in } Q, \quad \bar{y}^n = 0 \text{ on } \Sigma, \quad \bar{y}^n(0) = 0 \tag{1.385}$$

and

$$\begin{cases} \bar{g}^n \in H_0^1(\Omega), \quad \forall \phi \in H_0^1(\Omega), \\ \int_\Omega \nabla \bar{g}^n \cdot \nabla \phi \, \mathrm{d}x = k^{-1} \int_\Omega \nabla w^n \cdot \nabla \phi \, \mathrm{d}x + \langle \bar{y}^n(T), \phi \rangle. \end{cases} \tag{1.386}$$

Compute then

$$\rho_n = \int_\Omega |\nabla g^n|^2 \, \mathrm{d}x \Big/ \int_\Omega \nabla \bar{g}^n \cdot \nabla w^n \, \mathrm{d}x \,, \tag{1.387}$$

and update f^n and g^n by

$$f^{n+1} = f^n - \rho_n w^n, \tag{1.388}$$

and

$$g^{n+1} = g^n - \rho_n \bar{g}^n, \tag{1.389}$$

respectively. If $\|g^{n+1}\|_{H_0^1(\Omega)} \Big/ \|g^0\|_{H_0^1(\Omega)} \le \varepsilon$, take $f = f^{n+1}$; otherwise, compute

$$\gamma_n = \int_\Omega \left|\nabla g^{n+1}\right|^2 \mathrm{d}x \Big/ \int_\Omega |\nabla g^n|^2 \, \mathrm{d}x, \tag{1.390}$$

and update w^n by

$$w^{n+1} = g^{n+1} + \gamma_n w^n. \tag{1.391}$$

Do $n = n+1$ and go to (1.384).

Remark 1.62 Concerning the *speed of convergence* of Algorithm 1.61 it follows from Section 1.8.2, relation (1.133), that the convergence will be obtained for some n such that

$$n \leq n_0 \sim \ln\left(\frac{1}{\varepsilon}\nu_k^{\frac{3}{2}}\right) \Big/ \ln\left(\frac{\sqrt{\nu_k}+1}{\sqrt{\nu_k}-1}\right), \tag{1.392}$$

where

$$\nu_k = \left\|k^{-1}I + \tilde{\Lambda}\right\| \left\|\left(k^{-1}I + \tilde{\Lambda}\right)^{-1}\right\| \quad \text{(with } \tilde{\Lambda} = (-\Delta)^{-1}\Lambda\text{)}. \tag{1.393}$$

Since

$$\left\|k^{-1}I + \tilde{\Lambda}\right\| = k^{-1} + \left\|\tilde{\Lambda}\right\| \text{ and } \left\|\left(k^{-1}I + \tilde{\Lambda}\right)^{-1}\right\| = k,$$

it follows from (1.392) and (1.393) that for *large values* of k we have

$$n \leq n_0 \sim \left\|\tilde{\Lambda}\right\|^{\frac{1}{2}} k^{\frac{1}{2}} \ln\left(\frac{1}{\sqrt{\varepsilon}} k^{\frac{3}{4}}\right). \tag{1.394}$$

Similarly, we could have shown that an inequality such as (1.394) holds, concerning the convergence of Algorithm 1.60.

From a practical point of view we shall implement finite-dimensional variants of the above algorithms; these variants will be discussed in the following section.

1.10.6 Space–time discretizations of problems (1.336) and (1.357)

1.10.6.1 Generalities

We shall discuss in this section the *numerical solution* of the pointwise control problems discussed in Sections 1.10.2–1.10.5. The approximation methods to be discussed are closely related to those which have been employed in Section 1.8, namely, they will combine *time discretizations* by *finite difference* methods with *space discretizations by finite element methods*. Since the solution to the control problem (1.336) can be reduced to a sequence of problems such as (1.357), we shall focus our discussion on this last problem.

1.10.6.2 Approximations of the control problem (1.357)

We shall employ the *finite element* spaces V_h and V_{0h} defined in Section 1.8.4 (the notation of which is mostly kept). We approximate then the control problem (1.357) by

$$\min_{\mathbf{v}\in\mathbb{R}^N} J_h^{\Delta t}(\mathbf{v}), \tag{1.395}$$

where, in (1.395), we have $\Delta t = \frac{T}{N}$, $\mathbf{v} = \{v^n\}_{n=1}^N$ and

$$J_h^{\Delta t}(\mathbf{v}) = \frac{1}{2}\Delta t \sum_{n=1}^{N} |v^n|^2 + \frac{k}{2}\int_\Omega \left|\nabla \Phi_h^N\right|^2 dx, \tag{1.396}$$

with Φ_h^N obtained from $\mathbf{v}$ via the solution of the following *discrete parabolic problem*:

$$y_h^0 = 0; \tag{1.397}$$

then for $n = 1, \ldots, N$, assuming that y_h^{n-1} is known, we solve

$$\begin{cases} y_h^n \in V_{0h}, \\ \displaystyle\int_\Omega \frac{y_h^n - y_h^{n-1}}{\Delta t} z_h \, dx + a(y_h^n, z_h) = v^n z_h(b), \quad \forall z_h \in V_{0h}, \end{cases} \tag{1.398}$$

and finally

$$\begin{cases} \Phi_h^N \in V_{0h}, \\ \displaystyle\int_\Omega \nabla \Phi_h^N \cdot \nabla z_h \, dx = \langle y_T - y_h^N, z_h \rangle, \quad \forall z_h \in V_{0h}. \end{cases} \tag{1.399}$$

Problems (1.398) (for $n = 1, \ldots, N$) and (1.399) are *well-posed discrete Dirichlet problems* (we recall that $a(z_1, z_2) = \langle \mathcal{A} z_1, z_2 \rangle, \forall z_1, z_2 \in H_0^1(\Omega)$).

The discrete control problem (1.395) is *well-posed*; its unique solution – denoted by $\mathbf{u}_h^{\Delta t} = \{u^n\}_{n=1}^N$ – is *characterized* by

$$\nabla J_h^{\Delta t}(\mathbf{u}_h^{\Delta t}) = \mathbf{0}, \tag{1.400}$$

where, in (1.400), $\nabla J_h^{\Delta t}$ denotes the *gradient* of $J_h^{\Delta t}$.

Remark 1.63 The *convergence* of $\mathbf{u}_h^{\Delta t}$, and of the corresponding state vector, to their continuous counterparts is a fairly technical issue. It will not be addressed in this book. On the other hand, we shall address the solution of problem (1.395), via the solution of the *equivalent* linear problem (1.400); this will be the task of the following Sections 1.10.6.3 and 1.10.6.4.

Remark 1.64 The approximate control problem (1.395) relies on a time discretization by an *implicit Euler scheme*. Actually, we can improve accuracy by using, as in Section 1.8.6, a *second-order accurate two-step implicit time-discretization scheme*. By merging the techniques described in the present section with those in Section 1.8.6 we can easily derive a variant of the approximate problem (1.395) relying on the above second-order accurate time-discretization scheme.

1.10.6.3 Iterative solution of the discrete control problem (1.395) I: Calculation of $\nabla J_h^{\Delta t}$

In order to solve – via (1.400) – the discrete control problem (1.395) by a *conjugate gradient algorithm*, we need to be able to compute $\nabla J_h^{\Delta t}(\mathbf{v}), \forall \mathbf{v} \in \mathbb{R}^N$. To compute $\nabla J_h^{\Delta t}(\mathbf{v})$, we observe that

$$\lim_{\substack{\theta \to 0 \\ \theta \neq 0}} \frac{J_h^{\Delta t}(\mathbf{v} + \theta \mathbf{w}) - J_h^{\Delta t}(\mathbf{v})}{\theta} = \left(\nabla J_h^{\Delta t}(\mathbf{v}), \mathbf{w}\right)_{\Delta t}, \quad \forall \mathbf{v}, \mathbf{w} \in \mathbb{R}^N, \tag{1.401}$$

where

$$(\mathbf{v}, \mathbf{w})_{\Delta t} = \Delta t \sum_{n=1}^{N} v^n w^n, \quad \forall \mathbf{v}, \mathbf{w} \in \mathbb{R}^N, \quad (\text{and } \|\mathbf{v}\|_{\Delta t} = (\mathbf{v}, \mathbf{v})_{\Delta t}^{\frac{1}{2}}).$$

Combining (1.396)–(1.399) with (1.401) we can prove that

$$\left(\nabla J_h^{\Delta t}(\mathbf{v}), \mathbf{w}\right)_{\Delta t} = \Delta t \sum_{n=1}^{N} \left(v^n - p_h^n(b)\right) w^n, \quad \forall \mathbf{v}, \mathbf{w} \in \mathbb{R}^N, \tag{1.402}$$

where the family $\{p_h^n\}_{n=1}^N$ is obtained as the solution to the following *adjoint discrete parabolic problem*:

$$p_h^{N+1} = k\, \Phi_h^N; \tag{1.403}$$

then, for $n = N, \ldots, 1$, *assuming that* p_h^{n+1} *is known, solve (the well-posed discrete elliptic problem)*

$$\begin{cases} p_h^n \in V_{0h}, \\ \displaystyle\int_\Omega \frac{p_h^n - p_h^{n+1}}{\Delta t} z_h \, dx + a(z_h, p_h^n) = 0, \quad \forall z_h \in V_{0h}. \end{cases} \tag{1.404}$$

Owing to the importance of relation (1.402), we shall give a short proof of it (of the “engineer/physicist” type) based on a (formal) *perturbation analysis*. Hence, let us consider a perturbation $\delta\mathbf{v}$ of $\mathbf{v}$; we have then, from (1.396),

$$\delta J_h^{\Delta t}(\mathbf{v}) = \left(\nabla J_h^{\Delta t}(\mathbf{v}), \delta\mathbf{v}\right)_{\Delta t} = \Delta t \sum_{n=1}^{N} v^n \delta v^n + k \int_\Omega \nabla \Phi_h^N \cdot \nabla \delta \Phi_h^N \, dx, \tag{1.405}$$

where, in (1.405), $\delta\Phi_h^N$ is obtained from $\delta\mathbf{v}$ via

$$\delta y_h^0 = 0, \tag{1.406}$$

then for $n = 1, \ldots, N$, we have

$$\begin{cases} \delta y_h^n \in V_{0h}, \\ \displaystyle\int_\Omega \frac{\delta y_h^n - \delta y_h^{n-1}}{\Delta t} z_h \, dx + a\left(\delta y_h^n, z_h\right) = \delta v^n z_h(b), \quad \forall z_h \in V_{0h}, \end{cases} \tag{1.407}$$

and finally

$$\begin{cases} \delta \Phi_h^N \in V_{0h}, \\ \displaystyle\int_\Omega \nabla \delta \Phi_h^N \cdot \nabla z_h \, dx = -\int_\Omega \delta y_h^N z_h \, dx, \quad \forall z_h \in V_{0h}. \end{cases} \tag{1.408}$$

Taking $z_h = p_h^n$ in (1.407) we obtain, by summation from $n = 1$ to $n = N$,

$$\Delta t \sum_{n=1}^{N} p_h^n(b)\delta v^n = \Delta t \sum_{n=1}^{N} \int_\Omega \frac{\delta y_h^n - \delta y_h^{n-1}}{\Delta t} p_h^n \, dx + \Delta t \sum_{n=1}^{N} a\left(\delta y_h^n, p_h^n\right)$$
$$= \int_\Omega p_h^{N+1} \delta y_h^N \, dx$$
$$+ \Delta t \sum_{n=1}^{N} \left(\int_\Omega \frac{p_h^n - p_h^{n+1}}{\Delta t} \delta y_h^n \, dx + a\left(\delta y_h^n, p_h^n\right) \right). \tag{1.409}$$

Since $\{p_h^n\}_{n=1}^{N+1}$ satisfies (1.403), (1.404), it follows from (1.409) that

$$\int_\Omega p_h^{N+1} \delta y_h^N \, dx = \Delta t \sum_{n=1}^{N} p_h^n(b)\delta v^n. \tag{1.410}$$

Taking $z_h = \Phi_h^N$ in (1.408) we obtain, from (1.403),

$$k \int_\Omega \nabla \Phi_h^N \cdot \nabla \delta \Phi_h^N \, dx = -k \int_\Omega \delta y_h^N \Phi_h^N \, dx = -\int_\Omega p_h^{N+1} \delta y_h^N \, dx,$$

which combined with (1.405) and (1.408) implies

$$\left(\nabla J_h^{\Delta t}(\mathbf{v}), \delta \mathbf{v}\right)_{\Delta t} = \Delta t \sum_{n=1}^{N} (v^n - p_h^n(b))\delta v^n. \tag{1.411}$$

Since $\delta \mathbf{v}$ is "arbitrary," relation (1.411) implies (1.402).

1.10.6.4 Iterative solution of the discrete control (1.395). II: Conjugate gradient solution of problems (1.395) and (1.400)

The discrete control problem (1.395) is equivalent to a linear system (namely, (1.400)) which is associated with an $N \times N$ *symmetric* and *positive definite* matrix. Such a system can be solved therefore by a *conjugate gradient* algorithm which is a particular case of algorithm (1.124)–(1.131) (see Section 1.8.2) and a variant of algorithm (1.172)–(1.186) (see Section 1.8.5). This algorithm reads as follows.

Algorithm 1.65 *(Conjugate gradient solution of problem (1.395))*

$$\mathbf{u}_0 = \{u_0^n\}_{n=1}^{N} \textit{ is given in } \mathbb{R}^N; \tag{1.412}$$

take then

$$y_0^0 = 0, \tag{1.413}$$

and, for $n = 1, \ldots, N$, assuming that y_0^{n-1} is known, solve

$$\begin{cases} y_0^n \in V_{0h}, \\ \displaystyle\int_\Omega \frac{y_0^n - y_0^{n-1}}{\Delta t} z_h \, dx + a\left(y_0^n, z_h\right) = u_0^n z_h(b), \quad \forall z_h \in V_{0h}. \end{cases} \tag{1.414}$$

Solve next

$$\begin{cases} \Phi_0^N \in V_{0h}, \\ \displaystyle\int_\Omega \nabla \Phi_0^N \cdot \nabla z_h \, dx = \langle y_T - y_0^N, z_h \rangle, \quad \forall z_h \in V_{0h}. \end{cases} \tag{1.415}$$

Finally, take

$$p_0^{N+1} = k\, \Phi_0^N, \tag{1.416}$$

and, for $n = N, \ldots, 1$, assuming that p_0^{n+1} is known, solve

$$\begin{cases} p_0^n \in V_{0h}, \\ \displaystyle\int_\Omega \frac{p_0^n - p_0^{n+1}}{\Delta t} z_h \, dx + a\left(z_h, p_0^n\right) = 0, \quad \forall z_h \in V_{0h}. \end{cases} \tag{1.417}$$

Set

$$\mathbf{g}_0 = \{u_0^n - p_0^n(b)\}_{n=1}^N, \tag{1.418}$$

and

$$\mathbf{w}_0 = \mathbf{g}_0. \tag{1.419}$$

Then for $m \geq 0$, assuming that $\mathbf{u}_m$, $\mathbf{g}_m$, and $\mathbf{w}_m$ are known (with the last two different from $\mathbf{0}$) compute $\mathbf{u}_{m+1}$, $\mathbf{g}_{m+1}$, and $\mathbf{w}_{m+1}$ as follows:
Take

$$\bar{y}_m^0 = 0; \tag{1.420}$$

for $n = 1, \ldots, N$, assuming that $\bar{y}_m^{n-1}$ is known, solve

$$\begin{cases} \bar{y}_m^n \in V_{0h}, \\ \displaystyle\int_\Omega \frac{\bar{y}_m^n - \bar{y}_m^{n-1}}{\Delta t} z_h \, dx + a\left(\bar{y}_m^n, z_h\right) = w_m^n z_h(b), \quad \forall z_h \in V_{0h}. \end{cases} \tag{1.421}$$

Solve next

$$\begin{cases} \bar{\Phi}_m^N \in V_{0h}, \\ \displaystyle\int_\Omega \nabla \bar{\Phi}_m^N \cdot \nabla z_h \, dx = -\langle \bar{y}_m^N, z_h \rangle, \quad \forall z_h \in V_{0h}. \end{cases} \tag{1.422}$$

Finally, take

$$\bar{p}_m^{N+1} = k \bar{\Phi}_m^N, \tag{1.423}$$

and, for $n = N, \ldots, 1$, assuming that $\bar{p}_m^{n+1}$ is known, solve

$$\begin{cases} \bar{p}_m^n \in V_{0h}, \\ \displaystyle\int_\Omega \frac{\bar{p}_m^n - \bar{p}_m^{n+1}}{\Delta t} z_h \, dx + a\left(z_h, \bar{p}_m^n\right) = 0, \quad \forall z_h \in V_{0h}. \end{cases} \tag{1.424}$$

Set

$$\bar{\mathbf{g}}_m = \{w_m^n - \bar{p}_m^n(b)\}_{n=1}^N. \tag{1.425}$$

Compute

$$\rho_m = \frac{\|\mathbf{g}_m\|_{\Delta t}^2}{(\bar{\mathbf{g}}_m, \mathbf{w}_m)_{\Delta t}}, \tag{1.426}$$

and update $\mathbf{u}_m$ and $\mathbf{g}_m$ by

$$\mathbf{u}_{m+1} = \mathbf{u}_m - \rho_m \mathbf{w}_m, \tag{1.427}$$

$$\mathbf{g}_{m+1} = \mathbf{g}_m - \rho_m \bar{\mathbf{g}}_m, \tag{1.428}$$

respectively. If $\|\mathbf{g}_{m+1}\|_{\Delta t} / \|\mathbf{g}_0\|_{\Delta t} \leq \varepsilon$ take $\mathbf{u}_h^{\Delta t} = \mathbf{u}_{m+1}$; otherwise, compute

$$\gamma_m = \frac{\|\mathbf{g}_{m+1}\|_{\Delta t}^2}{\|\mathbf{g}_m\|_{\Delta t}^2}, \tag{1.429}$$

and update $\mathbf{w}_m$ by

$$\mathbf{w}_{m+1} = \mathbf{g}_{m+1} + \gamma_m \mathbf{w}_m. \tag{1.430}$$

Do $m = m + 1$ and go to (1.420).

Remark 1.66 Algorithm (1.412)–(1.430) is a discrete analogue of algorithm (1.365)–(1.377).

1.10.6.5 Approximation of the dual problem (1.362)

It was shown in Section 1.10.5 that there is *equivalence* between the *primal* control problem (1.357) and its *dual* problem (1.362). We shall discuss now the approximation of problem (1.362). There is no difficulty in adapting to problem (1.362) the (backward Euler scheme based) approximations methods discussed in Sections 1.8.3–1.8.5 for the solution of problem (1.118). Therefore, to avoid tedious repetitions we shall focus our discussion on an approximation of problem (1.362) which is based on a time discretization by the two-step backward implicit scheme considered in Section 1.8.6 (whose notation is kept); for simplicity, we shall take $H = h$ and $E_{0h} = V_{0h}$. We approximate then the dual problem (1.362), (1.378) by

$$\begin{cases} f_h^{\Delta t} \in V_{0h}, \quad \forall g_h \in V_{0h}, \\ k^{-1} \displaystyle\int_\Omega \nabla f_h^{\Delta t} \cdot \nabla g_h \, dx + \int_\Omega \left(\Lambda_h^{\Delta t} f_h^{\Delta t}\right) g_h \, dx = \langle y_T, g_h \rangle, \end{cases} \tag{1.431}$$

where, in (1.431), $\Lambda_h^{\Delta t}$ denotes the *linear operator* from V_{0h} into V_{0h} defined as follows:

$$\Lambda_h^{\Delta t} g_h = 2\varphi_{gh}^{N-1} - \varphi_{gh}^{N-2}, \tag{1.432}$$

where, to obtain φ_{gh}^{N-1} and φ_{gh}^{N-2}, we solve for $n = N-1, \ldots, 1$, the well-posed discrete elliptic problem

$$\begin{cases} \psi_{gh}^n \in V_{0h}, \quad \forall z_h \in V_{0h}, \\ \displaystyle\int_\Omega \frac{\frac{3}{2}\psi_{gh}^n - 2\psi_{gh}^{n+1} + \frac{1}{2}\psi_{gh}^{n+2}}{\Delta t} z_h \, dx + a\left(z_h, \psi_{gh}^n\right) = 0, \end{cases} \tag{1.433}$$

with

$$\psi_{gh}^N = 2g_h, \quad \psi_{gh}^{N+1} = 4g_h, \tag{1.434}$$

then, with $\varphi_{gh}^0 = 0$,

$$\begin{cases} \varphi_{gh}^1 \in V_{0h}, \quad \forall z_h \in V_{0h}, \\ \displaystyle\int_\Omega \frac{\varphi_{gh}^1 - \varphi_{gh}^0}{\Delta t} z_h \, dx + a\left(\frac{2}{3}\varphi_{gh}^1 + \frac{1}{3}\varphi_{gh}^0, z_h\right) = \frac{2}{3}\psi_{gh}^1(b) z_h(b), \end{cases} \tag{1.435}$$

and, finally, for $n = 2, \ldots, N-1$,

$$\begin{cases} \varphi_{gh}^n \in V_{0h}, \quad \forall z_h \in V_{0h}, \\ \displaystyle\int_\Omega \frac{\frac{3}{2}\varphi_{gh}^n - 2\varphi_{gh}^{n-1} + \frac{1}{2}\varphi_{gh}^{n-2}}{\Delta t} z_h \, dx + a\left(\varphi_{gh}^n, z_h\right) = \psi_{gh}^n(b) z_h(b). \end{cases} \tag{1.436}$$

It follows from (1.432)–(1.436) that (with obvious notation)

$$\int_\Omega \left(\Lambda_h^{\Delta t} g_1\right) g_2 \, dx = \Delta t \sum_{n=1}^{N-1} \psi_1^n(b) \psi_2^n(b), \quad \forall g_1, g_2 \in V_{0h},$$

that is, operator $\Lambda_h^{\Delta t}$ is *symmetric* and *positive semidefinite*, which implies in turn that the approximate dual problem (1.431) has a *unique* solution.

Remark 1.67 The discrete problem (1.431) is actually the *dual problem* of the following *discrete control problem* (a variant of problem (1.395); see Section 1.10.6.2):

$$\min_{\mathbf{v} \in \mathbb{R}^{N-1}} J_h^{\Delta t}(\mathbf{v}), \tag{1.437}$$

where, in (1.437), we have $\mathbf{v} = \{v^n\}_{n=1}^{N-1}$ and

$$J_h^{\Delta t}(\mathbf{v}) = \frac{1}{2} \Delta t \sum_{n=1}^{N-1} \left|v^n\right|^2 + \frac{k}{2} \int_\Omega \left|\nabla \Phi_h^N\right|^2 dx, \tag{1.438}$$

with Φ_h^N obtained from $\mathbf{v}$ via the solution of the following discrete parabolic problem:

$$y_h^0 = 0, \tag{1.439}$$

$$\begin{cases} y_h^1 \in V_{0h}, \quad \forall z_h \in V_{0h}, \\ \displaystyle\int_\Omega \frac{y_h^1 - y_h^0}{\Delta t} z_h \, dx + a\left(\frac{2}{3}y_h^1 + \frac{1}{3}y_h^0, z_h\right) = \frac{2}{3}v^1 z_h(b), \end{cases} \tag{1.440}$$

then, for $n = 2, \ldots, N-1$, assuming that y_h^{n-1} and y_h^{n-2} are known, solve

$$\begin{cases} y_h^n \in V_{0h}, \quad \forall z_h \in V_{0h}, \\ \displaystyle\int_\Omega \frac{\frac{3}{2}y_h^n - 2y_h^{n-1} + \frac{1}{2}y_h^{n-2}}{\Delta t} z_h \, dx + a\left(y_h^n, z_h\right) = v^n z_h(b), \end{cases} \tag{1.441}$$

and finally

$$\begin{cases} \Phi_h^N \in V_{0h}, \quad \forall z_h \in V_{0h}, \\ \displaystyle\int_\Omega \nabla \Phi_h^N \cdot \nabla z_h \, dx = \langle y_T - 2y_h^{N-1} + y_h^{N-2}, z_h \rangle. \end{cases} \tag{1.442}$$

Back to problem (1.431), it follows from the properties of operator $\Lambda_h^{\Delta t}$ that this problem can be solved by the following conjugate gradient algorithm (which is a discrete analogue of algorithm (1.379)–(1.391); see Section 1.10.5.3).

Algorithm 1.68 *(Conjugate gradient solution of problem (1.431))*

$$f_0 \in V_{0h} \text{ is given;} \tag{1.443}$$

take

$$p_0^N = 2f_0 \text{ and } p_0^{N+1} = 4f_0, \tag{1.444}$$

and solve, for $n = N-1, \ldots, 1$, the following discrete elliptic problem:

$$\begin{cases} p_0^n \in V_{0h}, \quad \forall z_h \in V_{0h}, \\ \displaystyle\int_\Omega \frac{\frac{3}{2}p_0^n - 2p_0^{n+1} + \frac{1}{2}p_0^{n+2}}{\Delta t} z_h \, dx + a\left(z_h, p_0^n\right) = 0. \end{cases} \tag{1.445}$$

Take now

$$y_0^0 = 0, \tag{1.446}$$

and solve

$$\begin{cases} y_0^1 \in V_{0h}, \quad \forall z_h \in V_{0h}, \\ \displaystyle\int_\Omega \frac{y_0^1 - y_0^0}{\Delta t} z_h \, dx + a\left(\frac{2}{3}y_0^1 + \frac{1}{3}y_0^0, z_h\right) = \frac{2}{3}p_0^1(b) z_h(b); \end{cases} \tag{1.447}$$

solve next, for $n = 2, \ldots, N-1$,

$$\begin{cases} y_0^n \in V_{0h}, \quad \forall z_h \in V_{0h}, \\ \displaystyle\int_\Omega \frac{\frac{3}{2}y_0^n - 2y_0^{n-1} + \frac{1}{2}y_0^{n-2}}{\Delta t} z_h \, dx + a\left(y_0^n, z_h\right) = p_0^n(b) z_h(b). \end{cases} \tag{1.448}$$

Solve next

$$\begin{cases} g_0 \in V_{0h}, \quad \forall z_h \in V_{0h}, \\ \displaystyle\int_\Omega \nabla g_0 \cdot \nabla z_h \, dx = k^{-1} \int_\Omega \nabla f_0 \cdot \nabla z_h \, dx \\ \qquad + \langle 2y_0^{N-1} - y_0^{N-2} - y_T, z_h \rangle, \end{cases} \tag{1.449}$$

and set

$$w_0 = g_0. \tag{1.450}$$

Then for $m \geq 0$, *assuming that* f_m, g_m, *and* w_m *are known, compute* f_{m+1}, g_{m+1}, *and* w_{m+1} *as follows:*
Take

$$\bar{p}_m^N = 2w_m \text{ and } \bar{p}_m^{N+1} = 4w_m, \tag{1.451}$$

and solve for $n = N-1, \ldots, 1$,

$$\begin{cases} \bar{p}_m^n \in V_{0h}, \quad \forall z_h \in V_{0h}, \\ \displaystyle\int_\Omega \frac{\frac{3}{2}\bar{p}_m^n - 2\bar{p}_m^{n+1} + \frac{1}{2}\bar{p}_m^{n+2}}{\Delta t} z_h \, dx + a\left(y_0^n, z_h\right) = p_0^n(b) z_h(b). \end{cases} \tag{1.452}$$

Take now

$$\bar{y}_m^0 = 0, \tag{1.453}$$

solve

$$\begin{cases} \bar{y}_m^1 \in V_{0h}, \quad \forall z_h \in V_{0h}, \\ \displaystyle\int_\Omega \frac{\bar{y}_m^1 - \bar{y}_m^0}{\Delta t} z_h \, dx + a\left(\frac{2}{3}\bar{y}_m^1 + \frac{1}{3}\bar{y}_m^0, z_h\right) = \frac{2}{3}\bar{p}_m^1(b) z_h(b); \end{cases} \tag{1.454}$$

and then for $n = 2, \ldots, N-1$,

$$\begin{cases} \bar{y}_m^n \in V_{0h}, \quad \forall z_h \in V_{0h}, \\ \displaystyle\int_\Omega \frac{\frac{3}{2}\bar{y}_m^n - 2\bar{y}_m^{n-1} + \frac{1}{2}\bar{y}_m^{n-2}}{\Delta t} z_h \, dx + a\left(\bar{y}_m^n, z_h\right) = \bar{p}_m^n(b) z_h(b). \end{cases} \tag{1.455}$$

Solve next

$$\begin{cases} \bar{g}_m \in V_{0h}, \\ \displaystyle\int_\Omega \nabla \bar{g}_m \cdot \nabla z_h \, dx = k^{-1} \int_\Omega \nabla w_m \cdot \nabla z_h \, dx \\ \qquad + \displaystyle\int_\Omega (2\bar{y}_m^{N-1} - \bar{y}_m^{N-2}) z_h \, dx, \quad \forall z_h \in V_{0h}, \end{cases} \tag{1.456}$$

and compute

$$\rho_m = \int_\Omega |\nabla g_m|^2 \, dx \bigg/ \int_\Omega \nabla \bar{g}_m \cdot \nabla w_m \, dx; \tag{1.457}$$

then update f_m and g_m by

$$f_{m+1} = f_m - \rho_m w_m, \tag{1.458}$$

$$g_{m+1} = g_m - \rho_m \bar{g}_m, \tag{1.459}$$

respectively. If $\int_\Omega |\nabla g_{m+1}|^2 \, dx / \int_\Omega |\nabla g_0|^2 \, dx \le \varepsilon^2$, take $f_h^{\Delta t} = f_{m+1}$; otherwise, compute

$$\gamma_m = \int_\Omega |\nabla g_{m+1}|^2 \, dx \bigg/ \int_\Omega |\nabla g_m|^2 \, dx, \tag{1.460}$$

and update w_m by

$$w_{m+1} = g_{m+1} + \gamma_m w_m. \tag{1.461}$$

Do $m = m + 1$ and go to (1.451).

Algorithm 1.68 is fairly easy to implement. It requires, essentially, *finite element* based *linear elliptic solvers* to compute the solution to problems (1.445), (1.447)–(1.449), (1.452), and (1.454)–(1.456); such solvers are easily available.

1.10.7 Numerical experiments

1.10.7.1 Generalities. Synopsis

In order to illustrate the various results and methods from Sections 1.10.1–1.10.6 we shall discuss in this section the solution of some *pointwise control problems*; these problems will be particular cases and variants of the *penalized* problem (1.357). We suppose for simplicity that $d = 1$ (that is, $\Omega \subset \mathbb{R}$); it follows then from Remark 1.58 that the solution of the parabolic problem (1.323) satisfies

$$y \in C^0\left([0, T]; L^2(\Omega)\right),$$

which implies that, in (1.357) it makes sense to replace $\|y(T) - y_T\|_{-1}$ by $\|y(T) - y_T\|_{L^2(\Omega)}$, a significant simplification, indeed. Also, for some of the test problems we will replace $\frac{1}{2}\int_0^T |v|^2 \, dt$ by $\frac{1}{s}\int_0^T |v|^s \, dt$, with $s > 2$, including some *very large* values of s for which the control u (in fact its discrete analog) has clearly a *bang-bang* behavior; this was expected from Section 1.7.

1.10.7.2 First test problems

What we have considered here is a family of test problems parametrized by T, y_T, k, and by the "support" b of the pointwise control. These test problems can be formulated as follows:

$$\min_{v \in L^2(0,T)} J_k(v), \tag{1.462}$$

where

$$J_k(v) = \frac{1}{2}\int_0^T |v|^2 \,\mathrm{d}t + \frac{k}{2}\int_0^1 |y(T) - y_T|^2 \,\mathrm{d}x, \tag{1.463}$$

with y the solution of the following *diffusion problem*:

$$\frac{\partial y}{\partial t} - \nu \frac{\partial^2 y}{\partial x^2} = v(t)\delta(x - b) \quad \text{in } (0,1)\times(0,T), \tag{1.464}$$

$$y(0,t) = y(1,t) = 0 \quad \text{on } (0,T), \tag{1.465}$$

$$y(0) = 0. \tag{1.466}$$

In equation (1.464) we have $\nu > 0$ and $b \in (0,1)$; we clearly have $\Omega = (0,1)$.

We have considered, first, test problems where the *target function* y_T is *even* with respect to the variable $x - \frac{1}{2}$. These target functions are given by

$$y_T(x) = 4x(1-x), \tag{1.467}$$

$$y_T(x) = \begin{cases} 8\left(x - \frac{1}{4}\right) & \text{if } \frac{1}{4} \le x \le \frac{1}{2}, \\ 8\left(\frac{3}{4} - x\right) & \text{if } \frac{1}{2} \le x \le \frac{3}{4}, \\ 0 & \text{elsewhere on } (0,1), \end{cases} \tag{1.468}$$

$$y_T(x) = \begin{cases} 1 & \text{if } \frac{1}{4} \le x \le \frac{3}{4}, \\ 0 & \text{elsewhere on } (0,1), \end{cases} \tag{1.469}$$

respectively. We have taken $\nu = \frac{1}{10}$ in equation (1.464), and $T = 3$ for all the three target functions given above. The continuous problem (1.462)–(1.466) has been approximated using the methods described in Sections 1.10.6.2–1.10.6.4 (that is, we have solved *directly* the control problems, taking into account the fact that for these problems we used a penalty term associated with the L^2-norm, instead of the H^{-1}-norm used in the general case). The *time discretization* has been obtained using the *backward Euler scheme* described in Section 1.10.6.2 with $\Delta t = 10^{-2}$, while the *space discretization* was obtained using a *uniform* mesh on $(0,1)$ with $h = 10^{-2}$. The discrete control problems have been solved by a variant of the *conjugate gradient* algorithm (1.412)–(1.430); we have taken $\mathbf{u}_0 = \mathbf{0}$ as *initializer* for the above algorithm and $\|\mathbf{g}_m\|_{\Delta t}/\|\mathbf{g}_0\|_{\Delta t} \le 10^{-6}$ as the *stopping criterion*. The corresponding numerical results have been summarized in Tables 1.1–1.3, where u^c and $y^c(T)$ denote the computed optimal control and the corresponding computed final state, respectively, and where *Nit* denotes the number of iterations necessary to achieve convergence.

In Figures 1.1–1.9 we have visualized, for $k = 10^4$, the computed optimal control and compared the corresponding computed value of $y(T)$ (that is, $y^c(T)$) to the target function y_T.

The above results deserve several comments:

(i) Since operator $\mathcal{A} = -\nu \frac{\mathrm{d}^2}{\mathrm{d}x^2}$ is *self-adjoint* for the *homogeneous Dirichlet boundary conditions* we can apply the controllability results of Section 1.10.2. The

Table 1.1. *Summary of numerical results (target function defined by (1.467)): $T = 3$, $h = \Delta t = 10^{-2}$*

b	k	Nit	$\|u^c\|_{L^2(0,T)}$	$\frac{\|y^c(T)-y_T\|_{L^2(0,1)}}{\|y_T\|_{L^2(0,1)}}$
$\frac{\sqrt{2}}{3}$	10^2	5	0.921	6.0×10^{-2}
	10^3	7	1.14	2.3×10^{-2}
	10^4	9	1.39	9.1×10^{-3}
	10^5	9	1.66	5.6×10^{-3}
	10^6	12	2.22	4.1×10^{-3}
$\frac{1}{2}$	10^2	5	0.909	5.5×10^{-2}
	10^3	7	1.09	2.1×10^{-2}
	10^4	7	1.30	9.3×10^{-3}
	10^5	9	1.63	5.1×10^{-3}
	10^6	10	2.18	3.2×10^{-3}
$\frac{\pi}{6}$	10^2	5	0.918	5.9×10^{-2}
	10^3	7	1.13	2.3×10^{-2}
	10^4	7	1.37	9.1×10^{-3}
	10^5	9	1.65	5.4×10^{-3}
	10^6	12	2.22	3.9×10^{-3}

Table 1.2. *Summary of numerical results (target function defined by (1.468)): $T = 3$, $h = \Delta t = 10^{-2}$.*

b	k	Nit	$\|u^c\|_{L^2(0,T)}$	$\frac{\|y^c(T)-y_T\|_{L^2(0,1)}}{\|y_T\|_{L^2(0,1)}}$
$\frac{\sqrt{2}}{3}$	10^2	6	1.23	2.2×10^{-1}
	10^3	7	1.96	1.9×10^{-1}
	10^4	10	5.54	1.6×10^{-1}
	10^5	11	11.2	1.4×10^{-1}
	10^6	13	37.1	1.3×10^{-1}
	10^{10}	21	585.0	6.7×10^{-2}
$\frac{1}{2}$	10^2	5	1.27	1.1×10^{-1}
	10^3	7	1.74	6.3×10^{-2}
	10^4	7	2.42	5.2×10^{-2}
	10^5	9	6.26	3.8×10^{-2}
	10^6	9	12.7	3.0×10^{-2}
	10^{10}	14	66.7	2.5×10^{-2}
$\frac{\pi}{6}$	10^2	5	1.24	1.9×10^{-1}
	10^3	7	1.84	1.6×10^{-1}
	10^4	9	4.84	1.4×10^{-1}
	10^5	12	9.87	1.2×10^{-1}
	10^6	15	29.7	1.1×10^{-1}
	10^{10}	21	534.0	5.9×10^{-2}

Table 1.3. *Summary of numerical results (target function defined by (1.469)): $T = 3$, $h = \Delta t = 10^{-2}$.*

b	k	Nit	$\|u^c\|_{L^2(0,T)}$	$\frac{\|y^c(T)-y_T\|_{L^2(0,1)}}{\|y_T\|_{L^2(0,1)}}$
$\frac{\sqrt{2}}{3}$	10^2	5	0.92	3.47×10^{-1}
	10^3	8	2.3	3.11×10^{-1}
	10^4	10	7.3	2.72×10^{-1}
	10^5	11	16.0	2.46×10^{-1}
	10^6	13	26.0	2.40×10^{-1}
	10^{10}	28	2200.0	2.27×10^{-1}
$\frac{1}{2}$	10^2	5	0.99	3.32×10^{-1}
	10^3	7	2.08	2.47×10^{-1}
	10^4	9	6.3	2.36×10^{-1}
	10^5	9	10.0	2.27×10^{-1}
	10^6	12	36.0	2.16×10^{-1}
	10^{10}	15	540.0	1.82×10^{-1}
$\frac{\pi}{6}$	10^2	5	0.94	3.43×10^{-1}
	10^3	7	2.4	3.02×10^{-1}
	10^4	9	6.9	2.66×10^{-1}
	10^5	10	16.0	2.40×10^{-1}
	10^6	13	29.0	2.34×10^{-1}
	10^{10}	28	2200.0	2.17×10^{-1}

eigenfunctions of operator $\mathcal{A}$, that is, the solutions of

$$-\nu \frac{d^2}{dx^2} w_j = \lambda_j w_j \text{ on } (0,1), \quad w_j(0) = w_j(1) = 0, \quad w_j \neq 0,$$

are clearly given by

$$w_j(x) = \sin j\pi x, \quad j = 1, 2, \ldots,$$

the corresponding *spectrum* being $\{\nu j^2 \pi^2\}_{j=1}^{+\infty}$. Since each eigenvalue is *simple*, b will be strategic if

$$\sin j\pi b \neq 0, \quad \forall j = 1, 2, \ldots,$$

that is,

$$b \notin (0,1) \cap \mathbb{Q}, \tag{1.470}$$

where, in (1.470), $\mathbb{Q}$ is the field of the *rational real numbers*. Clearly, $b = \frac{\sqrt{3}}{2}$ and $b = \frac{\pi}{6}$ being nonrational are strategic. On the other hand, $b = \frac{1}{2}$ is far from being strategic since $\sin \frac{1}{2} j\pi = 0$ for *any* even integer j; indeed, $b = \frac{1}{2}$ is *generically* the worst choice which can be made. However, if one takes $b = \frac{1}{2}$

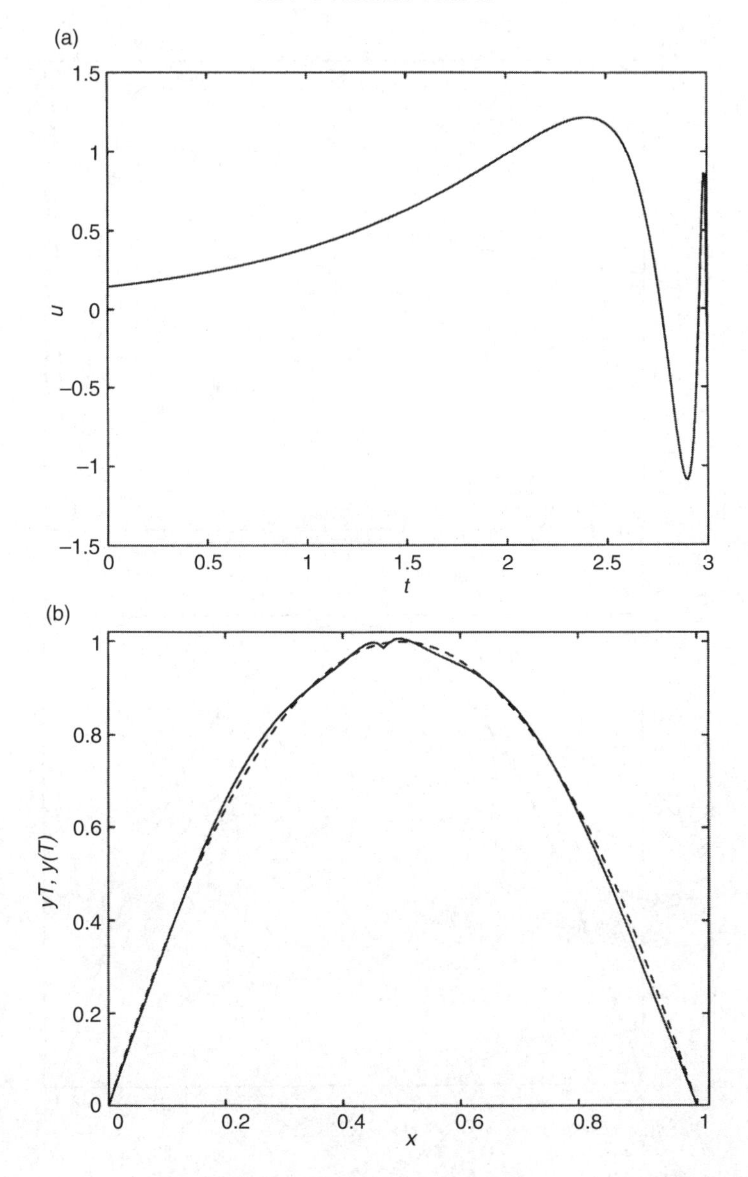

Fig. 1.1. (a) Variation of the computed optimal control, and (b) comparison between $y^c(T)$ and y_T (target function defined by (1.467)): $T = 3$, $b = \frac{\sqrt{2}}{3}$, $k = 10^4$, $h = \Delta t = 10^{-2}$.

the solution y of problem (1.464)–(1.466) satisfies

$$\forall t \in [0, T], y(t) \text{ is an } \textit{even} \text{ function of } x - \frac{1}{2}; \tag{1.471}$$

property (1.471) implies that the coefficients of w_j in the Fourier expansion of y are zero for j even. This property implies in turn that $b = \frac{1}{2}$ is strategic if the target y_T is also an even function of $x - \frac{1}{2}$; this is precisely the case for the target functions defined by (1.467)–(1.469). Actually, for target functions y_T which are

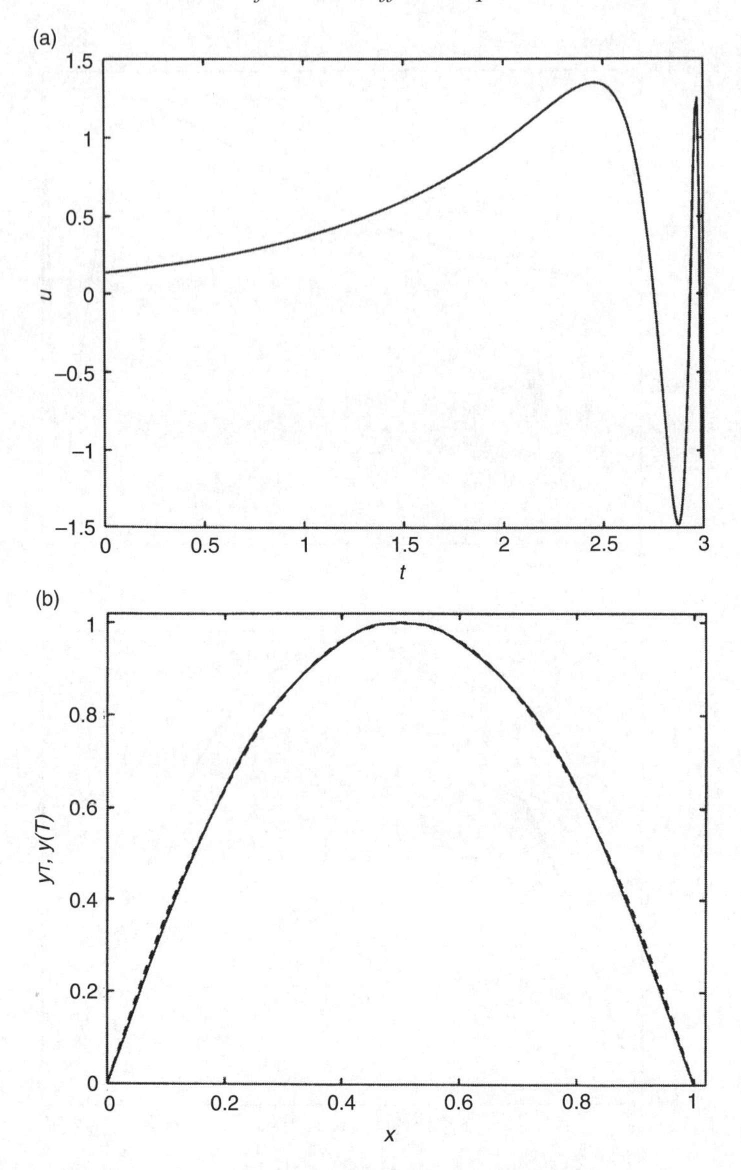

Fig. 1.2. (a) Variation of the computed optimal control, and (b) comparison between $y^c(T)$ and y_T (target function defined by (1.467)): $T = 3$, $b = \frac{1}{2}$, $k = 10^4$, $h = \Delta t = 10^{-2}$.

even with respect to $x - \frac{1}{2}$, $b = \frac{1}{2}$ is the best strategic point; this appears clearly in Tables 1.1–1.3 where the smallest *control norms* and *controllability errors* are obtained for $b = \frac{1}{2}$. In Section 1.10.7.3 we shall consider target functions which are not even with respect to $x - \frac{1}{2}$; $b = \frac{1}{2}$ will not be strategic at all for these test problems.

(ii) A *digital computer* "knows" only rational numbers; this means that for the particular test problems considered in this section, there is no – strictly speaking – strategic point for pointwise control. However, if b is the computer

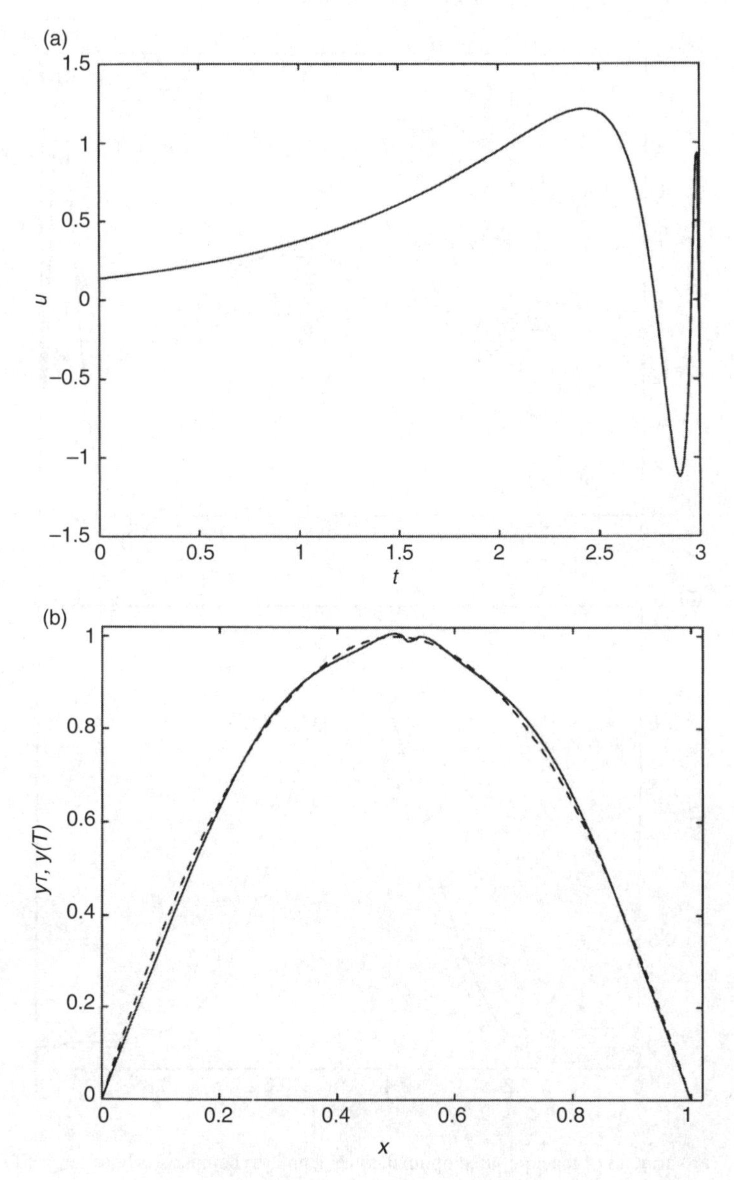

Fig. 1.3. (a) Variation of the computed optimal control, and (b) comparison between $y^c(T)$ and y_T (target function defined by (1.467)): $T = 3$, $b = \frac{\pi}{6}$, $k = 10^4$, $h = \Delta t = 10^{-2}$.

approximation of a nonrational number, the integers j such that bj is also an integer are *very large*. This last property implies that, unless h is extremely small and/or y_T is quite pathological (that is, its Fourier coefficients converge very slowly to zero as $j \to +\infty$), such a b is strategic in practice.

(iii) Tables 1.1–1.3 strongly suggest that the smoother is y_T, the smaller is the controllability error, a result we could expect. We observe also that the number of conjugate gradient iterations necessary to achieve convergence increases slowly with k, but remains in all circumstances significantly smaller than the dimension

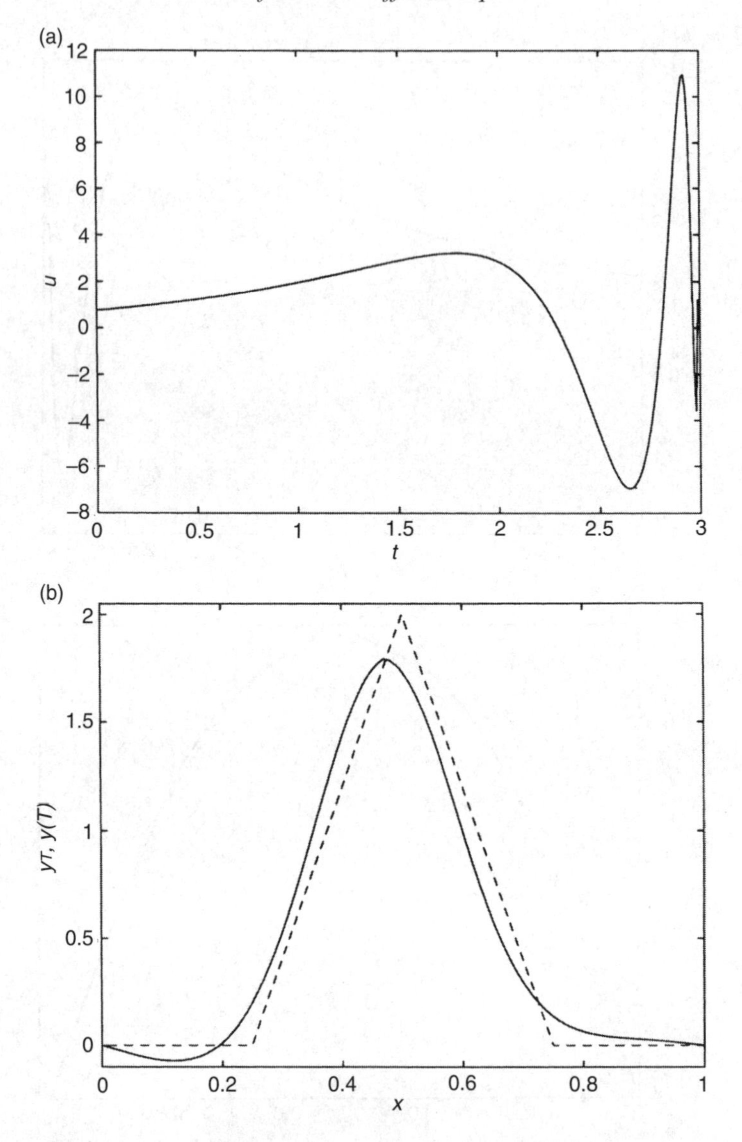

Fig. 1.4. (a) Variation of the computed optimal control, and (b) comparison between $y^c(T)$ and y_T (target function defined by (1.468)): $T = 3$, $b = \frac{\sqrt{2}}{3}$, $k = 10^4$, $h = \Delta t = 10^{-2}$.

of the discrete control space (300, here); this suggests that for the reasonably smooth target functions considered here the number of active modes – practically – involved in the control problems is much smaller than the dimension of the discrete control space.

1.10.7.3 Further test problems

The test problems in this section are still defined by (1.462)–(1.466) the main difference being that the target functions y_T are not even with respect to the variable $x - \frac{1}{2}$.

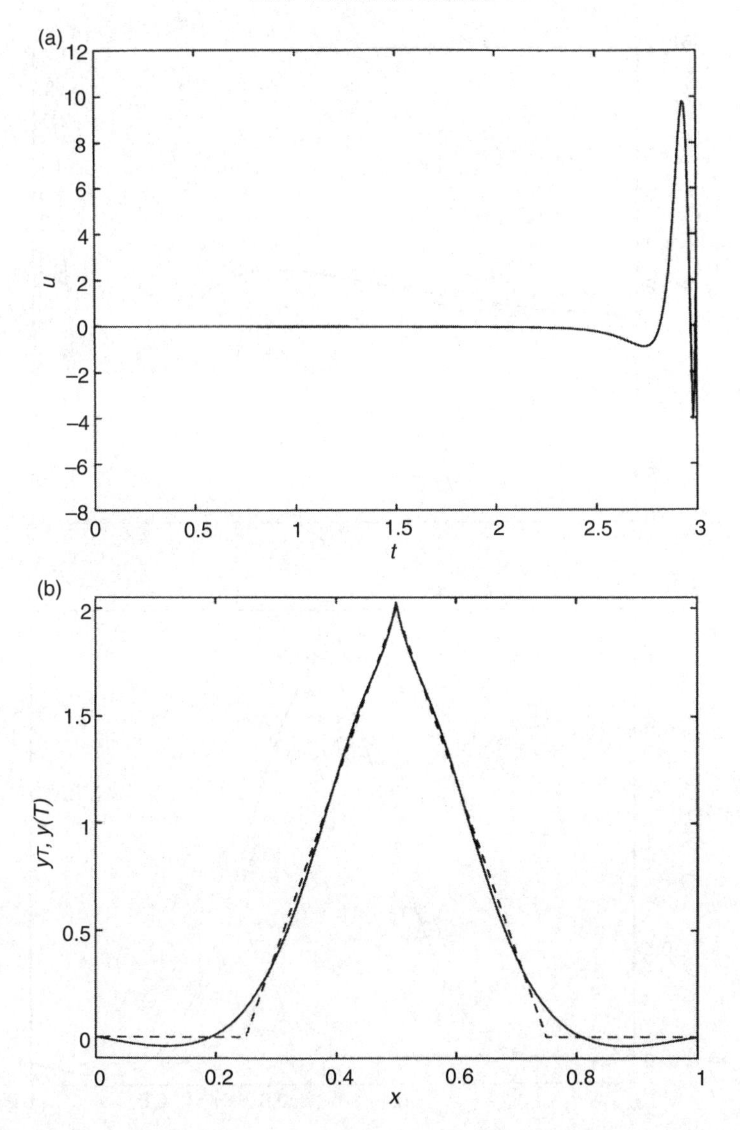

Fig. 1.5. (a) Variation of the computed optimal control, and (b) comparison between $y^c(T)$ and y_T (target function defined by (1.468)): $T = 3$, $b = \frac{1}{2}$, $k = 10^4$, $h = \Delta t = 10^{-2}$.

Indeed, the two target functions considered here are defined by

$$y_T(x) = \frac{27}{4}x^2(1-x), \tag{1.472}$$

and

$$y_T(x) = \begin{cases} 0 & \text{on } [0, \frac{1}{2}], \\ 8\left(x - \frac{1}{2}\right) & \text{on } [\frac{1}{2}, \frac{3}{4}], \\ 8(1-x) & \text{on } [\frac{3}{4}, 1]; \end{cases} \tag{1.473}$$

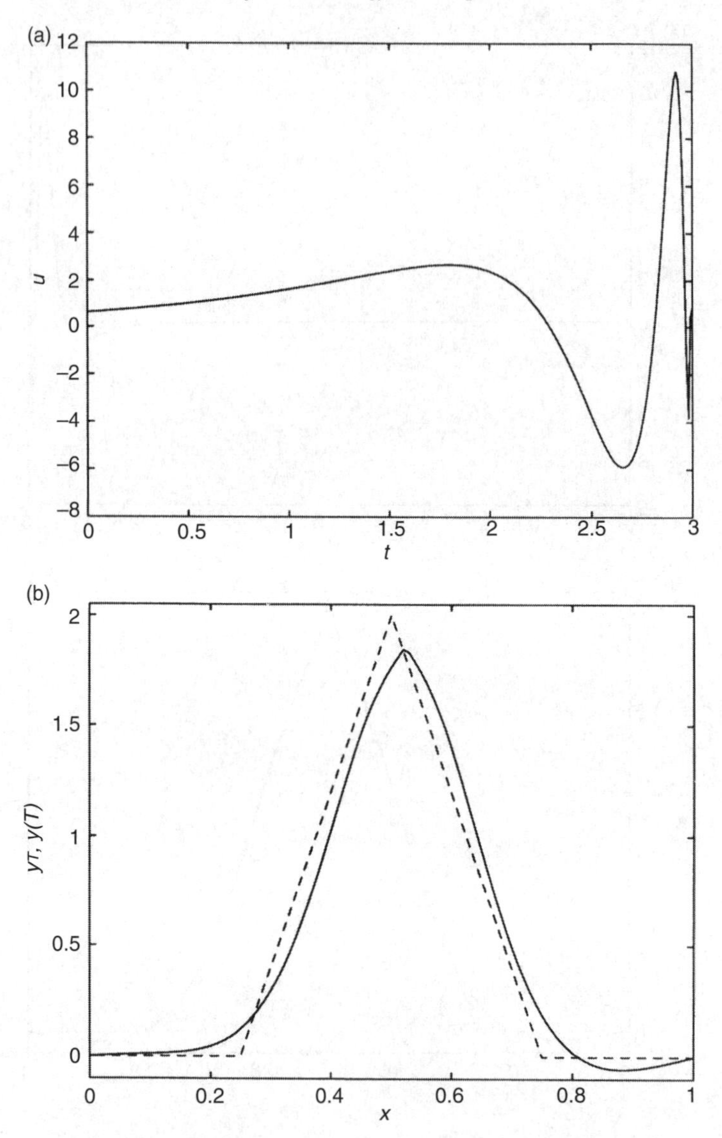

Fig. 1.6. (a) Variation of the computed optimal control, and (b) comparison between $y^c(T)$ and y_T (target function defined by (1.468)): $T = 3$, $b = \frac{\pi}{6}$, $k = 10^4$, $h = \Delta t = 10^{-2}$.

we have taken $T = 3$ for both target functions. The approximations and solution methods being those of Section 1.10.7.2, still with $\nu = \frac{1}{10}$, $h = \Delta t = 10^{-2}$, we have obtained the results summarized in Tables 1.4 and 1.5 and Figures 1.10–1.15 (the notation is the same as in Section 1.10.7.2). These results clearly show that $b = \frac{1}{2}$ is not strategic for the test problems considered here; this was expected since none of the functions y_T is even with respect to $x - \frac{1}{2}$. On the other hand, "small" *irrational* shifts of $\frac{1}{2}$, either to the right or to the left, produce strategic values of b. The other comments made in Section 1.10.7.2 still holds for the examples considered here.

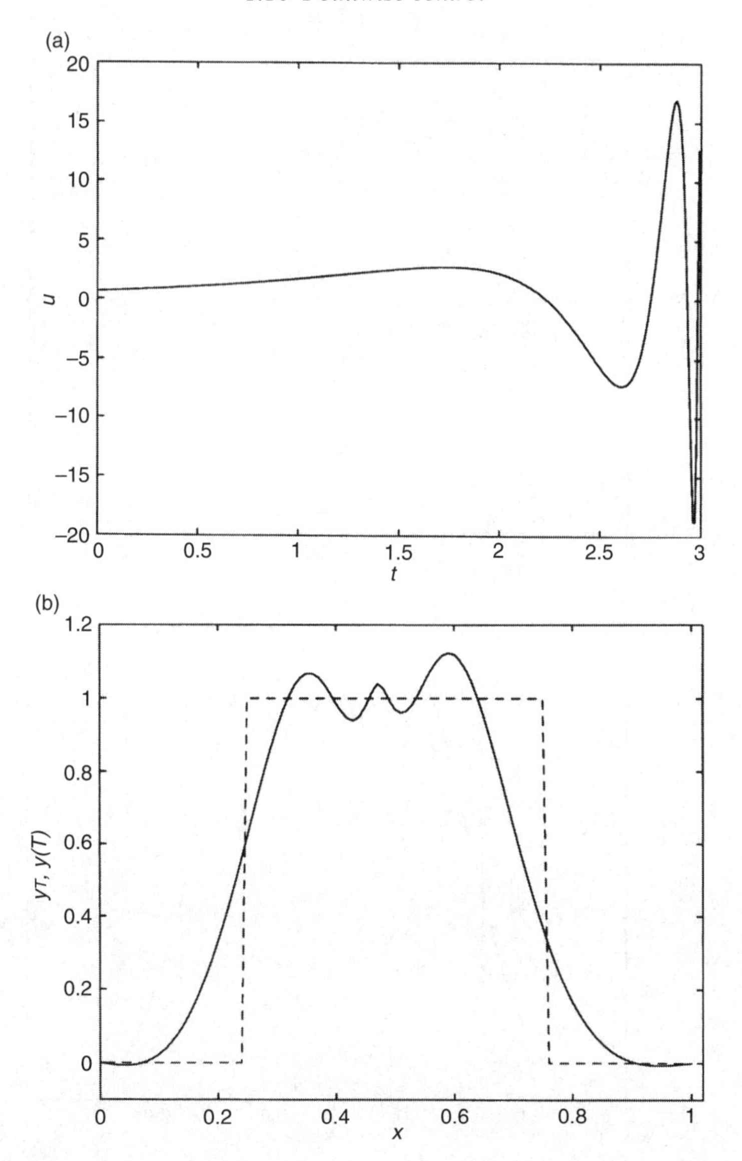

Fig. 1.7. (a) Variation of the computed optimal control, and (b) comparison between $y^c(T)$ and y_T (target function defined by (1.469)): $T = 3$, $b = \frac{\sqrt{2}}{3}$, $k = 10^4$, $h = \Delta t = 10^{-2}$.

1.10.7.4 Test problems for nonquadratic cost functions

Motivated by Section 1.9 we have been considering pointwise control problems defined by

$$\min_{v \in L^s(0,T)} \left(\frac{1}{2} \|v\|^2_{L^s(0,T)} + \frac{1}{2} k \|y(T) - y_T\|^2_{L^2(0,1)} \right), \tag{1.474}$$

with y still defined from v by (1.464)–(1.466), and s "large." It seems, unfortunately, that for $s > 2$, problem (1.474) is poorly conditioned, implying that the various

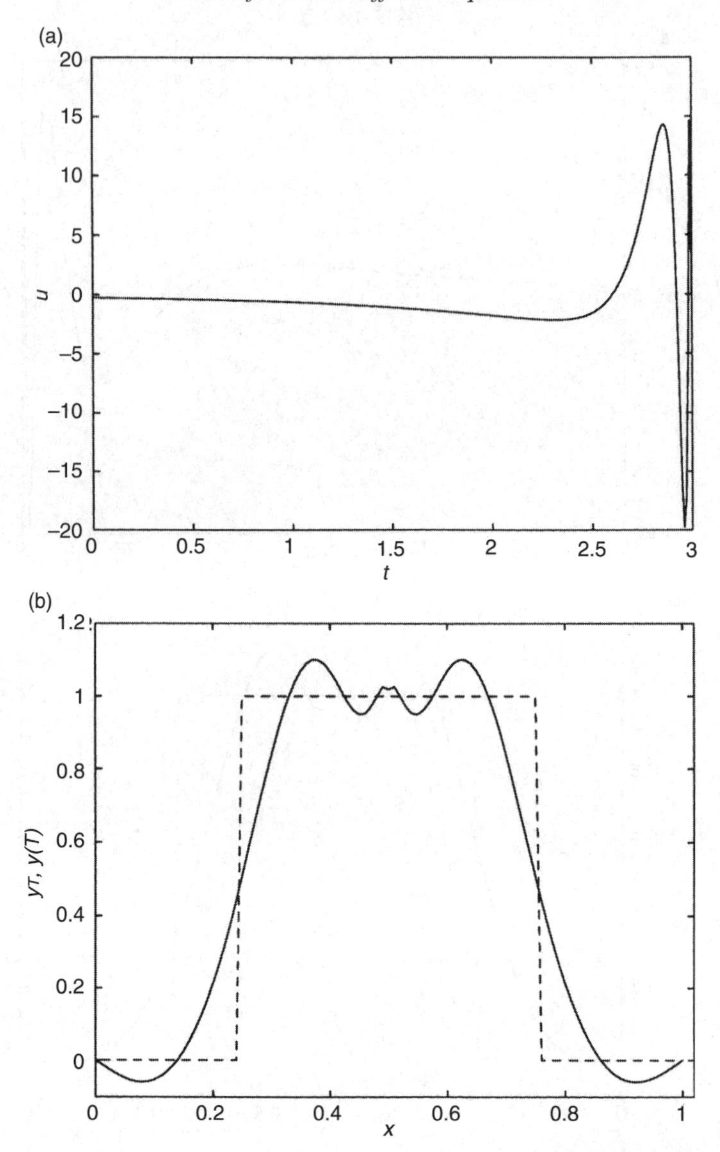

Fig. 1.8. (a) Variation of the computed optimal control, and (b) comparison between $y^c(T)$ and y_T (target function defined by (1.469)): $T = 3$, $b = \frac{1}{2}$, $k = 10^4$, $h = \Delta t = 10^{-2}$.

iterative methods we employed to solve it (conjugate gradient, Newton's, quasi-Newton, ...) failed to converge (or even worse, get "stuck" on some wrong solution). From these facts, it was quite natural to consider the variant of problem (1.474) defined by

$$\min_{v \in L^s(0,T)} \left(\frac{1}{s} \|v\|^s_{L^s(0,T)} + \frac{1}{2} k \|y(T) - y_T\|^2_{L^2(0,1)} \right), \tag{1.475}$$

with y defined from v as above. The cost function in (1.475) has better differentiability properties than the one in (1.474).

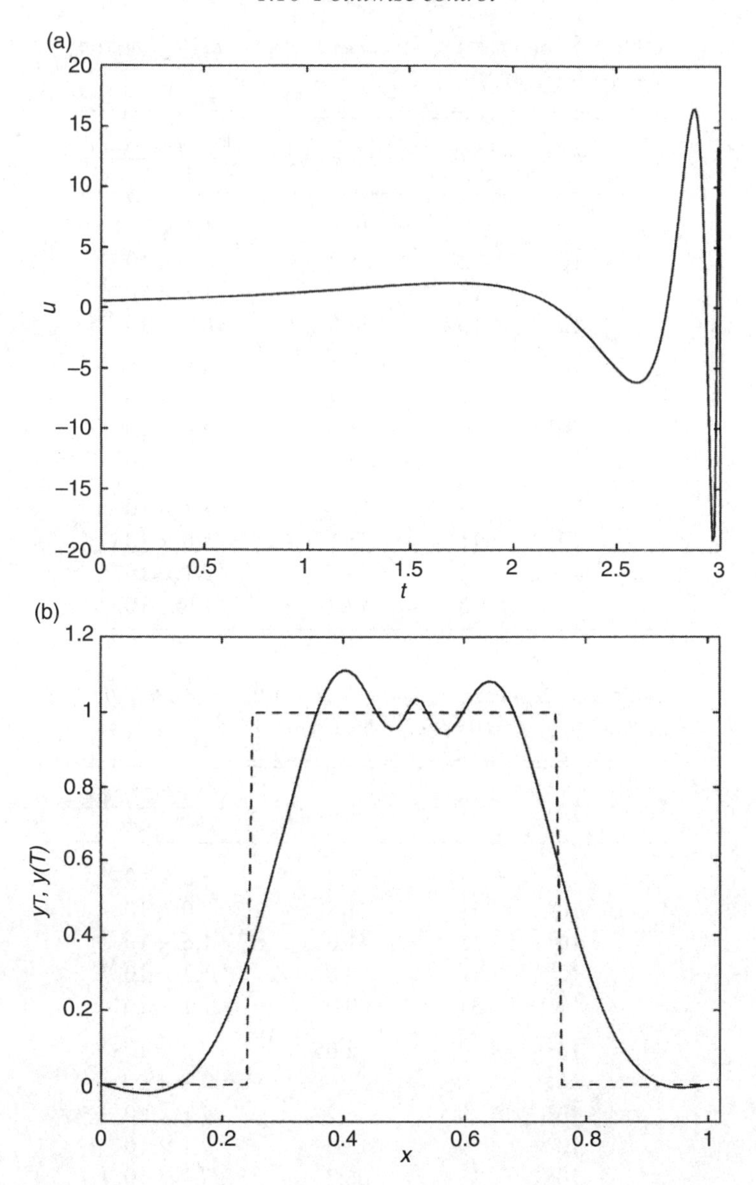

Fig. 1.9. (a) Variation of the computed optimal control, and (b) comparison between $y^c(T)$ and y_T (target function defined by (1.469)): $T = 3$, $b = \frac{\pi}{6}$, $k = 10^4$, $h = \Delta t = 10^{-2}$.

Let us denote by u the solution of the control problem (1.475); assuming that b in (1.464) is strategic we can expect that for s fixed $y(u; T)$ will get closer to y_T as k increases. If, on the other hand, k is fixed we can expect the distance between $y(u; T)$ and y_T to increase with s, since in that case the relative importance of the term $s^{-1} \int_0^T |v(t)|^s \, dt$ in the cost function increases also with s. These predictions are fully confirmed by the numerical experiments whose results are shown below. For these experiments, we have used essentially the same approximation methods as in

Table 1.4. *Summary of numerical results (target function defined by (1.472)):* $T = 3$, $h = \Delta t = 10^{-2}$.

b	k	*Nit*	$\|u^c\|_{L^2(0,T)}$	$\frac{\|y^c(T)-y_T\|_{L^2(0,1)}}{\|y_T\|_{L^2(0,1)}}$
$\frac{\sqrt{2}}{3}$	10^4	9	10.6	1.5×10^{-1}
	10^5	12	18.4	4.0×10^{-2}
	10^6	15	21.6	1.5×10^{-2}
	10^{10}	16	34.0	8.3×10^{-3}
$\frac{1}{2}$	10^4	7	1.10	3.5×10^{-1}
	10^5	9	1.38	3.5×10^{-1}
	10^6	9	1.92	3.5×10^{-1}
	10^{10}	9	3.28	3.5×10^{-1}
$\frac{\pi}{6}$	10^4	6	10.1	1.8×10^{-1}
	10^5	11	21.1	5.6×10^{-2}
	10^6	12	26.7	2.1×10^{-2}
	10^{10}	15	49.7	1.0×10^{-2}

Table 1.5. *Summary of numerical results (target function defined by (1.473)):* $T = 3$, $h = \Delta t = 10^{-2}$.

b	k	*Nit*	$\|u^c\|_{L^2(0,T)}$	$\frac{\|y^c(T)-y_T\|_{L^2(0,1)}}{\|y_T\|_{L^2(0,1)}}$
$\frac{\sqrt{2}}{3}$	10^3	8	6.53	5.2×10^{-1}
	10^4	11	21.8	3.0×10^{-1}
	10^5	12	41.6	1.6×10^{-1}
	10^6	16	58.8	1.3×10^{-1}
	10^{10}	31	1900.0	7.1×10^{-2}
$\frac{1}{2}$	10^3	7	2.05	7.1×10^{-1}
	10^4	9	2.68	7.1×10^{-1}
	10^5	9	6.27	7.1×10^{-1}
	10^6	9	12.6	7.1×10^{-1}
	10^{10}	15	66.7	7.1×10^{-1}
$\frac{\pi}{6}$	10^3	8	5.99	6.7×10^{-1}
	10^4	10	27.5	4.2×10^{-1}
	10^5	12	57.6	2.0×10^{-1}
	10^6	16	89.9	1.3×10^{-1}
	10^{10}	16	326.0	6.2×10^{-2}

Sections 1.10.7.2 and 1.10.7.3, with $h = \Delta t = 10^{-2}$, and taken $b = \frac{\sqrt{2}}{3}$, $T = 3$, $\nu = \frac{1}{10}$ and y_T defined by (1.472). The discrete control problems have been solved by *quasi-Newton methods* à la *BFGS*, like those discussed, for example, in the classical text book by Dennis and Schnabel (1983) (see also Liu and Nocedal, 1989; Nocedal,

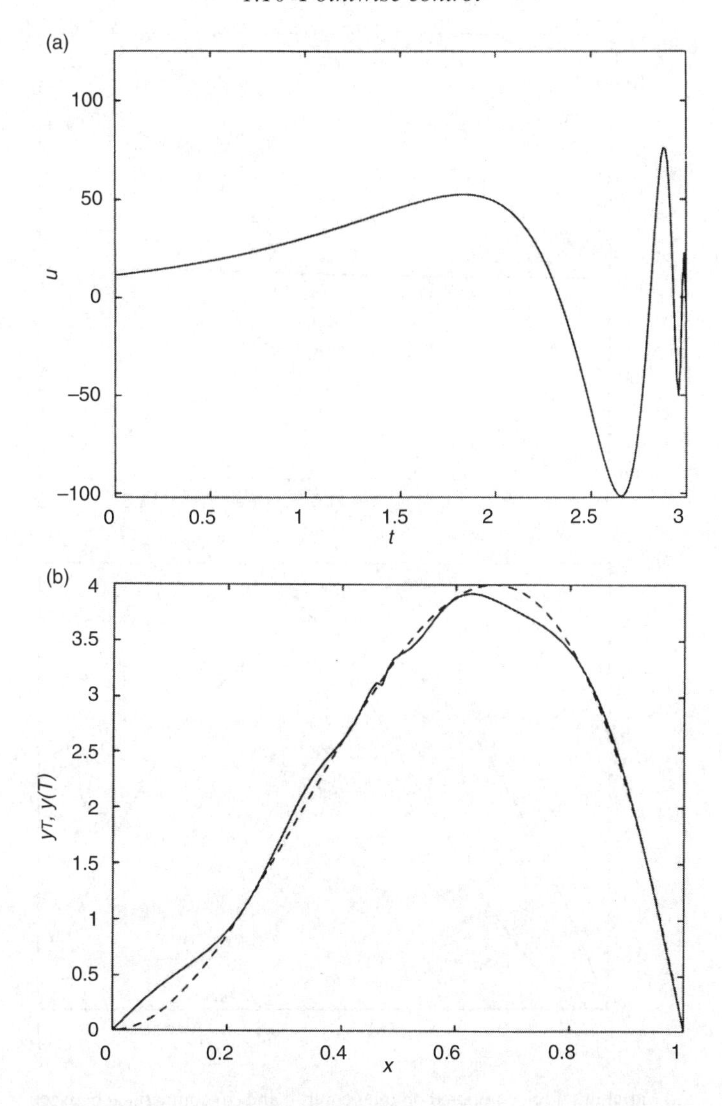

Fig. 1.10. (a) Variation of the computed optimal control, and (b) comparison between $y^c(T)$ and y_T (target function defined by (1.472)): $T = 3$, $b = \frac{\sqrt{2}}{3}$, $k = 10^5$, $h = \Delta t = 10^{-2}$.

1992); we shall return on these methods – and briefly describe them – in Chapter 9, Section 9.6, where they have been applied to the solution of *flow control problems*. For the problems discussed here these quasi-Newton methods appear to be much more efficient than conjugate gradient methods if $s > 2$.

On Figures 1.16–1.21 we have visualized – for $k = 10^7$ and $s = 2, 4, 6, 10, 20, 30$ – the computed optimal control u^c and compared the corresponding final state $y^c(T)$ with the target function y_T. From these figures, we clearly see that the distance of $y^c(T)$ to y_T increases with s; we also see the *bang-bang* character of the optimal control for large values of s.

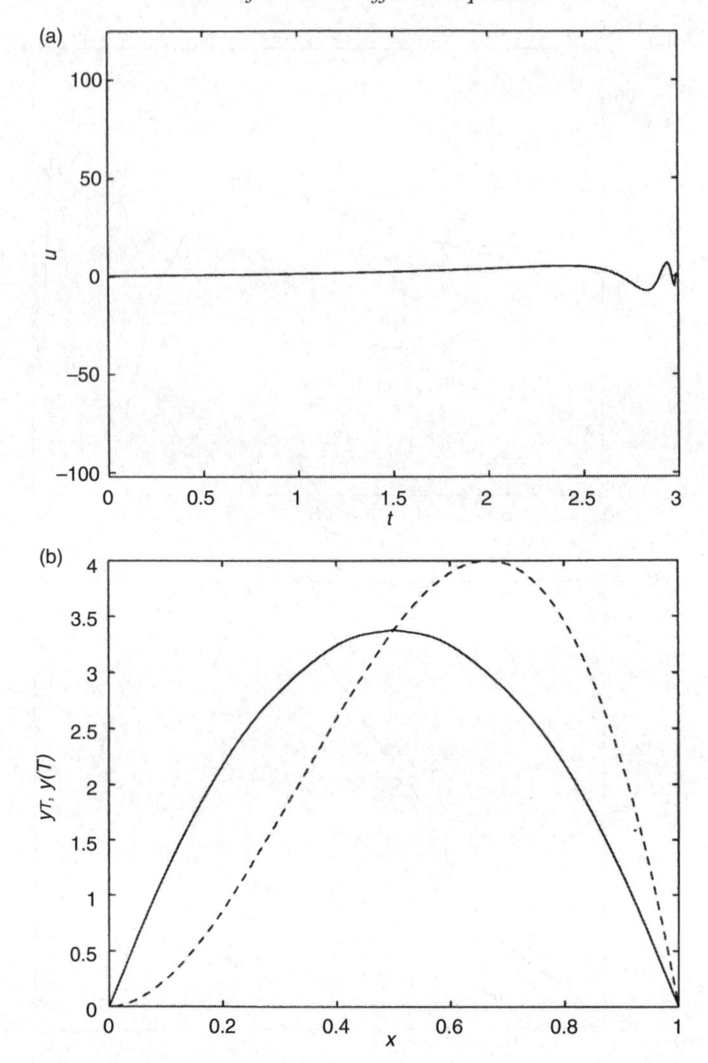

Fig. 1.11. (a) Variation of the computed optimal control, and (b) comparison between $y^c(T)$ and y_T (target function defined by (1.472)): $T = 3$, $b = \frac{1}{2}$, $k = 10^5$, $h = \Delta t = 10^{-2}$.

Finally, on Figures 1.22–1.25 we have shown some of the results obtained for large values of s and very large values of k; comparing with Figures 1.19–1.21 we observe that if for a given s we increase k, then $y^c(T)$ gets closer to y_T and $\|u^c\|_{L^2(0,T)}$ increases, which makes sense. We observe again that for very large values of s the optimal control is very close to bang-bang.

1.11 Further remarks (I): Additional constraints on the state function

In Control, there are many situations where the *linear state equations* modeling the system evolution (we have seen such equations in the preceding sections) have been obtained by *linearization* of a nonlinear mathematical model. For these situations,

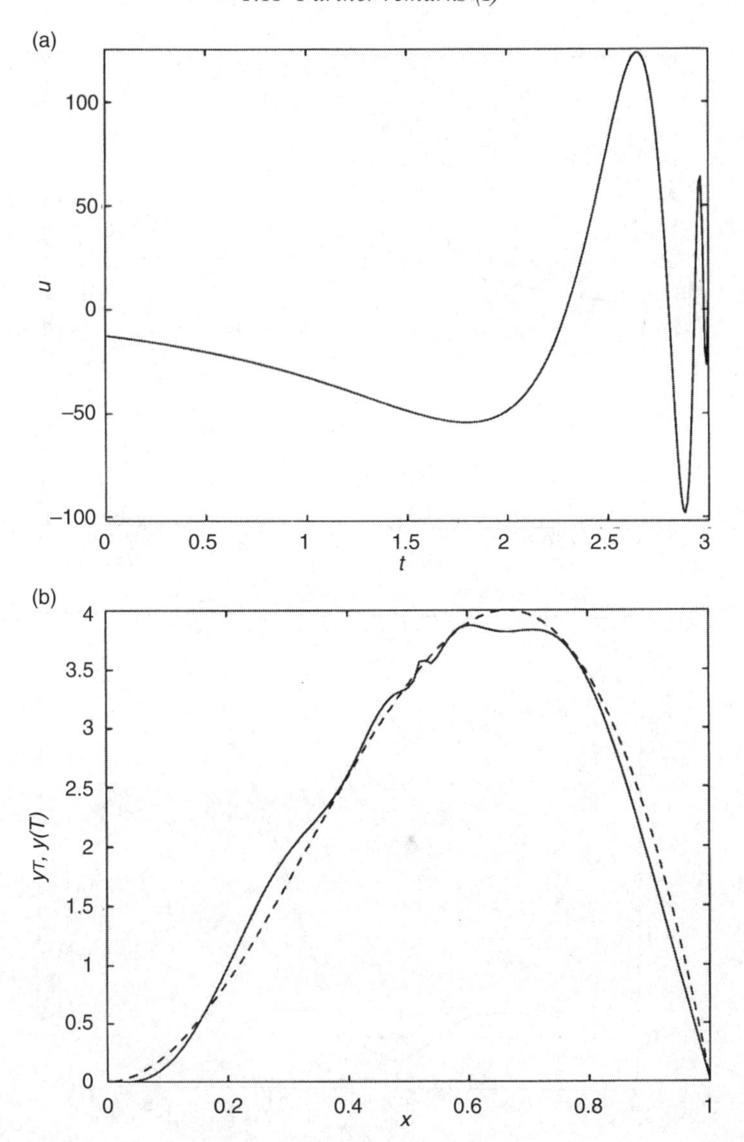

Fig. 1.12. (a) Variation of the computed optimal control, and (b) comparison between $y^c(T)$ and y_T (target function defined by (1.472)): $T = 3, b = \frac{\pi}{6}, k = 10^5, h = \Delta t = 10^{-2}$.

y is a perturbation that it is important to keep "small" on the full interval $(0, T)$ in order to preserve the validity of the linearization approach. Such a requirement leads to prescribe a *global state constraint* such as

$$\int_0^T \|y(t)\|_X^2 \, dt \leq \gamma, \tag{1.476}$$

where, in (1.476), γ is an arbitrary small positive number and where the norm $\|\cdot\|_X$ is chosen as "strong" as possible. In this section, the notation and terminology are those

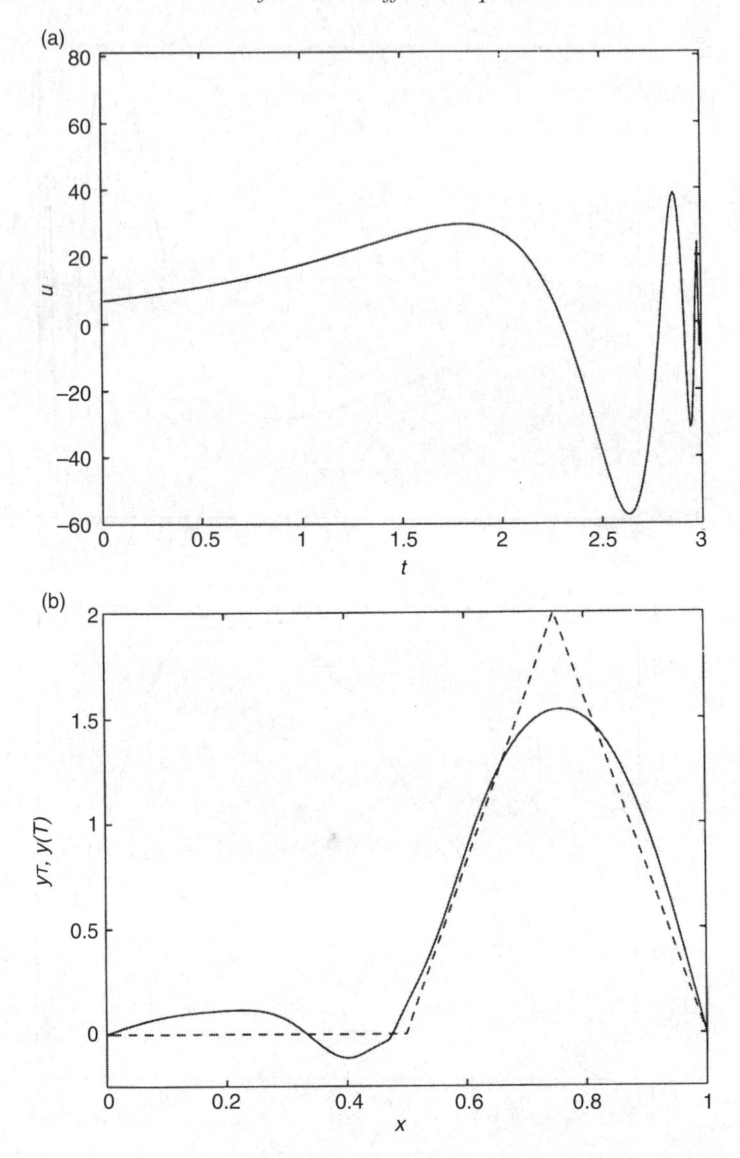

Fig. 1.13. (a) Variation of the computed optimal control, and (b) comparison between $y^c(T)$ and y_T (target function defined by (1.473)): $T = 3$, $b = \frac{\sqrt{2}}{3}$, $k = 10^5$, $h = \Delta t = 10^{-2}$.

of J.L. Lions (1997a). Indeed, following the above reference we consider a controlled distributed system whose evolution is described by

$$\begin{cases} \dfrac{\partial y}{\partial t} - \Delta y = v\chi_{\mathcal{O}\times(0,T)} \quad \text{in } \Omega \times (0, T), \\ y(0) = 0, \quad y = 0 \quad \text{on } \Gamma \times (0, T), \end{cases} \tag{1.477}$$

where, Ω is a bounded domain of $\mathbb{R}^d$ and Γ is its boundary; $\mathcal{O}$ is an arbitrary open subset of Ω; $\chi_{\mathcal{O}\times(0,T)}$ is the characteristic function of the set $\mathcal{O} \times (0, T)$; the control

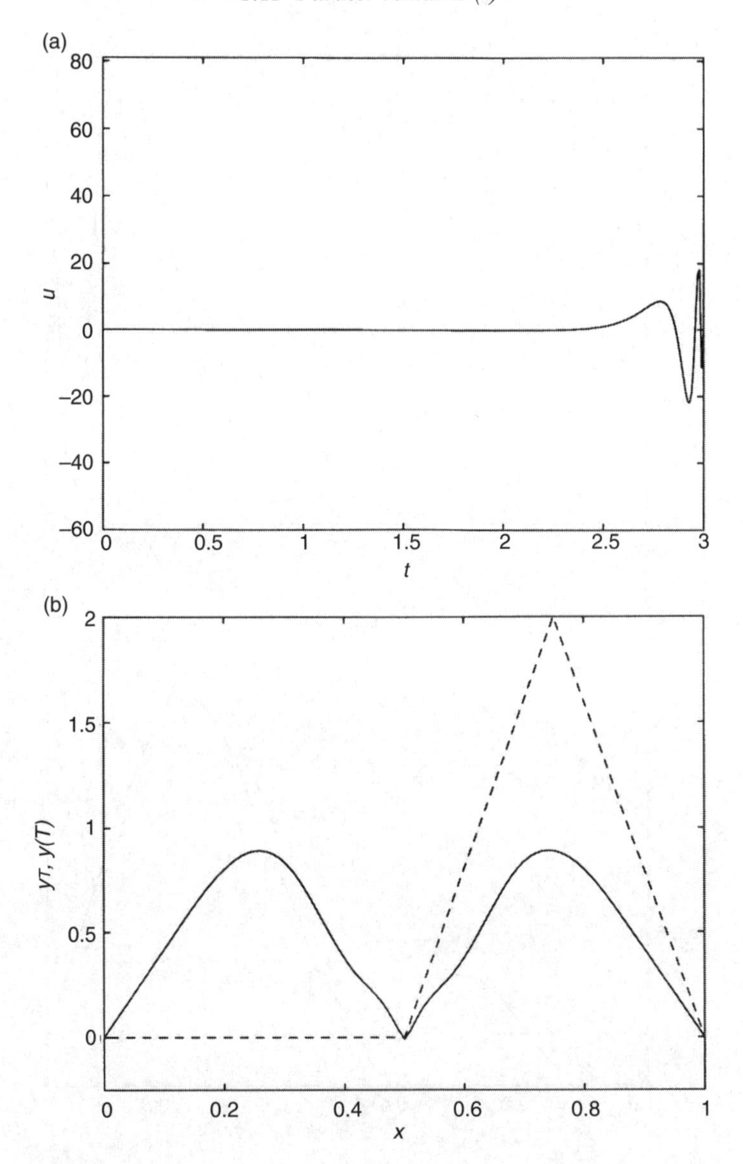

Fig. 1.14. (a) Variation of the computed optimal control, and (b) comparison between $y^c(T)$ and y_T (target function defined by (1.473)): $T = 3, b = \frac{1}{2}, k = 10^5, h = \Delta t = 10^{-2}$.

function v verifies

$$v \in L^2\big(\mathcal{O} \times (0, T)\big). \tag{1.478}$$

Next, we associate with $\mu > 0$ the *finite-dimensional* space E_μ defined by

$$E_\mu = \underset{j:\,\lambda_j \leq \mu}{\text{span}} \{w_j\}, \tag{1.479}$$

where

$$-\Delta w_j = \lambda_j w_j \text{ in } \Omega, \quad w_j = 0 \text{ on } \Gamma, \quad \|w_j\|_{L^2(\Omega)} = 1.$$

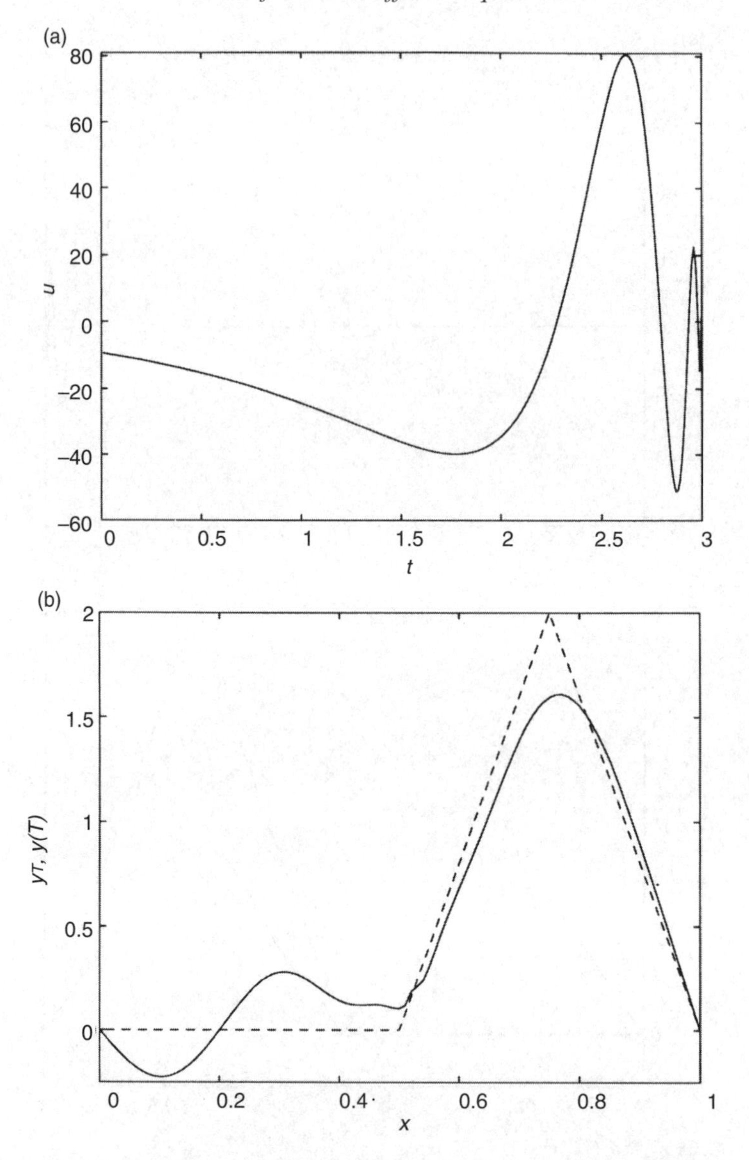

Fig. 1.15. (a) Variation of the computed optimal control, and (b) comparison between $y^c(T)$ and y_T (target function defined by (1.473)): $T = 3$, $b = \frac{\pi}{6}$, $k = 10^5$, $h = \Delta t = 10^{-2}$.

If μ is sufficiently large, then the space E_μ is nonempty and we denote by Π_μ the *orthogonal projector* from $L^2(\Omega)$ onto E_μ. Finally, we consider a target function $y_T \in L^2(\Omega)$ and β another positive number. It follows then from J.L. Lions (1997a) that there exists a pair $\{v, y\}$ verifying (1.477), (1.478) and

$$\left\| \Pi_\mu \big(y(T) - y_T\big) \right\|_{L^2(\Omega)} \leq \beta, \tag{1.480}$$

$$\int_0^T \|y(t)\|^2_{H^\alpha} \, \mathrm{d}t \leq \gamma, \tag{1.481}$$

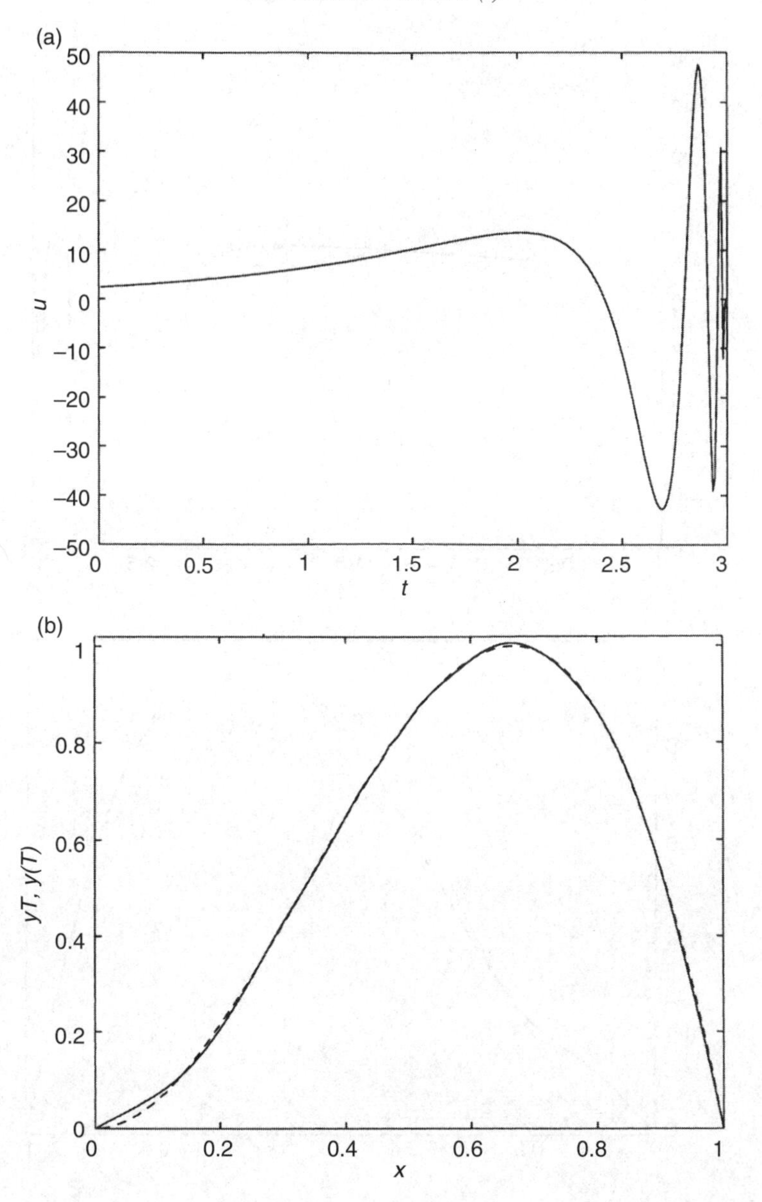

Fig. 1.16. (a) Variation of the computed optimal control, and (b) comparison between $y^c(T)$ and y_T (target function defined by (1.472)): $s = 2$, $T = 3$, $b = \frac{\sqrt{2}}{3}$, $k = 10^7$, $h = \Delta t = 10^{-2}$.

where, in (1.481), H^α is for $\alpha \in [0, 1)$ the *interpolate* between $L^2(\Omega)$ $(= H^0)$ and $H_0^1(\Omega)$ $(= H^1)$. The above existence result implies in turn the existence and uniqueness of a control u solution of the following *optimal control problem*:

$$\min_v \frac{1}{2} \iint_{\mathcal{O}\times(0,T)} |v|^2 \, \mathrm{d}x\mathrm{d}t, \tag{1.482}$$

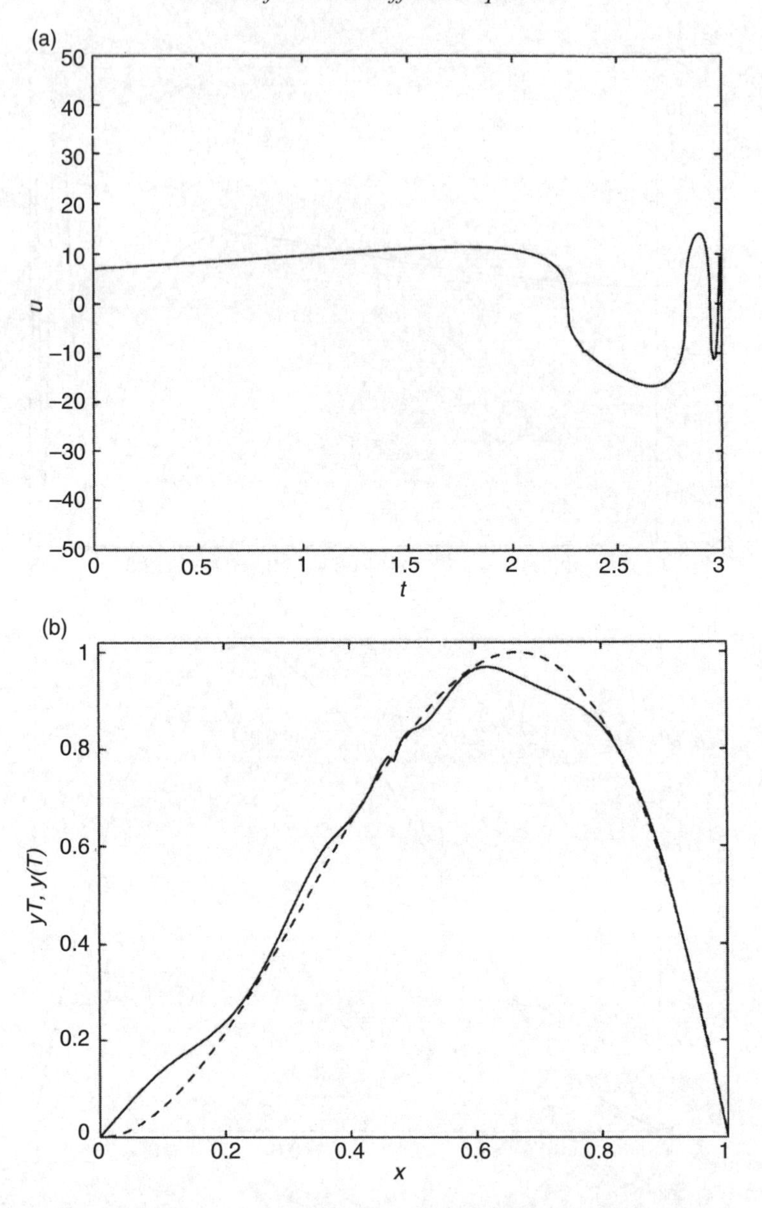

Fig. 1.17. (a) Variation of the computed optimal control, and (b) comparison between $y^c(T)$ and y_T (target function defined by (1.472)): $s = 4$, $T = 3$, $b = \frac{\sqrt{2}}{3}$, $k = 10^7$, $h = \Delta t = 10^{-2}$.

where, in (1.482), the control v describes the following closed convex subset of $L^2(\mathcal{O} \times (0,T))$:

$$\{\, v \,|\, v \in L^2(\mathcal{O} \times (0,T)),\ \{v,y\} \text{ verifies (1.477), (1.480), and (1.481)} \,\}.$$

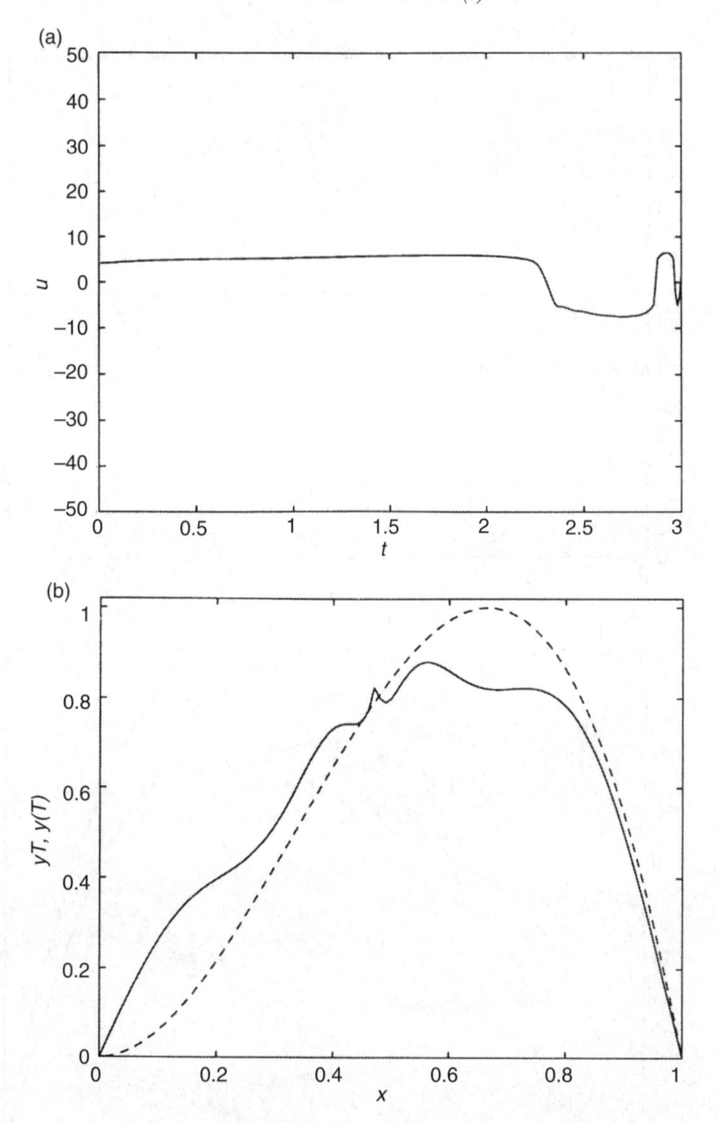

Fig. 1.18. (a) Variation of the computed optimal control, and (b) comparison between $y^c(T)$ and y_T (target function defined by (1.472)): $s = 6$, $T = 3$, $b = \frac{\sqrt{2}}{3}$, $k = 10^7$, $h = \Delta t = 10^{-2}$.

From a practical point of view, it is quite natural to consider the following *penalized* variant of problem (1.482):

$$\min_{v \in L^2(\mathcal{O}\times(0,T))} \frac{1}{2}\Bigg(\iint_{\mathcal{O}\times(0,T)} |v|^2 \,\mathrm{d}x\mathrm{d}t + k_1 \left\| \Pi_\mu\big(y(T) - y_T\big)\right\|^2_{L^2(\Omega)} + k_2 \int_0^T \|y(t)\|^2_{H^\alpha}\,\mathrm{d}t\Bigg); \quad (1.483)$$

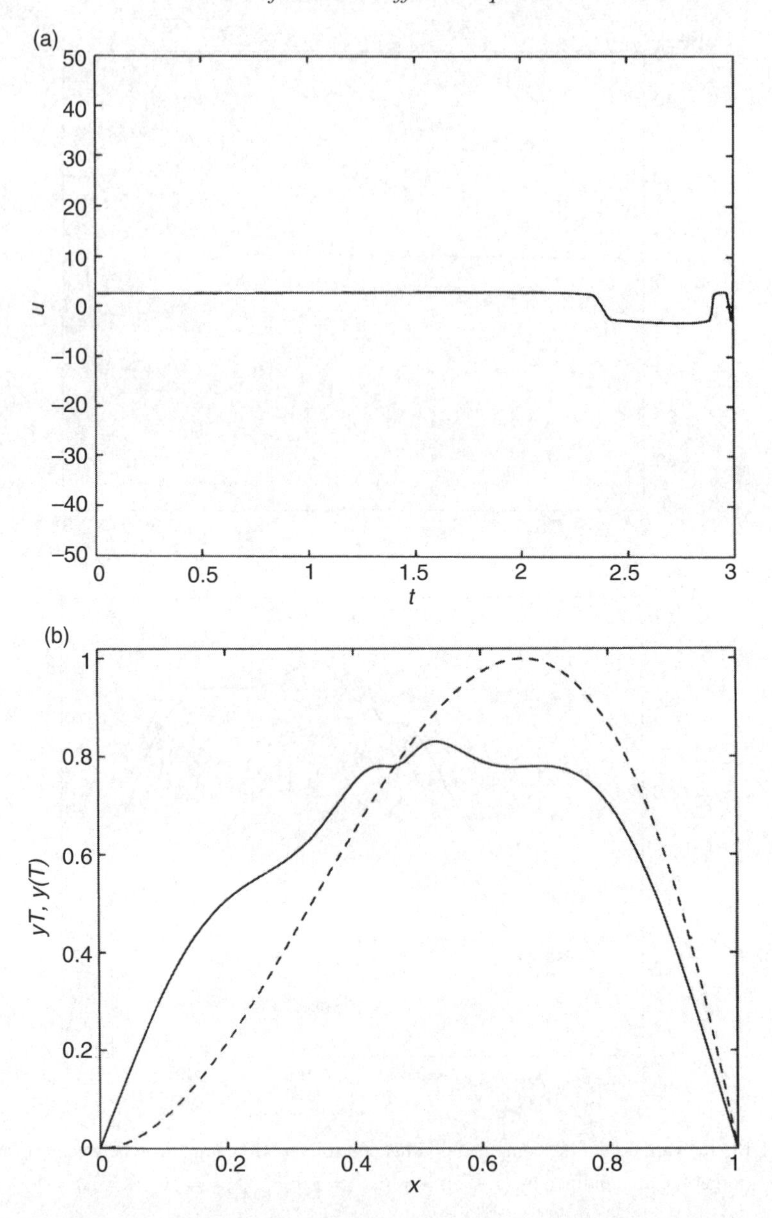

Fig. 1.19. (a) Variation of the computed optimal control, and (b) comparison between $y^c(T)$ and y_T (target function defined by (1.472)): $s = 10$, $T = 3$, $b = \frac{\sqrt{2}}{3}$, $k = 10^7$, $h = \Delta t = 10^{-2}$.

in (1.483), the (penalty) coefficients k_1 and k_2 are both positive, and y is a function of v via the solution of the state equation (1.477); using *convexity* arguments and other techniques, discussed in, for example, J.L. Lions (1968, 1971), we can easily prove that problem (1.483) has a unique solution. Suppose, for simplicity, that $\alpha = 0$;

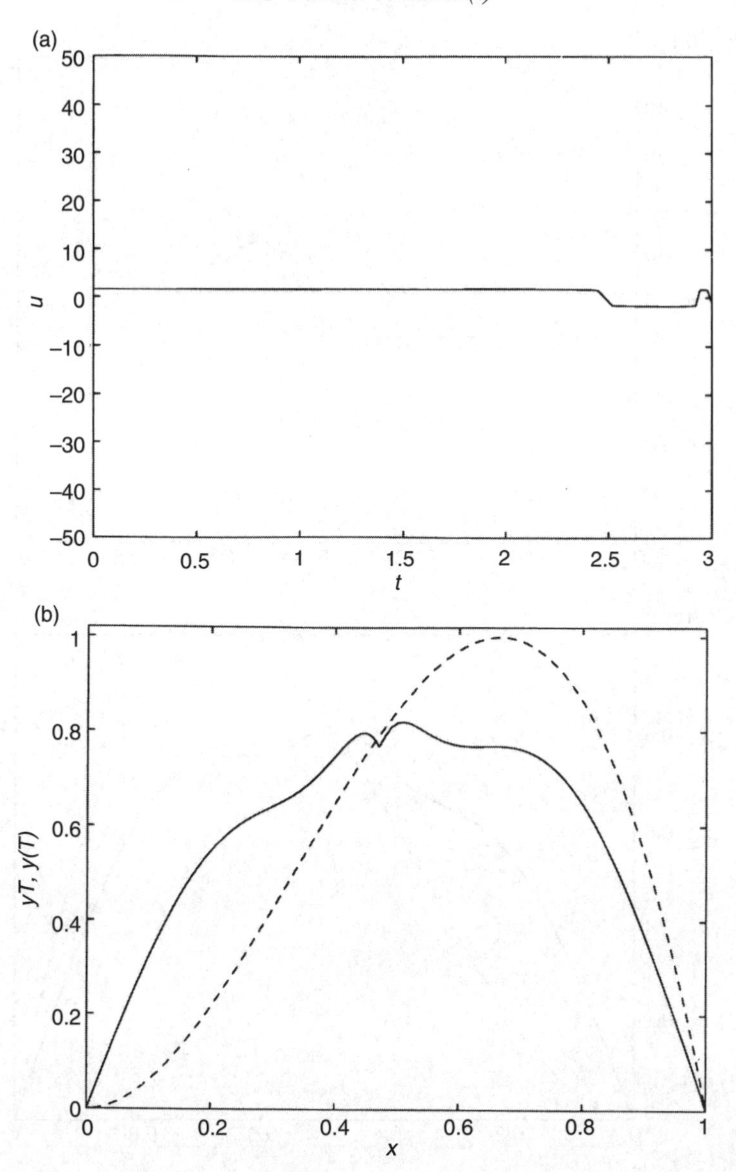

Fig. 1.20. (a) Variation of the computed optimal control, and (b) comparison between $y^c(T)$ and y_T (target function defined by (1.472)): $s = 20$, $T = 3$, $b = \frac{\sqrt{2}}{3}$, $k = 10^7$, $h = \Delta t = 10^{-2}$.

problem (1.483) reads then as follows:

$$\min_{v \in L^2(\mathcal{O}\times(0,T))} \frac{1}{2}\Bigg(\iint_{\mathcal{O}\times(0,T)} |v|^2 \,\mathrm{d}x\mathrm{d}t + k_1 \left\| \Pi_\mu\big(y(T) - y_T\big)\right\|^2_{L^2(\Omega)} + k_2 \iint_{\Omega\times(0,T)} |y|^2 \,\mathrm{d}x\mathrm{d}t\Bigg). \quad (1.484)$$

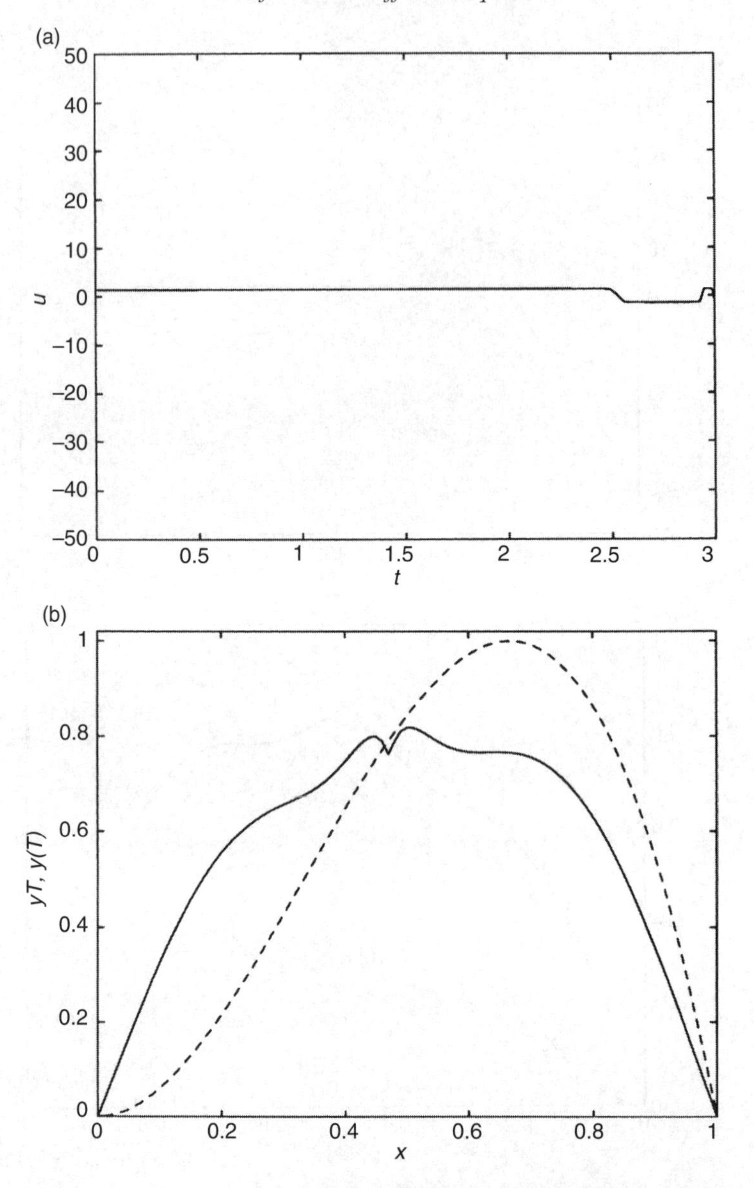

Fig. 1.21. (a) Variation of the computed optimal control, and (b) comparison between $y^c(T)$ and y_T (target function defined by (1.472)): $s = 30$, $T = 3$, $b = \frac{\sqrt{2}}{3}$, $k = 10^7$, $h = \Delta t = 10^{-2}$.

Let us denote by $J_{\mathbf{k}}(\cdot)$ the functional of v occurring in (1.484), with $\mathbf{k} = \{k_1, k_2\}$. The unique solution u of problem (1.484) is characterized by

$$J'_{\mathbf{k}}(u) = 0, \tag{1.485}$$

with $J'_{\mathbf{k}}(\cdot)$ the *differential* of the cost function $J_{\mathbf{k}}(\cdot)$. In order to compute $J'_{\mathbf{k}}(v)$, $\forall v \in L^2(\mathcal{O} \times (0,T))$, it is convenient to introduce the (Hilbert) space $\mathcal{U}$ and the

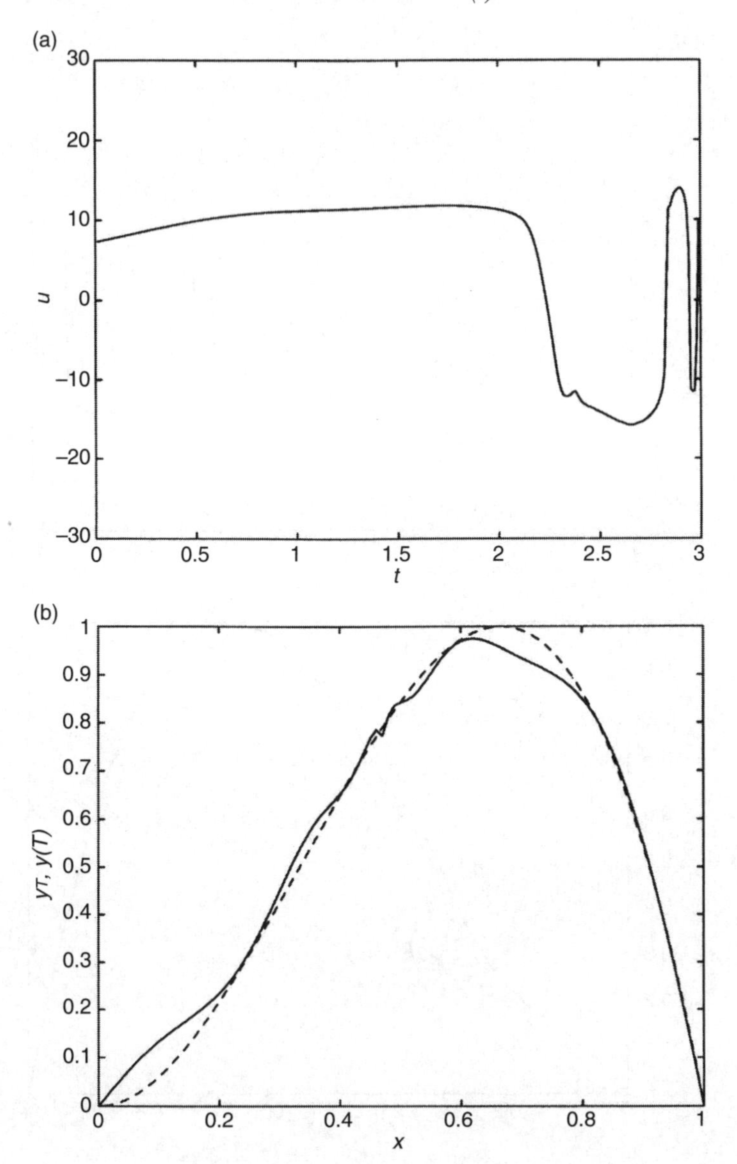

Fig. 1.22. (a) Variation of the computed optimal control, and (b) comparison between $y^c(T)$ and y_T (target function defined by (1.472)): $s = 6$, $T = 3$, $b = \frac{\sqrt{2}}{3}$, $k = 2 \times 10^9$, $h = \Delta t = 10^{-2}$.

scalar-product $(\cdot, \cdot)_{\mathcal{U}}$, defined by

$$\mathcal{U} = L^2(\mathcal{O} \times (0, T)), \tag{1.486}$$

and

$$(v, w)_{\mathcal{U}} = \iint_{\mathcal{O} \times (0,T)} vw \,\mathrm{d}x\mathrm{d}t, \quad \forall v, w \in \mathcal{U}, \tag{1.487}$$

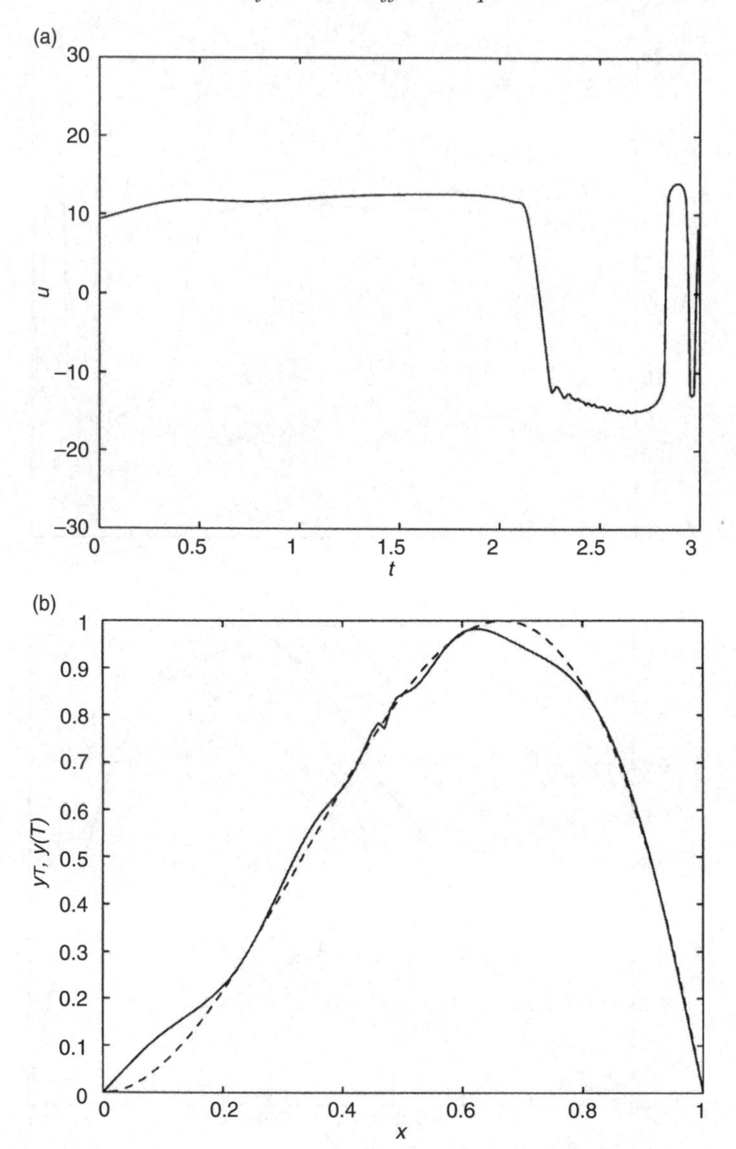

Fig. 1.23. (a) Variation of the computed optimal control, and (b) comparison between $y^c(T)$ and y_T (target function defined by (1.472)): $s = 10$, $T = 3$, $b = \frac{\sqrt{2}}{3}$, $k = 10^{15}$, $h = \Delta t = 10^{-2}$.

respectively. To compute $J'_{\mathbf{k}}(v)$ we are going to use a (formal) *perturbation method*: we have thus

$$\left(J'_{\mathbf{k}}(v), \delta v\right)_{\mathcal{U}} = \iint_{\mathcal{O}\times(0,T)} J'_{\mathbf{k}}(v)\delta v \,\mathrm{d}x\mathrm{d}t = \iint_{\mathcal{O}\times(0,T)} v\delta v \,\mathrm{d}x\mathrm{d}t$$
$$+ k_1 \int_\Omega \Pi_\mu\big(y(T) - y_T\big)\Pi_\mu \delta y(T)\,\mathrm{d}x + k_2 \iint_{\Omega\times(0,T)} y\delta y \,\mathrm{d}x\mathrm{d}t, \quad (1.488)$$

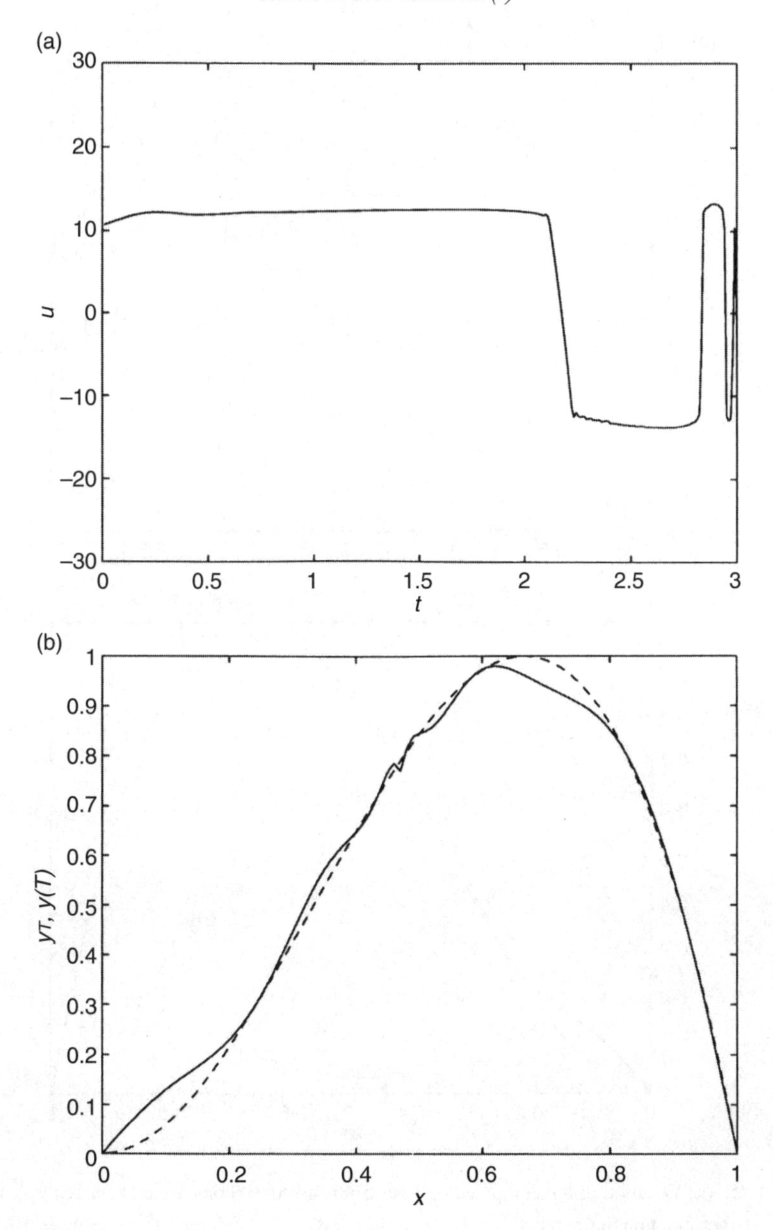

Fig. 1.24. (a) Variation of the computed optimal control, and (b) comparison between $y^c(T)$ and y_T (target function defined by (1.472)): $s = 20$, $T = 3$, $b = \frac{\sqrt{2}}{3}$, $k = 10^{25}$, $h = \Delta t = 10^{-2}$.

where, in (1.488), δy is the solution of

$$\begin{cases} \dfrac{\partial \delta y}{\partial t} - \Delta \delta y = \delta v \chi_{\mathcal{O}\times(0,T)} \text{ in } \Omega \times (0, T), \\ \delta y(0) = 0, \quad \delta y = 0 \text{ on } \Gamma \times (0, T). \end{cases} \tag{1.489}$$

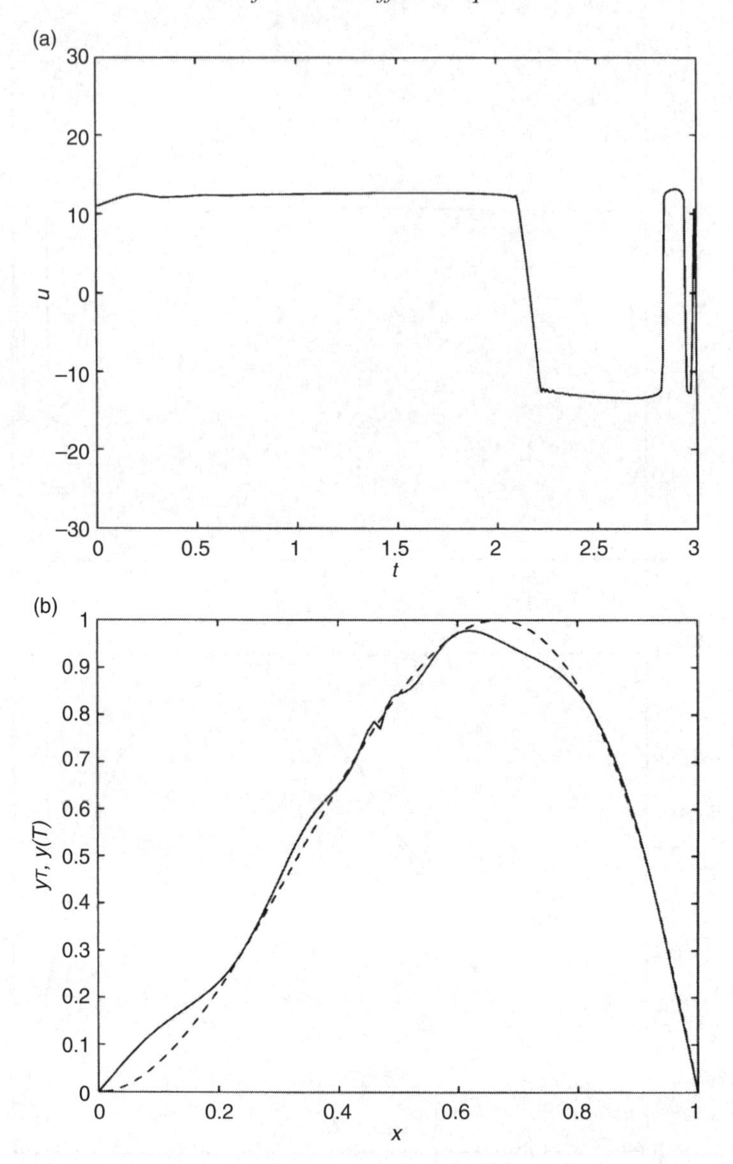

Fig. 1.25. (a) Variation of the computed optimal control, and (b) comparison between $y^c(T)$ and y_T (target function defined by (1.472)): $s = 30$, $T = 3$, $b = \frac{\sqrt{2}}{3}$, $k = 10^{30}$, $h = \Delta t = 10^{-2}$.

Since $\Pi_\mu = \Pi_\mu^*$ and $\Pi_\mu = \Pi_\mu^2$ we have, in (1.488),

$$\int_\Omega \Pi_\mu\big(y(T) - y_T\big)\Pi_\mu \delta y(T)\,\mathrm{d}x = \int_\Omega \Pi_\mu\big(y(T) - y_T\big)\delta y(T)\,\mathrm{d}x. \tag{1.490}$$

We introduce now a "smooth" function p of $\{x, t\}$ verifying

$$p = 0 \quad \text{on } \Gamma \times (0, T); \tag{1.491}$$

multiplying both sides of the first equation in (1.489) by p, and integrating by parts, we obtain

$$\iint_{\Omega\times(0,T)} p\frac{\partial\delta y}{\partial t}\,\mathrm{d}x\mathrm{d}t - \iint_{\Omega\times(0,T)} p\Delta\delta y\,\mathrm{d}x\mathrm{d}t = \int_\Omega p(T)\delta y(T)\,\mathrm{d}x$$
$$+\iint_{\Omega\times(0,T)}\left(-\frac{\partial p}{\partial t}-\Delta p\right)\delta y\,\mathrm{d}x\mathrm{d}t = \iint_{\mathcal{O}\times(0,T)} p\delta v\,\mathrm{d}x\mathrm{d}t. \quad (1.492)$$

Suppose that, in addition to (1.491), the function p verifies

$$-\frac{\partial p}{\partial t} - \Delta p = k_2 y \text{ in } \Omega\times(0,T), \quad p(T) = k_1\Pi_\mu\big(y(T)-y_T\big); \quad (1.493)$$

combining (1.493) with (1.488), (1.490), (1.492), and (1.493) yields

$$J'_{\mathbf{k}}(v) = v + p|_{\mathcal{O}\times(0,T)}. \quad (1.494)$$

It follows from (1.485) and (1.494) that the unique solution u of problem (1.484) is characterized by the existence of a pair $\{y,p\}$ such that

$$u + p|_{\mathcal{O}\times(0,T)} = 0, \quad (1.495)$$

$$\begin{cases} \dfrac{\partial y}{\partial t} - \Delta y = u\chi_{\mathcal{O}\times(0,T)} \text{ in } \Omega\times(0,T), \\ y(0) = 0, \quad y = 0 \text{ on } \Gamma\times(0,T), \end{cases} \quad (1.496)$$

and

$$\begin{cases} -\dfrac{\partial p}{\partial t} - \Delta p = k_2 y \text{ in } \Omega\times(0,T), \\ p(T) = k_1\Pi_\mu\big(y(T)-y_T\big), \quad p = 0 \text{ on } \Gamma\times(0,T). \end{cases} \quad (1.497)$$

Problem (1.484) can be solved by a *conjugate gradient algorithm* operating in $\mathcal{U}$; this algorithm reads as follows.

Algorithm 1.69 *(Conjugate gradient solution of problem (1.484))*

$$u^0 \text{ is given in } \mathcal{U}; \quad (1.498)$$

solve

$$\begin{cases} \dfrac{\partial y^0}{\partial t} - \Delta y^0 = u^0\chi_{\mathcal{O}\times(0,T)} \text{ in } \Omega\times(0,T), \\ y^0(0) = 0, \quad y^0 = 0 \text{ on } \Gamma\times(0,T), \end{cases} \quad (1.499)$$

$$\begin{cases} -\dfrac{\partial p^0}{\partial t} - \Delta p^0 = k_2 y^0 \text{ in } \Omega\times(0,T), \\ p^0(T) = k_1\Pi_\mu\big(y^0(T)-y_T\big), \quad p^0 = 0 \text{ on } \Gamma\times(0,T), \end{cases} \quad (1.500)$$

and

$$\begin{cases} g^0 \in \mathcal{U}, \ \forall v \in \mathcal{U}, \\ \iint_{\mathcal{O}\times(0,T)} g^0 v \, dx dt = \iint_{\mathcal{O}\times(0,T)} \left(u^0 + p^0\right) v \, dx dt, \end{cases} \tag{1.501}$$

and set

$$w^0 = g^0. \tag{1.502}$$

Then, for $n \geq 0$, assuming that u^n, g^n, and w^n are known (with the last two different from 0), solve

$$\begin{cases} \dfrac{\partial \bar{y}^n}{\partial t} - \Delta \bar{y}^n = w^n \chi_{\mathcal{O}\times(0,T)} \ \textit{in } \Omega \times (0,T), \\ \bar{y}^n(0) = 0, \quad \bar{y}^n = 0 \ \textit{on } \Gamma \times (0,T), \end{cases} \tag{1.503}$$

$$\begin{cases} -\dfrac{\partial \bar{p}^n}{\partial t} - \Delta \bar{p}^n = k_2 \bar{y}^n \ \textit{ in } \Omega \times (0,T), \\ \bar{p}^n(T) = k_1 \Pi_\mu \bar{y}^n(T), \quad \bar{p}^n = 0 \ \textit{ on } \Gamma \times (0,T), \end{cases} \tag{1.504}$$

and

$$\begin{cases} \bar{g}^n \in \mathcal{U}, \ \forall v \in \mathcal{U}, \\ \iint_{\mathcal{O}\times(0,T)} \bar{g}^n v \, dx dt = \iint_{\mathcal{O}\times(0,T)} \left(w^n + \bar{p}^n\right) v \, dx dt. \end{cases} \tag{1.505}$$

Compute

$$\rho_n = \iint_{\mathcal{O}\times(0,T)} |g^n|^2 \, dx dt \Big/ \iint_{\mathcal{O}\times(0,T)} \bar{g}^n w^n \, dx dt, \tag{1.506}$$

$$u^{n+1} = u^n - \rho_n w^n, \tag{1.507}$$

$$g^{n+1} = g^n - \rho_n \bar{g}^n. \tag{1.508}$$

If $\iint_{\mathcal{O}\times(0,T)} |g^{n+1}|^2 \, dx dt \big/ \iint_{\mathcal{O}\times(0,T)} |g^0|^2 \, dx dt \leq \varepsilon^2$ *take* $u = u^{n+1}$*; otherwise, update the search direction* w^n *via*

$$\gamma_n = \iint_{\mathcal{O}\times(0,T)} \left|g^{n+1}\right|^2 dx dt \Big/ \iint_{\mathcal{O}\times(0,T)} |g^n|^2 \, dx dt, \tag{1.509}$$

$$w^{n+1} = g^{n+1} + \gamma_n w^n. \tag{1.510}$$

Do $n = n + 1$ *and return to* (1.503).

1.12 Further remarks (II): A bisection based memory saving method for the solution of time dependent control problems by adjoint equation based methodologies

1.12.1 Motivation. Synopsis

A superficial inspection suggests that applying *adjoint equation* based methods to the solution of control (or inverse) problems modeled by time dependent linear or nonlinear partial differential equations will require the storage of huge quantities of information, particularly in the nonlinear case. One may be led to believe that the solution of the state equation has to be stored at each time step. If such was the case, it is clear that the adjoint equation approach may not be applicable, for example, for time dependent problems in three-space dimension. Actually, very substantial storage memory saving can be achieved through a *bi section based storage method*, the price to be paid being reasonable additional computational time in the sense that the state equation will have to be integrated more than once (3–5 times, typically). In Section 1.12.2 we shall consider a model optimal control problem and use an adjoint equation technique to compute the gradient of the related cost function after time discretization. Then in Section 1.12.3 we will describe the bi section method.

1.12.2 A model optimal control problem and its time discretization

Let us consider the following *optimal control problem*:

$$\begin{cases} \mathbf{u} \in \mathcal{U}\,(= L^2(0,T;\mathbb{R}^c)\,), \\ J(\mathbf{u}) \leq J(\mathbf{v}), \quad \forall \mathbf{v} \in \mathcal{U}, \end{cases} \tag{1.511}$$

with, in (1.511), the *cost function* J defined by

$$J(\mathbf{v}) = \frac{1}{2}\int_0^T \mathbf{S}\mathbf{v}(t)\cdot\mathbf{v}(t)\,\mathrm{d}t + \frac{k_1}{2}\int_0^T \|\mathbf{C}_1\mathbf{y}(t) - \mathbf{z}_1(t)\|_1^2\,\mathrm{d}t + \frac{k_2}{2}\|\mathbf{C}_2\mathbf{y}(T) - \mathbf{z}_2\|_2^2, \tag{1.512}$$

$\mathbf{y}$ being a function of $\mathbf{v}$ through the solution of the following *state equation*:

$$\begin{cases} \dfrac{\mathrm{d}\mathbf{y}}{\mathrm{d}t} + \mathbf{A}(\mathbf{y},t) = \mathbf{f} + \mathbf{B}\mathbf{v} \quad \text{in } (0,T), \\ \mathbf{y}(0) = \mathbf{y}_0. \end{cases} \tag{1.513}$$

We suppose that in (1.511)–(1.513):

- $T \in (0,+\infty)$.
- $\mathbf{S}$ is a time independent $c \times c$ matrix, symmetric, and positive definite.
- k_1 and k_2 are both nonnegative with $k_1 + k_2 > 0$.
- $\mathbf{C}_i$ is a time independent $M_i \times d$ matrix, $\mathbf{z}_1 \in L^2(0,T;\mathbb{R}^{M_1})$, $\mathbf{z}_2 \in \mathbb{R}^{M_2}$, and $\|\cdot\|_i$ denotes the canonical Euclidean norm of $\mathbb{R}^{M_i}$.
- $\mathbf{y}(t) \in \mathbb{R}^d, \forall t \in [0,T]$, $\mathbf{f} \in L^2(0,T;\mathbb{R}^d)$, $\mathbf{B}$ is a time independent $c \times d$ matrix, $\mathbf{y}_0 \in \mathbb{R}^d$, $\mathbf{A} : \mathbb{R}^d \times [0,T] \to \mathbb{R}^d$; we shall assume that the operator $\mathbf{A}$ is differentiable with respect to its first argument.

Assuming that problem (1.511) has a solution $\mathbf{u}$, this solution will verify

$$J'(\mathbf{u}) = \mathbf{0}, \tag{1.514}$$

where J' denotes the differential of J. We can easily show (by a *perturbation analysis*, for example) that

$$J'(\mathbf{v}) = \mathbf{S}\mathbf{v} + \mathbf{B}^t\mathbf{p}, \quad \forall \mathbf{v} \in \mathcal{U}, \tag{1.515}$$

where, in (1.515), the vector-valued function $\mathbf{p}$ is solution of the following *adjoint system*:

$$\begin{cases} -\dfrac{d\mathbf{p}}{dt} + \left(\dfrac{\partial \mathbf{A}}{\partial y}(\mathbf{y}, t)\right)^t \mathbf{p} = k_1\mathbf{C}_1^t(\mathbf{C}_1\mathbf{y} - \mathbf{z}_1) \text{ in } (0, T), \\ \mathbf{p}(T) = k_2\mathbf{C}_2^t(\mathbf{C}_2\mathbf{y}(T) - \mathbf{z}_2). \end{cases} \tag{1.516}$$

Let us briefly discuss now the *time discretization* of the control problem (1.511). For simplicity we shall time discretize (1.513) by the *forward Euler scheme* with $\Delta t = \frac{T}{N}$, N being a positive integer. We obtain then as *discrete control problem*

$$\begin{cases} \mathbf{u}^{\Delta t}\,(= \{\mathbf{u}^n\}_{n=1}^N\,) \in \mathcal{U}^{\Delta t}\,(= (\mathbb{R}^c)^N\,), \\ J_{\Delta t}(\mathbf{u}^{\Delta t}) \leq J_{\Delta t}(\mathbf{v}), \quad \forall \mathbf{v}\,(= \{\mathbf{v}^n\}_{n=1}^N\,) \in \mathcal{U}^{\Delta t}, \end{cases} \tag{1.517}$$

with

$$J_{\Delta t}(\mathbf{v}) = \frac{1}{2}\Delta t \sum_{n=}^{N} \mathbf{S}\mathbf{v}^n \cdot \mathbf{v}^n + \frac{1}{2}\Delta t\, k_1 \sum_{n=1}^{N} \left\|\mathbf{C}_1\mathbf{y}^n - \mathbf{z}_1^n\right\|_1^2 + \frac{1}{2}k_2 \left\|\mathbf{C}_2\mathbf{y}^N - \mathbf{z}_2\right\|_2^2, \tag{1.518}$$

where, in (1.518), $\{\mathbf{y}^n\}_{n=1}^N$ is obtained via the solution of the following *discrete state equation*:

$$\begin{cases} \mathbf{y}^0 = \mathbf{y}_0; \\ \text{for } n = 1, \ldots, N \text{ compute } \mathbf{y}^n \text{ from} \\ \dfrac{\mathbf{y}^n - \mathbf{y}^{n-1}}{\Delta t} + \mathbf{A}(\mathbf{y}^{n-1}, (n-1)\Delta t) = \mathbf{f}^n + \mathbf{B}\mathbf{v}^n. \end{cases} \tag{1.519}$$

If $\mathbf{u}^{\Delta t}$ is a solution of the above discrete control problem we have then

$$J'_{\Delta t}(\mathbf{u}^{\Delta t}) = \mathbf{0}, \tag{1.520}$$

where, in (1.520), the differential $J'_{\Delta t}$ of $J_{\Delta t}$ is obtained as follows:

$$J'_{\Delta t}(\mathbf{v}) = \{\mathbf{S}\mathbf{v}^n + \mathbf{B}^t\mathbf{p}^n\}_{n=1}^N, \quad \forall \mathbf{v} \in \mathcal{U}^{\Delta t}, \tag{1.521}$$

with $\{\mathbf{p}^n\}_{n=1}^N$ the solution of the following *discrete adjoint equation*:

$$\begin{cases} \mathbf{p}^{N+1} = k_2 \mathbf{C}_2^t(\mathbf{C}_2\mathbf{y}^N - \mathbf{z}_2), \\ \dfrac{\mathbf{p}^N - \mathbf{p}^{N+1}}{\Delta t} = k_1 \mathbf{C}_1^t(\mathbf{C}_1\mathbf{y}^N - \mathbf{z}_1^N), \\ \text{for } n = N-1, \ldots, 1 \text{ compute } \mathbf{p}^n \text{ from} \\ \dfrac{\mathbf{p}^n - \mathbf{p}^{n+1}}{\Delta t} + \left(\dfrac{\partial \mathbf{A}}{\partial y}(\mathbf{y}^n, n\Delta t)\right)^t \mathbf{p}^{n+1} = k_1 \mathbf{C}_1^t(\mathbf{C}_1\mathbf{y}^n - \mathbf{z}_1^n). \end{cases} \tag{1.522}$$

A superficial inspection of the relations (1.521) and (1.522) suggests that to compute $J'_{\Delta t}(\mathbf{v})$ it is necessary, in general, to store the vector $\{\mathbf{y}^n\}_{n=1}^N$ which for large values of d and N makes the above approach not very practical. We will show in the following paragraph that in fact this difficulty can be easily overcome, the price to pay being the necessity to solve the discrete state equation (1.519) more than once (3–5 times, typically).

1.12.3 Description of the bi section memory saving method

Suppose that $N = 2^M$ with $M > 1$ (we have then $\Delta t = 2^{-M}T$). We shall store the components of the discrete state vector $\{\mathbf{y}^n\}_{n=0}^N$ at $t = T_0, T_1, \ldots, T_q, \ldots, T_Q$, with

$$T_q = \left(1 - 2^{-q}\right)T, \quad 0 \le q \le Q \le M, \tag{1.523}$$

and at all the discrete time instants on interval $(T_Q, T]$; we have thus stored S_Q snapshots with

$$S_Q = Q + 1 + 2^{M-Q}; \tag{1.524}$$

see Figure 1.26 for the time location of the stored snapshots.

Consider the function $S : \mathbb{R}_+ \to \mathbb{R}$ defined by $S(\xi) = \xi + 1 + 2^{M-\xi}$. Function S is minimal at $\xi = \xi^*$ such that $S'(\xi^*) = 0$, with the derivative S' of S given by

$$S'(\xi) = 1 - \ln 2\, 2^{M-\xi}.$$

We have then

$$\xi^* = M + \frac{\ln(\ln 2)}{\ln 2}. \tag{1.525}$$

Since $\ln 2 = 0.69\cdots \approx \frac{1}{\sqrt{2}}$ we have, from (1.525),

$$\xi^* \approx M - \frac{1}{2}. \tag{1.526}$$

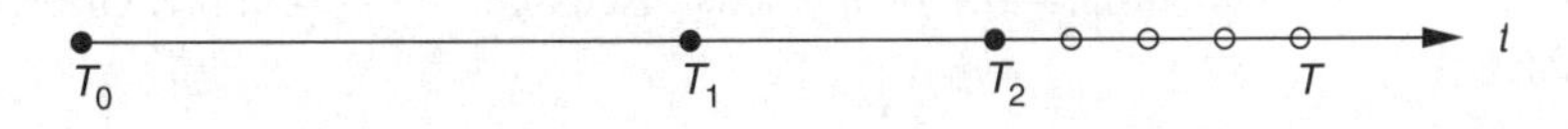

Fig. 1.26. Time location of the stored snapshots ($M = 4$, $Q = 2$).

It follows from (1.526) that ξ^* is not an integer, implying that in terms of memory saving an optimal choice is provided by $Q = M$ ($Q = M - 1$ is another possibility since ξ^* is "almost" the mid-point of interval $[M-1, M]$). We shall suppose from now on that that $Q = M$, implying that $T_Q = T_M = T - \Delta t$.

In order to evaluate the cost of computing $J'_{\Delta t}(\mathbf{v})$, from relations (1.519), (1.521), and (1.522), we are going to proceed by induction over Q.

- Suppose that $Q = 0$, that is, we store the full state vector $\{\mathbf{y}^n\}_{n=0}^N$. To obtain $J'_{\Delta t}(\mathbf{v})$ we just have to solve *once* equation (1.519) and the adjoint equation (1.522).
- Suppose now that $Q = 1$. In order to compute $J'_{\Delta t}(\mathbf{v})$ we have to solve (1.519) and store $\mathbf{y}^0$ and $\{\mathbf{y}^n\}_{n=\frac{N}{2}}^N$. Then we solve (1.522) from $n = N+1$ to $n = \frac{N}{2}$ and compute $\{\mathbf{S}\mathbf{v}^n + \mathbf{B}^t\mathbf{p}^n\}_{n=\frac{N}{2}}^N$ "on the fly." Next, we solve (1.519) from $n = 0$ to $n = \frac{N}{2} - 1$ and store the corresponding snapshots in the storage space previously occupied by $\{\mathbf{y}^n\}_{n=\frac{N}{2}+1}^N$. Finally, we solve (1.522) from $n = \frac{N}{2}$ to $n = 1$ and compute $\{\mathbf{S}\mathbf{v}^n + \mathbf{B}^t\mathbf{p}^n\}_{n=1}^{\frac{N}{2}-1}$. The state equation has been solved 1.5 times and the adjoint equation only once.
- Generalizing the above procedure for $Q > 1$ is straightforward; proceeding by induction we can easily show that we shall have to solve the state equation (1.519) "$1+\frac{Q}{2}$" times and the adjoint equation (1.522) only once. Thus, if $Q = M$ we shall have to solve the state equation (1.519) "$1 + \frac{M}{2}$" times and the adjoint equation (1.522) only once.

Let us summarize:

Assuming that $Q = M$ *we shall store* $M + 2 = \log_2 N + 2 = \log_2 4N$ *snapshots extracted from* $\{\mathbf{y}^n\}_{n=0}^N$ *and shall have to solve the discrete state equation* (1.519) $1 + \frac{M}{2} = \log_2 2\sqrt{N}$ *times and the adjoint equation* (1.522) *only once.*

To illustrate the above procedure, suppose that $N = 1024 = 2^{10}$. We have then to store 12 snapshots (instead of 1025 if $Q = 0$) and solve the state equation (1.519) six times. Since we have to include the cost of integrating the adjoint equation (1.522) we can say that using $Q = M$ instead of $Q = 0$ implies that:

- The required memory is divided by 85.
- We have to solve 7 discrete differential equations instead of 2, that is, a factor of 3.5.

Remark 1.70 The above memory saving method has been discussed in Glowinski (2003, Chapter 10); it is a variant of the one described in, for example, Berggren, Glowinski, and Lions (1996a, 1996b) and He and Glowinski (1998). Both methods are related to the *automatic differentiation methods* discussed in, for example, Griewank (1992).

1.13 Further remarks (III): A brief introduction to Riccati equations based control methods

1.13.1 Generalities

The main drawback of the control methods discussed in the previous sections is that they require the solution of a *two-point boundary value problem* with respect to the time variable, in order to compute the optimal control (the two points are $t = 0$ and $t = T$). Actually, what we would like is to be able to obtain the control at time t from the state function vector at the same time t, via a "simple" linear transformation. It follows from Kalman (see, for example, Kalman, 1960, 1963; Kalman and Bucy, 1961; Kalman, Falb, and Arbib, 1969; J.L. Lions, 1968, 1971; Faurre, 1971) that the above objective can be reached if the *state equation is linear* and the *cost function is quadratic*; there is, however, a price to pay which is the backward integration of a matrix or operator *Riccati equation* and the storage of the gain matrix or operator (of smaller size, in general). Our goal here is to give a very brief and simple introduction on this most important issue; some applications to *Neumann boundary control* will be given in Chapter 2 (an application to the *pointwise control* of diffusion equations is discussed in Barck-Holst, 1996).

Remark 1.71 The Kalman/Riccati approach, to be discussed in the following paragraphs, has motivated such a huge literature that we have decided to limit ourselves to the few references given just above, since they are at the foundation of modern *System Theory*.

1.13.2 A class of control problems

Focusing on *finite-dimensional systems* (see, for example, J.L. Lions, 1968, 1971 for infinite-dimensional situations associated with the control of distributed parameter systems) we consider a control problem close to the one introduced in Section 1.12.2, namely,

$$\begin{cases} \mathbf{u} \in \mathcal{U}\,(= L^2(0, T; \mathbb{R}^c)\,), \\ J(\mathbf{u}) \le J(\mathbf{v}), \quad \forall \mathbf{v} \in \mathcal{U}, \end{cases} \tag{1.527}$$

the *cost function* J in (1.527) being defined by

$$J(\mathbf{v}) = \frac{1}{2}\int_0^T \mathbf{S}\mathbf{v}(t)\cdot\mathbf{v}(t)\,\mathrm{d}t \\ + \frac{k_1}{2}\int_0^T \|\mathbf{C}_1\mathbf{y}(t) - \mathbf{z}_1(t)\|_1^2\,\mathrm{d}t + \frac{k_2}{2}\|\mathbf{C}_2\mathbf{y}(T) - \mathbf{z}_2\|_2^2, \tag{1.528}$$

$\mathbf{y}$ being a function of $\mathbf{v}$ through the solution of the following *state equation*:

$$\begin{cases} \dfrac{\mathrm{d}\mathbf{y}}{\mathrm{d}t} + \mathbf{A}\mathbf{y} = \mathbf{f} + \mathbf{B}\mathbf{v} \quad \text{in } (0, T), \\ \mathbf{y}(0) = \mathbf{y}_0; \end{cases} \tag{1.529}$$

the notation and assumptions on the various mathematical objects occurring in (1.527)–(1.529)) are those of Section 1.12.2, except that now $\mathbf{A}$ can be identified with a $d \times d$ matrix that we will assume time independent for simplicity.

1.13.3 Optimality conditions

It follows from Section 1.12.2 that the solution $\mathbf{u}$ of the control problem (1.527) is characterized by the existence of a pair $\{\mathbf{y}, \mathbf{p}\}$ such that the following *optimality conditions* (*necessary* and *sufficient* here) hold

$$\mathbf{u} = -\mathbf{S}^{-1}\mathbf{B}^t\mathbf{p}, \tag{1.530}$$

$$\begin{cases} \dfrac{d\mathbf{y}}{dt} + \mathbf{A}\mathbf{y} = \mathbf{f} + \mathbf{B}\mathbf{u} \quad \text{in } (0, T), \\ \mathbf{y}(0) = \mathbf{y}_0; \end{cases} \tag{1.531}$$

$$\begin{cases} -\dfrac{d\mathbf{p}}{dt} + \mathbf{A}^t\mathbf{p} = k_1\mathbf{C}_1^t(\mathbf{C}_1\mathbf{y} - \mathbf{z}_1) \quad \text{in } (0, T), \\ \mathbf{p}(T) = k_2\mathbf{C}_2^t(\mathbf{C}_2\mathbf{y}(T) - \mathbf{z}_2). \end{cases} \tag{1.532}$$

In previous sections, we used relations such as (1.530)–(1.532) to compute (after an appropriate time discretization) the optimal control by conjugate gradient algorithms. The idea in this section is to go further and to look for a methodology allowing the *online* solution of the optimal control problem. This will be achieved by using (1.530) to eliminate $\mathbf{u}$ from relations (1.531), (1.532).

1.13.4 Riccati equation (I): The continuous-time case

It follows from the optimality system (1.530)–(1.532) that $\mathbf{p}$ is an *affine function* of $\mathbf{y}$ (and conversely). We are going to verify if the relation between $\mathbf{y}$ and $\mathbf{p}$ is of the following form:

$$\mathbf{p} = \mathbf{P}\mathbf{y} + \mathbf{r}, \tag{1.533}$$

where in (1.533), $\mathbf{P}(t)$ is $d \times d$ matrix and $\mathbf{r}(t) \in \mathbb{R}^d$. Combining (1.530) with (1.531) yields

$$\frac{d\mathbf{y}}{dt} + \mathbf{A}\mathbf{y} + \mathbf{B}\mathbf{S}^{-1}\mathbf{B}^t\mathbf{p} = \mathbf{f}. \tag{1.534}$$

Differentiating (1.533) and combining with (1.534) we obtain (using $\dot{\mathbf{X}}$ to denote $d\mathbf{X}/dt$)

$$\dot{\mathbf{p}} = \dot{\mathbf{P}}\mathbf{y} + \mathbf{P}\dot{\mathbf{y}} + \dot{\mathbf{r}} = \dot{\mathbf{P}}\mathbf{y} + \mathbf{P}\left(\mathbf{f} - \mathbf{A}\mathbf{y} - \mathbf{B}\mathbf{S}^{-1}\mathbf{B}^t\mathbf{p}\right) + \dot{\mathbf{r}}. \tag{1.535}$$

Using (1.532) and (1.533) to eliminate **p** from (1.535) we obtain

$$\left(-\dot{\mathbf{P}} + \mathbf{PA} + \mathbf{A}^t\mathbf{P} + \mathbf{PBS}^{-1}\mathbf{B}^t\mathbf{P} - k_1\mathbf{C}_1^t\mathbf{C}_1\right)\mathbf{y}$$
$$- \dot{\mathbf{r}} + \left(\mathbf{A}^t + \mathbf{PBS}^{-1}\mathbf{B}^t\right)\mathbf{r} + k_1\mathbf{C}_1^t\mathbf{z}_1 - \mathbf{Pf} = \mathbf{0}. \tag{1.536}$$

Relation (1.536) will hold, $\forall \mathbf{y}_0 \in \mathbb{R}^d$, if the matrix-valued function **P** and the vector-valued function **r** verify

$$-\dot{\mathbf{P}} + \mathbf{PA} + \mathbf{A}^t\mathbf{P} + \mathbf{PBS}^{-1}\mathbf{B}^t\mathbf{P} - k_1\mathbf{C}_1^t\mathbf{C}_1 = \mathbf{0} \quad \text{in } (0, T), \tag{1.537}$$

$$-\dot{\mathbf{r}} + \left(\mathbf{A}^t + \mathbf{PBS}^{-1}\mathbf{B}^t\right)\mathbf{r} + k_1\mathbf{C}_1^t\mathbf{z}_1 - \mathbf{Pf} = \mathbf{0} \quad \text{in } (0, T), \tag{1.538}$$

respectively; equation (1.537) is a *matrix-valued Riccati equation*. In order to solve equation (1.537) and then (1.538), once **P** is known, further information is needed. It follows from (1.532) and (1.533) that

$$\mathbf{p}(T) = \mathbf{P}(T)\mathbf{y}(T) + \mathbf{r}(T) = k_2\mathbf{C}_2^t(\mathbf{C}_2\mathbf{y}(T) - \mathbf{z}_2), \quad \forall \mathbf{y}_0 \in \mathbb{R}^d.$$

The above relations imply

$$\mathbf{P}(T) = k_2\mathbf{C}_2^t\mathbf{C}_2, \tag{1.539}$$

and

$$\mathbf{r}(T) = -k_2\mathbf{C}_2^t\mathbf{z}_2; \tag{1.540}$$

relations (1.539) and (1.540) allow the backward integration of the differential equations (1.537) and (1.538). We observe that the matrix-valued function **P** is independent of $\mathbf{y}_0$, $\mathbf{z}_1$, $\mathbf{z}_2$, and **f**, while the vector-valued function **r** is independent of $\mathbf{y}_0$. Additional properties of the function **P** are provided by the following.

Theorem 1.72 *The matrix* $\mathbf{P}(t)$ *is symmetric and positive semidefinite,* $\forall t \in [0, T]$.

Proof. Our objectives are to show that, $\forall t \in [0, T]$, one has

$$\mathbf{P}(t)\boldsymbol{\zeta}_1 \cdot \boldsymbol{\zeta}_2 = \mathbf{P}(t)\boldsymbol{\zeta}_2 \cdot \boldsymbol{\zeta}_1, \quad \forall \boldsymbol{\zeta}_1, \boldsymbol{\zeta}_2 \in \mathbb{R}^d, \tag{1.541}$$

and

$$\mathbf{P}(t)\boldsymbol{\zeta} \cdot \boldsymbol{\zeta} \geq 0, \quad \forall \boldsymbol{\zeta} \in \mathbb{R}^d. \tag{1.542}$$

To achieve these goals, we consider the following control problem:

$$\min_{\mathbf{v} \in L^2(s,T;\mathbb{R}^d)} J_s(\mathbf{v}), \tag{1.543}$$

with $s \in [0, T)$ and

$$J_s(\mathbf{v}) = \frac{1}{2}\int_s^T \mathbf{S}\mathbf{v}(t) \cdot \mathbf{v}(t)\, dt + \frac{k_1}{2}\int_s^T \|\mathbf{C}_1\mathbf{y}_s(t)\|_1^2\, dt + \frac{k_2}{2}\|\mathbf{C}_2\mathbf{y}_s(T)\|_2^2, \tag{1.544}$$

the vector-valued function $\mathbf{y}_s$ being the solution of the following initial value problem:

$$\begin{cases} \dfrac{d\,\mathbf{y}_s}{d\,t} + \mathbf{A}\mathbf{y}_s = \mathbf{B}\mathbf{v} \quad \text{in } (s, T), \\ \mathbf{y}_s(s) = \boldsymbol{\zeta}. \end{cases} \tag{1.545}$$

The optimality conditions associated to (1.544), (1.545) read as follows (with obvious notation):

$$\mathbf{u}_s = -\mathbf{S}^{-1}\mathbf{B}^t\mathbf{p}_s, \tag{1.546}$$

$$\begin{cases} \dfrac{d\,\mathbf{y}_s}{d\,t} + \mathbf{A}\mathbf{y}_s = \mathbf{B}\mathbf{u}_s \quad \text{in } (s, T), \\ \mathbf{y}_s(s) = \boldsymbol{\zeta}, \end{cases} \tag{1.547}$$

$$\begin{cases} -\dfrac{d\,\mathbf{p}_s}{d\,t} + \mathbf{A}^t\mathbf{p}_s = k_1\mathbf{C}_1^t\mathbf{C}_1\mathbf{y}_s \quad \text{in } (s, T), \\ \mathbf{p}_s(T) = k_2\mathbf{C}_2^t\mathbf{C}_2\mathbf{y}_s(T). \end{cases} \tag{1.548}$$

It follows from (1.548) that $\mathbf{p}_s$ depends linearly from $\mathbf{y}_s$, namely $\mathbf{p}_s = \mathbf{P}_s\mathbf{y}_s$ with matrix $\mathbf{P}_s$ the solution of the matrix Riccati final value problem

$$\begin{cases} -\dot{\mathbf{P}}_s + \mathbf{P}_s\mathbf{A} + \mathbf{A}^t\mathbf{P}_s \\ \qquad\qquad + \mathbf{P}_s\mathbf{B}\mathbf{S}^{-1}\mathbf{B}^t\mathbf{P}_s - k_1\mathbf{C}_1^t\mathbf{C}_1 = \mathbf{0} \quad \text{in } (s, T), \\ \mathbf{P}_s(T) = k_2\mathbf{C}_2^t\mathbf{C}_2. \end{cases} \tag{1.549}$$

On interval $[s, T]$ the matrix $\mathbf{P}_s$ verifies the same differential equation and condition at time $t = T$ than matrix $\mathbf{P}$ in (1.537) and (1.539); *assuming* the uniqueness of the solution we have coincidence of $\mathbf{P}$ and $\mathbf{P}_s$ on the time interval $[s, T]$, $\forall s \in [0, T]$; we also have

$$\mathbf{p}_s(s) = \mathbf{P}_s(s)\boldsymbol{\zeta} = \mathbf{P}(s)\boldsymbol{\zeta}. \tag{1.550}$$

Let us denote by $\{\mathbf{y}_i, \mathbf{p}_i, \mathbf{u}_i\}$ the solution of the optimality system (1.546)–(1.548) associated with $\boldsymbol{\zeta}_i$. By elimination of $\mathbf{u}_1$ in the related system (1.546)–(1.548) we obtain

$$\begin{cases} \dfrac{d\,\mathbf{y}_1}{d\,t} + \mathbf{A}\mathbf{y}_1 + \mathbf{B}\mathbf{S}^{-1}\mathbf{B}^t\mathbf{p}_1 = \mathbf{0} \quad \text{in } (s, T), \\ \mathbf{y}_1(s) = \boldsymbol{\zeta}_1. \end{cases} \tag{1.551}$$

"Multiplying" by $\mathbf{p}_2$ both sides of the differential equation in (1.551) associated with $\boldsymbol{\zeta}_1$ and integrating by parts over $[s, T]$ yields

$$\mathbf{y}_1(T)\cdot\mathbf{p}_2(T) - \mathbf{y}_1(s)\cdot\mathbf{p}_2(s) - \int_s^T \mathbf{y}_1\cdot\frac{d\,\mathbf{p}_2}{d\,t}\,dt \\ + \int_s^T \mathbf{A}\mathbf{y}_1\cdot\mathbf{p}_2\,dt + \int_s^T \mathbf{S}^{-1}\mathbf{B}^t\mathbf{p}_1\cdot\mathbf{B}^t\mathbf{p}_2\,dt = 0. \tag{1.552}$$

Combining (1.552) with (1.546)–(1.548) and (1.550) we obtain

$$\boldsymbol{\zeta}_1\cdot\mathbf{P}(s)\boldsymbol{\zeta}_2 = k_2\mathbf{C}_2\mathbf{y}_1(T)\cdot\mathbf{C}_2\mathbf{y}_2(T) \\ + \int_s^T \left(k_1\mathbf{C}_1\mathbf{y}_1\cdot\mathbf{C}_1\mathbf{y}_2 + \mathbf{S}^{-1}\mathbf{B}^t\mathbf{p}_1\cdot\mathbf{B}^t\mathbf{p}_2\right)dt, \quad \forall\boldsymbol{\zeta}_1,\boldsymbol{\zeta}_2\in\mathbb{R}^d. \tag{1.553}$$

It follows from relation (1.553) that, $\forall s\in[0,T]$, the matrix $\mathbf{P}(s)$ is *symmetric* and *positive semidefinite.* □

Assuming that it is implementable the above *Riccati equation* based approach provides the control $\mathbf{u}$ from relations (1.530), (1.531), and (1.533), assuming that the matrix $\mathbf{P}$ and the function $\mathbf{r}$ are known (via the solution of (1.537)–(1.540)). From a practical point of view a time-discrete analogue of the above derivation is a natural next step to investigate; this will be the object of the following subsection.

1.13.5 Riccati equation (II): The discrete-time case

In order to derive the discrete analogue of the *Riccati equation* based approach, discussed in Section 1.13.4, we will rely on an analysis following closely the one used in the continuous case. We consider again the model optimal control problem (1.527); in order to discretize the above problem, we introduce first a *time-discretization step* $\Delta t = T/N$, with N a positive integer, and approximate the *control space* $\mathcal{U}$ $(= L^2(0,T;\mathbb{R}^c))$ by $\mathcal{U}^{\Delta t}$ $(= (\mathbb{R}^c)^N)$; we equip the above space with the following Euclidean scalar product:

$$(\mathbf{v},\mathbf{w})_{\Delta t} = \Delta t\sum_{n=1}^{N}\mathbf{v}^n\cdot\mathbf{w}^n, \quad \forall\mathbf{v}=\{\mathbf{v}^n\}_{n=1}^N,\ \mathbf{w}=\{\mathbf{w}^n\}_{n=1}^N\in\mathcal{U}^{\Delta t}.$$

We approximate then the control problem (1.527) by

$$\begin{cases} \mathbf{u}^{\Delta t}(=\{\mathbf{u}^n\}_{n=1}^N)\in\mathcal{U}^{\Delta t}, \\ J_{\Delta t}(\mathbf{u}^{\Delta t})\le J_{\Delta t}(\mathbf{v}), \quad \forall\mathbf{v}(=\{\mathbf{v}^n\}_{n=1}^N)\in\mathcal{U}^{\Delta t}, \end{cases} \tag{1.554}$$

with

$$J_{\Delta t}(\mathbf{v}) = \frac{1}{2}\Delta t \sum_{n=1}^{N} \mathbf{S}\mathbf{v}^n \cdot \mathbf{v}^n + \frac{k_1}{2}\Delta t \sum_{n=1}^{N} \left\|\mathbf{C}_1\mathbf{y}^n - \mathbf{z}_1^n\right\|_1^2 + \frac{k_2}{2}\left\|\mathbf{C}_2\mathbf{y}^N - \mathbf{z}_2\right\|_2^2, \tag{1.555}$$

where, in (1.555), $\{\mathbf{y}^n\}_{n=1}^N$ is obtained via the solution of the following *discrete state equation*:

$$\begin{cases} \mathbf{y}^0 = \mathbf{y}_0; \\ \text{for } n = 1, \ldots, N, \ \mathbf{y}^{n-1} \text{ being known, compute } \mathbf{y}^n \text{ from} \\ \dfrac{\mathbf{y}^n - \mathbf{y}^{n-1}}{\Delta t} + \mathbf{A}\mathbf{y}^{n-1} = \mathbf{f}^n + \mathbf{B}\mathbf{v}^n, \end{cases} \tag{1.556}$$

with $\mathbf{f}^n = \mathbf{f}(n\Delta t)$. The gradient $\nabla J_{\Delta t}(\mathbf{v})$ of the cost function $J_{\Delta t}(\cdot)$ at $\mathbf{v}$ is given by

$$\nabla J_{\Delta t}(\mathbf{v}) = \left\{\mathbf{S}\mathbf{v}^n + \mathbf{B}^t\mathbf{p}^n\right\}_{n=1}^N, \tag{1.557}$$

where, in (1.557), $\{\mathbf{p}^n\}_{n=1}^N$ is the solution of the following *discrete adjoint equation*:

$$\begin{cases} \mathbf{p}^N = k_2\mathbf{C}_2^t\left(\mathbf{C}_2\mathbf{y}^N - \mathbf{z}_2\right) + \Delta t\, k_1\mathbf{C}_1^t\left(\mathbf{C}_1\mathbf{y}^N - \mathbf{z}_1^N\right), \\ \text{for } n = N-1, \ldots, 1, \ \mathbf{p}^{n+1} \text{ being known, compute } \mathbf{p}^n \text{ from} \\ \dfrac{\mathbf{p}^n - \mathbf{p}^{n+1}}{\Delta t} + \mathbf{A}^t\mathbf{p}^{n+1} = k_1\mathbf{C}_1^t\left(\mathbf{C}_1\mathbf{y}^n - \mathbf{z}_1^n\right). \end{cases} \tag{1.558}$$

The discrete control problem (1.554) has a unique solution characterized by:

$$\mathbf{u}^n = -\mathbf{S}^{-1}\mathbf{B}^t\mathbf{p}^n, \quad n = 1, \ldots, N, \tag{1.559}$$

$$\begin{cases} \mathbf{y}^0 = \mathbf{y}_0; \\ \text{for } n = 1, \ldots, N, \ \mathbf{y}^{n-1} \text{ being known, compute } \mathbf{y}^n \text{ from} \\ \dfrac{\mathbf{y}^n - \mathbf{y}^{n-1}}{\Delta t} + \mathbf{A}\mathbf{y}^{n-1} = \mathbf{f}^n + \mathbf{B}\mathbf{u}^n, \end{cases} \tag{1.560}$$

$$\begin{cases} \mathbf{p}^N = k_2\mathbf{C}_2^t\left(\mathbf{C}_2\mathbf{y}^N - \mathbf{z}_2\right) + \Delta t\, k_1\mathbf{C}_1^t\left(\mathbf{C}_1\mathbf{y}^N - \mathbf{z}_1^N\right), \\ \text{for } n = N-1, \ldots, 1, \ \mathbf{p}^{n+1} \text{ being known, compute } \mathbf{p}^n \text{ from} \\ \dfrac{\mathbf{p}^n - \mathbf{p}^{n+1}}{\Delta t} + \mathbf{A}^t\mathbf{p}^{n+1} = k_1\mathbf{C}_1^t\left(\mathbf{C}_1\mathbf{y}^n - \mathbf{z}_1^n\right) \end{cases} \tag{1.561}$$

(for simplicity, we left some notation unchanged). Combining (1.559) with (1.561), we obtain

$$\begin{cases} \mathbf{y}^0 = \mathbf{y}_0; \\ \text{for } n = 1, \ldots, N, \ \mathbf{y}^{n-1} \text{ being known, compute } \mathbf{y}^n \text{ from} \\ \dfrac{\mathbf{y}^n - \mathbf{y}^{n-1}}{\Delta t} + \mathbf{A}\mathbf{y}^{n-1} + \mathbf{B}\mathbf{S}^{-1}\mathbf{B}^t\mathbf{p}^n = \mathbf{f}^n. \end{cases} \tag{1.562}$$

It follows of the above relations, that $\{\mathbf{p}^n\}_{n=1}^N$ is an *affine function* of $\{\mathbf{y}^n\}_{n=1}^N$ and conversely. From a practical point of view, the ideal situation would be the one where:

$$\mathbf{p}^n = \mathbf{P}^n\mathbf{y}^{n-1} + \mathbf{r}^n, \quad \forall n. \tag{1.563}$$

If (1.563) was taking place it would make, from (1.562), the computation of $\mathbf{y}^n$ explicit. Unfortunately, the discrete Riccati equation associated to (1.563) is nonlinear (quadratic nonlinearity) and relatively costly to solve. An alternative to (1.563) is provided by

$$\mathbf{p}^n = \mathbf{P}^n\mathbf{y}^n + \mathbf{r}^n, \quad \forall n; \tag{1.564}$$

the corresponding discrete Riccati equation is easier to solve, but it will make implicit the computation of $\mathbf{y}^n$ from (1.562) (indeed, we shall have to solve the following linear system:

$$\left(\mathbf{I} + \Delta t\, \mathbf{B}\mathbf{S}^{-1}\mathbf{B}^t\mathbf{P}^n\right)\mathbf{y}^n = \mathbf{y}^{n-1} + \Delta t\left(\mathbf{f}^n - \mathbf{A}\mathbf{y}^{n-1} - \mathbf{B}\mathbf{S}^{-1}\mathbf{B}^t\mathbf{r}^n\right).$$

Assuming that (1.564) holds, we can show (after tedious calculations, to be skipped here) that $\mathbf{P}^n$ and $\mathbf{r}^n$ verify

$$\begin{cases} \mathbf{P}^n = \left(\mathbf{I} - \Delta t\, \mathbf{A}^t\right)\left(\mathbf{I} + \Delta t\, \mathbf{P}^{n+1}\mathbf{B}\mathbf{S}^{-1}\mathbf{B}^t\right)^{-1}\mathbf{P}^{n+1}\left(\mathbf{I} - \Delta t\, \mathbf{A}\right) \\ \qquad + \Delta t\, k_1 \mathbf{C}_1^t\mathbf{C}_1, \quad n = N-1, \ldots, 1, \\ \mathbf{P}^N = k_2\mathbf{C}_2^t\mathbf{C}_2 + \Delta t\, k_1\mathbf{C}_1^t\mathbf{C}_1, \end{cases} \tag{1.565}$$

and

$$\begin{cases} \mathbf{r}^n = \left(\mathbf{I} - \Delta t\, \mathbf{A}^t\right)\left(\mathbf{I} + \Delta t\, \mathbf{P}^{n+1}\mathbf{B}\mathbf{S}^{-1}\mathbf{B}^t\right)^{-1} \\ \qquad \times \left(\mathbf{r}^{n+1} + \Delta t\, \mathbf{P}^{n+1}\mathbf{f}^{n+1}\right) - \Delta t\, k_1\mathbf{C}_1^t\mathbf{z}_1^n, \\ \qquad n = N-1, \ldots, 1, \\ \mathbf{r}^N = -k_2\mathbf{C}_2^t\mathbf{z}_2 - \Delta t\, k_1\mathbf{C}_1^t\mathbf{z}_1^N, \end{cases} \tag{1.566}$$

respectively.

2

Boundary control

2.1 Dirichlet control (I): Formulation of the control problem

We consider again the *state equation*

$$\frac{\partial y}{\partial t} + \mathcal{A}y = 0 \text{ in } Q, \tag{2.1}$$

where the *second-order elliptic operator* $\mathcal{A}$ is as in Section 1.1, and where the control v is now a *boundary control of Dirichlet* type, namely,

$$y = \begin{cases} v \text{ on } \Sigma_0 = \Gamma_0 \times (0, T), \\ 0 \text{ on } \Sigma \backslash \Sigma_0, \end{cases} \tag{2.2}$$

where Γ_0 is a (regular) subset of Γ.

The *initial condition* is (for simplicity)

$$y(0) = 0. \tag{2.3}$$

In (2.2) we assume that

$$v \in L^2(\Sigma_0). \tag{2.4}$$

Then, assuming that the coefficients of operator $\mathcal{A}$ are smooth enough (see, for example, Lions and Magenes, 1968 for precise statements), the parabolic problem (2.1)–(2.3) has a *unique solution* such that

$$y \in L^2(0, T; L^2(\Omega)) \; (= L^2(Q)), \quad \frac{\partial y}{\partial t} \in L^2(0, T; H^{-2}(\Omega)), \tag{2.5}$$

so that

$$y \in C^0([0, T]; H^{-1}(\Omega)). \tag{2.6}$$

Remark 2.1 The solution y to (2.1)–(2.3) is defined, as usual, by *transposition*. Properties (2.5) and (2.6) still hold true if $v \in L^2(0, T; H^{-\frac{1}{2}}(\Gamma_0))$ (the notation is that used in Lions and Magenes, 1968).

Concerning *controllability*, the key result is given by the following.

Proposition 2.2 *When v spans $L^2(\Sigma_0)$, the function $y(T;v)$ spans a dense subspace of $H^{-1}(\Omega)$.*

Proof. We shall give a (nonconstructive) proof based on the *Hahn–Banach theorem*. Consider thus $f \in H_0^1(\Omega)$ such that

$$\langle y(T;v), f\rangle = 0, \quad \forall v \in L^2(\Sigma_0), \tag{2.7}$$

where, in (2.7), $\langle\cdot,\cdot\rangle$ denotes the duality pairing between $H^{-1}(\Omega)$ and $H_0^1(\Omega)$; next, define ψ by

$$-\frac{\partial\psi}{\partial t} + \mathcal{A}^*\psi = 0 \text{ in } Q, \quad \psi(T) = f, \quad \psi = 0 \text{ on } \Sigma. \tag{2.8}$$

Multiplying both sides of the first equation in (2.8) by the solution $\{x,t\} \to y(x,t;v)$ of problem (2.1)–(2.3) we obtain after integration by parts

$$\langle y(T;v), f\rangle = -\int_{\Sigma_0} \frac{\partial\psi}{\partial n_{\mathcal{A}^*}} v \, d\Gamma \, dt, \tag{2.9}$$

where $\frac{\partial}{\partial n_{\mathcal{A}^*}}$ denotes the *conormal derivative* operator associated with $\mathcal{A}^*$ (if $\mathcal{A} = \mathcal{A}^* = -\Delta$, then $\frac{\partial}{\partial n_{\mathcal{A}}} = \frac{\partial}{\partial n_{\mathcal{A}^*}} = \frac{\partial}{\partial n}$ where $\frac{\partial}{\partial n}$ is the usual outward normal derivative operator at Γ). Then (2.7) is equivalent to

$$\frac{\partial\psi}{\partial n_{\mathcal{A}^*}} = 0 \quad \text{on } \Sigma_0. \tag{2.10}$$

It follows from (2.8), (2.10) that the *Cauchy data* of ψ vanish on Σ_0; using again the *Mizohata's uniqueness theorem*, we obtain that $\psi = 0$ in Q, so that the function f in (2.7) verifies $f = 0$, which completes the proof of the proposition. □

We can formulate, now, the following *approximate controllability problems* (where $d\Sigma = d\Gamma \, dt$):

Problem 1 It is defined by

$$\inf_{\substack{v \in L^2(\Sigma_0) \\ y(T;v) \in y_T + \beta B_{-1}}} \frac{1}{2}\int_{\Sigma_0} v^2 \, d\Sigma, \tag{2.11}$$

where, in (2.11), the target function y_T is given in $H^{-1}(\Omega)$, $\beta > 0$, B_{-1} denotes the unit ball of $H^{-1}(\Omega)$ and the function $t \to y(t;v)$ is the solution of (2.1)–(2.3) associated with the control v.

Problem 2 It is the variant of problem (2.11) defined by

$$\inf_{v \in L^2(\Sigma_0)} \left(\frac{1}{2}\int_{\Sigma_0} v^2 \, d\Sigma + \frac{k}{2}\|y(T;v) - y_T\|_{-1}^2 \right), \tag{2.12}$$

where, in (2.12), $k > 0$, y_T and $y(T;v)$ are as in (2.11), and where $\forall z \in H^{-1}(\Omega)$, we have $\|z\|_{-1} = \left(\int_\Omega |\nabla\phi|^2\,dx\right)^{1/2}$ with ϕ the unique solution in $H_0^1(\Omega)$ of the Dirichlet problem

$$\int_\Omega \nabla\phi\cdot\nabla\theta\,dx = \langle z,\theta\rangle, \quad \forall\theta\in H_0^1(\Omega).$$

Using relatively simple *convexity arguments* it can be shown that *problems* (2.11) *and* (2.12) *have both a unique solution.*

2.2 Dirichlet control (II): Optimality conditions and dual formulations

We discuss first problem (2.12) since it is simpler than problem (2.11). Let us denote by $J_k(\cdot)$ the cost function in (2.12); using the relation

$$(J_k'(v), w)_{L^2(\Sigma_0)} = \lim_{\substack{\theta\to 0\\ \theta\neq 0}} \frac{J_k(v+\theta w) - J_k(v)}{\theta}, \quad \forall v, w \in L^2(\Sigma_0), \tag{2.13}$$

we can show that

$$(J_k'(v), w)_{L^2(\Sigma_0)} = \int_{\Sigma_0}\left(v - \frac{\partial p}{\partial n_{A^*}}\right) w\,d\Sigma, \quad \forall v, w\in L^2(\Sigma_0), \tag{2.14}$$

where, in (2.14), the *adjoint state function p* is obtained from v via the solution of (2.1)–(2.3) and of the *adjoint state equation*

$$\begin{cases} -\dfrac{\partial p}{\partial t} + A^*p = 0 \text{ in } Q, \quad p = 0 \text{ on } \Sigma, \\ p(T)\in H_0^1(\Omega) \text{ with } -\Delta p(T) = k(y(T) - y_T) \text{ in } \Omega. \end{cases} \tag{2.15}$$

Suppose now that u is *the* solution of the control problem (2.12); since $J_k'(u) = 0$, we have then the following optimality system satisfied by u and the corresponding state and adjoint state functions:

$$u = \frac{\partial p}{\partial n_{A^*}}|_{\Sigma_0},$$

$$\frac{\partial y}{\partial t} + Ay = 0 \text{ in } Q, \quad y(0) = 0, \quad y = 0 \text{ on } \Sigma\backslash\Sigma_0, \; y = \frac{\partial p}{\partial n_{A^*}} \text{ on } \Sigma_0,$$

$$-\frac{\partial p}{\partial t} + A^*p = 0 \text{ in } Q, \quad p = 0 \text{ on } \Sigma, \quad p(T) = f,$$

where f is the *unique* solution in $H_0^1(\Omega)$ of the Dirichlet problem

$$-\Delta f = k(y(T) - y_T) \text{ in } \Omega, \quad f = 0 \text{ on } \Gamma. \tag{2.16}$$

In order to identify a dual problem of (2.12), we proceed as in Chapter 1 by introducing (in the spirit of the *Hilbert Uniqueness Method* (HUM)) the operator $\Lambda \in \mathcal{L}(H_0^1(\Omega), H^{-1}(\Omega))$ defined by

$$\Lambda g = -\phi_g(T), \quad \forall g \in H_0^1(\Omega), \tag{2.17}$$

where the function ϕ_g is obtained from g as follows:
Solve first

$$-\frac{\partial \psi_g}{\partial t} + \mathcal{A}^* \psi_g = 0 \text{ in } Q, \;\; \psi_g = 0 \text{ on } \Sigma, \;\; \psi_g(T) = g, \tag{2.18}$$

and then,

$$\begin{cases} \dfrac{\partial \phi_g}{\partial t} + \mathcal{A}\phi_g = 0 \text{ in } Q, \;\; \phi_g = 0 \text{ on } \Sigma \backslash \Sigma_0, \;\; \phi_g = \dfrac{\partial \psi_g}{\partial n_{\mathcal{A}^*}} \text{ on } \Sigma_0, \\ \phi_g(0) = 0. \end{cases} \tag{2.19}$$

We can easily show that (with obvious notation)

$$\langle \Lambda g_1, g_2 \rangle = \int_{\Sigma_0} \frac{\partial \psi_{g_1}}{\partial n_{\mathcal{A}^*}} \frac{\partial \psi_{g_2}}{\partial n_{\mathcal{A}^*}} \, d\Sigma, \quad \forall g_1, \, g_2 \in H_0^1(\Omega). \tag{2.20}$$

It follows from (2.20) that the operator Λ is *self-adjoint* and *positive semi-definite*; indeed, it follows from the Mizohata's uniqueness theorem that the operator Λ is *positive definite*. However, the above operator Λ *is not an isomorphism* from $H_0^1(\Omega)$ *onto* $H^{-1}(\Omega)$ (implying that, in general, *we do not have exact boundary controllability here*).

Back to (2.16) we observe that from the definition of Λ we have $y(T) = -\Lambda f$, which implies in turn that f is the unique solution in $H_0^1(\Omega)$ of

$$-k^{-1}\Delta f + \Lambda f = -y_T. \tag{2.21}$$

Problem (2.21) is precisely the *dual problem* we are looking for. From the properties of operator $-k^{-1}\Delta + \Lambda$, problem (2.21) can be solved by a *conjugate gradient algorithm* operating in the space $H_0^1(\Omega)$; we shall return to this issue in Section 2.3.

Let us consider the control problem (2.11); using the *Fenchel–Rockafellar convex duality theory* as in Chapter 1, we can show that the solution u of problem (2.11) is characterized by the following optimality system:

$$u = \frac{\partial p}{\partial n_{\mathcal{A}^*}} |_{\Sigma_0}, \tag{2.22}$$

$$\frac{\partial y}{\partial t} + \mathcal{A}y = 0 \text{ in } Q, \, y(0) = 0, \, y = 0 \text{ on } \Sigma \backslash \Sigma_0, \text{ and } y = \frac{\partial p}{\partial n_{\mathcal{A}^*}} \text{ on } \Sigma_0, \tag{2.23}$$

$$-\frac{\partial p}{\partial t} + \mathcal{A}^* p = 0 \text{ in } Q, \quad p = 0 \text{ on } \Sigma, \, p(T) = f, \tag{2.24}$$

where f is the *unique* solution of the following *variational inequality*:

$$\begin{cases} f \in H_0^1(\Omega), \\ \langle \Lambda f, g - f \rangle + \beta \left(\int_\Omega |\nabla g|^2 \, dx \right)^{\frac{1}{2}} - \beta \left(\int_\Omega |\nabla f|^2 \, dx \right)^{\frac{1}{2}} \\ \qquad + \langle y_T, g - f \rangle \geq 0, \quad \forall g \in H_0^1(\Omega). \end{cases} \tag{2.25}$$

Problem (2.25) is precisely the *dual problem* to (2.11). The solution of problem (2.25) will be discussed in Section 2.3.

2.3 Dirichlet control (III): Iterative solution of the control problems

2.3.1 Conjugate gradient solution of problem (2.12)

It follows from Section 2.2 that solving the control problem (2.12) is equivalent to solving the linear equation

$$J_k'(u) = 0, \tag{2.26}$$

where operator J_k' is defined by (2.1)–(2.3), (2.14), (2.15). It is fairly easy to show that the *linear part* of operator J_k', namely,

$$v \to J_k'(v) - J_k'(0),$$

is *symmetric* and *strongly elliptic* over $L^2(\Sigma_0)$. From these properties, problem (2.26) can be solved by a *conjugate gradient algorithm* operating in the space $L^2(\Sigma_0)$. It follows from Chapter 1, Section 1.8.2 that this algorithm is as follows.

Algorithm 2.3 *(Conjugate gradient solution of problem (2.26))*

$$u^0 \textit{ is given in } L^2(\Sigma_0); \tag{2.27}$$

solve

$$\begin{cases} \dfrac{\partial y^0}{\partial t} + \mathcal{A} y^0 = 0 \;\; in \; Q, \quad y^0 = u^0 \;\; on \; \Sigma_0, \quad y^0 = 0 \;\; on \; \Sigma \backslash \Sigma_0, \\ y^0(0) = 0, \end{cases} \tag{2.28}$$

and then

$$\begin{cases} f^0 \in H_0^1(\Omega), \\ -\Delta f^0 = k(y^0(T) - y_T) \quad in \; \Omega, \end{cases} \tag{2.29}$$

and finally

$$-\frac{\partial p^0}{\partial t} + \mathcal{A}^* p^0 = 0 \;\; in \; Q, \quad p^0 = 0 \;\; on \; \Sigma, \quad p^0(T) = f^0. \tag{2.30}$$

Set

$$g^0 = u^0 - \left. \frac{\partial p^0}{\partial n_{\mathcal{A}^*}} \right|_{\Sigma_0}, \tag{2.31}$$

and then

$$w^0 = g^0. \tag{2.32}$$

For $n \geq 0$, assuming that u^n, g^n, w^n are known, compute u^{n+1}, g^{n+1}, w^{n+1} as follows:
Solve

$$\begin{cases} \dfrac{\partial \bar{y}^n}{\partial t} + \mathcal{A}\bar{y}^n = 0 \text{ in } Q, \quad \bar{y}^n = w^n \text{ on } \Sigma_0, \quad \bar{y}^n = 0 \text{ on } \Sigma \backslash \Sigma_0, \\ \bar{y}^n(0) = 0, \end{cases} \tag{2.33}$$

and then

$$\begin{cases} \bar{f}^n \in H_0^1(\Omega), \\ -\Delta \bar{f}^n = k\bar{y}^n(T) \quad \text{in } \Omega, \end{cases} \tag{2.34}$$

and finally

$$-\frac{\partial \bar{p}^n}{\partial t} + \mathcal{A}^*\bar{p}^n = 0 \text{ in } Q, \quad \bar{p}^n = 0 \text{ on } \Sigma, \quad \bar{p}^n(T) = \bar{f}^n. \tag{2.35}$$

Compute

$$\bar{g}^n = w^n - \left.\frac{\partial \bar{p}^n}{\partial n_{\mathcal{A}^*}}\right|_{\Sigma_0}, \tag{2.36}$$

and then

$$\rho_n = \int_{\Sigma_0} |g^n|^2 \, d\Sigma \bigg/ \int_{\Sigma_0} \bar{g}^n w^n \, d\Sigma, \tag{2.37}$$

$$u^{n+1} = u^n - \rho_n w^n, \tag{2.38}$$

$$g^{n+1} = g^n - \rho_n \bar{g}^n. \tag{2.39}$$

If $\|g^{n+1}\|_{L^2(\Sigma_0)}/\|g^0\|_{L^2(\Sigma_0)} \leq \epsilon$ take $u = u^{n+1}$; else compute

$$\gamma_n = \frac{\|g^{n+1}\|^2_{L^2(\Sigma_0)}}{\|g^n\|^2_{L^2(\Sigma_0)}}, \tag{2.40}$$

and update w^n by

$$w^{n+1} = g^{n+1} + \gamma_n w^n. \tag{2.41}$$

Do $n = n + 1$ and go to (2.33).

Remark 2.4 Here too, the number of iterations sufficient to obtain convergence varies, generically, like $\sqrt{k} \ln \epsilon^{-1}$.

2.3.2 Conjugate gradient solution of the dual problem (2.21)

We mentioned in Section 2.2 that the *dual problem* (2.21), namely,

$$-k^{-1}\Delta f + \Lambda f = -y_T,$$

can be solved by a *conjugate gradient algorithm* operating in $H_0^1(\Omega)$; from the definition of operator Λ (see (2.17)–(2.19)), and from Chapter 1, Section 1.8.2, this algorithm takes the following form.

Algorithm 2.5 *(Conjugate gradient solution of problem (2.21))*

$$f^0 \text{ is given in } H_0^1(\Omega); \tag{2.42}$$

solve

$$\begin{cases} -\dfrac{\partial p^0}{\partial t} + \mathcal{A}^* p^0 = 0 \quad \text{in } Q, \\ p^0 = 0 \quad \text{on } \Sigma, \\ p^0(T) = f^0, \end{cases} \tag{2.43}$$

and

$$\begin{cases} \dfrac{\partial y^0}{\partial t} + \mathcal{A} y^0 = 0 \text{ in } Q, \\ y^0 = \dfrac{\partial p^0}{\partial n_{\mathcal{A}^*}} \text{ on } \Sigma_0, \quad y^0 = 0 \text{ on } \Sigma\backslash\Sigma_0, \\ y^0(0) = 0. \end{cases} \tag{2.44}$$

Solve now

$$\begin{cases} g^0 \in H_0^1(\Omega), \\ \displaystyle\int_\Omega \nabla g^0 \cdot \nabla z \, dx = k^{-1} \int_\Omega \nabla f^0 \cdot \nabla z \, dx \\ \qquad + \langle y_T - y^0(T), z \rangle, \quad \forall z \in H_0^1(\Omega), \end{cases} \tag{2.45}$$

and set

$$w^0 = g^0. \tag{2.46}$$

Then, for $n \geq 0$, assuming that f^n, g^n, w^n are known, compute f^{n+1}, g^{n+1}, w^{n+1} as follows:
Solve

$$\begin{cases} -\dfrac{\partial \bar{p}^n}{\partial t} + \mathcal{A}^* \bar{p}^n = 0 \quad \text{in } Q, \\ \bar{p}^n = 0 \quad \text{on } \Sigma, \\ \bar{p}^n(T) = w^n, \end{cases} \tag{2.47}$$

and

$$\begin{cases} \dfrac{\partial \bar{y}^n}{\partial t} + \mathcal{A}\bar{y}^n = 0 \quad \text{in } Q, \\ \bar{y}^n = \dfrac{\partial \bar{p}^n}{\partial n_{\mathcal{A}^*}} \quad \text{on } \Sigma_0, \quad \bar{y}^n = 0 \ \text{ on } \Sigma \backslash \Sigma_0, \\ \bar{y}^n(0) = 0. \end{cases} \tag{2.48}$$

Solve now

$$\begin{cases} \bar{g}^n \in H_0^1(\Omega), \ \forall z \in H_0^1(\Omega), \\ \displaystyle\int_\Omega \nabla \bar{g}^n \cdot \nabla z \, dx = k^{-1} \int_\Omega \nabla w^n \cdot \nabla z \, dx - \langle \bar{y}^n(T), z \rangle. \end{cases} \tag{2.49}$$

Compute

$$\rho_n = \int_\Omega |\nabla g^n|^2 \, dx \Big/ \int_\Omega \nabla \bar{g}^n \cdot \nabla w^n \, dx, \tag{2.50}$$

and then

$$f^{n+1} = f^n - \rho_n w^n, \tag{2.51}$$

$$g^{n+1} = g^n - \rho_n \bar{g}^n. \tag{2.52}$$

If $\int_\Omega |\nabla g^{n+1}|^2 \, dx \Big/ \int_\Omega |\nabla g^0|^2 \, dx \le \epsilon^2$ *take* $f = f^{n+1}$ *and solve (2.24) to obtain* $u = \left.\frac{\partial p}{\partial n_{\mathcal{A}^*}}\right|_{\Sigma_0}$*; if the above stopping test is not satisfied, compute*

$$\gamma_n = \int_\Omega |\nabla g^{n+1}|^2 \, dx \Big/ \int_\Omega |\nabla g^n|^2 \, dx, \tag{2.53}$$

and then

$$w^{n+1} = g^{n+1} + \gamma_n w^n. \tag{2.54}$$

Do $n = n + 1$ *and go to (2.47).*

Remark 2.6 Remark 2.4 still holds for algorithm (2.42)–(2.54).

The *finite element* implementation of the above algorithm will be discussed in Section 2.5, while the results of numerical experiments will be presented in Section 2.6.

2.3.3 Iterative solution of problem (2.25)

Problem (2.25) can also be written as

$$-y_T \in \Lambda f + \beta \partial j(f), \tag{2.55}$$

which is a *multivalued* equation in $H^{-1}(\Omega)$, the unknown function f belonging to $H_0^1(\Omega)$; in (2.55), $\partial j(f)$ denotes the *subgradient* at f of the convex function

$j : H_0^1(\Omega) \to \mathbb{R}$ defined by

$$j(g) = \left(\int_\Omega |\nabla g|^2 \, dx \right)^{\frac{1}{2}}, \quad \forall g \in H_0^1(\Omega).$$

Problem (2.25), (2.55) is clearly a variant of problem (1.240) (see Chapter 1, Section 1.8.8) and as such it can be solved by those *operator splitting methods* advocated in Chapter 1, Section 1.8.8. To derive these methods we associate with the "*elliptic problem*" (2.55) the following *initial value problem* (*flow* in the *Dynamical System* terminology):

$$\begin{cases} \dfrac{\partial}{\partial \tau}(-\Delta f) + \Lambda f + \beta \partial j(f) \ni -y_T, \\ \\ f(0) = f_0 \; (\in H_0^1(\Omega)), \end{cases} \tag{2.56}$$

where, in (2.56), τ is a *pseudotime* variable.

To capture the steady-state solution of (2.56) (that is, the solution of problem (2.25), (2.55)) we can approximately integrate (2.56) from $\tau = 0$ to $\tau = +\infty$ by a *Peaceman–Rachford scheme*, like the one described just below.

$$f^0 = f_0 \text{ given in } H_0^1(\Omega); \tag{2.57}$$

then, for $m \geq 0$, *compute* $f^{m+\frac{1}{2}}$ *and* f^{m+1}, *from* f^m, *by solving in* $H_0^1(\Omega)$ *the following problems*:

$$\frac{(-\Delta f^{m+\frac{1}{2}}) - (-\Delta f^m)}{\frac{1}{2}\Delta \tau} + \beta \partial j(f^{m+\frac{1}{2}}) + \Lambda f^m \ni -y_T, \tag{2.58}$$

and

$$\frac{(-\Delta f^{m+1}) - (-\Delta f^{m+\frac{1}{2}})}{\frac{1}{2}\Delta \tau} + \beta \partial j(f^{m+\frac{1}{2}}) + \Lambda f^{m+1} \ni -y_T, \tag{2.59}$$

where $\Delta\tau (> 0)$ is a (pseudo) time-discretization step.

As in Chapter 1, Section 1.8.8, for problem (1.240), the *convergence* of $\{f^m\}_{m \geq 0}$ to the solution of (2.25), (2.55) is a direct consequence of P.L. Lions and B. Mercier (1979), Gabay (1982, 1983), and Glowinski and Le Tallec (1989); the convergence results proved in the above references apply to the present problem since operator Λ (respectively functional $j(\cdot)$) is *linear, continuous, and positive definite* (respectively *convex* and *continuous*) over $H_0^1(\Omega)$. As in Chapter 1, Section 1.8.8, we can also use a θ-scheme to solve problem (2.25), (2.55); we shall not describe this scheme here since it is a straightforward variant of algorithm (1.245)–(1.248) (actually, such an algorithm is described in Carthel, Glowinski, and Lions (1994), where it has been applied to the solution of the boundary control problem (2.11), (2.25) in the particular case where $\Gamma_0 = \Gamma$).

Back to algorithm (2.57)–(2.59) we observe that problem (2.59) can also be written as

$$\frac{-\Delta f^{m+1} + 2\Delta f^{m+\frac{1}{2}} - \Delta f^m}{\frac{1}{2}\Delta\tau} + \Lambda f^{m+1} = \Lambda f^m. \tag{2.60}$$

Problem (2.60) being a particular case of problem (2.21) can be solved by the conjugate gradient algorithm described in Section 2.3.2. Concerning the solution of problem (2.58), we observe that the solution of a closely related problem (namely, problem (1.346) in Chapter 1, Section 1.10.4) has already been discussed; since the solution methods for problem (1.346) and (2.58) are essentially the same we shall not discuss the solution of (2.58) any further.

2.4 Dirichlet control (IV): Approximation of the control problems

2.4.1 Generalities and synopsis

It follows from Section 2.3.3 that the solution of the *state constrained* control problem (2.11) (in fact of its dual problem (2.25)) can be reduced to a sequence of problems similar to (2.21), which is itself the dual problem of the control problem (2.12) (where the closeness of $y(T)$ to the target y_T is forced via *penalty*); we shall therefore concentrate our discussion on the approximation of the control problem (2.12), only. We shall address both the "direct" solution of problem (2.12) and the solution of the dual problem (2.21). The notation will be essentially as in Chapter 1, Sections 1.8 and 1.10.6.

2.4.2 Time discretization of problems (2.12) and (2.21)

The *time discretization* of problems (2.12) and (2.21) can be achieved using either first-order or second-order accurate time-discretization schemes, very close to those already discussed in Chapter 1, Sections 1.8 and 1.10.6 (see also Carthel, Glowinski, and Lions, 1994, Sections 5 and 6). Instead of essentially repeating the discussion which took place in the above sections and reference, we shall describe another *second-order accurate* time discretization scheme, introduced by Carthel (1994); actually, the numerical results shown in Section 2.6 have been obtained using this new scheme.

The time discretization of the control problem (2.12) is defined as follows (where $\Delta t = T/N$, N being a positive integer):

$$\min_{\mathbf{v}\in\mathcal{U}^{\Delta t}} J_k^{\Delta t}(\mathbf{v}), \tag{2.61}$$

where $\mathbf{v} = \{v^n\}_{n=1}^{N-1}$, $\mathcal{U}^{\Delta t} = (L^2(\Gamma_0))^{N-1}$ and

$$J_k^{\Delta t}(\mathbf{v}) = \frac{\Delta t}{2}\sum_{n=1}^{N-1} a_n \left\|v^n\right\|^2_{L^2(\Gamma_0)} + \frac{k}{2}\left\|y^N - y_T\right\|^2_{-1}; \tag{2.62}$$

in (2.62) we have $a_n = 1$ for $n = 1, 2, \ldots, N-2$, $a_{N-1} = \frac{3}{2}$ and y^N obtained from $\mathbf{v}$ as follows:

$$y^0 = 0; \tag{2.63}$$

to obtain y^1 (respectively y^n, $n = 2, \ldots, N-1$) we solve the following *elliptic* problem:

$$\begin{cases} \dfrac{y^1 - y^0}{\Delta t} + \mathcal{A}\left(\dfrac{2}{3}y^1 + \dfrac{1}{3}y^0\right) = 0 \quad \text{in } \Omega, \\ y^1 = v^1 \text{ on } \Gamma_0, \quad y^1 = 0 \text{ on } \Gamma\backslash\Gamma_0, \end{cases} \tag{2.64}$$

(respectively

$$\begin{cases} \dfrac{\frac{3}{2}y^n - 2y^{n-1} + \frac{1}{2}y^{n-2}}{\Delta t} + \mathcal{A}y^n = 0 \quad \text{in } \Omega, \\ y^n = v^n \text{ on } \Gamma_0, \quad y^n = 0 \text{ on } \Gamma\backslash\Gamma_0); \end{cases} \tag{2.65}$$

finally y^N is defined via

$$\frac{2y^N - 3y^{N-1} + y^{N-2}}{\Delta t} + \mathcal{A}y^{N-1} = 0. \tag{2.66}$$

The discrete control problem (2.61) has a *unique* solution.

In order to discretize the *dual problem* (2.21) we look for the dual problem of the discrete control problem (2.61). The simplest way to derive the dual of problem (2.61) is to start from the *optimality condition*

$$\nabla J_k^{\Delta t}(\mathbf{u}^{\Delta t}) = \mathbf{0}, \tag{2.67}$$

where, in (2.67), $\mathbf{u}^{\Delta t} = \{u^n\}_{n=1}^{N-1}$ is *the* solution of the discrete control problem (2.61), and where $\nabla J_k^{\Delta t}$ denotes the gradient of the discrete cost function $J_k^{\Delta t}$. Suppose that the discrete control space $\mathcal{U}^{\Delta t} = (L^2(\Gamma_0))^{N-1}$ is equipped with the scalar product

$$(\mathbf{v}, \mathbf{w})_{\Delta t} = \Delta t \sum_{n=1}^{N-1} a_n \int_{\Gamma_0} v^n w^n \, d\Gamma, \quad \forall \mathbf{v}, \mathbf{w} \in \mathcal{U}^{\Delta t}; \tag{2.68}$$

then a tedious calculation will show that $\forall \mathbf{v}, \mathbf{w} \in \mathcal{U}^{\Delta t}$,

$$\begin{aligned} (\nabla J_k^{\Delta t}(\mathbf{v}), \mathbf{w})_{\Delta t} = {} & \Delta t \sum_{n=1}^{N-2} \int_{\Gamma_0} \left(v^n - \frac{\partial p^n}{\partial n_{\mathcal{A}^*}}\right) w^n \, d\Gamma \\ & + \frac{3}{2}\Delta t \int_{\Gamma_0} \left[v^{N-1} - \left(\frac{2}{3}\frac{\partial p^{N-1}}{\partial n_{\mathcal{A}^*}} + \frac{1}{3}\frac{\partial p^N}{\partial n_{\mathcal{A}^*}}\right)\right] w^{N-1} \, d\Gamma, \end{aligned} \tag{2.69}$$

where, in (2.69), the *adjoint state vector* $\{p^n\}_{n=1}^N$ belongs to $(H_0^1(\Omega))^N$ and is obtained as follows:
First, compute p^N as the solution in $H_0^1(\Omega)$ of the elliptic problem

$$-\Delta p^N = k(y^N - y_T) \text{ in } \Omega, \quad p^N = 0 \text{ on } \Gamma, \tag{2.70}$$

then p^{N-1} (respectively p^n, $n = N-2, \ldots, 2, 1$) as the solution in $H_0^1(\Omega)$ of the elliptic problem

$$\frac{p^{N-1} - p^N}{\Delta t} + \mathcal{A}^*\left(\frac{2}{3}p^{N-1} + \frac{1}{3}p^N\right) = 0 \text{ in } \Omega, \quad p^{N-1} = 0 \text{ on } \Gamma, \tag{2.71}$$

(respectively

$$\frac{\frac{3}{2}p^n - 2p^{n+1} + \frac{1}{2}p^{n+2}}{\Delta t} + \mathcal{A}^* p^n = 0 \text{ in } \Omega, \quad p^n = 0 \text{ on } \Gamma). \tag{2.72}$$

Combining (2.67) and (2.69) shows that the *optimal triple*

$$\{\mathbf{u}^{\Delta t}, \{y^n\}_{n=1}^N, \{p^n\}_{n=1}^N\},$$

is *characterized* by

$$\begin{cases} u^n = \left.\dfrac{\partial p^n}{\partial n_{\mathcal{A}^*}}\right|_{\Gamma_0} & \text{if } n = 1, \ldots, N-2, \\ u^{N-1} = \left.\left(\dfrac{2}{3}\dfrac{\partial p^{N-1}}{\partial n_{\mathcal{A}^*}} + \dfrac{1}{3}\dfrac{\partial p^N}{\partial n_{\mathcal{A}^*}}\right)\right|_{\Gamma_0}, \end{cases} \tag{2.73}$$

to be completed by (2.70)–(2.72) and by

$$y^0 = 0, \tag{2.74}$$

$$\begin{cases} \dfrac{y^1 - y^0}{\Delta t} + \mathcal{A}\left(\dfrac{2}{3}y^1 + \dfrac{1}{3}y^0\right) = 0 \quad \text{in } \Omega, \\ y^1 = u^1 \text{ on } \Gamma_0, \quad y^1 = 0 \text{ on } \Gamma \backslash \Gamma_0, \end{cases} \tag{2.75}$$

$$\begin{cases} \dfrac{\frac{3}{2}y^n - 2y^{n-1} + \frac{1}{2}y^{n-2}}{\Delta t} + \mathcal{A}y^n = 0 \quad \text{in } \Omega, \\ y^n = u^n \text{ on } \Gamma_0, \quad y^n = 0 \text{ on } \Gamma \backslash \Gamma_0, \end{cases} \tag{2.76}$$

if $n = 2, \ldots, N-1$, and

$$\frac{2y^N - 3y^{N-1} + y^{N-2}}{\Delta t} + \mathcal{A}y^{N-1} = 0. \tag{2.77}$$

Following Section 2.2 we define $\Lambda^{\Delta t} \in \mathcal{L}(H_0^1(\Omega), H^{-1}(\Omega))$ by

$$\Lambda^{\Delta t} g = -\phi_g^N, \quad \forall g \in H_0^1(\Omega), \tag{2.78}$$

where ϕ_g^N is obtained from g via the solution of the *discrete backward parabolic problem*

$$\psi_g^N = g, \tag{2.79}$$

$$\frac{\psi_g^{N-1} - \psi_g^N}{\Delta t} + \mathcal{A}^*\left(\frac{2}{3}\psi_g^{N-1} + \frac{1}{3}\psi_g^N\right) = 0 \text{ in } \Omega, \quad \psi_g^{N-1} = 0 \text{ on } \Gamma, \tag{2.80}$$

$$\frac{\frac{3}{2}\psi_g^n - 2\psi_g^{n+1} + \frac{1}{2}\psi_g^{n+2}}{\Delta t} + \mathcal{A}^*\psi_g^n = 0 \text{ in } \Omega, \quad \psi_g^n = 0 \text{ on } \Gamma, \tag{2.81}$$

for $n = N-2, \ldots, 1$, and then of the *discrete forward parabolic problem*

$$\phi_g^0 = 0, \tag{2.82}$$

$$\begin{cases} \dfrac{\phi_g^1 - \phi_g^0}{\Delta t} + \mathcal{A}\left(\dfrac{2}{3}\phi_g^1 + \dfrac{1}{3}\phi_g^0\right) = 0 \text{ in } \Omega, \\ \phi_g^1 = \dfrac{\partial \psi_g^1}{\partial n_{\mathcal{A}^*}} \text{ on } \Gamma_0, \quad \phi_g^1 = 0 \text{ on } \Gamma\backslash\Gamma_0, \end{cases} \tag{2.83}$$

$$\begin{cases} \dfrac{\frac{3}{2}\phi_g^n - 2\phi_g^{n-1} + \frac{1}{2}\phi_g^{n-2}}{\Delta t} + \mathcal{A}\phi_g^n = 0 \text{ in } \Omega, \\ \phi_g^n = \dfrac{\partial \psi_g^n}{\partial n_{\mathcal{A}^*}} \text{ on } \Gamma_0, \quad \phi_g^n = 0 \text{ on } \Gamma\backslash\Gamma_0, \end{cases} \tag{2.84}$$

if $n = 2, \ldots, N-2$,

$$\begin{cases} \dfrac{\frac{3}{2}\phi_g^{N-1} - 2\phi_g^{N-2} + \frac{1}{2}\phi_g^{N-3}}{\Delta t} + \mathcal{A}\phi_g^{N-1} = 0 \text{ in } \Omega, \\ \phi_g^{N-1} = \dfrac{2}{3}\dfrac{\partial \psi_g^{N-1}}{\partial n_{\mathcal{A}^*}} + \dfrac{1}{3}\dfrac{\partial \psi_g^N}{\partial n_{\mathcal{A}^*}} \text{ on } \Gamma_0, \quad \phi_g^{N-1} = 0 \text{ on } \Gamma\backslash\Gamma_0, \end{cases} \tag{2.85}$$

$$\frac{2\phi_g^N - 3\phi_g^{N-1} + \phi_g^{N-2}}{\Delta t} + \mathcal{A}\phi_g^{N-1} = 0. \tag{2.86}$$

We can show that (with obvious notation) we have, $\forall f_1, f_2 \in H_0^1(\Omega)$,

$$\begin{aligned} \langle \Lambda^{\Delta t} f_1, f_2 \rangle = {} & \Delta t \sum_{n=1}^{N-2} \int_{\Gamma_0} \frac{\partial \psi_1^n}{\partial n_{\mathcal{A}^*}} \frac{\partial \psi_2^n}{\partial n_{\mathcal{A}^*}}\, d\Gamma \\ & + \frac{3}{2}\Delta t \int_{\Gamma_0} \left(\frac{2}{3}\frac{\partial \psi_1^{N-1}}{\partial n_{\mathcal{A}^*}} + \frac{1}{3}\frac{\partial \psi_1^N}{\partial n_{\mathcal{A}^*}}\right)\left(\frac{2}{3}\frac{\partial \psi_2^{N-1}}{\partial n_{\mathcal{A}^*}} + \frac{1}{3}\frac{\partial \psi_2^N}{\partial n_{\mathcal{A}^*}}\right) d\Gamma, \end{aligned} \tag{2.87}$$

where, in (2.87), $\langle \cdot, \cdot \rangle$ denotes the duality pairing between $H^{-1}(\Omega)$ and $H_0^1(\Omega)$.

It follows from (2.87) that operator $\Lambda^{\Delta t}$ is *self-adjoint* and *positive semidefinite* over $H_0^1(\Omega)$.

Back to the optimality system (2.70)–(2.77), let us denote by $f^{\Delta t}$ the function p^N; it follows then from the definition of $\Lambda^{\Delta t}$ that (2.70) can be reformulated as

$$-k^{-1}\Delta f^{\Delta t} + \Lambda^{\Delta t} f^{\Delta t} = -y_T, \tag{2.88}$$

which is precisely the *dual problem* we have been looking for. The full space/time discretization of problems (2.12) and (2.21) will be discussed in the following paragraph.

2.4.3 Full space/time discretization of problems (2.12) and (2.21)

The *full discretization* of control problems, related to (2.12) and (2.21), has been already discussed in Chapter 1, Sections 1.8.4 and 1.10.6. Despite many similarities, the *boundary control* problems discussed here are substantially more complicated to fully discretize than the above distributed and pointwise control problems. The main reason for this increased complexity arises from the fact that we still intend to employ *low-order finite element* approximations – as in Chapter 1, Sections 1.8 and 1.10 – to space discretize the parabolic state problem (2.1)–(2.3) and the corresponding adjoint system (2.15). With such a choice the "obvious" approximations of $\frac{\partial}{\partial n_{A^*}}\Big|_{\Gamma_0}$ will be fairly inaccurate. In order to obtain a (formal, at least) second-order accurate approximations of $\frac{\partial}{\partial n_{A^*}}\Big|_{\Gamma_0}$, we shall rely on a *discrete Green's formula*, following a strategy which has been successfully used in, for example, Glowinski, Li, and Lions (1990), Glowinski (1992a) (for the boundary control of the *wave equation*), and Carthel, Glowinski, and Lions (1994) (for the boundary control of the *heat equation*).

We suppose for simplicity that Ω is a *bounded* polygonal domain of $\mathbb{R}^2$. We introduce then, as in Chapter 1, Sections 1.8.4. and 1.10.6, a *triangulation* $\mathcal{T}_h$ of Ω (h: largest length of the edges of $\mathcal{T}_h$). Next, we approximate $H^1(\Omega)$, $L^2(\Omega)$, and $H_0^1(\Omega)$ by

$$H_h^1 = \{z_h \,|\, z_h \in C^0(\bar{\Omega}),\ z_h|_T \in P_1,\ \forall T \in \mathcal{T}_h\}, \tag{2.89}$$

$$H_{0h}^1 = \{z_h \,|\, z_h \in H_h^1,\ z_h = 0 \text{ on } \Gamma\} \quad (= H_0^1(\Omega) \cap H_h^1), \tag{2.90}$$

respectively (with, as usual, P_1 the space of polynomials in x_1, x_2 of degree ≤ 1). Another important finite element space is

$$V_{0h} = \{z_h \mid z_h \in H_h^1, z_h = 0 \text{ on } \Gamma\backslash\Gamma_0\}; \tag{2.91}$$

if $\int_{\Gamma\backslash\Gamma_0} d\Gamma > 0$ we shall assume that those boundary points at the interface of Γ_0 and $\Gamma\backslash\Gamma_0$ are vertices of $\mathcal{T}_h$. Finally, the role of $L^2(\Gamma_0)$ will be played by the space $M_h (\subset V_{0h})$ defined as follows:

$$\begin{cases} M_h \oplus H_{0h}^1 = V_{0h}, \\ \mu_h \in M_h \Rightarrow \mu_h|_T = 0,\ \forall T \in \mathcal{T}_h, \ \text{ such as } \ \partial T \cap \Gamma = \emptyset. \end{cases} \tag{2.92}$$

The space M_h is clearly *isomorphic* to the boundary space consisting of the traces on Γ of those functions belonging to V_{0h}; also, $\dim(M_h)$ is equal to the number of $\mathcal{T}_h$ boundary vertices interior to Γ_0 and the following bilinear form:

$$\{\mu_{1h}, \mu_{2h}\} \to \int_{\Gamma_0} \mu_{1h}\mu_{2h}\, d\Gamma,$$

defines a scalar product over M_h.

Since the full space/time discretization of problems (2.12) and (2.21) will rely on *variational* techniques, it is convenient to introduce the bilinear functional a : $H^1(\Omega) \times H_0^1(\Omega) \to \mathbb{R}$ defined by

$$a(y,z) = \langle \mathcal{A}y, z\rangle, \quad \forall y \in H^1(\Omega), \forall z \in H_0^1(\Omega), \tag{2.93}$$

where $\langle \cdot, \cdot \rangle$ denotes the duality pairing between $H^{-1}(\Omega)$ and $H_0^1(\Omega)$. Assuming that the coefficients of the second-order elliptic operator $\mathcal{A}$ are sufficiently smooth we also have

$$\begin{cases} a(y,z) = \displaystyle\int_\Omega (\mathcal{A}^* z)y\, dx + \int_\Gamma \frac{\partial z}{\partial n_{\mathcal{A}^*}} y\, d\Gamma, \\ \forall y \in H^1(\Omega), \forall z \in H_0^1(\Omega) \cap H^2(\Omega), \end{cases} \tag{2.94}$$

which is definitely a generalization of the well-known *Green's formula*

$$\begin{cases} \displaystyle\int_\Omega \nabla y \cdot \nabla z\, dx = -\int_\Omega \Delta z y\, dx + \int_\Gamma \frac{\partial z}{\partial n} y\, d\Gamma, \\ \forall y \in H^1(\Omega), \forall z \in H_0^1(\Omega) \cap H^2(\Omega). \end{cases}$$

Following Section 2.4.2 we approximate the control problem (2.12) by

$$\min_{\mathbf{v} \in \mathcal{U}_h^{\Delta t}} J_h^{\Delta t}(\mathbf{v}), \tag{2.95}$$

where, in (2.95), we have $\mathcal{U}_h^{\Delta t} = (M_h)^{N-1}$, $\mathbf{v} = \{v^n\}_{n=1}^{N-1}$, and

$$J_h^{\Delta t}(\mathbf{v}) = \frac{\Delta t}{2} \sum_{n=1}^{N-1} a_n \int_{\Gamma_0} |v^n|^2\, d\Gamma + \frac{k}{2} \int_\Omega |\nabla \phi^N|^2\, dx, \tag{2.96}$$

with, in (2.96), ϕ^N obtained from $\mathbf{v}$ via the solution of the following well-posed *discrete parabolic* and *elliptic* problems:

Parabolic problem:

$$y^0 = 0; \tag{2.97}$$

compute y^1 from

$$\begin{cases} y^1 \in V_{0h},\ y^1 = v^1 \ \text{ on } \Gamma_0, \\ \displaystyle\int_\Omega \frac{y^1 - y^0}{\Delta t} z\, dx + a\left(\frac{2}{3}y^1 + \frac{1}{3}y^0, z\right) = 0, \quad \forall z \in H^1_{0h}, \end{cases} \tag{2.98}$$

then y^n from

$$\begin{cases} y^n \in V_{0h},\ y^n = v^n \ \text{ on } \Gamma_0, \\ \displaystyle\int_\Omega \frac{\frac{3}{2}y^n - 2y^{n-1} + \frac{1}{2}y^{n-2}}{\Delta t} z\, dx + a(y^n, z) = 0, \quad \forall z \in H^1_{0h}, \end{cases} \tag{2.99}$$

for $n = 2, \ldots, N-1$, and y^N from

$$\begin{cases} y^N \in H^1_{0h}, \\ \displaystyle\int_\Omega \frac{2y^N - 3y^{N-1} + y^{N-2}}{\Delta t} z\, dx + a(y^{N-1}, z) = 0, \quad \forall z \in H^1_{0h}. \end{cases} \tag{2.100}$$

Elliptic problem:

$$\begin{cases} \phi^N \in H^1_{0h}, \\ \displaystyle\int_\Omega \nabla\phi^N \cdot \nabla z\, dx = \int_\Omega (y^N - y_T) z\, dx, \quad \forall z \in H^1_{0h}. \end{cases} \tag{2.101}$$

We then have the following.

Proposition 2.7 *The discrete control problem* (2.95) *has a unique solution* $\mathbf{u}_h^{\Delta t} = \{u^n\}_{n=1}^{N-1}$. *If we denote by* $\mathbf{y}_h^{\Delta t} = \{y^n\}_{n=0}^{N}$ *the solution of* (2.97)–(2.100) *associated with* $\mathbf{v} = \mathbf{u}_h^{\Delta t}$, *the optimal pair* $\{\mathbf{u}_h^{\Delta t},\ \mathbf{y}_h^{\Delta t}\}$ *is characterized by the existence of* $\mathbf{p}_h^{\Delta t} = \{p^n\}_{n=1}^{N} \in (H^1_{0h})^N$ *such that*

$$\begin{cases} p^N \in H^1_{0h}, \\ \displaystyle\int_\Omega \nabla p^N \cdot \nabla z\, dx = k\left(\int_\Omega y^N z\, dx - \langle y_T, z\rangle\right), \quad \forall z \in H^1_{0h}, \end{cases} \tag{2.102}$$

($\langle \cdot, \cdot \rangle$*: duality pairing between* $H^{-1}(\Omega)$ *and* $H^1_0(\Omega)$),

$$\begin{cases} p^{N-1} \in H^1_{0h}, \\ \displaystyle\int_\Omega \frac{p^{N-1} - p^N}{\Delta t} z\, dx + a\left(z, \frac{2}{3}p^{N-1} + \frac{1}{3}p^N\right) = 0, \quad \forall z \in H^1_{0h}, \end{cases} \tag{2.103}$$

$$\begin{cases} p^n \in H^1_{0h}, \\ \displaystyle\int_\Omega \frac{\frac{3}{2}p^n - 2p^{n+1} + \frac{1}{2}p^{n+2}}{\Delta t} z\, dx + a(z, p^n) = 0, \quad \forall z \in H^1_{0h}, \end{cases} \tag{2.104}$$

for $n = N-2, \dots, 1$, *and also*

$$\begin{cases} u^n \in M_h, \ \forall \mu \in M_h, \\ \displaystyle\int_{\Gamma_0} u^n \mu \, d\Gamma = \int_\Omega \frac{\frac{3}{2}p^n - 2p^{n+1} + \frac{1}{2}p^{n+2}}{\Delta t} \mu \, dx + a(\mu, p^n), \end{cases} \tag{2.105}$$

if $n = 1, 2, \dots, N-2$, *and finally*

$$\begin{cases} u^{N-1} \in M_h, \ \forall \mu \in M_h, \\ \displaystyle\int_{\Gamma_0} u^{N-1} \mu \, d\Gamma = \int_\Omega \frac{p^{N-1} - p^N}{\Delta t} \mu \, dx + a\left(\mu, \frac{2}{3}p^{N-1} + \frac{1}{3}p^N\right). \end{cases} \tag{2.106}$$

Proof. The *existence* and *uniqueness* properties are obvious. Concerning now the relations characterizing the optimal pair $\{\mathbf{u}_h^{\Delta t}, \mathbf{y}_h^{\Delta t}\}$ they follow from the *optimality condition*

$$\nabla J_h^{\Delta t}(\mathbf{u}_h^{\Delta t}) = \mathbf{0}, \tag{2.107}$$

where $\nabla J_h^{\Delta t}$ is the gradient of the functional $J_h^{\Delta t}$. Indeed, if we use

$$(\mathbf{v}, \mathbf{w})_{\Delta t} = \Delta t \sum_{n=1}^{N-1} a_n \int_{\Gamma_0} v^n w^n \, d\Gamma,$$

as the scalar product over $\mathcal{U}_h^{\Delta t}$, it can be shown that, $\forall \mathbf{v}, \mathbf{w} \in \mathcal{U}_h^{\Delta t}$, we have

$$\begin{aligned} &(\nabla J_h^{\Delta t}(\mathbf{v}), \mathbf{w})_{\Delta t} \\ &= \Delta t \sum_{n=1}^{N-2} \left(\int_{\Gamma_0} v^n w^n \, d\Gamma - \int_\Omega \frac{\frac{3}{2}p^n - 2p^{n+1} + \frac{1}{2}p^{n+2}}{\Delta t} w^n \, dx - a(w^n, p^n) \right) \\ &\quad + \frac{3}{2}\Delta t \left(\int_{\Gamma_0} v^{N-1} w^{N-1} \, d\Gamma - \int_\Omega \frac{p^{N-1} - p^N}{\Delta t} w^{N-1} \, dx \right. \\ &\quad \left. - a\left(w^{N-1}, \frac{2}{3}p^{N-1} + \frac{1}{3}p^N\right)\right), \end{aligned} \tag{2.108}$$

where, in (2.108), $\{p^n\}_{n=1}^{N-1}$ is obtained from $\mathbf{v} = \{v^n\}_{n=1}^{N-1}$ via the solution of the discrete parabolic and elliptic problems (2.97)–(2.100), (2.102), and (2.103), (2.104). Relations (2.107) and (2.108) clearly imply (2.102)–(2.106). □

Remark 2.8 Relations (2.105), (2.106) are not that mysterious. For the continuous problem (2.12), we know (see Section 2.2) that the optimal control u satisfies

$$u = \frac{\partial p}{\partial n_{\mathcal{A}^*}} \text{ on } \Sigma_0, \tag{2.109}$$

where p is the solution of the corresponding adjoint system (2.15). We have thus $\frac{\partial p}{\partial t} = \mathcal{A}^* p$, which combined with Green's formula (2.94), implies that a.e. on $(0, T)$

we have

$$\int_{\Gamma_0} u\mu \, d\Gamma = -\int_\Omega \frac{\partial p}{\partial t}\mu \, dx + a(\mu, p), \ \forall \mu \in H^1(\Omega), \ \mu = 0 \ \text{ on } \Gamma\backslash\Gamma_0. \quad (2.110)$$

Relations (2.105), (2.106) are clearly discrete analogues of (2.110).

To obtain the *fully discrete* analogue of the *dual problems* (2.21) and (2.88) we introduce the operator $\Lambda_h^{\Delta t} \in \mathcal{L}(H_{0h}^1, H_{0h}^1)$ defined as follows:

$$\Lambda_h^{\Delta t} g = -\phi_g^N, \quad \forall g \in H_{0h}^1, \quad (2.111)$$

where ϕ_g^N is obtained from g via the solution of the *fully discrete backward parabolic problem*

$$\psi_g^N = g, \quad (2.112)$$

$$\begin{cases} \psi_g^{N-1} \in H_{0h}^1, \\ \displaystyle\int_\Omega \frac{\psi_g^{N-1} - \psi_g^N}{\Delta t} z \, dx + a\left(z, \frac{2}{3}\psi_g^{N-1} + \frac{1}{3}\psi_g^N\right) = 0, \quad \forall z \in H_{0h}^1, \end{cases} \quad (2.113)$$

$$\begin{cases} \psi_g^n \in H_{0h}^1, \\ \displaystyle\int_\Omega \frac{\frac{3}{2}\psi_g^n - 2\psi_g^{n+1} + \frac{1}{2}\psi_g^{n+2}}{\Delta t} z \, dx + a(z, \psi_g^n) = 0, \quad \forall z \in H_{0h}^1, \end{cases} \quad (2.114)$$

for $n = N-2, \ldots, 1$, and then of the *fully discrete forward parabolic problem*

$$\phi_g^0 = 0, \quad (2.115)$$

$$\begin{cases} \phi_g^1 \in V_{0h}, \ \phi_g^1 = u_g^1 \ \text{ on } \Gamma_0, \\ \displaystyle\int_\Omega \frac{\phi_g^1 - \phi_g^0}{\Delta t} z \, dx + a\left(\frac{2}{3}\phi_g^1 + \frac{1}{3}\phi_g^0, z\right) = 0, \quad \forall z \in H_{0h}^1, \end{cases} \quad (2.116)$$

$$\begin{cases} \phi_g^n \in V_{0h}, \ \phi_g^n = u_g^n \ \text{ on } \Gamma_0, \\ \displaystyle\int_\Omega \frac{\frac{3}{2}\phi_g^n - 2\phi_g^{n-1} + \frac{1}{2}\phi_g^{n-2}}{\Delta t} z \, dx + a(\phi_g^n, z) = 0, \quad \forall z \in H_{0h}^1, \end{cases} \quad (2.117)$$

for $n = 2, \ldots, N-1$, and finally

$$\begin{cases} \phi_g^N \in H_{0h}^1, \\ \displaystyle\int_\Omega \frac{2\phi_g^N - 3\phi_g^{N-1} + \phi_g^{N-2}}{\Delta t} z \, dx + a(\phi_g^{N-1}, z) = 0, \quad \forall z \in H_{0h}^1; \end{cases} \quad (2.118)$$

in (2.116), (2.117) the vector $\{u_g^n\}_{n=1}^{N-1}$ is defined from $\{\psi_g^n\}_{n=1}^N$ as follows:

$$\begin{cases} u_g^n \in M_h, \\ \displaystyle\int_{\Gamma_0} u_g^n \mu \, d\Gamma = \int_\Omega \frac{\frac{3}{2}\psi_g^n - 2\psi_g^{n+1} + \frac{1}{2}\psi_g^{n+2}}{\Delta t} \mu \, dx + a(\mu, \psi_g^n), \quad \forall \mu \in M_h, \end{cases} \quad (2.119)$$

if $n = 1, 2, \ldots, N-2$, and

$$
\begin{cases}
u_g^{N-1} \in M_h, \ \forall \mu \in M_h, \\
\displaystyle\int_{\Gamma_0} u_g^{N-1} \mu \, d\Gamma = \int_\Omega \frac{\psi_g^{N-1} - \psi_g^N}{\Delta t} \mu \, dx + a\left(\mu, \frac{2}{3}\psi_g^{N-1} + \frac{1}{3}\psi_g^N\right).
\end{cases}
\tag{2.120}
$$

We can show that

$$
\int_\Omega \left(\Lambda_h^{\Delta t} g_1\right) g_2 \, dx = \Delta t \sum_{n=1}^{N-1} a_n \int_{\Gamma_0} u_1^n u_2^n \, d\Gamma, \quad \forall g_1, g_2 \in H_{0h}^1, \tag{2.121}
$$

where, in (2.121), $\{u_i^n\}_{n=1}^{N-1}$, $i = 1, 2$ is obtained from g_i via (2.112)–(2.114), (2.119), (2.120).

It follows from (2.121) that operator $\Lambda_h^{\Delta t}$ is *symmetric* and *positive semidefinite* over H_{0h}^1.

Let us consider now the *optimal triple* $\{\mathbf{u}_h^{\Delta t}, \mathbf{y}_h^{\Delta t}, \mathbf{p}_h^{\Delta t}\}$ and define $f_h^{\Delta t} \in H_h^1$ by

$$
f_h^{\Delta t} = p^N. \tag{2.122}
$$

It follows then from Proposition 2.7 and from the definition of $\Lambda_h^{\Delta t}$ that

$$
\Lambda_h^{\Delta t} f_h^{\Delta t} = -y^N. \tag{2.123}
$$

Combining (2.123) with (2.102) we obtain

$$
\begin{cases}
f_h^{\Delta t} \in H_{0h}^1, \\
k^{-1} \displaystyle\int_\Omega \nabla f_h^{\Delta t} \cdot \nabla z \, dx + \int_\Omega \left(\Lambda_h^{\Delta t} f_h^{\Delta t}\right) z \, dx = -\langle y_T, z \rangle, \quad \forall z \in H_{0h}^1.
\end{cases}
\tag{2.124}
$$

Problem (2.124) is precisely the *fully discrete dual problem* we were looking for. From the properties of $\Lambda_h^{\Delta t}$ (*symmetry* and *positive semidefiniteness*), problem (2.124) can be solved by a *conjugate gradient* algorithm operating in H_{0h}^1 (a fully discrete analogue of algorithm (2.42)–(2.54)); we shall describe this algorithm in Section 2.5.

Remark 2.9 From a practical point of view, it makes sense to use the *trapezoidal rule* to (approximately) compute the various $L^2(\Omega)$ and $L^2(\Gamma_0)$ scalar products occurring in the definition of the approximate control problem (2.95), and of its dual problem (2.124). If this approach is retained, the corresponding operator $\Lambda_h^{\Delta t}$ has the same basic properties as the one defined by (2.111), namely, *symmetry* and *positive semidefiniteness*, implying that the corresponding variant of problem (2.121) can also be solved by a *conjugate gradient algorithm* operating in the space H_{0h}^1.

2.5 Dirichlet control (V): Iterative solution of the fully discrete dual problem (2.124)

We have described in Section 2.3 two *conjugate gradient algorithms* for solving the control problem (2.12), either *directly* (by algorithm (2.27)–(2.41); see Section 2.3.1) or via the solution of the *dual problem* (2.21) (by algorithm (2.42)–(2.54); see Section 2.3.2). Since the numerical results presented in the following section were obtained via the solution of the dual problem we shall focus our discussion on the *iterative solution* of the *fully discrete* approximation of problem (2.21) (that is problem (2.124)). From the properties of the operator $\Lambda_h^{\Delta t}$ problem (2.124) can be solved by a *conjugate gradient algorithm* operating in the finite-dimensional space H_{0h}^1. From Chapter 1, Section 1.8.2, and from Section 2.3.2 this algorithm takes the following form:

$$f_0 \text{ is given in } H_{0h}^1; \tag{2.125}$$

set

$$p_0^N = f_0, \tag{2.126}$$

and solve first

$$\begin{cases} p_0^{N-1} \in H_{0h}^1, \\ \displaystyle\int_\Omega \frac{p_0^{N-1} - p_0^N}{\Delta t} z \, dx + a\left(z, \frac{2}{3} p_0^{N-1} + \frac{1}{3} p_0^N\right) = 0, \quad \forall z \in H_{0h}^1, \end{cases} \tag{2.127}$$

and

$$\begin{cases} u_0^{N-1} \in M_h, \ \forall \mu \in M_h, \\ \displaystyle\int_{\Gamma_0} u_0^{N-1} \mu \, d\Gamma = \int_\Omega \frac{p_0^{N-1} - p_0^N}{\Delta t} \mu \, dx + a\left(\mu, \frac{2}{3} p_0^{N-1} + \frac{1}{3} p_0^N\right), \end{cases} \tag{2.128}$$

and then for $n = N-2, \ldots, 1$

$$\begin{cases} p_0^n \in H_{0h}^1, \\ \displaystyle\int_\Omega \frac{\frac{3}{2} p_0^n - 2 p_0^{n+1} + \frac{1}{2} p_0^{n+2}}{\Delta t} z \, dx + a(z, p_0^n) = 0, \quad \forall z \in H_{0h}^1, \end{cases} \tag{2.129}$$

$$\begin{cases} u_0^n \in M_h, \ \forall \mu \in M_h, \\ \displaystyle\int_{\Gamma_0} u_0^n \mu \, d\Gamma = \int_\Omega \frac{\frac{3}{2} p_0^n - 2 p_0^{n+1} + \frac{1}{2} p_0^{n+2}}{\Delta t} \mu \, dx + a(\mu, p_0^n). \end{cases} \tag{2.130}$$

Solve next the following fully discrete forward parabolic problem

$$y_0^0 = 0, \tag{2.131}$$

$$\begin{cases} y_0^1 \in V_{0h}, \ y_0^1 = u_0^1 \text{ on } \Gamma_0, \\ \displaystyle\int_\Omega \frac{y_0^1 - y_0^0}{\Delta t} z \, dx + a\left(\frac{2}{3} y_0^1 + \frac{1}{3} y_0^0, z\right) = 0, \quad \forall z \in H_{0h}^1, \end{cases} \tag{2.132}$$

$$\begin{cases} y_0^n \in V_{0h},\ y_0^n = u_0^n \ \text{on}\ \Gamma_0, \\ \displaystyle\int_\Omega \frac{\frac{3}{2}y_0^n - 2y_0^{n-1} + \frac{1}{2}y_0^{n-2}}{\Delta t} z\,dx + a(y_0^n, z) = 0, \quad \forall z \in H_{0h}^1, \end{cases} \tag{2.133}$$

for $n = 2, \ldots, N-1$, and finally

$$\begin{cases} y_0^N \in H_{0h}^1, \\ \displaystyle\int_\Omega \frac{2y_0^N - 3y_0^{N-1} + y_0^{N-2}}{\Delta t} z\,dx + a(y_0^{N-1}, z) = 0, \quad \forall z \in H_{0h}^1. \end{cases} \tag{2.134}$$

Solve now

$$\begin{cases} g_0 \in H_{0h}^1,\ \forall z \in H_{0h}^1, \\ \displaystyle\int_\Omega \nabla g_0 \cdot \nabla z\,dx = k^{-1}\int_\Omega \nabla f_0 \cdot \nabla z\,dx + \langle y_T, z\rangle - \int_\Omega y_0^N z\,dx, \end{cases} \tag{2.135}$$

and set

$$w_0 = g_0. \tag{2.136}$$

Then for $m \geq 0$, assuming that f_m, g_m, and w_m are known (the last two different from 0), compute f_{m+1}, g_{m+1}, and w_{m+1} as follows:
Take

$$\bar{p}_m^N = w_m, \tag{2.137}$$

and solve

$$\begin{cases} \bar{p}_m^{N-1} \in H_{0h}^1, \\ \displaystyle\int_\Omega \frac{\bar{p}_m^{N-1} - \bar{p}_m^N}{\Delta t} z\,dx + a\left(z, \frac{2}{3}\bar{p}_m^{N-1} + \frac{1}{3}\bar{p}_m^N\right) = 0, \quad \forall z \in H_{0h}^1, \end{cases} \tag{2.138}$$

and

$$\begin{cases} \bar{u}_m^{N-1} \in M_h,\ \forall \mu \in M_h, \\ \displaystyle\int_{\Gamma_0} \bar{u}_m^{N-1}\mu\,d\Gamma = \int_\Omega \frac{\bar{p}_m^{N-1} - \bar{p}_m^N}{\Delta t}\mu\,dx + a\left(\mu, \frac{2}{3}\bar{p}_m^{N-1} + \frac{1}{3}\bar{p}_m^N\right), \end{cases} \tag{2.139}$$

and then for $n = N-2, \ldots, 1$

$$\begin{cases} \bar{p}_m^n \in H_{0h}^1, \\ \displaystyle\int_\Omega \frac{\frac{3}{2}\bar{p}_m^n - 2\bar{p}_m^{n+1} + \frac{1}{2}\bar{p}_m^{n+2}}{\Delta t} z\,dx + a(z, \bar{p}_m^n) = 0, \forall z \in H_{0h}^1, \end{cases} \tag{2.140}$$

and

$$\begin{cases} \bar{u}_m^n \in M_h,\ \forall \mu \in M_h, \\ \displaystyle\int_{\Gamma_0} \bar{u}_m^n \mu\,d\Gamma = \int_\Omega \frac{\frac{3}{2}\bar{p}_m^n - 2\bar{p}_m^{n+1} + \frac{1}{2}\bar{p}_m^{n+2}}{\Delta t}\mu\,dx + a(\mu, \bar{p}_m^n). \end{cases} \tag{2.141}$$

Solve next the following discrete forward parabolic problem

$$\bar{y}_m^0 = 0, \tag{2.142}$$

$$\begin{cases} \bar{y}_m^1 \in V_{0h},\ \bar{y}_m^1 = \bar{u}_m^1 \ \text{on}\ \Gamma_0, \\ \displaystyle\int_\Omega \frac{\bar{y}_m^1 - \bar{y}_m^0}{\Delta t} z\,\mathrm{d}x + a\left(\frac{2}{3}\bar{y}_m^1 + \frac{1}{3}\bar{y}_m^0, z\right) = 0, \quad \forall z \in H_{0h}^1, \end{cases} \tag{2.143}$$

$$\begin{cases} \bar{y}_m^n \in V_{0h},\ \bar{y}_m^n = \bar{u}_m^n \ \text{on}\ \Gamma_0, \\ \displaystyle\int_\Omega \frac{\frac{3}{2}\bar{y}_m^n - 2\bar{y}_m^{n-1} + \frac{1}{2}\bar{y}_m^{n-2}}{\Delta t} z\,\mathrm{d}x + a(\bar{y}_m^n, z) = 0, \quad \forall z \in H_{0h}^1, \end{cases} \tag{2.144}$$

for $n = 2, \ldots, N-1$, *and finally*

$$\begin{cases} \bar{y}_m^N \in H_{0h}^1,\ \forall z \in H_{0h}^1, \\ \displaystyle\int_\Omega (2\bar{y}_m^N - 3\bar{y}_m^{N-1} + \bar{y}_m^{N-2}) z\,\mathrm{d}x + \Delta t a(\bar{y}_m^{N-1}, z) = 0. \end{cases} \tag{2.145}$$

Solve now

$$\begin{cases} \bar{g}_m \in H_{0h}^1,\ \forall z \in H_{0h}^1, \\ \displaystyle\int_\Omega \nabla \bar{g}_m \cdot \nabla z\,\mathrm{d}x = k^{-1}\int_\Omega \nabla w_m \cdot \nabla z\,\mathrm{d}x - \int_\Omega \bar{y}_m^N z\,\mathrm{d}x, \end{cases} \tag{2.146}$$

and compute

$$\rho_m = \int_\Omega |\nabla g_m|^2\,\mathrm{d}x \Big/ \int_\Omega \nabla \bar{g}_m \cdot \nabla w_m\,\mathrm{d}x, \tag{2.147}$$

$$f_{m+1} = f_m - \rho_m w_m, \tag{2.148}$$

$$g_{m+1} = g_m - \rho_m \bar{g}_m. \tag{2.149}$$

If $\int_\Omega |\nabla g_{m+1}|^2\,\mathrm{d}x / \int_\Omega |\nabla g_0|^2\,\mathrm{d}x \le \epsilon^2$ *take* $f_h^{\Delta t} = f_{m+1}$ *and solve* (2.112)–(2.120) *with* $g = f_h^{\Delta t}$ *to obtain* $\mathbf{u}_h^{\Delta t} = \{u^n\}_{n=1}^{N-1}$*; if the above stopping test is not satisfied, compute*

$$\gamma_m = \int_\Omega |\nabla g_{m+1}|^2\,\mathrm{d}x \Big/ \int_\Omega |\nabla g_m|^2\,\mathrm{d}x, \tag{2.150}$$

and then

$$w_{m+1} = g_{m+1} + \gamma_m w_m. \tag{2.151}$$

Do $m = m + 1$ *and go to* (2.137).

Remark 2.10 Algorithm (2.125)–(2.151) may seem complicated (almost thirty instructions); in fact, it is quite easy to implement since it requires, essentially, a *fast elliptic solver*. For the calculations presented in Section 2.6, hereafter, we have been using a *multigrid*-based elliptic solver (see, for example, Hackbush, 1985; Yserentant, 1993; Brenner and Scott, 2002 and the references therein for a thorough discussion of the solution of discrete elliptic problems by multigrid methods; see also Glowinski, 2003, Chapter 5).

Remark 2.11 If h and Δt are sufficiently small Remarks 2.4 and 2.6 still hold for algorithm (2.125)–(2.151).

2.6 Dirichlet control (VI): Numerical experiments

2.6.1 *First test problem*

The first test problem is one for which the *exact controllability property holds*; indeed, in order to construct more easily a test problem whose exact solution is known we have taken a *nonzero source term* in the right-hand side of the state equation (2.1), obtaining thus

$$\frac{\partial y}{\partial t} + \mathcal{A}y = s \text{ in } Q, \tag{SE}$$

and also replaced the initial condition (2.3) by

$$y(0) = y_0, \tag{IC}$$

with, possibly, $y_0 \neq 0$. For these numerical experiments we have taken $\Omega = (0,1) \times (0,1)$, $\Gamma_0 = \Gamma$, $T = 1$, and $\mathcal{A} = -\nu\Delta$, with $\nu > 0$ ((SE) is therefore a *heat equation*); the *source term* s, the *initial value* y_0 and the *target function* y_T are defined by

$$s(x_1,x_2,t) = 3\pi^3\nu e^{2\pi^2\nu t}(\sin \pi x_1 + \sin \pi x_2), \tag{2.152}$$

$$y_0(x_1,x_2) = \pi(\sin \pi x_1 + \sin \pi x_2), \tag{2.153}$$

$$y_T(x_1,x_2) = \pi e^{2\pi^2\nu}(\sin \pi x_1 + \sin \pi x_2), \tag{2.154}$$

respectively.

With these data the (unique) solution u of the optimal control problem

$$\min_{v\in\mathcal{U}_f} J(v), \tag{2.155}$$

(with

$$J(v) = \frac{1}{2}\int_\Sigma |v|^2 \, d\Gamma \, dt,$$

$$\mathcal{U}_f = \{v \mid v \in L^2(\Sigma), \text{ the pair } \{v,y\} \text{ satisfies (SE), (2.2), (IC), and } y(T) = y_T\}),$$

is given by

$$\begin{cases} u(x_1,x_2,t) = \pi e^{2\pi^2\nu t} \sin \pi x_1 & \text{if } 0 < x_1 < 1 \text{ and } x_2 = 0 \text{ or } 1, \\ u(x_1,x_2,t) = \pi e^{2\pi^2\nu t} \sin \pi x_2 & \text{if } 0 < x_2 < 1 \text{ and } x_1 = 0 \text{ or } 1, \end{cases} \tag{2.156}$$

the corresponding function y being defined by

$$y(x_1,x_2,t) = \pi e^{2\pi^2\nu t}(\sin \pi x_1 + \sin \pi x_2). \tag{2.157}$$

Concerning now the *dual problem* of (2.155) we can easily show that it is defined by

$$\Lambda f = Y_0(T) - y_T, \tag{2.158}$$

where the operator Λ is still defined by (2.17)–(2.19) and where the function Y_0 is the solution of the following parabolic problem:

$$\frac{\partial Y_0}{\partial t} + \mathcal{A}Y_0 = s \text{ on } Q, \quad Y_0(0) = y_0, \quad Y_0 = 0 \text{ on } \Sigma. \tag{2.159}$$

Since the data have been chosen so that we have exact controllability, the dual problem (2.158) has a unique solution which, in the particular case discussed here, is given by

$$f(x_1, x_2) = \pi e^{2\pi^2\nu} \sin \pi x_1 \sin \pi x_2. \tag{2.160}$$

To approximate problem (2.158) (and therefore problem (2.155)) we have used the method described in Sections 2.3–2.5, namely, *time discretization* by a *second-order accurate scheme, space discretization* by *finite element methods* (using *regular* triangulations $\mathcal{T}_h$ like the one in Figure 2.1) and *iterative solution* by a trivial variant of algorithm (2.125)–(2.151) (with $k = +\infty$) with $\epsilon = 10^{-4}$ for the stopping criterion.

The above solution methodology has been tested for various values of h and Δt; for all of them, we have taken $\nu = \frac{1}{2\pi^2}$ $(= 5.066059\ldots \times 10^{-2})$. On Table 2.1 we have summarized the results which have been obtained (we have used a $*$ to indicate a *computed* quantity). All the calculations have been done with $f_0 = 0$ as initializer for the conjugate gradient algorithms.

The results presented in Table 2.1 deserve some comments, such as,

(i) The convergence of the conjugate gradient algorithm is fairly fast if we keep in mind that the solution $f_h^{\Delta t}$ of the discrete problem which has been solved can be viewed as a vector with $(31)^2 = 961$ components if $h = \Delta t = 1/32$ (respectively $(63)^2 = 3969$ components if $h = \Delta t = 1/64$).

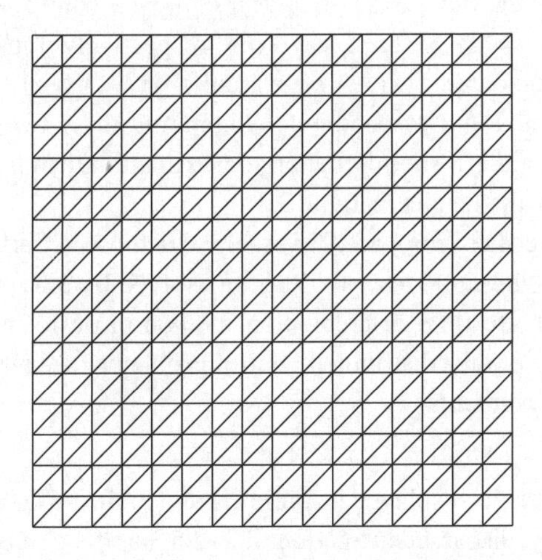

Fig. 2.1. A regular triangulation of $\Omega = (0, 1) \times (0, 1)$.

Table 2.1. *Summary of numerical results (first test problem).*

	$h = \Delta t = \frac{1}{32}$	$h = \Delta t = \frac{1}{64}$
Number of iterations	10	11
$\dfrac{\|y_T^* - y_T\|_{-1}}{\|y_T\|_{-1}}$	2.24×10^{-5}	1.78×10^{-5}
$\|u^*\|_{L^2(\Sigma)}$	7.791	7.863
$\dfrac{\|u^* - u\|_{L^2(\Sigma)}}{\|u\|_{L^2(\Sigma)}}$	2.50×10^{-3}	1.21×10^{-3}
$\|f^*\|_{H_0^1(\Omega)}$	6.07	6.041
$\dfrac{\|f^* - f\|_{H_0^1(\Omega)}}{\|f\|_{H_0^1(\Omega)}}$	2.44×10^{-2}	2.85×10^{-2}
$\dfrac{\|f^* - f\|_{L^2(\Omega)}}{\|f\|_{L^2(\Omega)}}$	6.53×10^{-3}	7.02×10^{-3}

(ii) The target function y_T has been reached within a good accuracy, similar comments holding for the approximation of the optimal control u and of the solution f of the dual problem (2.158).

(iii) For information, we have $\|u\|_{L^2(\Sigma)} = \pi\sqrt{e^2 - 1} = 7.94087251\ldots$ and $\|f\|_{H_0^1(\Omega)} = \pi e/\sqrt{2} = 6.03850398$.

On Figures 2.2 and 2.3 we have compared $y_T(x_1, \frac{1}{2})$ (−−) and $y_T^*(x_1, \frac{1}{2})$ (—) for $x_1 \in (0, 1)$ and $h = \Delta t = 1/32$ and $1/64$, respectively; we recall that $y_T^* = y_h^{\Delta t}(T)$ and that our methodology forces $y_h^{\Delta t}(T)$ to belong to H_{0h}^1, explaining the observed behavior of the above function in the neighborhood of Γ. On Figures 2.4 and 2.5 we have represented the functions $t \to \|u(t)\|_{L^2(\Gamma)}$ (−−) and $t \to \|u^*(t)\|_{L^2(\Gamma)}$ (—) for $t \in (0, T)$ and $h = \Delta t = 1/32$ and $1/64$, respectively. Finally, on Figures 2.6 and 2.7 we have compared $f(x_1, \frac{1}{2})$ (−−) and $f^*(x_1, \frac{1}{2})$ (—) for $x_1 \in (0, 1)$ and $h = \Delta t = 1/32$ and $1/64$, respectively. Comparing these two figures shows that $h = \Delta t = 1/32$ provides a slightly better approximation than $h = \Delta t = 1/64$; this is in agreement with the results in Table 2.1.

The results obtained here compared favorably with those in Carthel, Glowinski, and Lions (1994) where the same test problem was solved by other methods, including a second-order accurate time discretization method close to the one discussed in Chapter 1, Section 1.8.6 for distributed control problems (see also Carthel, 1994 for further results and comments).

Remark 2.12 The results displayed in Table 2.1 and Figures 2.2–2.5 strongly suggest that the methodology, discussed in Sections 2.2–2.5, has done a better "job" at solving the *primal problem* (2.155) than its *dual* counterpart (2.158) (the approximation errors

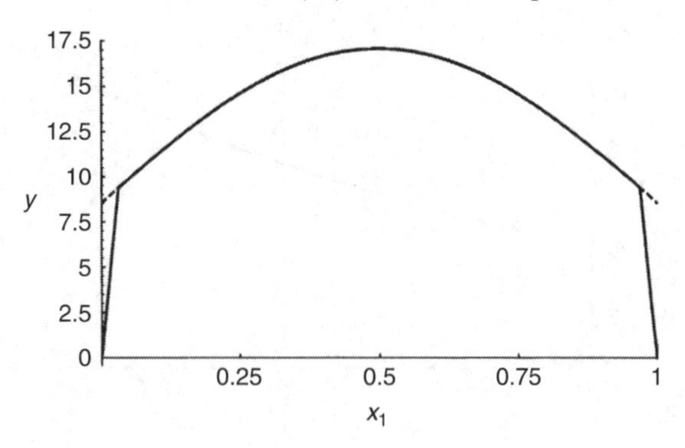

Fig. 2.2. First test problem: comparison between y_T (−−) and y_T^* (—) ($h = \Delta t = 1/32$).

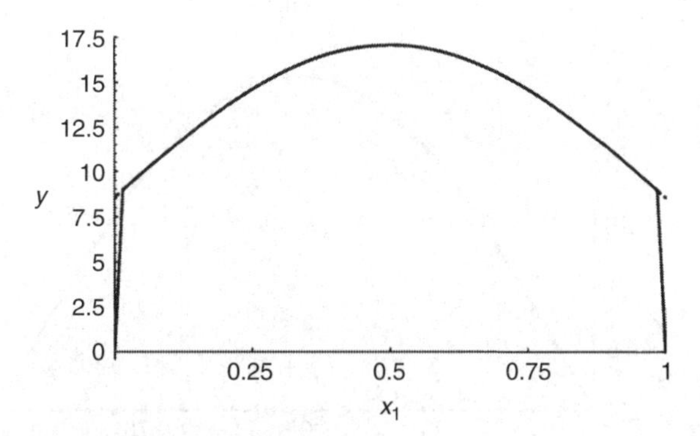

Fig. 2.3. First test problem: comparison between y_T (−−) and y_T^* (—) ($h = \Delta t = 1/64$).

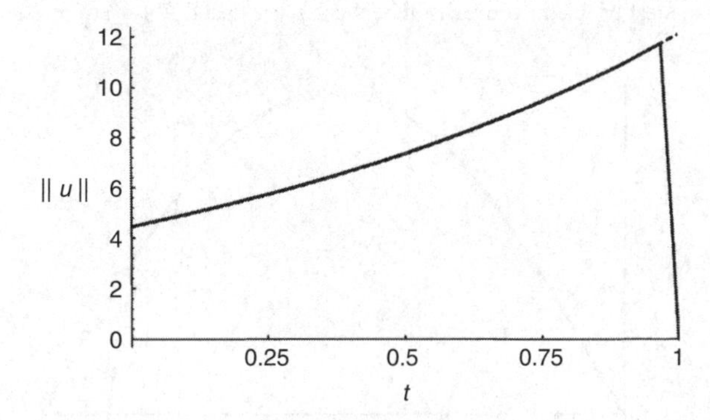

Fig. 2.4. First test problem: comparison between $\|u(t)\|_{L^2(\Gamma)}$ (−−) and $\|u^*(t)\|_{L^2(\Gamma)}$ (—) ($h = \Delta t = 1/32$).

on y_T and u are roughly of the order of $O(h)$). In order to explain this phenomenon let us observe that to discretize the dual problem (2.158) we took the following approach:

(i) First, we approximated the primal problem (2.155).
(ii) Second, we derived the dual of the fully discretized primal problem, obtaining thus the finite-dimensional problem (2.124).

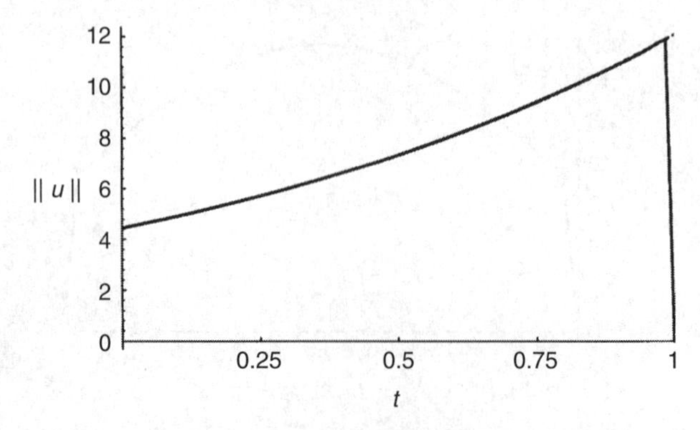

Fig. 2.5. First test problem: comparison between $\|u(t)\|_{L^2(\Gamma)}$ (−−) and $\|u^*(t)\|_{L^2(\Gamma)}$ (—) ($h = \Delta t = 1/64$).

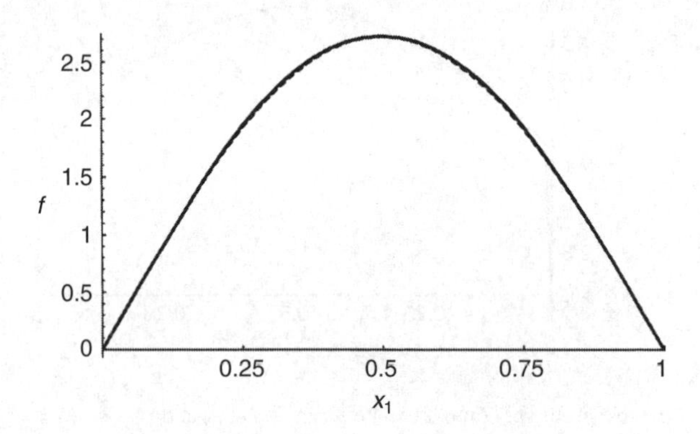

Fig. 2.6. First test problem: comparison between f (−−) and f^* (—) ($h = \Delta t = 1/32$).

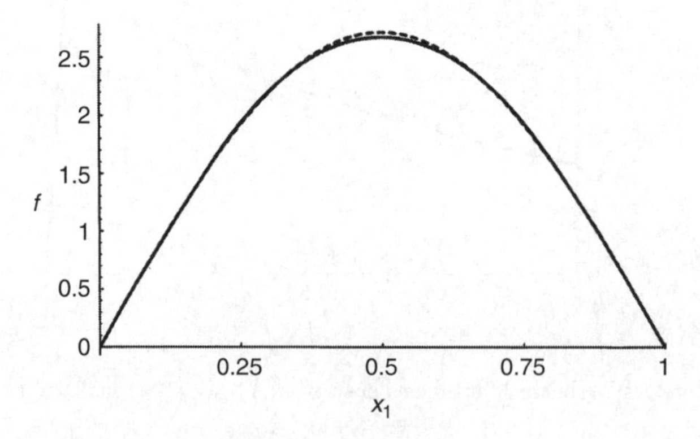

Fig. 2.7. First test problem: comparison between f (−−) and f^* (—) ($h = \Delta t = 1/64$).

It follows from (i) and (ii) that algorithm (2.125)–(2.151) is nothing but an iterative method for solving the fully discretized primal problem. With the strategy we followed, our "driver" was the approximation of the primal problem, the discrete dual being essentially a convenient way to solve, let us say *indirectly*, the discrete primal.

Concerning now the fact that the approximation of the solution f of the dual problem deteriorates (slightly) as h $(= \Delta t)$ decreases, let us observe that (for this first test problem) f does not contain high frequency modes with respect to the space variables, implying that, in principle, reducing h does not bring additional accuracy. Moreover, the operator Λ not being strongly elliptic, the condition number of $\Lambda_h^{\Delta t}$ deteriorates as $h(= \Delta t)$ decreases making the conjugate gradient algorithm less efficient.

2.6.2 Second test problem

If one uses the notation of Section 2.6.1 we have for this test problem $\Omega = (0,1) \times (0,1)$, $\mathcal{A} = -\nu\Delta$, $T = 1$, $s = 0$, $y_0 = 0$, $y_T(x_1,x_2) = \min(x_1,x_2,1-x_1,1-x_2)$, and $\nu = \frac{1}{2\pi^2}$; unlike the test problem of Section 2.6.1, for which $\Gamma_0 = \Gamma$, we have here $\Gamma_0 \neq \Gamma$ since here we took

$$\Gamma_0 = \big\{\{x_1,x_2\} \mid 0 < x_1 < 1,\ x_2 = 0\big\}.$$

The function y_T is *Lipschitz continuous*, but not smooth enough to have (see the discussion in Carthel, Glowinski, and Lions 1994, Section 2.3.3) exact controllability. This implies that problem (2.21) has no solution if $k = +\infty$; on the other hand, problems (2.11), (2.12), (2.21), and (2.25) are *well-posed* for any positive value of k or β. Focusing on the solution of problem (2.21) we have used the same space–time-discretization method as for the first test problem, with $h = \Delta t = 1/32$ and $1/64$. We have taken $k = 10^5$ and 10^7 for the penalty parameter and used $\epsilon = 10^{-3}$ for the stopping criterion of the conjugate gradient algorithm (2.125)–(2.151) (which has been initialized with $f^0 = 0$).

The numerical results have been summarized in Table 2.2.

The above results suggest the following comments: first, we observe that $\|y_T^* - y(T)\|_{-1}$ varies like $k^{-1/4}$, approximately. Second, we observe that the number of iterations necessary for convergence, increases as h $(= \Delta t)$ and k^{-1} decrease; there is no mystery here, since – from Chapter 1, Section 1.8.2, relation (2.133) – the key factor controlling the speed of convergence is the *condition number* of the *bilinear*

Table 2.2. *Summary of numerical results (second test problem).*

$h = \Delta t$	1/32	1/64	1/32	1/64
k	10^5	10^5	10^7	10^7
Number of iterations	56	60	292	505
$\dfrac{\|y_T^* - y_T\|_{-1}}{\|y_T\|_{-1}}$	1.31×10^{-1}	1.28×10^{-1}	4.15×10^{-2}	3.93×10^{-2}
$\|u^*\|_{L^2(\Sigma_0)}$	8.18	8.12	25.59	24.78
$\|f^*\|_{H_0^1(\Omega)}$	600.4	584.2	18 960	17 950
$\|f^*\|_{L^2(\Omega)}$	75.95	73.63	1 632	1 525

functional in the left-hand side of equation (2.124). This condition number, denoted by $\nu_h^{\Delta t}(k)$, is defined by

$$\nu_h^{\Delta t}(k) = \max_{z \in H_{0h}^1 \setminus \{0\}} R_h(z) \Big/ \min_{z \in H_{0h}^1 \setminus \{0\}} R_h(z), \tag{2.161}$$

where, in (2.161), $R_h(z)$ is the *Rayleigh quotient* defined by

$$R_h(z) = \frac{k^{-1} \int_\Omega |\nabla z|^2 \, dx + \int_\Omega (\Lambda_h^{\Delta t} z) z \, dx}{\int_\Omega |\nabla z|^2 \, dx}; \tag{2.162}$$

it can be shown that

$$\lim_{\substack{h \to 0 \\ \Delta t \to 0}} \nu_h^{\Delta t}(k) = k \|\Lambda\|_{\mathcal{L}(H_0^1(\Omega); H^{-1}(\Omega))}, \tag{2.163}$$

implying that for small values of h, Δt, and k^{-1}, problem (2.124) is *badly conditioned*. Indeed, we can expect from (2.163) and from Chapter 1, Section 1.8.2, relation (1.133), that for h and Δt sufficiently small the number of iterations necessary to obtain convergence will vary like $k^{\frac{1}{2}}$, approximately; this prediction is confirmed by the results in Table 2.2 (and will be further confirmed by the results in Section 2.6.3, Table 2.3, concerning the solution of our third test problem). Third, and finally, we observe that $\|u^*\|_{L^2(\Sigma_0)}$ (respectively $\|f^*\|_{H_0^1(\Omega)}$) varies like $k^{1/4}$ (respectively $k^{3/4}$); it can be shown that the behavior of $\|f^*\|_{H_0^1(\Omega)}$ follows from that of $\|y_T^* - y_T\|_{-1}$ since we have (see, for example, Carthel, Glowinski, and Lions, 1994, Remark 4.3)

$$k^{-1} = \frac{\|y_T - y(T)\|_{-1}}{\|f\|_{H_0^1(\Omega)}}, \tag{2.164}$$

where y is the state function obtained from the optimal control u via (2.1)–(2.3).

On the following figures, we have represented or shown the following information and results: A view of the target function y_T on Figure 2.8. On Figures 2.9 (top)–2.12 (top) (respectively 2.9 (bottom) to 2.12(bottom)) the graph of the function $y_T^*(= y_h^{\Delta t}(T))$ (respectively a comparison between y_T (– –) and the actually reached state function y_T^* (—)) for various values of h, Δt, and k (we have shown the graphs of the functions $x_2 \to y_T(\frac{1}{2}, x_2)$ and $x_2 \to y_T^*(\frac{1}{2}, x_2)$ for $x_2 \in (0,1)$). The graphs of the computed solution $f_h^{\Delta t}(= f^*)$ and of the function $x_2 \to f_h^{\Delta t}(\frac{1}{2}, x_2)$ on Figures 2.13–2.16. On Figures 2.17–2.20 the graphs of the functions $t \to \|u^*(t)\|_{L^2(\Gamma_0)}$ and $\{x_1, t\} \to u^*(x_1, t)$. Finally, we have visualized on Figures 2.21–2.24 (using a log-scale) the convergence to zero of the conjugate gradient residual $\|g_m\|_{H_0^1(\Omega)}$; the observed behavior (highly oscillatory, particularly for $k = 10^7$) is typical of a *badly conditioned* problem.

Table 2.3. *Summary of numerical results (third test problem).*

$h = \Delta t$	1/32	1/64	1/32	1/64
k	10^5	10^5	10^7	10^7
Number of iterations	55	56	361	569
$\dfrac{\lVert y_T^* - y_T \rVert_{-1}}{\lVert y_T \rVert_{-1}}$	1.64×10^{-1}	1.57×10^{-1}	1.05×10^{-1}	9.88×10^{-2}
$\lVert u^* \rVert_{L^2(\Sigma_0)}$	14.68	15.07	56.80	58.53
$\lVert f^* \rVert_{H_0^1(\Omega)}$	1,407	1,410	90,010	88,510
$\lVert f^* \rVert_{L^2(\Omega)}$	120.7	122.5	5,608	5,566

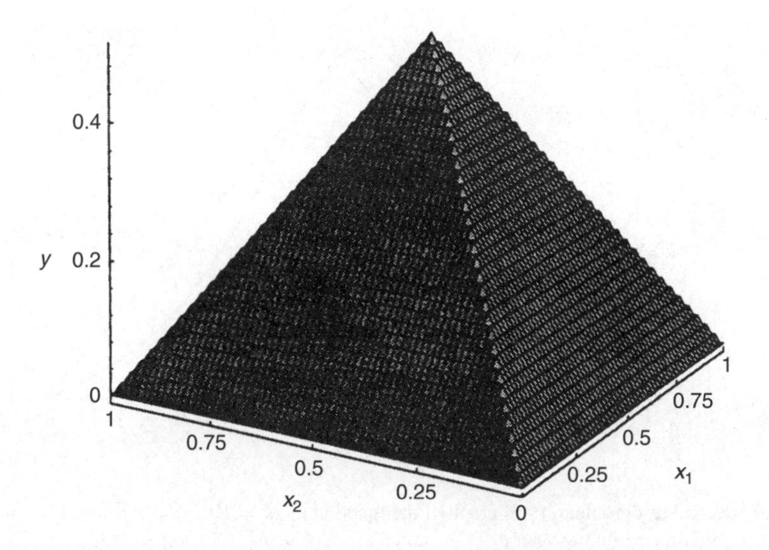

Fig. 2.8. Second test problem: graph of the target function y_T ($y_T(x_1, x_2) = \min(x_1, x_2, 1 - x_1, 1 - x_2)$, $0 \leq x_1, x_2 \leq 1$).

2.6.3 Third test problem

For this test problem Ω, T, Γ_0, y_0, s, $\mathcal{A}$, and ν are as in Section 2.6.2, namely, $\Omega = (0.1) \times (0.1)$, $T = 1$, $y_0 = 0$, $s = 0$, $\mathcal{A} = -\nu\Delta$ with $\nu = 1/2\pi^2$; the only difference is that this time y_T is the *discontinuous* function defined by

$$y_T(x_1, x_2) = \begin{cases} 1 & \text{if } \frac{1}{4} < x_1, x_2 < \frac{3}{4}, \\ 0 & \text{otherwise.} \end{cases} \tag{2.165}$$

We have applied to this problem the solution methods considered in Section 2.6.2; their behavior here is essentially the same as that for the test problem of Section 2.6.2 (where y_T was Lipschitz continuous). We have shown in the following Table 2.3 the results of our numerical experiments (the notation is as in Section 2.6.2).

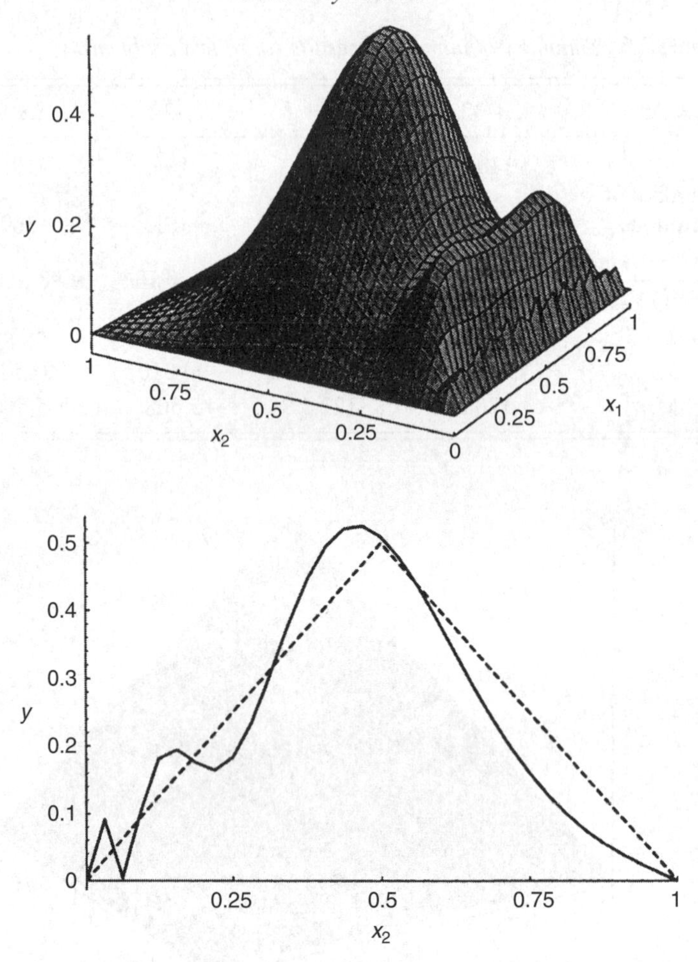

Fig. 2.9. Second test problem: (top) graph of the function y_T^* ($k = 10^5$, $h = \Delta t = 1/32$); (bottom) comparison between y_T (−−) and y_T^* (—) ($k = 10^5$, $h = \Delta t = 1/32$).

Comparing to Table 2.2 we observe that the convergence properties of the conjugate gradient algorithm are essentially the same, despite the fact that the target function y_T is much less smooth here; on the other hand, we observe that $\|y_T^* - y_T\|_{-1}$ varies like $k^{-1/3}$, approximately, implying in turn (from (2.164)) that $\|f^*\|_{H_0^1(\Omega)}$ varies like $k^{7/8}$, approximately. The dependence of $\|u^*\|_{L^2(\Sigma_0)}$ is less clear (to us at least); it looks "faster," however, than $k^{1/4}$.

On Figure 2.25 we have visualized the graph of the target function y_T, then on Figures 2.26 and 2.27 we have compared the functions $x_2 \to y_T(\frac{1}{2}, x_2)$ (−−) and $x_2 \to y_T^*(\frac{1}{2}, x_2)$ (—) for various values of k and for $h = \Delta t = 1/64$; on Figures 2.28 and 2.29 we have shown the graph of the corresponding functions y_T^*. Finally, for the above values of k, h, and Δt, we have shown, on Figures 2.30–2.35, further information concerning $u_h^{\Delta t}$, $f_h^{\Delta t}$, and the convergence of the conjugate gradient algorithm (2.125)–(2.151).

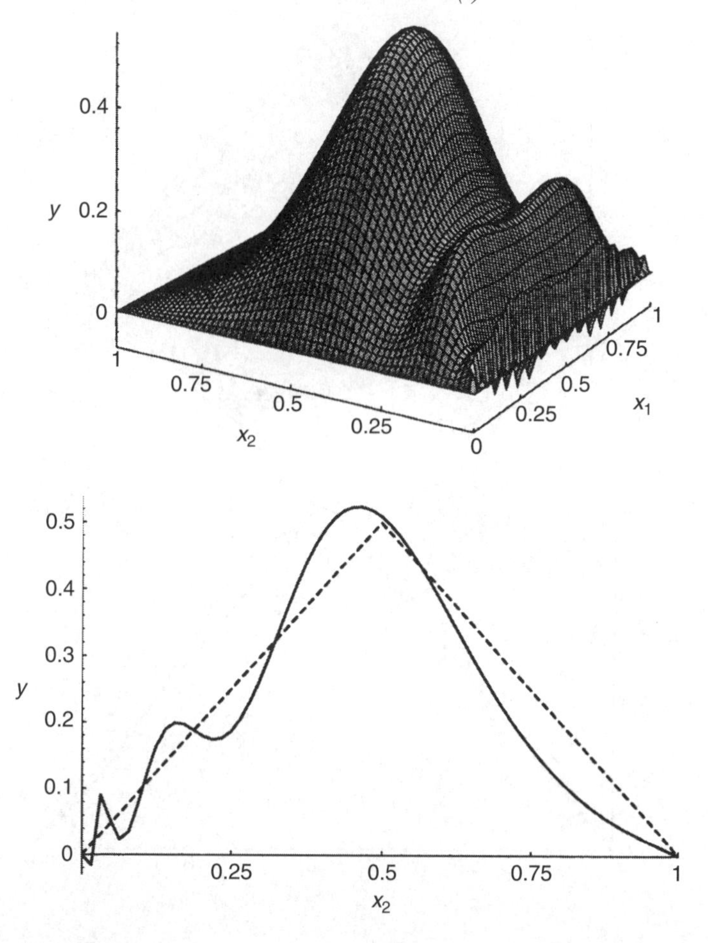

Fig. 2.10. Second test problem: (top) graph of the function y_T^* ($k = 10^5$, $h = \Delta t = 1/64$); (bottom) comparison between y_T (−−) and y_T^* (—) ($k = 10^5$, $h = \Delta t = 1/64$).

2.7 Neumann control (I): Formulation of the control problems and synopsis

2.7.1 Formulation of the control problems

We consider again the state equation (2.1) in Q and the initial condition (2.3). We suppose this time that the *boundary control* is of the *Neumann's type*. To be more precise, the state function y is defined now by

$$\begin{cases} \dfrac{\partial y}{\partial t} + \mathcal{A}y = 0 \quad \text{in } Q, \\ \\ y(0) = 0, \\ \\ \dfrac{\partial y}{\partial n_{\mathcal{A}}} = v \ \text{ on } \Sigma_0, \quad \dfrac{\partial y}{\partial n_{\mathcal{A}}} = 0 \ \text{ on } \Sigma \backslash \Sigma_0. \end{cases} \tag{2.166}$$

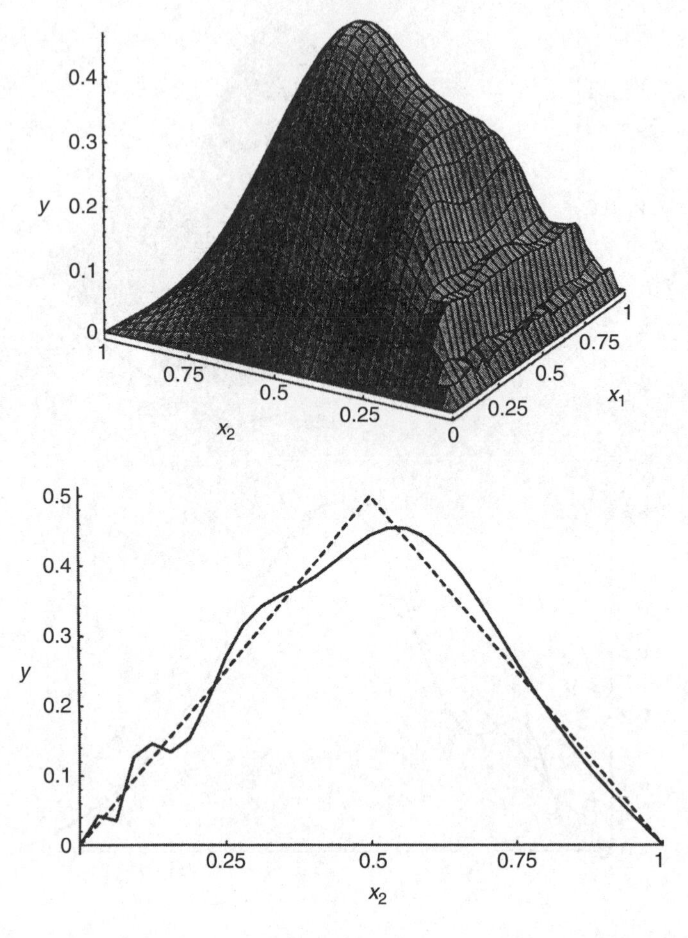

Fig. 2.11. Second test problem: (top) graph of the function y_T^* ($k = 10^7$, $h = \Delta t = 1/32$); (bottom) comparison between y_T (-- --) and y_T^* (—) ($k = 10^7$, $h = \Delta t = 1/32$).

In (2.166), $\frac{\partial}{\partial n_{\mathcal{A}}}$ denotes the *conormal derivative* operator; if operator $\mathcal{A}$ is defined by

$$\mathcal{A}\varphi = -\sum_{i=1}^{d}\sum_{j=1}^{d}\frac{\partial}{\partial x_i}a_{ij}\frac{\partial \varphi}{\partial x_j}, \tag{2.167}$$

then $\frac{\partial}{\partial n_{\mathcal{A}}}$ is defined by

$$\frac{\partial \varphi}{\partial n_{\mathcal{A}}} = \sum_{i=1}^{d}\sum_{j=1}^{d} a_{ij}\frac{\partial \varphi}{\partial x_j}n_i, \tag{2.168}$$

where $\mathbf{n} = \{n_i\}_{i=1}^d$ is the *unit vector* of the *outward normal* at Γ.

We assume that

$$v \in L^2(\Sigma_0). \tag{2.169}$$

There are slight (and subtle) technical differences between Neumann and Dirichlet boundary controls. Indeed, suppose that operator $\mathcal{A}$ is defined by (2.167) with the

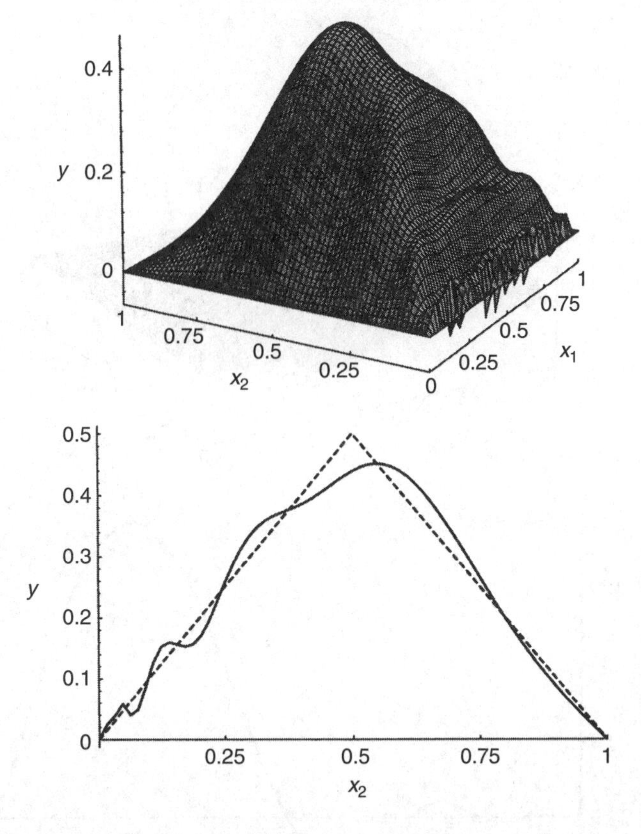

Fig. 2.12. Second test problem: (top) graph of the function y_T^* ($k = 10^7$, $h = \Delta t = 1/64$); (bottom) comparison between y_T (−−) and y_T^* (—) ($k = 10^7$, $h = \Delta t = 1/64$).

following additional properties:

$$a_{ij} \in L^\infty(\Omega),\ \forall 1 \leq i,j \leq d, \tag{2.170}$$

$$\sum_{i=1}^{d}\sum_{j=1}^{d} a_{ij}(x)\xi_i\xi_j \geq \alpha|\boldsymbol{\xi}|^2,\ \forall \boldsymbol{\xi} \in \mathbb{R}^d,\ \text{ a.e. in } \Omega, \text{ with } \alpha > 0, \tag{2.171}$$

(in (2.171), $|\boldsymbol{\xi}|^2 = \sum_{i=1}^{d} |\xi_i|^2$, $\forall \boldsymbol{\xi} = \{\xi_i\}_{i=1}^{d} \in \mathbb{R}^d$); then problem (2.166) can be expressed in *variational* form as follows:

$$\begin{cases} \left(\dfrac{\partial y}{\partial t}, z\right) + a(y,z) = \displaystyle\int_{\Gamma_0} vz \, d\Gamma, \quad \forall z \in H^1(\Omega), \\ \\ y(t) \in H^1(\Omega) \text{ a.e. on } (0,T), \quad y(0) = 0, \end{cases} \tag{2.172}$$

where

$$a(y,z) = \sum_{i=1}^{d}\sum_{j=1}^{d} \int_\Omega a_{ij} \frac{\partial y}{\partial x_j}\frac{\partial z}{\partial x_i}\, dx, \tag{2.173}$$

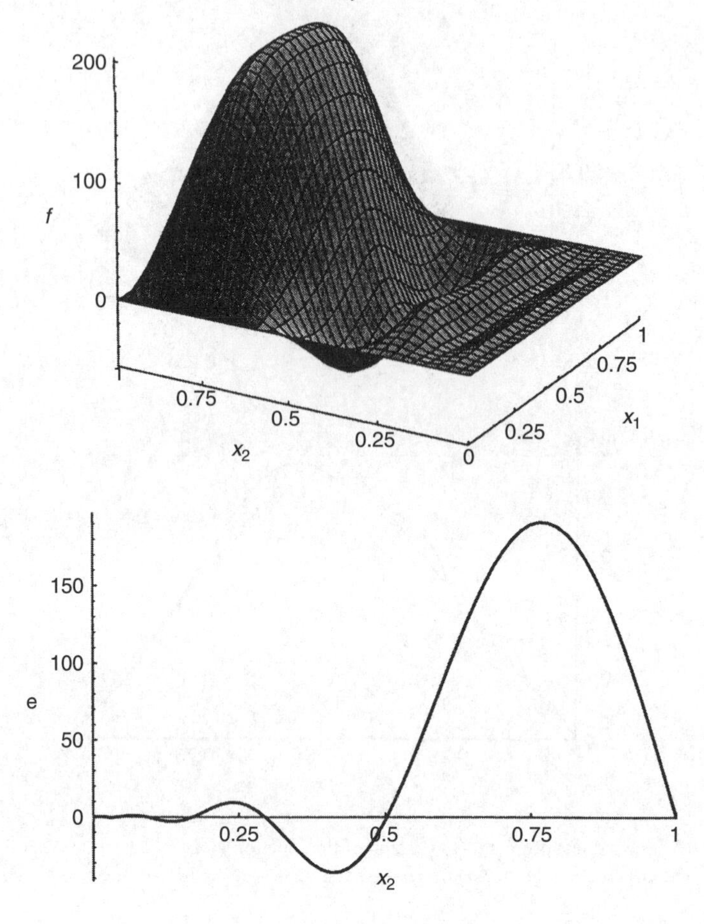

Fig. 2.13. Second test problem: (top) graph of the function $f_h^{\Delta t}$ $(k = 10^5,\ h = \Delta t = 1/32)$; (bottom) graph of the function $x_2 \to f_h^{\Delta t}(0.5, x_2)$ $(k = 10^5,\ h = \Delta t = 1/32)$.

(all this applies, actually, to the case where the coefficients a_{ij} depend on x *and* t and verify $a_{ij}(x,t) \in L^\infty(Q)$, $\forall i,j,\ 1 \leq i,j \leq d$, and

$$\sum_{i=1}^{d}\sum_{j=1}^{d} a_{ij}(x,t)\xi_i\xi_j \geq \alpha|\boldsymbol{\xi}|^2, \quad \forall \boldsymbol{\xi} \in \mathbb{R}^d, \text{ a.e. in } Q,$$

with $\alpha > 0$). Therefore, without any further hypothesis on the coefficients a_{ij}, problem (2.166) admits a unique solution $y(v)$ $(= \{x,t\} \to y(x,t;v))$ such that

$$y(v) \in L^2(0,T;H^1(\Omega)) \cap C^0([0,T];L^2(\Omega)). \tag{2.174}$$

To obtain the *approximate controllability* property we shall assume further *regularity* properties for the a_{ij}s, more specifically we shall assume that

$$a_{ij} \in C^1(\bar{\Omega}), \quad \forall i,j,\ 1 \leq i,j \leq d. \tag{2.175}$$

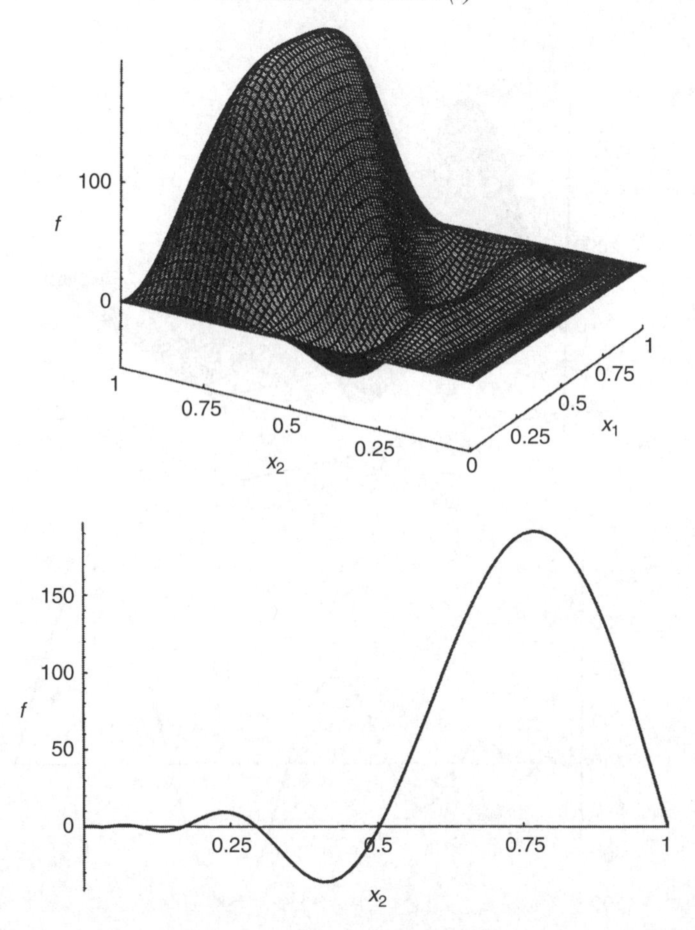

Fig. 2.14. Second test problem: (top) graph of the function $f_h^{\Delta t}$ ($k = 10^5$, $h = \Delta t = 1/64$); (bottom) graph of the function $x_2 \to f_h^{\Delta t}(0.5, x_2)$ ($k = 10^5$, $h = \Delta t = 1/64$).

We have then the following.

Proposition 2.13 *Suppose that the coefficients a_{ij} verify* (2.171) *and* (2.175). *Then $y(T; v)$ spans a dense subset of $L^2(\Omega)$ when v spans $L^2(\Sigma_0)$.*

Proof. The proof is similar to the proof of Proposition 2.1 (see Section 2.1). Let us assume therefore that $f \in L^2(\Omega)$ satisfies

$$\int_\Omega y(T; v) f \, dx = 0, \; \forall v \in L^2(\Sigma_0). \tag{2.176}$$

We associate with f the solution ψ of

$$-\frac{\partial \psi}{\partial t} + \mathcal{A}^* \psi = 0 \;\text{ in } Q, \quad \psi(T) = f, \quad \frac{\partial \psi}{\partial n_{\mathcal{A}^*}} = 0 \;\text{ on } \Sigma; \tag{2.177}$$

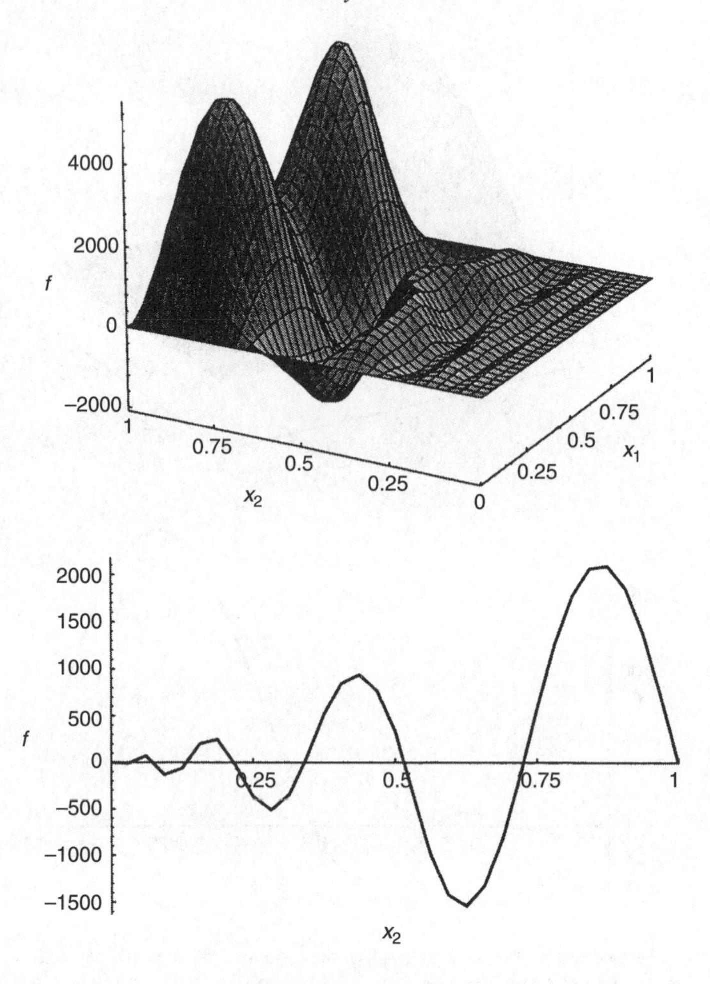

Fig. 2.15. Second test problem: (top) graph of the function $f_h^{\Delta t}$ ($k = 10^7$, $h = \Delta t = 1/32$); (bottom) graph of the function $x_2 \to f_h^{\Delta t}(0.5, x_2)$ ($k = 10^7$, $h = \Delta t = 1/32$).

then (2.176) is equivalent to

$$\psi = 0 \text{ on } \Sigma_0. \tag{2.178}$$

Thanks to the regularity hypothesis (2.175) we can use the *Mizohata's uniqueness theorem* (Mizohata, 1958; see also Saut and Schoerer, 1987); it follows then from (2.177) and (2.178) that $\psi = 0$, hence $f = 0$ and the proof is completed. □

Remark 2.14 The applicability of the Mizohata's uniqueness theorem under the only assumption that $a_{ij} \in L^\infty(Q)$ does not seem to have been completely settled, yet.

We will state now two basic *controllability problems*, both closely related to problems (2.11) and (2.12) in Section 2.1.

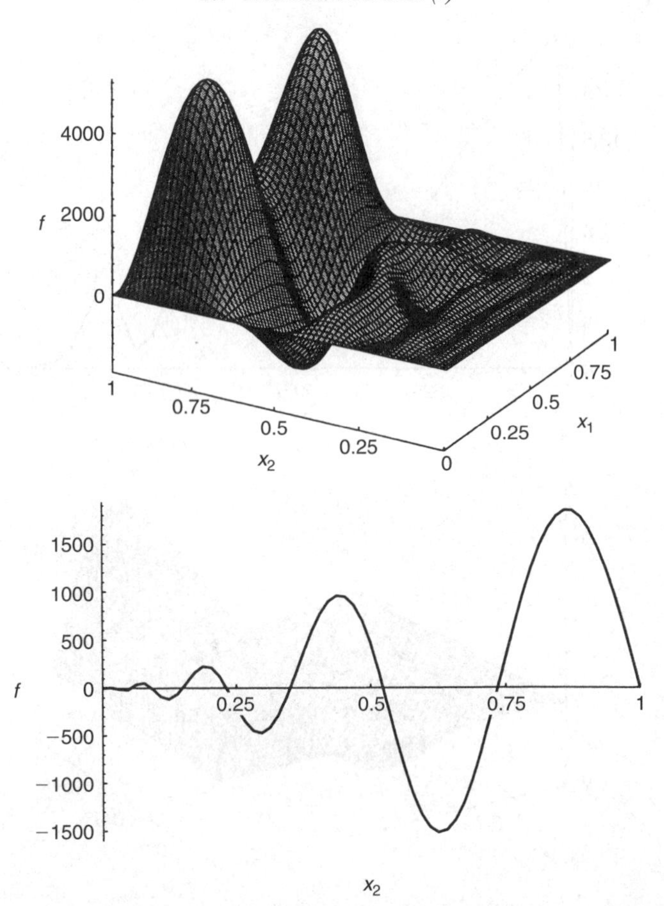

Fig. 2.16. Second test problem: (top) graph of the function $f_h^{\Delta t}$ ($k = 10^7$, $h = \Delta t = 1/64$); (bottom) graph of the function $x_2 \to f_h^{\Delta t}(0.5, x_2)(k = 10^7,\ h = \Delta t = 1/64)$.

The *first* Neumann control problem that we consider is defined by

$$\inf_{v} \frac{1}{2} \int_{\Sigma_0} v^2 \, d\Sigma, \tag{2.179}$$

where v is subjected to

$$y(T; v) \in y_T + \beta B; \tag{2.180}$$

in (2.180), $y(t; v)$ is the solution of problem (2.166), the target function y_T belongs to $L^2(\Omega)$, B denotes the closed unit ball of $L^2(\Omega)$ and β is a positive number, arbitrarily small.

The *second* Neumann control problem to be considered is defined by

$$\inf_{v} \left(\frac{1}{2} \int_{\Sigma_0} v^2 \, d\Sigma + \frac{1}{2} k \|y(T; v) - y_T\|^2_{L^2(\Omega)} \right), \tag{2.181}$$

where $v \in L^2(\Sigma_0)$ and k is a positive number, arbitrarily large.

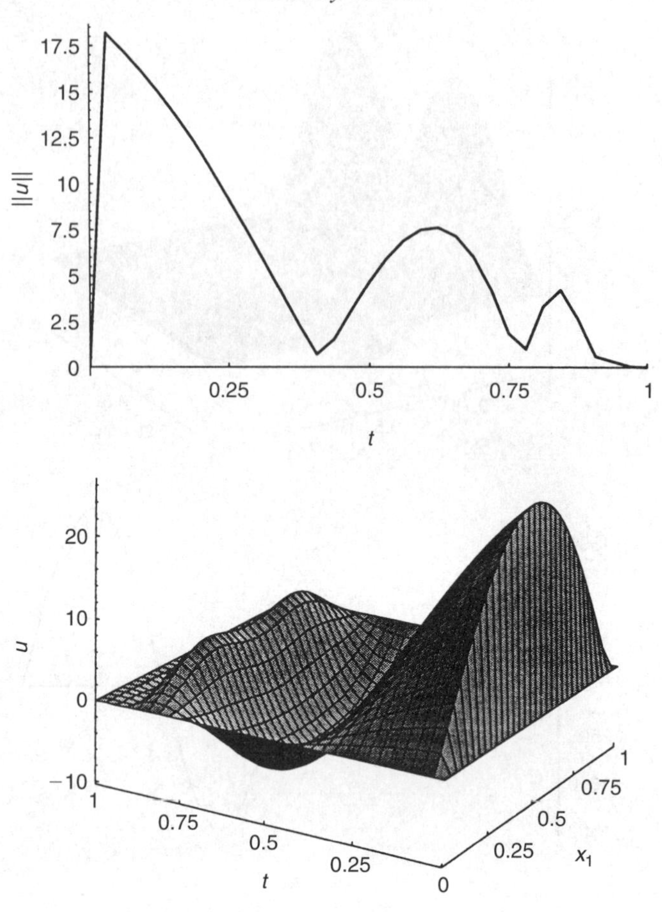

Fig. 2.17. Second test problem: (top) graph of $t \to \|u^*(t)\|_{L^2(\Gamma_0)}$ ($k = 10^5$, $h = \Delta t = 1/32$); (bottom) graph of the computed boundary control ($k = 10^5$, $h = \Delta t = 1/32$).

Both problems (2.179) and (2.181) admits a *unique* solution. There is, however, a technical difference between these two problems since problem (2.181) admits a unique solution under the only hypothesis $a_{ij} \in L^\infty(\Omega)$ (and of course the ellipticity property (2.171)), while the existence of a solution for problem (2.179), with β arbitrarily small, requires, so far, some regularity property (such as (2.175)) for the a_{ij}s. In the following we shall assume that property (2.175) holds, even if this hypothesis is not always necessary.

2.7.2 Synopsis

Sections 2.8–2.10, hereafter, will follow closely the text by Glowinski and J.L. Lions in *Acta Numerica 1995* (and also in J.L. Lions, 2003). On the other hand, new material has been included at the end of this chapter, namely,

(i) In Section 2.11, the stabilization and control, via Neumann boundary actions, of systems governed by some *unstable nonlinear parabolic equations* (or systems

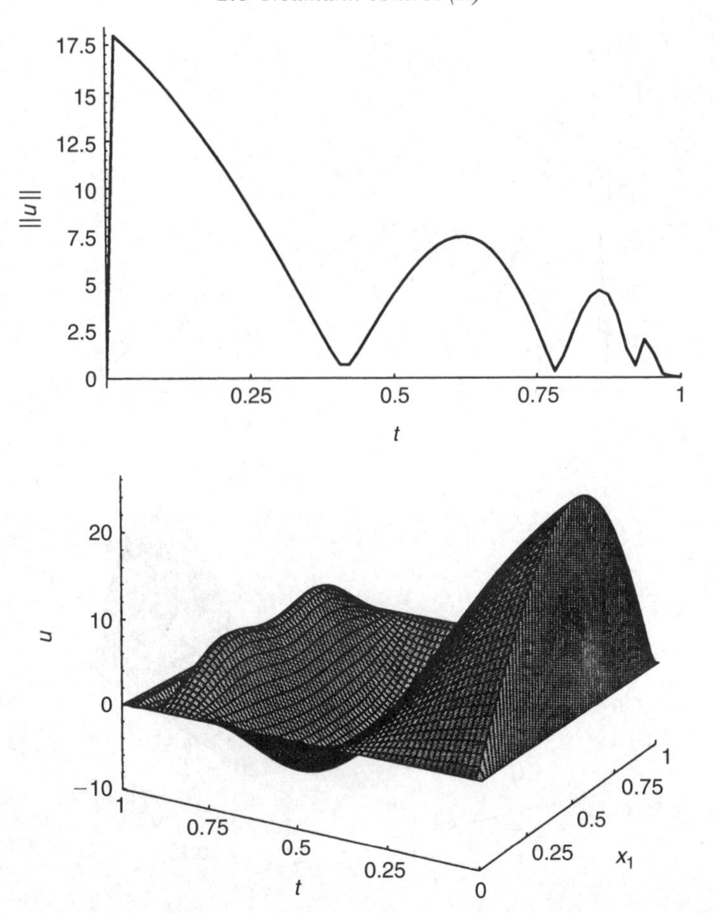

Fig. 2.18. Second test problem: (top) graph of $t \to \|u^*(t)\|_{L^2(\Gamma_0)}$ ($k = 10^5$, $h = \Delta t = 1/64$); (bottom) graph of the computed boundary control ($k = 10^5$, $h = \Delta t = 1/64$).

of such equations); the use of *chattering control* will be discussed also in this section which follows closely He and Glowinski (1998).

(ii) In Section 2.12, the solution of some of the problems discussed in Section 2.11 by those *recursive methods à la Kalman* briefly described in Chapter 1, Section 1.13; this section follows He, Glowinski, Metcalfe, and Périaux (1998).

2.8 Neumann control (II): Optimality conditions and dual formulations

The *optimality* system for problem (2.181) is obtained by arguments which are fairly classical (see, for example, J. L. Lions, 1968, 1971), as recalled in Section 2.2. Following, precisely, the approach taken in Section 2.2, we introduce the functional $J_k : L^2(\Sigma_0) \to \mathbb{R}$ defined by

$$J_k(v) = \frac{1}{2}\int_{\Sigma_0} v^2 \, d\Sigma + \frac{1}{2}k\|y(T;v) - y_T\|^2_{L^2(\Omega)}. \tag{2.182}$$

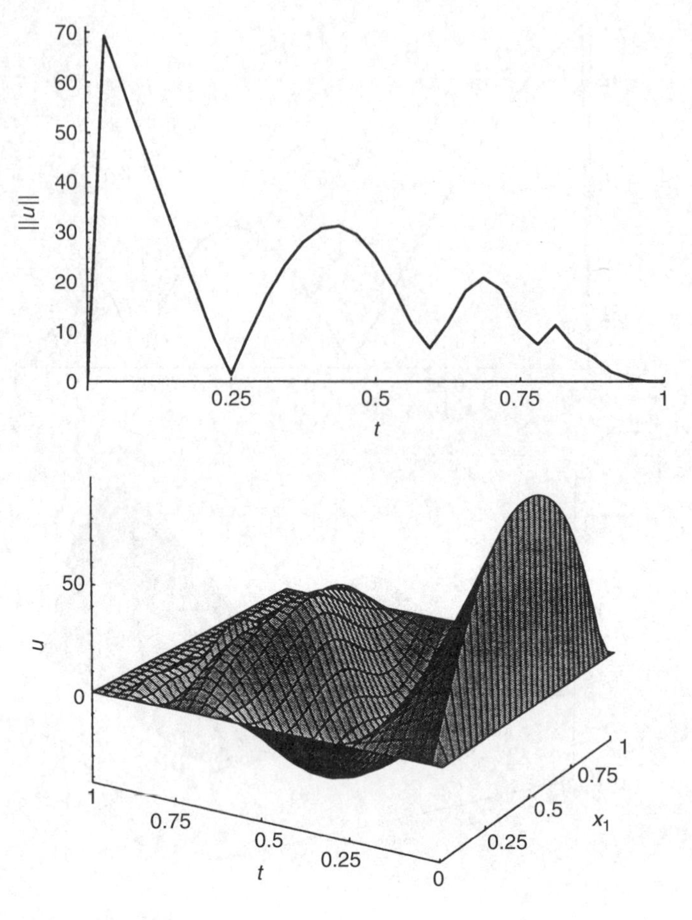

Fig. 2.19. Second test problem: (top) graph of $t \to \|u^*(t)\|_{L^2(\Gamma_0)}$ ($k = 10^7$, $h = \Delta t = 1/32$); (bottom) graph of the computed boundary control ($k = 10^7$, $h = \Delta t = 1/32$).

We can show that the derivative J_k' of J_k is defined by

$$(J_k'(v), w)_{L^2(\Sigma_0)} = \int_{\Sigma_0} (v + p)w \, d\Sigma, \quad \forall v, w \in L^2(\Sigma_0), \tag{2.183}$$

where, in (2.183), the *adjoint state function p* is obtained from v via the solution of (2.166) and of the *adjoint state equation*

$$-\frac{\partial p}{\partial t} + \mathcal{A}^* p = 0 \text{ in } Q, \quad \frac{\partial p}{\partial n_{\mathcal{A}^*}} = 0 \text{ on } \Sigma, \quad p(T) = k(y(T) - y_T). \tag{2.184}$$

Suppose now that u is *the* solution of the control problem (2.181); since $J_k'(u) = 0$, we have then the following optimality system satisfied by u and by the corresponding state and adjoint state functions:

$$u = -p|_{\Sigma_0}, \tag{2.185}$$

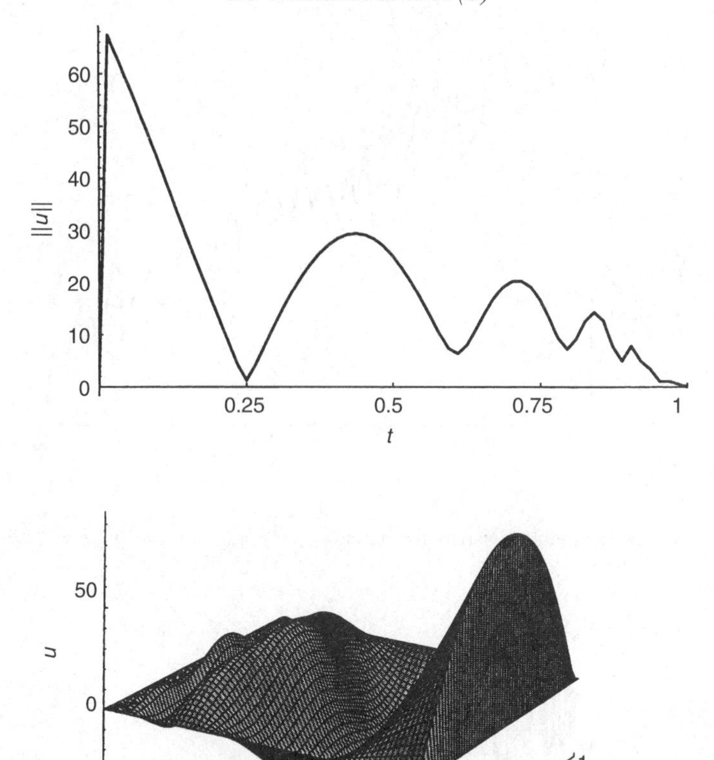

Fig. 2.20. Second test problem: (top) graph of $t \to \|u^*(t)\|_{L^2(\Gamma_0)}$ ($k = 10^7$, $h = \Delta t = 1/64$); (bottom) graph of the computed boundary control ($k = 10^7$, $h = \Delta t = 1/64$).

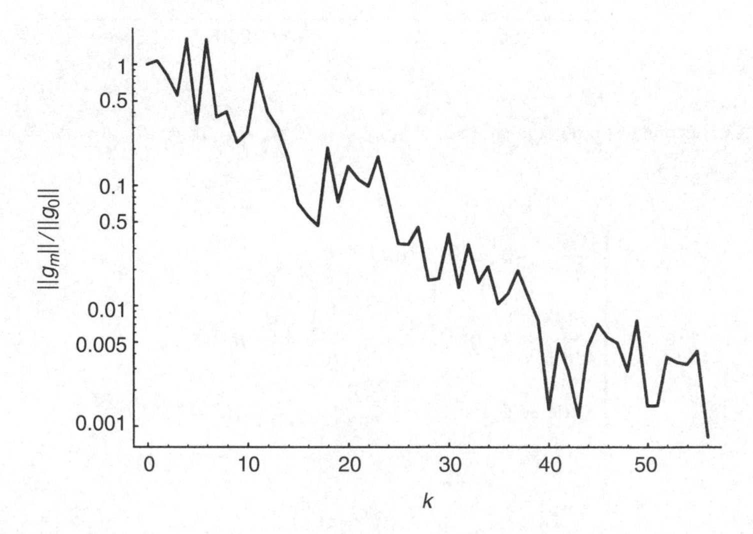

Fig. 2.21. Second test problem: variation of $\|g_m\|_{H_0^1(\Omega)}/\|g_0\|_{H_0^1(\Omega)}$ ($k = 10^5$, $h = \Delta t = 1/32$).

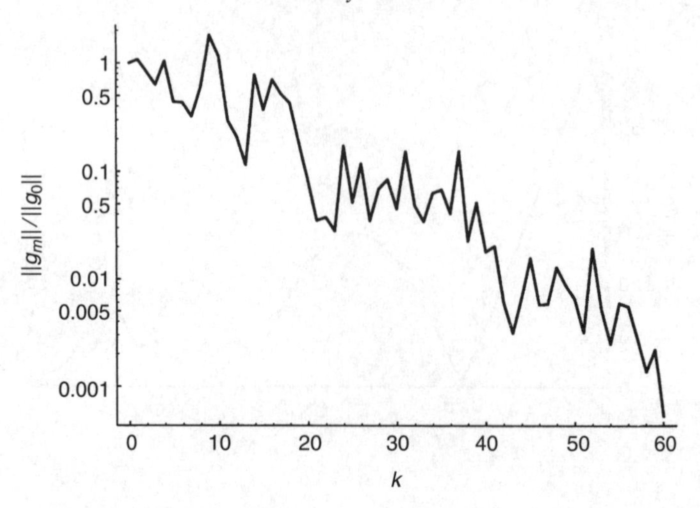

Fig. 2.22. Second test problem: variation of $\|g_m\|_{H_0^1(\Omega)}/\|g_0\|_{H_0^1(\Omega)}$ ($k = 10^5$, $h = \Delta t = 1/64$).

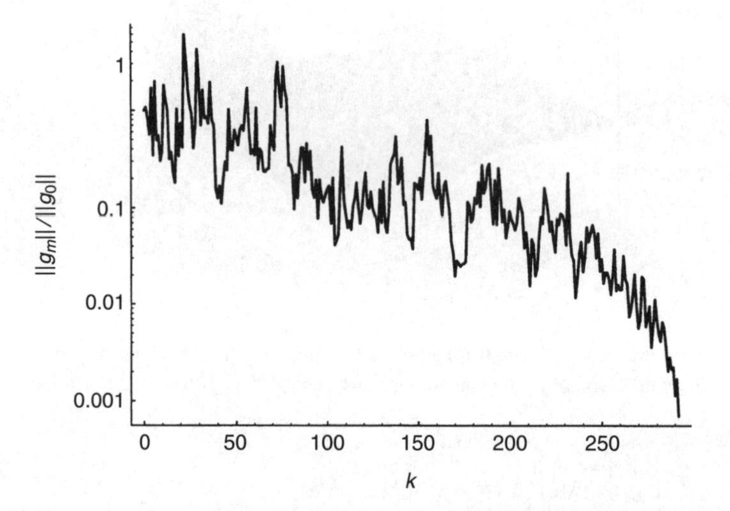

Fig. 2.23. Second test problem: variation of $\|g_m\|_{H_0^1(\Omega)}/\|g_0\|_{H_0^1(\Omega)}$ ($k = 10^7$, $h = \Delta t = 1/32$).

$$\begin{cases} \dfrac{\partial y}{\partial t} + \mathcal{A}y = 0 \quad \text{in } Q, \\ \dfrac{\partial y}{\partial n_{\mathcal{A}}} = 0 \ \text{ on } \Sigma \backslash \Sigma_0, \quad \dfrac{\partial y}{\partial n_{\mathcal{A}}} = -p \ \text{ on } \Sigma_0, \\ y(0) = 0, \end{cases} \tag{2.186}$$

$$-\frac{\partial p}{\partial t} + \mathcal{A}^* p = 0 \ \text{ in } Q, \quad \frac{\partial p}{\partial n_{\mathcal{A}^*}} = 0 \ \text{ on } \Sigma, \quad p(T) = k(y(T) - y_T). \tag{2.187}$$

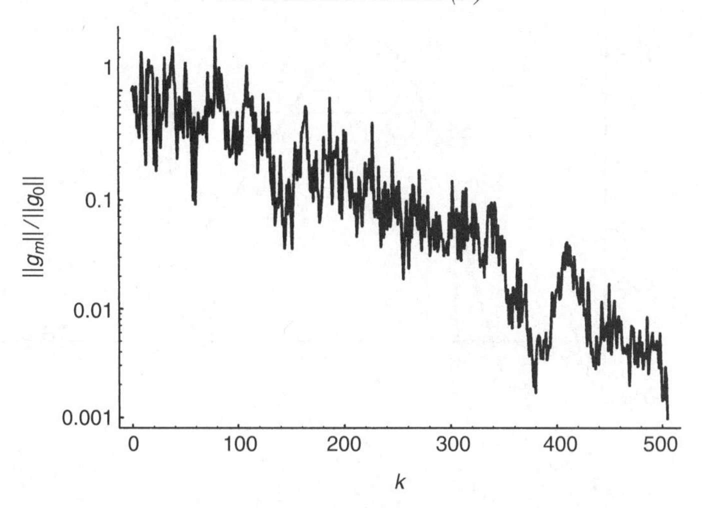

Fig. 2.24. Second test problem: variation of $\|g_m\|_{H_0^1(\Omega)}/\|g_0\|_{H_0^1(\Omega)}$ ($k = 10^7$, $h = \Delta t = 1/64$).

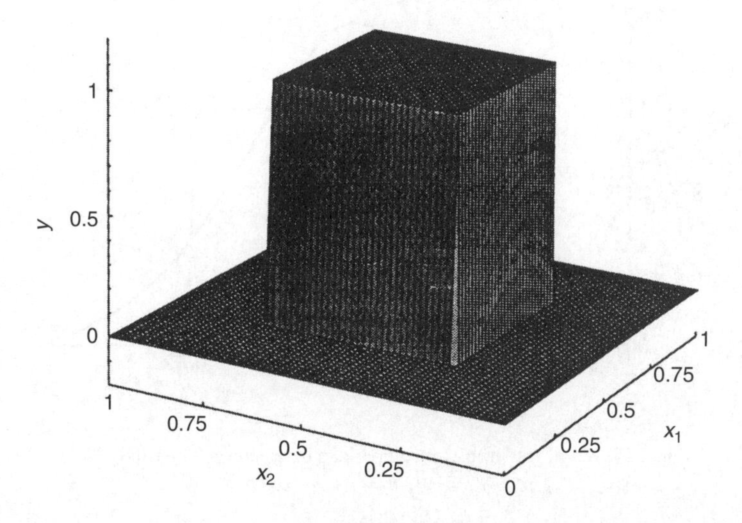

Fig. 2.25. Third test problem: graph of the target function y_T (y_T is the characteristic function of the square $\left(\frac{1}{4}, \frac{3}{4}\right) \times \left(\frac{1}{4}, \frac{3}{4}\right)$).

In order to identify the dual problem of (2.181) we proceed essentially as in Section 2.2. We introduce therefore the operator $\Lambda \in \mathcal{L}(L^2(\Omega), L^2(\Omega))$ defined by

$$\Lambda g = -\phi_g(T), \quad \forall g \in L^2(\Omega), \tag{2.188}$$

where, in (2.188), ϕ_g is obtained from g as follows:
Solve first

$$-\frac{\partial \psi_g}{\partial t} + \mathcal{A}^* \psi_g = 0 \text{ in } Q, \quad \frac{\partial \psi_g}{\partial n_{\mathcal{A}^*}} = 0 \text{ on } \Sigma, \quad \psi_g(T) = g, \tag{2.189}$$

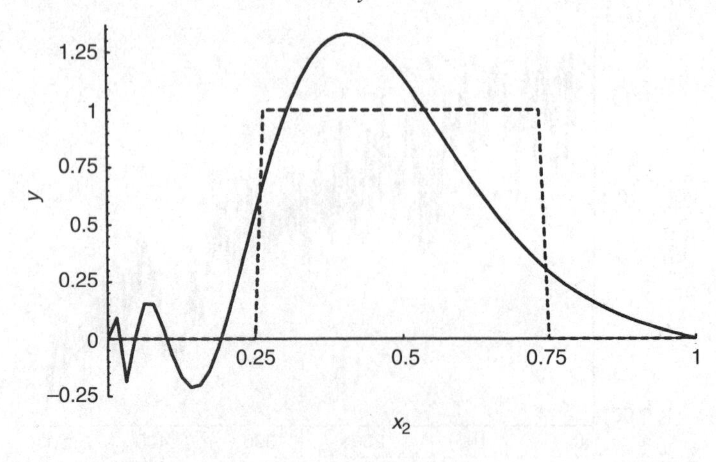

Fig. 2.26. Third test problem: comparison between y_T (−−) and y_T^* (—) ($k = 10^5$, $h = \Delta t = 1/64$).

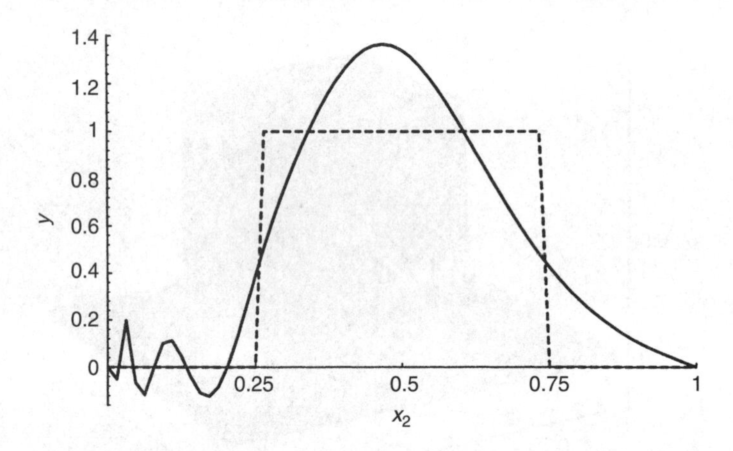

Fig. 2.27. Third test problem: comparison between y_T (−−) and y_T^* (—) ($k = 10^7$, $h = \Delta t = 1/64$).

and then

$$\begin{cases} \dfrac{\partial \phi_g}{\partial t} + \mathcal{A}\phi_g = 0 \text{ in } Q, \\ \phi_g(0) = 0, \\ \dfrac{\partial \phi_g}{\partial n_{\mathcal{A}}} = 0 \text{ on } \Sigma \backslash \Sigma_0, \quad \dfrac{\partial \phi_g}{\partial n_{\mathcal{A}}} = -\psi_g \text{ on } \Sigma_0. \end{cases} \tag{2.190}$$

We can easily show that (with obvious notation)

$$\int_\Omega (\Lambda g_1) g_2 \,\mathrm{d}x = \int_{\Sigma_0} \psi_{g_1} \psi_{g_2} \,\mathrm{d}\Sigma, \quad \forall g_1, g_2 \in L^2(\Omega). \tag{2.191}$$

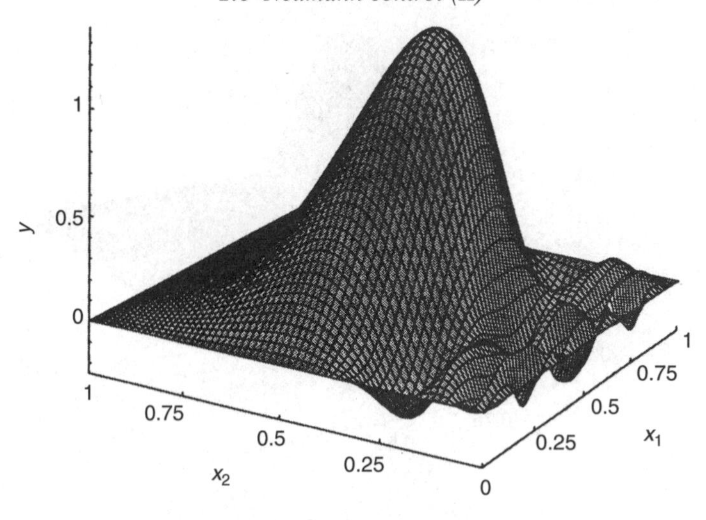

Fig. 2.28. Third test problem: graph of the function y_T^* ($k = 10^5$, $h = \Delta t = 1/64$).

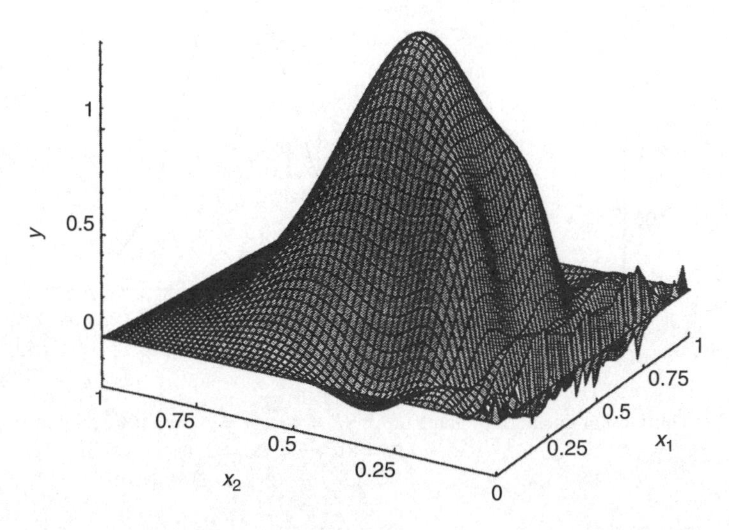

Fig. 2.29. Third test problem: graph of the function y_T^* ($k = 10^7$, $h = \Delta t = 1/64$).

It follows from (2.191) that operator Λ is *symmetric* and *positive semidefinite*; indeed, it follows from the *Mizohata's uniqueness theorem* that operator Λ is *positive definite* (if (2.175) holds, at least). However, operator Λ is not an isomorphism from $L^2(\Omega)$ onto $L^2(\Omega)$ (implying that, in general, we do not have here exact boundary controllability).

Back to (2.188), we observe that, if we denote by f the function $p(T)$ in (2.187), it follows from the definition of operator Λ that we have

$$k^{-1}f + \Lambda f = -y_T. \tag{2.192}$$

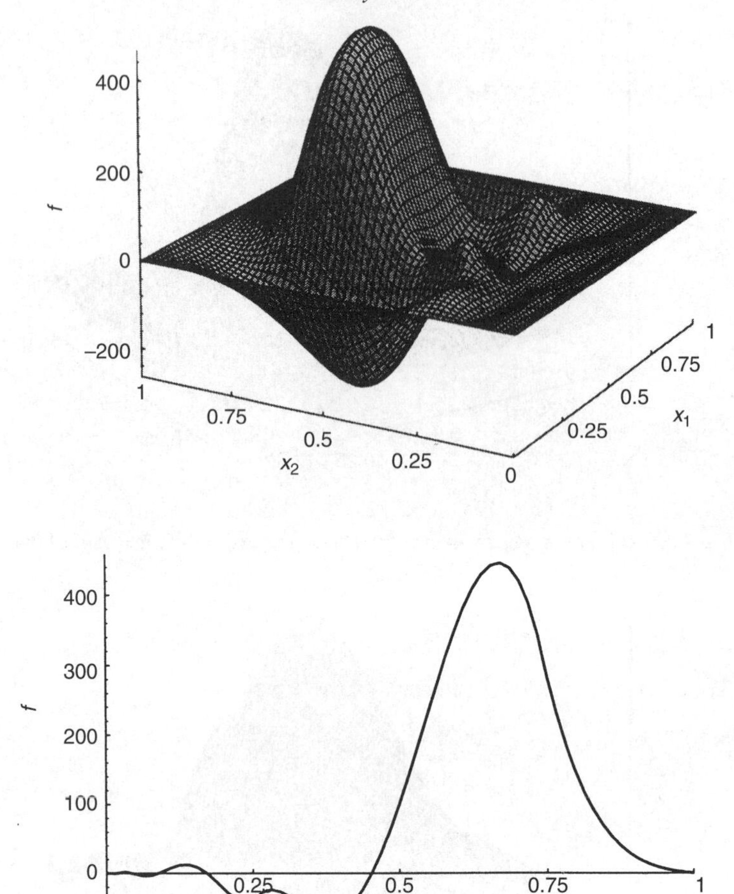

Fig. 2.30. Third test problem: (top) graph of $f_h^{\Delta t}$ ($k = 10^5$, $h = \Delta t = 1/64$); (bottom) graph of $x_2 \to f_h^{\Delta t}(0.5, x_2)$ ($k = 10^5$, $h = \Delta t = 1/64$).

Problem (2.192) is the dual problem of (2.181). From the properties of the operator $k^{-1}I + \Lambda$, problem (2.192) can be solved by a *conjugate gradient algorithm* operating in the space $L^2(\Omega)$; we shall return to this issue in Section 2.9.

The dual problem (2.192) has been obtained by a fairly simple method. Obtaining the dual problem of (2.179) is more complicated. We can use – as already done in previous sections – the *Fenchel–Rockafellar duality theory*; however, in order to introduce (possibly) our readers to other duality techniques we shall derive the dual problem of (2.179) through a *Lagrangian* approach (which is indeed closely related to the Fenchel–Rockafellar method, as shown in, for example, Rockafellar, 1970 and Ekeland and Temam, 1974).

Our starting point is to observe that problem (2.179) is equivalent to

$$\inf_{\{v,z\}} \frac{1}{2} \int_{\Sigma_0} v^2 \, d\Sigma, \tag{2.193}$$

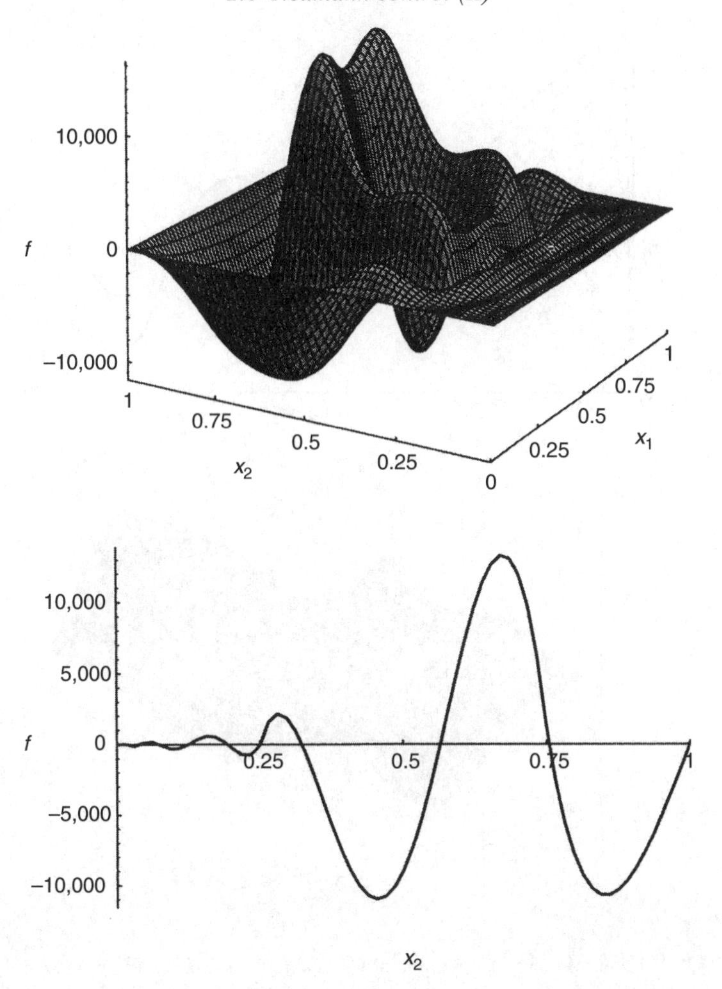

Fig. 2.31. Third test problem: (top) graph of $f_h^{\Delta t}$ ($k = 10^7$, $h = \Delta t = 1/64$); (bottom) graph of $x_2 \to f_h^{\Delta t}(0.5, x_2)$ ($k = 10^7$, $h = \Delta t = 1/64$).

where, in (2.193), the pair $\{v, z\}$ satisfies

$$v \in L^2(\Sigma_0), \tag{2.194}$$

$$z \in y_T + \beta B, \tag{2.195}$$

$$y(T) - z = 0, \tag{2.196}$$

$y(T)$ being obtained from v via the solution of (2.166). The idea here is to "dualize" the linear constraint (2.196) via an appropriate *Lagrangian functional* and then to compute the corresponding *dual functional*. A Lagrangian functional naturally associated with problem (2.193)–(2.196) is defined by

$$\mathcal{L}(v, z; \mu) = \frac{1}{2}\int_{\Sigma_0} v^2 \, d\Sigma + \int_\Omega \mu(y(T) - z) \, dx. \tag{2.197}$$

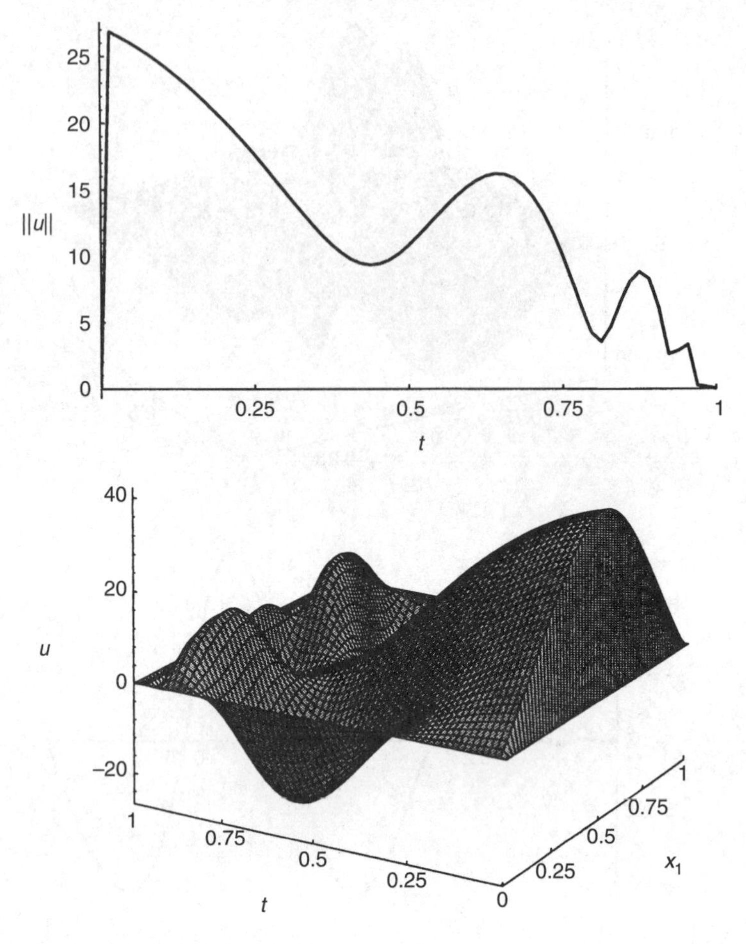

Fig. 2.32. Third test problem: (top) graph of $t \to \|u^*(t)\|_{L^2(\Gamma_0)}$ ($k = 10^5$, $h = \Delta t = 1/64$); (bottom) graph of the computed boundary control ($k = 10^5$, $h = \Delta t = 1/64$).

The dual problem associated with (2.193)–(2.197) is defined by

$$\inf_{\mu \in L^2(\Omega)} J^*(\mu), \tag{2.198}$$

where, in (2.198), the dual functional J^* is defined by

$$J^*(\mu) = -\inf_{\{v,z\}} \mathcal{L}(v, z; \mu), \tag{2.199}$$

where $\{v, z\}$ still satisfies (2.194), (2.195). We clearly have

$$\inf_{\{v,z\}} \mathcal{L}(v, z; \mu)$$
$$= \inf_{v \in L^2(\Sigma_0)} \left(\frac{1}{2} \int_{\Sigma_0} v^2 \, d\Sigma + \int_\Omega y(T)\mu \, dx \right) - \sup_{z \in y_T + \beta B} \int_\Omega \mu z \, dx, \tag{2.200}$$

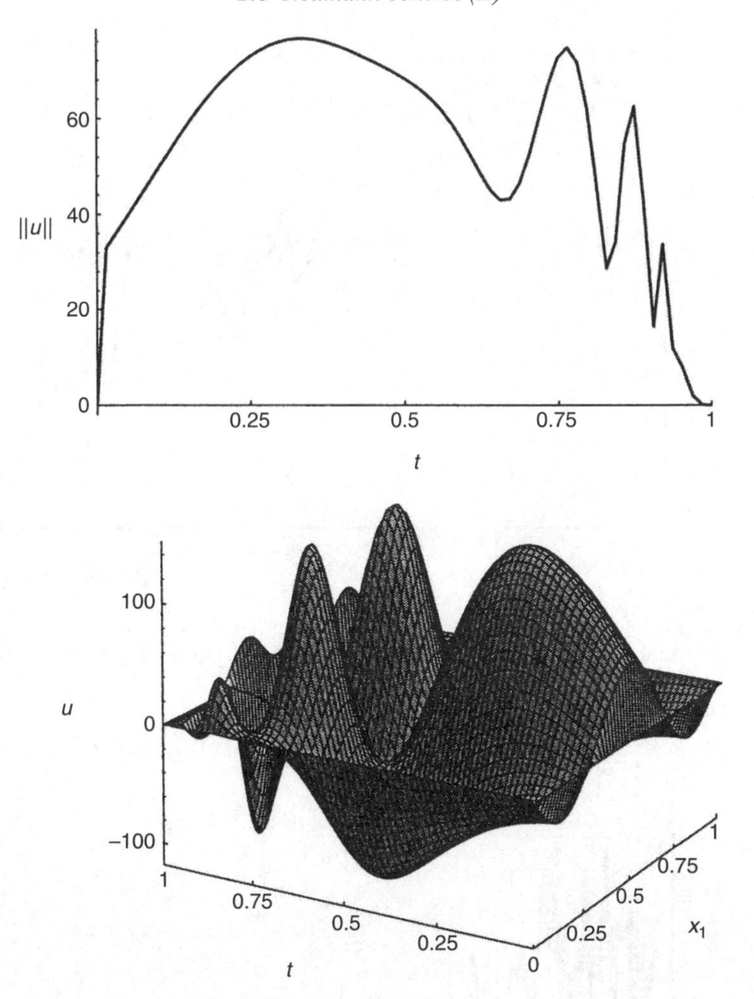

Fig. 2.33. Third test problem: (top) graph of $t \to \|u^*(t)\|_{L^2(\Gamma_0)}$ ($k = 10^7$, $h = \Delta t = 1/64$); (bottom) graph of the computed boundary control ($k = 10^7$, $h = \Delta t = 1/64$).

and then

$$\sup_{z \in y_T + \beta B} \int_\Omega \mu z \, dx = \sup_{z \in y_T + \beta B} \left(\int_\Omega \mu(z - y_T) \, dx + \int_\Omega \mu y_T \, dx \right)$$

$$= \beta \|\mu\|_{L^2(\Omega)} + \int_\Omega \mu y_T \, dx. \tag{2.201}$$

It remains to evaluate

$$\inf_{v \in L^2(\Sigma_0)} \left(\frac{1}{2} \int_{\Sigma_0} v^2 \, d\Sigma + \int_\Omega y(T) \mu \, dx \right); \tag{2.202}$$

indeed, solving the (linear) control problem (2.202) is quite easy since its unique solution u_μ is characterized (see, for example, J. L. Lions, 1968) by the existence of

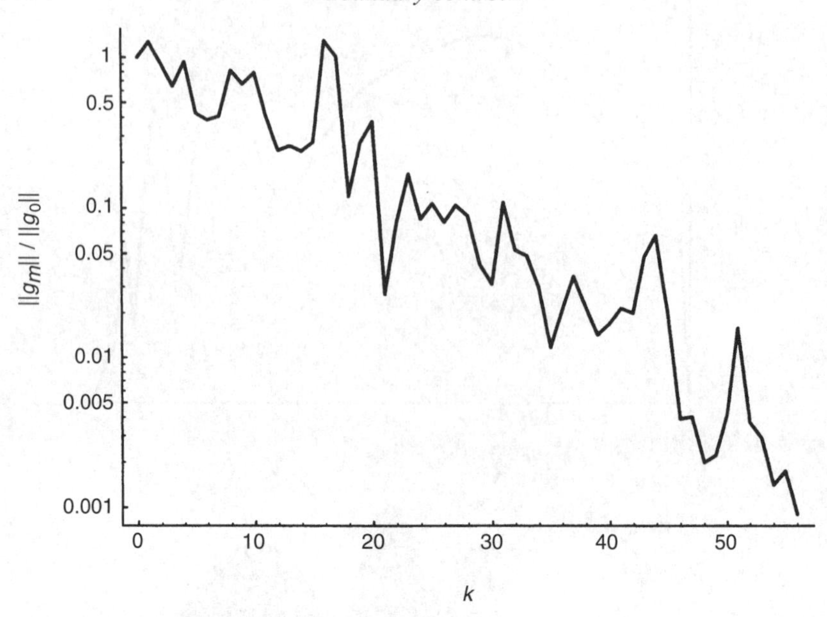

Fig. 2.34. Third test problem: variation of $\|g_m\|_{H_0^1(\Omega)}/\|g_0\|_{H_0^1(\Omega)}$ ($k = 10^5,\ h = \Delta t = 1/64$).

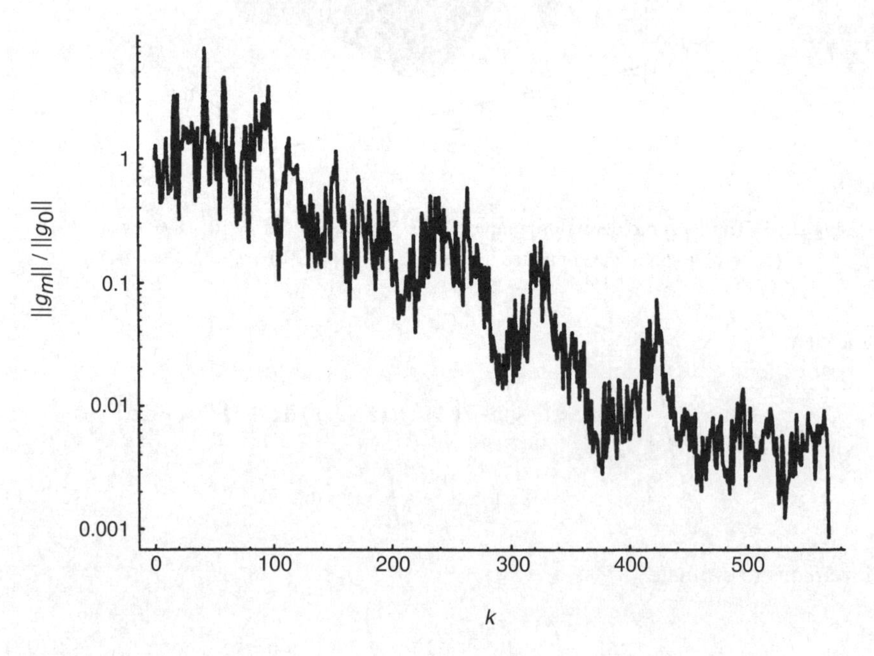

Fig. 2.35. Third test problem: variation of $\|g_m\|_{H_0^1(\Omega)}/\|g_0\|_{H_0^1(\Omega)}$ ($k = 10^7,\ h = \Delta t = 1/64$).

$\{y_\mu, p_\mu\}$ such that

$$u_\mu = -p_\mu|_{\Sigma_0}, \tag{2.203}$$

$$\begin{cases} \dfrac{\partial y_\mu}{\partial t} + \mathcal{A} y_\mu = 0 \text{ in } Q, \\[2ex] \dfrac{\partial y_\mu}{\partial n_\mathcal{A}} = -p_\mu \text{ on } \Sigma_0, \quad \dfrac{\partial y_\mu}{\partial n_\mathcal{A}} = 0 \text{ on } \Sigma \backslash \Sigma_0, \\[2ex] y_\mu(0) = 0, \end{cases} \tag{2.204}$$

$$-\frac{\partial p_\mu}{\partial t} + \mathcal{A}^* p_\mu = 0 \text{ in } Q, \quad \frac{\partial p_\mu}{\partial n_{\mathcal{A}^*}} = 0 \text{ on } \Sigma,\ p_\mu(T) = \mu. \tag{2.205}$$

We have then from (2.202)–(2.205) and from the definition and properties of the operator Λ

$$\inf_{v \in L^2(\Sigma_0)} \left(\frac{1}{2} \int_{\Sigma_0} v^2 \, d\Sigma + \int_\Omega y(T)\mu \, dx \right)$$
$$= \frac{1}{2} \int_{\Sigma_0} p_\mu^2 \, d\Sigma + \int_\Omega y_\mu(T)\mu \, dx = -\frac{1}{2} \int_\Omega (\Lambda\mu)\mu \, dx. \tag{2.206}$$

Combining (2.199), (2.200), (2.201) with (2.206) implies that

$$J^*(\mu) = \frac{1}{2} \int_\Omega (\Lambda\mu)\mu \, dx + \beta \|\mu\|_{L^2(\Omega)} + \int_\Omega y_T \mu \, dx. \tag{2.207}$$

The dual problem to (2.179) is defined then by

$$\inf_{g \in L^2(\Omega)} \left(\frac{1}{2} \int_\Omega (\Lambda g) g \, dx + \beta \|g\|_{L^2(\Omega)} + \int_\Omega y_T g \, dx \right), \tag{2.208}$$

or, equivalently, by the following *variational inequality*:

$$\begin{cases} f \in L^2(\Omega), \\[2ex] \displaystyle\int_\Omega (\Lambda f)(g - f) \, dx + \beta \|g\|_{L^2(\Omega)} - \beta \|f\|_{L^2(\Omega)} \\[2ex] \displaystyle\qquad + \int_\Omega y_T (g - f) \, dx \geq 0, \quad \forall g \in L^2(\Omega). \end{cases} \tag{2.209}$$

Once f is known, obtaining the solution u of problem (2.179) is quite easy, since

$$u = -p|_{\Sigma_0}, \tag{2.210}$$

where, in (2.210), p is the solution of

$$-\frac{\partial p}{\partial t} + \mathcal{A}^* p = 0 \text{ on } Q, \quad \frac{\partial p}{\partial n_{\mathcal{A}^*}} = 0 \text{ on } \Sigma, \quad p(T) = f. \tag{2.211}$$

The numerical solution of problem (2.208), (2.209) will be discussed in Section 2.10.

Remark 2.15 Proving *directly* the existence and uniqueness of the solution f of problem (2.208), (2.209) is not obvious. Actually, proving it without some *regularity* hypothesis on the a_{ij}s (like (2.175)) is still an *open* question.

2.9 Neumann control (III): Conjugate gradient solution of the dual problem (2.192)

We shall address in this section the *iterative solution* of the control problem (2.181), via the solution of its *dual problem* (2.192). From the properties of Λ (*symmetry* and *positive definiteness*) problem (2.192) can be solved by a *conjugate gradient algorithm* operating in the space $L^2(\Omega)$. Such an algorithm is given below; we will use there a *variational description* in order to facilitate *finite element implementations* of the algorithm.

Description of the algorithm

$$f^0 \text{ is given in } L^2(\Omega); \tag{2.212}$$

solve

$$\begin{cases} -\displaystyle\int_\Omega \frac{\partial \psi^0}{\partial t}(t) z \, dx + a(z, \psi^0(t)) = 0, \quad \forall z \in H^1(\Omega), \\ \psi^0(t) \in H^1(\Omega), \quad \textit{a.e. on } (0, T), \quad \psi^0(T) = f^0, \end{cases} \tag{2.213}$$

and then

$$\begin{cases} \displaystyle\int_\Omega \frac{\partial \varphi^0}{\partial t}(t) z \, dx + a(\varphi^0(t), z) = -\int_{\Gamma_0} \psi^0(t) z \, d\Gamma, \quad \forall z \in H^1(\Omega), \\ \varphi^0(t) \in H^1(\Omega), \quad \textit{a.e. on } (0, T), \quad \varphi^0(0) = 0. \end{cases} \tag{2.214}$$

Solve next

$$\begin{cases} g^0 \in L^2(\Omega), \ \forall v \in L^2(\Omega), \\ \displaystyle\int_\Omega g^0 v \, dx = k^{-1} \int_\Omega f^0 v \, dx + \int_\Omega (y_T - \varphi^0(T)) v \, dx, \end{cases} \tag{2.215}$$

and set

$$w^0 = g^0. \tag{2.216}$$

Then, for $n \geq 0$, assuming that f^n, g^n, w^n are known, compute f^{n+1}, g^{n+1}, w^{n+1} as follows:
Solve

$$\begin{cases} -\displaystyle\int_\Omega \frac{\partial \bar{\psi}^n}{\partial t}(t) z \, dx + a(z, \bar{\psi}^n(t)) = 0, \quad \forall z \in H^1(\Omega), \\ \bar{\psi}^n(t) \in H^1(\Omega), \ \text{a.e. on } (0, T), \quad \bar{\psi}^n(T) = w^n, \end{cases} \tag{2.217}$$

and then

$$\begin{cases} \displaystyle\int_\Omega \frac{\partial \bar{\varphi}^n}{\partial t}(t) z \, dx + a(\bar{\varphi}^n(t), z) = -\int_{\Gamma_0} \bar{\psi}^n(t) z \, d\Gamma, \quad \forall z \in H^1(\Omega), \\ \bar{\varphi}^n(t) \in H^1(\Omega), \ \text{a.e. on } (0, T), \quad \bar{\varphi}^n(0) = 0. \end{cases} \tag{2.218}$$

Solve next

$$\begin{cases} \bar{g}^n \in L^2(\Omega), \\ \displaystyle\int_\Omega \bar{g}^n v \, dx = k^{-1} \int_\Omega w^n v \, dx - \int_\Omega \bar{\varphi}^n(T) v \, dx, \quad \forall v \in L^2(\Omega), \end{cases} \tag{2.219}$$

and compute

$$\rho_n = \int_\Omega |g^n|^2 \, dx \Big/ \int_\Omega \bar{g}^n w^n \, dx. \tag{2.220}$$

Set then

$$f^{n+1} = f^n - \rho_n w^n, \tag{2.221}$$

$$g^{n+1} = g^n - \rho_n \bar{g}^n. \tag{2.222}$$

If $\|g^{n+1}\|_{L^2(\Omega)} / \|g^0\|_{L^2(\Omega)} \leq \epsilon$, take $f = f^{n+1}$; else, compute

$$\gamma_n = \int_\Omega |g^{n+1}|^2 \, dx \Big/ \int_\Omega |g^n|^2 \, dx, \tag{2.223}$$

and update w^n via

$$w^{n+1} = g^{n+1} + \gamma_n w^n. \tag{2.224}$$

Do $n = n + 1$ and go to (2.217).

In (2.212)–(2.224), the bilinear functional $a(\cdot, \cdot)$ is defined by (2.173).

It is fairly easy to derive a fully discrete analogue of algorithm (2.212)–(2.224), obtained by combining finite elements for the space discretization and finite differences for the time discretization. We shall then obtain a variation of algorithm (2.125)–(2.151) (see Section 2.5), which is itself the fully discrete analogue of algorithm (2.42)–(2.54) (see Section 2.3). Actually, algorithm (2.212)–(2.224) is easier to implement than (2.42)–(2.54) since it operates in $L^2(\Omega)$, instead of $H_0^1(\Omega)$; no preconditioning is required, thus.

2.10 Neumann control (IV): Iterative solution of the dual problem (2.208), (2.209)

Problem (2.208), (2.209) can also be formulated as

$$-y_T \in \Lambda f + \beta \partial j(f), \tag{2.225}$$

which is a *multivalued* equation in $L^2(\Omega)$. In (2.225), $\partial j(\cdot)$ is the subgradient of the *convex functional* $j(\cdot)$ defined by

$$j(g) = \left(\int_\Omega |g|^2 \, dx \right)^{1/2}, \quad \forall g \in L^2(\Omega).$$

As done in preceding sections we associate with the (kind of) *elliptic equation* (2.225) the *initial value problem*

$$\begin{cases} \dfrac{\partial f}{\partial \tau} + \Lambda f + \beta \partial j(f) + y_T \ni 0, \\ f(0) = f_0. \end{cases} \tag{2.226}$$

To obtain the *steady-state* solution of (2.226), that is, the solution of (2.225), we shall use the following algorithm obtained, from (2.226), by application of the *Peaceman–Rachford time-discretization scheme* (where $\Delta\tau (> 0)$ is a pseudotime-discretization step):

$$f^0 = f_0; \tag{2.227}$$

then, for $n \geq 0$, *compute* $f^{n+1/2}$ *and* f^{n+1}, *from* f^n, *via*

$$\frac{f^{n+1/2} - f^n}{\frac{\Delta\tau}{2}} + \Lambda f^n + \beta \partial j(f^{n+1/2}) + y_T \ni 0, \tag{2.228}$$

$$\frac{f^{n+1} - f^{n+1/2}}{\frac{\Delta\tau}{2}} + \Lambda f^{n+1} + \beta \partial j(f^{n+1/2}) + y_T \ni 0. \tag{2.229}$$

Problem (2.229) can be reformulated as

$$\frac{f^{n+1} - 2f^{n+1/2} + f^n}{\Delta\tau/2} + \Lambda f^{n+1} = \Lambda f^n; \tag{2.230}$$

problem (2.230) being a simple variation of problem (2.192) can be solved by an algorithm similar to (2.212)–(2.224). On the other hand, problem (2.228) can be (easily) solved by the methods used in Chapter 1, Section 1.8.8 to solve problems (1.243), (1.246), (1.248) which are simple variants of problem (2.228).

2.11 Neumann control of unstable parabolic systems: a numerical approach

2.11.1 Generalities. Synopsis

So far we have followed very closely our original article from *Acta Numerica*, the additions we made being relatively minor (in size, but not in importance).

We are going to encounter now our first major addition, namely, the *Neumann control* of systems governed by *unstable parabolic equations* (or systems of such equations). This section will follow closely He and Glowinski (1998), an article dedicated to H.B. Keller, on the occasion of his 70th birthday. It is well-known in the scientific community that H.B. Keller has given a lot of attention to the *Bratu problem*

$$-\Delta\phi = \lambda\, e^{\phi}, \tag{2.231}$$

and has greatly contributed to improve what was known about it (see, for example Keller and Langford, 1972; Keener and Keller, 1973, 1974; Keller, 1977; Decker and Keller, 1980; Chan and Keller, 1982; Keller, 1982; Glowinski, Keller, and Reinhart 1985); there are several reasons for this attention, a most important one being that (2.231) completed by appropriate boundary conditions will provide various types of *bifurcation phenomena* and also that (2.231) or variations of it play a significant role in the modeling of *combustion phenomena* as shown for example in the book by Bebernes and Eberly (1989). Our goal here is to control via *Neumann boundary actions* the evolution of systems modeled by time-dependent variants of (2.231) and in particular to avoid *blow-up phenomena*. The following part of this section is organized as follows. In Section 2.11.2, we consider a nonlinear model problem which will provide, after linearization, a motivation to address the solution of the control problems considered in this section and whose precise formulation is given in Section 2.11.3. In Section 2.11.4, we derive first the *optimality conditions* satisfied by the solution of the control problems and discuss then the *iterative solution* of these problems by *conjugate gradient algorithms*. In Section 2.11.5, we discuss the *time discretization* of the control problems and derive the corresponding optimality conditions. The *space approximation* by *finite-element* methods is discussed in Section 2.11.6, where we describe also the iterative solution of the *fully discrete* problems by a conjugate gradient algorithm. In Section 2.11.7, we analyze the results of some *numerical experiments* and, on the basis of these results, introduce *chattering boundary control* in order to enhance controllability, as shown by further experiments. In Section 2.11.8, we generalize the methods discussed in the previous sub-sections in order to solve boundary control problems for systems of unstable parabolic equations. Finally, the controls computed in Section 2.11.7 from linear models are applied in Section 2.11.9 on the original nonlinear models in order to validate the linearization based approached followed in this section.

2.11.2 Motivation

2.11.2.1 A family of unstable nonlinear parabolic equations

Let Ω be a bounded domain of $\mathbb{R}^d$ (typical values of d being 1, 2, and 3); as in the previous sections of this book we denote by Γ the boundary of Ω and by $x = \{x_i\}_{i=1}^d$ the generic point of $\mathbb{R}^d$. We suppose that during the time interval $(0, T)$ (with $0 < T \leq +\infty$) a physical phenomenon is taking place in Ω, modeled by the following time dependent variant of equation (2.231):

$$\frac{\partial \phi}{\partial t} - \nu \Delta \phi + \mathbf{V} \cdot \nabla \phi = \lambda e^{\phi} - C \qquad \text{in } Q \; (= \Omega \times (0, T)), \tag{2.232}$$

$$\phi(0) = \phi_0, \tag{2.233}$$

$$\nu \frac{\partial \phi}{\partial n} = 0 \qquad \text{on } \Sigma \; (= \Gamma \times (0, T)), \tag{2.234}$$

where ν, λ, and C are *positive constants*, and where $\mathbf{V}$ is a *divergence-free* vector-valued function (that is, $\nabla \cdot \mathbf{V} = 0$), independent of t (for simplicity) and such that $\mathbf{V} \cdot \mathbf{n} = 0$, where $\mathbf{n}$ is the unit outward normal vector at Γ. Problem (2.232)–(2.234) is of the *reaction-advection-diffusion* type. Variants of the above problem are discussed at length in Bebernes and Eberly (1989), since they are related to the modeling of *combustion phenomena*; as before, in relation (2.233) and elsewhere in this section, we use the notation $\phi(t)$ for the function $x \to \phi(x, t)$. Let us define the *constant* function ϕ_s by

$$\phi_s = \ln\left(\frac{C}{\lambda}\right); \tag{2.235}$$

the function ϕ_s is clearly a *steady-state solution* of the system (2.232)–(2.234). Let us denote by $\delta\phi_s$ a perturbation of ϕ_s; we suppose that $\delta\phi_s$ is constant in x. The perturbed solution of (2.232)–(2.234) corresponding to $\phi_0 = \phi_s + \delta\phi_s$ is also constant in x and verifies the following *ordinary differential equation* and *initial condition*

$$\frac{d\phi}{dt} = \lambda e^{\phi} - C, \quad \phi(0) = \phi_0 \; (= \phi_s + \delta\phi_s). \tag{2.236}$$

The *closed-form* solution of (2.236) is easy to compute and is given by

$$\phi(t) = \phi_s - Ct - \ln\left(e^{-Ct} + e^{\phi_s - \phi_0} - 1\right). \tag{2.237}$$

It follows from (2.237) that:

(i) If $\phi_0 - \phi_s \; (= \delta\phi_s) \; > 0$, the function ϕ is defined only for $t \in [0, t^*)$, where $t^* = -\frac{1}{C} \ln\left(1 - e^{\phi_s - \phi_0}\right)$, and

$$\lim_{t \to t^*} \phi(t) = +\infty; \tag{2.238}$$

we have therefore a finite time *blow-up phenomenon*.

(ii) If $\phi_0 - \phi_s \; (= \delta\phi_s) < 0$, the function ϕ is defined in $\mathbb{R}_+$ and we have

$$\lim_{t \to +\infty} \phi(t) = -\infty. \tag{2.239}$$

Our goal in this section is to stabilize the system modeled by (2.232)–(2.234) via *Neumann boundary* controls. The approach taken here is the following classical one, namely,

(a) Linearize the model (2.232)–(2.234) in the neighborhood of ϕ_s, assuming that it is feasible.

(b) Compute an optimal control for the linearized model.
(c) Apply the above control to the nonlinear system.

2.11.2.2 Linearization of problem (2.232)–(2.234) in the neighborhood of ϕ_s

Let us consider a "small" variation $\delta\phi_s$ of ϕ_s (possibly not constant in x, this time); since $\lambda e^{\phi_s} = C$, the perturbation $\delta\phi$ of the steady-state solution ϕ_s satisfies approximately the following linear model

$$\frac{\partial}{\partial t}\delta\phi - \nu\Delta\delta\phi + \mathbf{V}\cdot\nabla\delta\phi = C\delta\phi \quad \text{in } Q, \tag{2.240}$$

$$\delta\phi(0) = \delta\phi_s, \tag{2.241}$$

$$\nu\frac{\partial}{\partial n}\delta\phi = 0 \quad \text{on } \Sigma. \tag{2.242}$$

An analysis similar to the one done in Section 2.11.2.1 for the system (2.232)–(2.234) will show that the system (2.240)–(2.242) is unstable in general and can develop blow-up phenomena (in infinite time); it is clear that the model (2.240)–(2.242) is no longer valid if $\delta\phi$ becomes too large. The (simple) idea behind considering the linearized model (2.240)–(2.242) is to use it to compute a control action preventing $\delta\phi$ from becoming too large (and possibly driving $\delta\phi$ to zero) and hope that the computed control will also stabilize the original nonlinear system. The controlled variant of system (2.240)–(2.242) that we consider is defined by

$$\frac{\partial}{\partial t}\delta\phi - \nu\Delta\delta\phi + \mathbf{V}\cdot\nabla\delta\phi = C\delta\phi \quad \text{in } Q, \tag{2.243}$$

$$\delta\phi(0) = \delta\phi_s, \tag{2.244}$$

$$\nu\frac{\partial}{\partial n}\delta\phi = v \quad \text{on } \sigma\ (= \gamma\times(0,T)), \tag{2.245}$$

$$\nu\frac{\partial}{\partial n}\delta\phi = 0 \quad \text{on } \Sigma\setminus\sigma, \tag{2.246}$$

where γ is a subset (possibly multiconnected) of Γ, and where v belongs to an appropriate control space.

Remark 2.16 Nonlinear parabolic equations enjoying blow-up properties have motivated a very large number of publications. Concentrating on books only, let us mention again Bebernes and Eberly (1989), and also Samarskii, Galaktionov, Kurdyumov, and Mikhailov (1995) whose Chapter 7 includes finite difference methods for the numerical solution of those unstable parabolic equations; see also the related references in these two books.

Remark 2.17 Numerical methods for the control of systems modeled by *nonlinear parabolic equations* are discussed in the book by Neitaanmäki and Tiba (1994); see also the references therein.

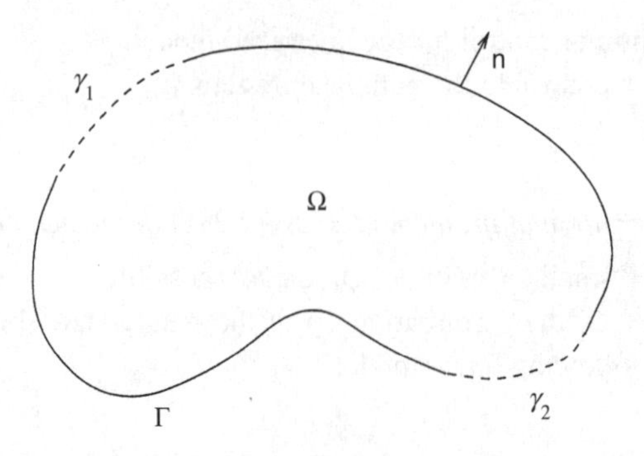

Fig. 2.36. A domain Ω with $\gamma = \gamma_1 \cup \gamma_2$.

2.11.3 Formulation of the Neumann boundary control problem for the linearized models

We shall suppose from now on that the number of connected components of γ is finite, implying (with obvious notation) that

$$\gamma = \bigcup_{m=1}^{M} \gamma_m; \tag{2.247}$$

we shall assume that

$$\int_{\gamma_m} \mathrm{d}\Gamma > 0, \quad \forall m = 1, \ldots, M;$$

see Figure 2.36, for a particular but typical geometrical configuration where $M = 2$.

Concentrating for the time being on *finite-horizon control*, we consider the evolution of the system modeled by (2.232)–(2.234) on the time interval $[0, T]$, with $0 < T < +\infty$. Using the notation $y = \delta\phi$, we shall (try to) stabilize (2.243)–(2.246) via the following control formulation:

$$\min_{\mathbf{v} \in \mathcal{U}} J(\mathbf{v}), \tag{2.248}$$

where $\mathbf{v} = \{v_m\}_{m=1}^{M}$ and

$$\mathcal{U} = L^2(0, T; \mathbb{R}^M), \tag{2.249}$$

$$J(\mathbf{v}) = \frac{1}{2}\int_0^T |\mathbf{v}|^2 \,\mathrm{d}t + \frac{1}{2}k_1 \int_Q y^2 \,\mathrm{d}x \,\mathrm{d}t + \frac{1}{2}k_2 \int_\Omega |y(T)|^2 \,\mathrm{d}x, \tag{2.250}$$

with $\mathrm{d}x = \mathrm{d}x_1 \cdots \mathrm{d}x_d$, $k_1 \geq 0$, $k_2 \geq 0$, $k_1 + k_2 > 0$, and

$$|\mathbf{v}| = \left(\sum_{m=1}^{M} |v_m|^2 \right)^{\frac{1}{2}}, \tag{2.251}$$

and finally y obtained from $\mathbf{v}$ via the solution of the following linear parabolic problem:

$$\frac{\partial y}{\partial t} - \nu \Delta y + \mathbf{V} \cdot \nabla y - Cy = 0 \quad \text{in } Q, \tag{2.252}$$

$$y(0) = y_0 \ (= \delta\phi_s), \tag{2.253}$$

$$\nu \frac{\partial y}{\partial n} = 0 \quad \text{on } \Sigma \setminus \sigma, \tag{2.254}$$

$$\forall\, m = 1, \ldots, M, \quad \nu \frac{\partial y}{\partial n} = v_m \ \text{ on } \sigma_m \ (= \gamma_m \times (0, T)). \tag{2.255}$$

Remark 2.18 Other norms than the one defined in (2.251) can be used for the vector $\mathbf{v}$.

Suppose that $\mathbf{V} \in (L^\infty(\Omega))^d$ and that $y_0 \in L^2(\Omega)$; using the results in J.L. Lions (1968, 1971), it can be shown that the optimal control problem (2.248) has a unique solution. The numerical solution of the above problem will be addressed in the following sections together with its relevance to the stabilization of the system (2.232)–(2.234).

Remark 2.19 The control space $\mathcal{U}$ consists of time-dependent functions which are constant in space on each γ_m. Actually, the above existence and uniqueness result still holds if in (2.248) we replace $L^2(0, T; \mathbb{R}^M)$ by $L^2(\sigma)$. Similarly, the computational techniques described in the following sections would be easily modified to handle the case $\mathcal{U} = L^2(\sigma)$.

2.11.4 Optimality conditions and conjugate gradient solution for problem (2.248)

2.11.4.1 Generalities and synopsis

Let us denote by $J'(\mathbf{v})$ the differential of J at $\mathbf{v} \in \mathcal{U}$. Since $\mathcal{U}$ is a *Hilbert space* for the scalar product defined by

$$(\mathbf{v}, \mathbf{w})_{\mathcal{U}} = \int_0^T \mathbf{v}(t) \cdot \mathbf{w}(t)\, dt, \tag{2.256}$$

with $\mathbf{v}(t) \cdot \mathbf{w}(t) = \sum_{m=1}^{M} v_m(t) w_m(t)$, the action $\langle J'(\mathbf{v}), \mathbf{w} \rangle$ of $J'(\mathbf{v})$ on $\mathbf{w} \in \mathcal{U}$ can also be written as

$$\langle J'(\mathbf{v}), \mathbf{w} \rangle = \int_0^T J'(\mathbf{v})(t) \cdot \mathbf{w}(t)\, dt, \quad \forall \mathbf{v}, \mathbf{w} \in \mathcal{U}, \tag{2.257}$$

with $J'(\mathbf{v}) \in \mathcal{U}$. If $\mathbf{u}$ is the solution of problem (2.248), it is characterized (from *convexity* arguments; see, for example, J.L. Lions, 1968, 1971) by

$$J'(\mathbf{u}) = \mathbf{0}. \tag{2.258}$$

Since the cost function J is *quadratic* and the state model (2.252)–(2.255) is *linear*, $J'(\mathbf{v})$ is in fact an *affine* function of $\mathbf{v}$, implying in turn from (2.258) that $\mathbf{u}$ is the solution of a *linear* equation in the control space $\mathcal{U}$. In abstract form, equation (2.258) can be written as

$$J'(\mathbf{u}) - J'(\mathbf{0}) = -J'(\mathbf{0}), \tag{2.259}$$

and from the properties (to be shown in Section 2.11.4.4) of the operator $\mathbf{v} \to J'(\mathbf{v}) - J'(\mathbf{0})$, problem (2.258), (2.259) can be solved by a *conjugate gradient algorithm* operating in the space $\mathcal{U}$; we shall describe this algorithm in Section 2.11.4.5. The practical implementation of the above algorithm requires the explicit knowledge of $J'(\mathbf{v})$; the calculation of $J'(\mathbf{v})$ will be discussed in the following paragraph.

2.11.4.2 Calculation of $J'(\mathbf{v})$

In order to compute $J'(\mathbf{v})$ we shall use a (formal) *perturbation analysis*, which can be make rigorous by using for example the methods in J.L. Lions (1968, 1971). Let us consider $\mathbf{v} \in \mathcal{U}$ and then a "small" perturbation $\delta\mathbf{v}$ of $\mathbf{v}$. We have then, with obvious notation

$$\delta J(\mathbf{v}) = \int_0^T J'(\mathbf{v}) \cdot \delta\mathbf{v}\, \mathrm{d}t = \int_0^T \mathbf{v} \cdot \delta\mathbf{v}\, \mathrm{d}t + k_1 \int_Q y\delta y\, \mathrm{d}x\, \mathrm{d}t + k_2 \int_\Omega y(T)\delta y(T)\, \mathrm{d}x, \tag{2.260}$$

where δy is the solution of

$$\frac{\partial}{\partial t}\delta y - \nu\Delta\delta y + \mathbf{V} \cdot \nabla\delta y - C\delta y = 0 \quad \text{in } Q, \tag{2.261}$$

$$\delta y(0) = 0, \tag{2.262}$$

$$\nu\frac{\partial}{\partial n}\delta y = 0 \quad \text{on } \Sigma \setminus \sigma, \tag{2.263}$$

$$\forall m = 1, \cdots, M, \quad \nu\frac{\partial}{\partial n}\delta y = \delta v_m \quad \text{on } \sigma_m. \tag{2.264}$$

Let us introduce now a "smooth" function p defined over $\overline{Q}$. Multiplying both sides of (2.261) by p and integrating by parts over Q, we obtain

$$\begin{aligned}\int_\Omega p(T)\,\delta y(T)\, \mathrm{d}x - \int_\Omega p(0)\,\delta y(0)\, \mathrm{d}x - \int_Q \frac{\partial p}{\partial t}\,\delta y\, \mathrm{d}x\, \mathrm{d}t - \nu\int_Q \Delta p\,\delta y\, \mathrm{d}x\, \mathrm{d}t \\ + \nu\int_\Sigma \left(\frac{\partial p}{\partial n}\,\delta y - p\frac{\partial}{\partial n}\delta y\right)\, \mathrm{d}\Sigma + \int_\Sigma p\mathbf{V} \cdot \mathbf{n}\,\delta y\, \mathrm{d}\Sigma \\ - \int_Q (\mathbf{V} \cdot \nabla p + p\nabla \cdot \mathbf{V})\,\delta y\, \mathrm{d}x\, \mathrm{d}t - C\int_Q p\,\delta y\, \mathrm{d}x\, \mathrm{d}t = 0.\end{aligned} \tag{2.265}$$

Taking into account (2.262), (2.264), and $\nabla \cdot \mathbf{V} = 0$ in Ω, $\mathbf{V} \cdot \mathbf{n} = 0$ on Γ, (2.265) reduces to

$$\int_\Omega p(T)\,\delta y(T)\,\mathrm{d}x + \int_Q \left(-\frac{\partial p}{\partial t} - \nu\Delta p - \mathbf{V}\cdot\nabla p - Cp\right)\delta y\,\mathrm{d}x\,\mathrm{d}t$$
$$+\nu\int_\Sigma \frac{\partial p}{\partial n}\delta y\;\mathrm{d}\Sigma = \int_0^T \sum_{m=1}^M \left(\int_{\gamma_m} p\;\mathrm{d}\Gamma\right)\delta v_m\,\mathrm{d}t. \quad (2.266)$$

Suppose now that the function p is chosen so that it satisfies the following *adjoint system*:

$$-\frac{\partial p}{\partial t} - \nu\Delta p - \mathbf{V}\cdot\nabla p - Cp = k_1\,y \quad \text{in } Q, \quad (2.267)$$

$$p(T) = k_2\,y(T), \quad (2.268)$$

$$\nu\frac{\partial p}{\partial n} = 0 \qquad \text{on } \Sigma; \quad (2.269)$$

this choice makes sense since problem (2.267)–(2.268) has a unique solution. Assuming that p verifies (2.267)–(2.269), we have then from (2.260) and (2.266)–(2.269) that

$$\int_0^T J'(\mathbf{v})\cdot\delta\mathbf{v}\,\mathrm{d}t = \int_0^T \sum_{m=1}^M \left(v_m + \int_{\gamma_m} p\;\mathrm{d}\Gamma\right)\delta v_m\,\mathrm{d}t, \quad (2.270)$$

which implies that $J'(\mathbf{v})$ can be identified with the vector-valued function of t defined by

$$J'(\mathbf{v}) = \left\{v_m + \int_{\gamma_m} p\;\mathrm{d}\Gamma\right\}_{m=1}^M. \quad (2.271)$$

2.11.4.3 Optimality conditions for problem (2.248)

Let $\mathbf{u} = \{u_m\}_{m=1}^M$ be the solution of problem (2.248) and let us denote by y (respectively, p) the corresponding solution of the state system (2.252)–(2.255) (respectively, of the adjoint system (2.267)–(2.269)). It follows from Section 2.11.4.2 that $J'(\mathbf{u}) = \mathbf{0}$ is equivalent to the following (optimality) system:

$$u_m = -\int_{\gamma_m} p\;\mathrm{d}\Gamma, \quad \forall\, m = 1,\ldots,M, \quad (2.272)$$

$$\frac{\partial y}{\partial t} - \nu\Delta y + \mathbf{V}\cdot\nabla y - Cy = 0 \quad \text{in } Q, \quad (2.273)$$

$$y(0) = y_0, \quad (2.274)$$

$$\nu \frac{\partial y}{\partial n} = 0 \quad \text{on } \Sigma \setminus \sigma, \tag{2.275}$$

$$\forall\, m = 1, \ldots, M, \quad \nu \frac{\partial y}{\partial n} = u_m \text{ on } \sigma_m, \tag{2.276}$$

$$-\frac{\partial p}{\partial t} - \nu \Delta p - \mathbf{V} \cdot \nabla p - Cp = k_1 y \quad \text{in } Q, \tag{2.277}$$

$$p(T) = k_2 y(T), \tag{2.278}$$

$$\nu \frac{\partial p}{\partial n} = 0 \quad \text{on } \Sigma. \tag{2.279}$$

Conversely, it can be shown (see, for example, J.L. Lions, 1968, 1971) that the system (2.272)–(2.279) characterizes $\mathbf{u} = \{u_m\}_{m=1}^M$ as the solution (necessarily unique here) of the control problem (2.248). The optimality conditions (2.272)–(2.279) will play a crucial role concerning the iterative solution of the control problem (2.248).

Remark 2.20 Actually, the above optimality system plays also a fundamental role in order to derive a *Riccati equation* based *recursive solution method à la Kalman* as shown in, for example, J.L. Lions (1968, 1971); this issue was briefly addressed in Chapter 1, Section 1.13, and we shall return to it in Section 2.12.

2.11.4.4 A functional equation satisfied by the optimal control

Our goal here is to show that the optimality condition $J'(\mathbf{u}) = \mathbf{0}$ can also be written as

$$\mathbf{Au} = \boldsymbol{\beta}, \tag{2.280}$$

where the linear operator $\mathbf{A}$ is a *strongly elliptic* and *symmetric isomorphism* from $\mathcal{U}$ onto itself (an automorphism of $\mathcal{U}$) and where $\boldsymbol{\beta} \in \mathcal{U}$. A "candidate" for $\mathbf{A}$ is the linear operator from $\mathcal{U}$ into itself, defined by

$$\mathbf{Av} = \left\{ v_m + \int_{\gamma_m} p(\mathbf{v})\, d\Gamma \right\}_{m=1}^M, \tag{2.281}$$

where $p(\mathbf{v})$ is obtained from $\mathbf{v}$ via the successive solution of the following two parabolic problems:

$$\frac{\partial}{\partial t} y(\mathbf{v}) - \nu \Delta y(\mathbf{v}) + \mathbf{V} \cdot \nabla y(\mathbf{v}) - Cy(\mathbf{v}) = 0 \quad \text{in } Q, \tag{2.282}$$

$$y(\mathbf{v})(0) = 0, \tag{2.283}$$

$$\nu \frac{\partial}{\partial n} y(\mathbf{v}) = 0 \quad \text{on } \Sigma \setminus \sigma, \tag{2.284}$$

$$\forall m = 1, \cdots, M, \quad \nu \frac{\partial}{\partial n} y(\mathbf{v}) = v_m \text{ on } \sigma_m, \tag{2.285}$$

which is *forward in time*, and

$$-\frac{\partial}{\partial t}p(\mathbf{v}) - \nu\Delta p(\mathbf{v}) - \mathbf{V}\cdot\nabla p(\mathbf{v}) - Cp(\mathbf{v}) = k_1 y(\mathbf{v}) \text{ in } Q, \tag{2.286}$$

$$p(\mathbf{v})(T) = k_2 y(\mathbf{v})(T), \tag{2.287}$$

$$\nu\frac{\partial}{\partial n}p(\mathbf{v}) = 0 \qquad \text{on } \Sigma, \tag{2.288}$$

which is *backward in time*.

Proposition 2.21 *The operator* $\mathbf{A}$ *is symmetric and is a strongly elliptic isomorphism from* $\mathcal{U}$ *onto itself.*

Proof. Consider $\mathbf{v}^1$ and $\mathbf{v}^2$ belonging to $\mathcal{U}$ and define y_i and p_i by

$$y_i = y(\mathbf{v}^i) \quad \text{and} \quad p_i = p(\mathbf{v}^i), \quad \forall i = 1,2; \tag{2.289}$$

we have then

$$\begin{aligned}\int_0^T (\mathbf{A}\mathbf{v}^1)\cdot\mathbf{v}^2\,\mathrm{d}t &= \int_0^T \left(\sum_{m=1}^M v_m^1 v_m^2\right)\mathrm{d}t \\ &\quad + \int_0^T \left(\sum_{m=1}^M \int_{\gamma_m} p_1 v_m^2\,\mathrm{d}\Gamma\right)\mathrm{d}t \\ &= \int_0^T \left(\sum_{m=1}^M v_m^1 v_m^2\right)\mathrm{d}t + \nu\int_\Sigma p_1\frac{\partial y_2}{\partial n}\,\mathrm{d}\Sigma.\end{aligned} \tag{2.290}$$

We have on the other hand, using integration by parts,

$$\begin{aligned}0 &= \int_Q p_1\left(\frac{\partial y_2}{\partial t} - \nu\Delta y_2 + \mathbf{V}\cdot\nabla y_2 - Cy_2\right)\mathrm{d}x\,\mathrm{d}t \\ &\quad - \int_Q y_2\left(-\frac{\partial p_1}{\partial t} - \nu\Delta p_1 - \mathbf{V}\cdot\nabla p_1 - Cp_1 - k_1 y_1\right)\mathrm{d}x\,\mathrm{d}t \\ &= k_1\int_Q y_1 y_2\,\mathrm{d}x\,\mathrm{d}t + k_2\int_\Omega y_1(T)y_2(T)\,\mathrm{d}x - \nu\int_\Sigma p_1\frac{\partial y_2}{\partial n}\,\mathrm{d}\Sigma.\end{aligned} \tag{2.291}$$

Combining relations (2.290) and (2.291), we obtain that

$$\begin{aligned}\int_0^T (\mathbf{A}\mathbf{v}^1)\cdot\mathbf{v}^2\,\mathrm{d}t &= \int_0^T \mathbf{v}^1\cdot\mathbf{v}^2\,\mathrm{d}t \\ &\quad + k_1\int_Q y_1 y_2\,\mathrm{d}x\,\mathrm{d}t + k_2\int_\Omega y_1(T)y_2(T)\,\mathrm{d}x.\end{aligned} \tag{2.292}$$

Relation (2.292) implies obviously the *symmetry* of $\mathbf{A}$; we also have

$$\int_0^T (\mathbf{A}\mathbf{v}) \cdot \mathbf{v}\, dt \geq \int_0^T |\mathbf{v}|^2\, dt, \quad \forall \mathbf{v} \in \mathcal{U}, \tag{2.293}$$

which implies the *strong ellipticity* of $\mathbf{A}$ over $\mathcal{U}$ (to prove the *continuity* of $\mathbf{A}$ over $\mathcal{U}$, we can use for example the results in J.L. Lions (1968, 1971) on the regularity properties of the solution of linear parabolic equations for Neumann boundary conditions). The linear operator $\mathbf{A}$, being continuous and strongly elliptic over $\mathcal{U}$, is an automorphism of $\mathcal{U}$. To identify the right-hand side $\boldsymbol{\beta}$ of (2.280), we introduce Y_0 and P_0 defined as the solutions of

$$\frac{\partial Y_0}{\partial t} - \nu \Delta Y_0 + \mathbf{V} \cdot \nabla Y_0 - C Y_0 = 0 \quad \text{in } Q, \tag{2.294}$$

$$Y_0(0) = y_0, \tag{2.295}$$

$$\nu \frac{\partial Y_0}{\partial n} = 0 \quad \text{on } \Sigma, \tag{2.296}$$

and

$$-\frac{\partial P_0}{\partial t} - \nu \Delta P_0 - \mathbf{V} \cdot \nabla P_0 - C P_0 = k_1\, Y_0 \quad \text{in } Q, \tag{2.297}$$

$$P_0(T) = k_2\, Y_0(T), \tag{2.298}$$

$$\nu \frac{\partial P_0}{\partial n} = 0 \quad \text{on } \Sigma. \tag{2.299}$$

Suppose that y and p verify the optimality system (2.272)–(2.278), and define $\overline{y}$ and $\overline{p}$ by

$$\overline{y} = y - Y_0 \quad \text{and} \quad \overline{p} = p - P_0. \tag{2.300}$$

By subtraction, we obtain

$$\frac{\partial}{\partial t}\overline{y} - \nu \Delta \overline{y} + \mathbf{V} \cdot \nabla \overline{y} - C \overline{y} = 0 \quad \text{in } Q, \tag{2.301}$$

$$\overline{y}(0) = 0, \tag{2.302}$$

$$\nu \frac{\partial}{\partial n}\overline{y} = 0 \quad \text{on } \Sigma \backslash \sigma, \tag{2.303}$$

$$\forall m = 1, \ldots, M, \quad \nu \frac{\partial}{\partial n}\overline{y} = u_m \quad \text{on } \sigma_m, \tag{2.304}$$

and

$$-\frac{\partial}{\partial t}\overline{p} - \nu \Delta \overline{p} - \mathbf{V} \cdot \nabla \overline{p} - C \overline{p} = k_1\, \overline{y} \quad \text{in } Q, \tag{2.305}$$

$$\overline{p}(T) = k_2\, \overline{y}(T), \tag{2.306}$$

$$\nu \frac{\partial}{\partial n}\overline{p} = 0 \quad \text{on } \Sigma. \tag{2.307}$$

It follows from the definition of the operator **A** that

$$\mathbf{Au} = \left\{ u_m + \int_{\gamma_m} \overline{p}\, \mathrm{d}\Gamma \right\}_{m=1}^{M}. \tag{2.308}$$

Since

$$\overline{p} = p - P_0 \quad \text{and} \quad u_m = -\int_{\gamma_m} p\, \mathrm{d}\Gamma, \quad \forall m = 1, \ldots, M,$$

it follows from (2.308) that

$$\mathbf{Au} = -\left\{ \int_{\gamma_m} P_0\, \mathrm{d}\Gamma \right\}_{m=1}^{M}; \tag{2.309}$$

the right-hand side of (2.309) is obviously the vector $\boldsymbol{\beta}$ that we were looking for. □

To summarize, we have shown that the optimal control **u** is solution of an equation such as (2.280). From the properties of operator **A**, problem (2.280), (2.309) can be solved by a *conjugate gradient algorithm* operating in the Hilbert space $\mathcal{U}$. Such an algorithm will be described in the following section.

2.11.4.5 Conjugate gradient solution of the control problem (2.248)

Writing problem (2.309) in variational form, we obtain

$$\begin{cases} \mathbf{u} \in \mathcal{U}\,(= L^2(0,T); \mathbb{R}^M), \quad \forall \mathbf{v}\,(= \{v_m\}_{m=1}^M) \in \mathcal{U}, \\ \displaystyle\int_0^T (\mathbf{Au}) \cdot \mathbf{v}\, dt = -\int_0^T \sum_{m=1}^{M} \left(\int_{\gamma_m} P_0\, \mathrm{d}\Gamma \right) v_m\, dt. \end{cases} \tag{2.310}$$

From the *symmetry* and $\mathcal{U}$*-ellipticity* of the bilinear functional

$$\{\mathbf{v}, \mathbf{w}\} \rightarrow \int_0^T (\mathbf{Av}) \cdot \mathbf{w}\, dt,$$

the *linear variational problem* (2.310) is a particular case of the variational problem (1.123) whose *conjugate gradient solution* is discussed in Chapter 1, Section 1.8.2. Applying the results of Section 1.8.2 to the solution of problem (2.310) we obtain the following iterative method:

$$\mathbf{u}^0\,(= \{u_m^0\}_{m=1}^M) \quad \text{is given in } \mathcal{U}. \tag{2.311}$$

Solve first

$$\frac{\partial}{\partial t}y^0 - \nu\Delta y^0 + \mathbf{V}\cdot\nabla y^0 - Cy^0 = 0 \quad \text{in } Q, \tag{2.312}$$

$$y^0(0) = 0, \tag{2.313}$$

$$\nu\frac{\partial}{\partial n}y^0 = 0 \quad \text{on } \Sigma\setminus\sigma, \tag{2.314}$$

$$\forall m = 1,\ldots,M, \quad \nu\frac{\partial}{\partial n}y^0 = u_m^0 \ \text{ on } \sigma_m. \tag{2.315}$$

Solve now

$$-\frac{\partial}{\partial t}p^0 - \nu\Delta p^0 - \mathbf{V}\cdot\nabla p^0 - Cp^0 = k_1 y^0 \quad \text{in } Q, \tag{2.316}$$

$$p^0(T) = k_2 y^0(T), \tag{2.317}$$

$$\nu\frac{\partial}{\partial n}p^0 = 0 \qquad \text{on } \Sigma. \tag{2.318}$$

Finally, solve

$$\begin{cases} \mathbf{g}^0 \in \mathcal{U}, \quad \forall\, \mathbf{v}\,(=\{v_m\}_{m=1}^M) \in \mathcal{U}, \\ \displaystyle\int_0^T \mathbf{g}^0\cdot\mathbf{v}\,\mathrm{d}t = \int_0^T \sum_{m=1}^M \left(u_m^0 + \int_{\gamma_m} p^0\,\mathrm{d}\Gamma\right) v_m\,\mathrm{d}t, \end{cases} \tag{2.319}$$

and set

$$\mathbf{w}^0 = \mathbf{g}^0. \tag{2.320}$$

Then, for $n \geq 0$, assuming that $\mathbf{u}^n$, $\mathbf{g}^n$, and $\mathbf{w}^n$ are known, compute $\mathbf{u}^{n+1}$, $\mathbf{g}^{n+1}$, and $\mathbf{w}^{n+1}$ as follows:
Solve first

$$\frac{\partial}{\partial t}\bar{y}^n - \nu\Delta\bar{y}^n + \mathbf{V}\cdot\nabla\bar{y}^n - C\bar{y}^n = 0 \quad \text{in } Q, \tag{2.321}$$

$$\bar{y}^n(0) = 0, \tag{2.322}$$

$$\nu\frac{\partial}{\partial n}\bar{y}^n = 0 \quad \text{on } \Sigma\setminus\sigma, \tag{2.323}$$

$$\forall\, m = 1,\ldots,M, \quad \nu\frac{\partial}{\partial n}\bar{y}^n = w_m^n \ \text{ on } \sigma_m. \tag{2.324}$$

Solve now

$$-\frac{\partial}{\partial t}\bar{p}^n - \nu\Delta\bar{p}^n - \mathbf{V}\cdot\nabla\bar{p}^n - C\bar{p}^n = k_1\bar{y}^n \quad \text{in } Q, \tag{2.325}$$

$$\bar{p}^n(T) = k_2\bar{y}^n(T), \tag{2.326}$$

$$\nu\frac{\partial}{\partial n}\bar{p}^n = 0 \quad \text{on } \Sigma, \tag{2.327}$$

and then

$$\begin{cases} \bar{\mathbf{g}}^n \in \mathcal{U}, \quad \forall\, \mathbf{v}\,(=\{v_m\}_{m=1}^M) \in \mathcal{U}, \\ \displaystyle\int_0^T \bar{\mathbf{g}}^n \cdot \mathbf{v}\,\mathrm{d}t = \int_0^T \sum_{m=1}^M \left(w_m^n + \int_{\gamma_m} \bar{p}^n\,\mathrm{d}\Gamma \right) v_m\,\mathrm{d}t. \end{cases} \tag{2.328}$$

Compute

$$\rho_n = \int_0^T |\mathbf{g}^n|^2\,\mathrm{d}t \Big/ \int_0^T \bar{\mathbf{g}}^n \cdot \mathbf{w}^n\,\mathrm{d}t, \tag{2.329}$$

$$\mathbf{u}^{n+1} = \mathbf{u}^n - \rho_n \mathbf{w}^n, \tag{2.330}$$

$$\mathbf{g}^{n+1} = \mathbf{g}^n - \rho_n \bar{\mathbf{g}}^n. \tag{2.331}$$

If $\int_0^T |\mathbf{g}^{n+1}|^2\,\mathrm{d}t / \int_0^T |\mathbf{g}^0|^2\,\mathrm{d}t \le \varepsilon^2$, *take* $\mathbf{u} = \mathbf{u}^{n+1}$; *else, compute*

$$\gamma_n = \int_0^T |\mathbf{g}^{n+1}|^2\,\mathrm{d}t \Big/ \int_0^T |\mathbf{g}^n|^2\,\mathrm{d}t, \tag{2.332}$$

and then

$$\mathbf{w}^{n+1} = \mathbf{g}^{n+1} + \gamma_n\,\mathbf{w}^n. \tag{2.333}$$

Do $n = n + 1$ *and go to* (2.321).

2.11.5 Time discretization of the control problem (2.248)

2.11.5.1 Formulation of the discrete control problem

Assuming that T is finite, we introduce a *time-discretization step* Δt, defined by $\Delta t = \frac{T}{N}$, where N is a positive integer. We approximate then the control problem (2.248) by the following finite-dimensional minimization problem:

$$\min_{\mathbf{v} \in \mathcal{U}^{\Delta t}} J^{\Delta t}(\mathbf{v}), \tag{2.334}$$

with

$$\mathcal{U}^{\Delta t} = \mathbb{R}^{M \times N}, \tag{2.335}$$

$$\mathbf{v} = \{\mathbf{v}^n\}_{n=1}^N, \quad \mathbf{v}^n = \{v_m^n\}_{m=1}^M, \tag{2.336}$$

$$J^{\Delta t}(\mathbf{v}) = \frac{1}{2}\Delta t \sum_{n=1}^N |\mathbf{v}^n|^2 + \frac{k_1}{2}\Delta t \sum_{n=1}^N \int_\Omega |y^n|^2\,\mathrm{d}x + \frac{k_2}{2}\int_\Omega |y^N|^2\,\mathrm{d}x, \tag{2.337}$$

where $|\mathbf{v}^n| = \left(\sum_{m=1}^M |v_m^n|^2\right)^{\frac{1}{2}}$ and $\{y^n\}_{n=1}^N$ is defined from the solution of the following semidiscrete parabolic problem:

$$y^0 = y_0, \tag{2.338}$$

and for $n = 1,\ldots,N$,

$$\frac{y^n - y^{n-1}}{\Delta t} - \nu \Delta y^n + \mathbf{V} \cdot \nabla y^{n-1} - C y^{n-1} = 0 \quad \text{in } \Omega, \tag{2.339}$$

$$\nu \frac{\partial}{\partial n} y^n = 0 \quad \text{on } \Gamma \backslash \gamma, \tag{2.340}$$

$$\forall m = 1,\ldots,M, \quad \nu \frac{\partial}{\partial n} y^n = v_m^n \text{ on } \gamma_m. \tag{2.341}$$

Remark 2.22 For simplicity, we have chosen a one-step semiexplicit scheme à la Euler to time discretize the parabolic system (2.252)–(2.255). This scheme is first-order accurate and reasonably robust, once combined to an appropriate space discretization. The application of second-order time discretization to the solution of control problems is discussed in, for example, Carthel, Glowinski, and Lions (1994) and in Chapter 1, Sections 1.8.6 and 1.10.6, and in Section 2.4.2.

Remark 2.23 At each step of scheme (2.338)–(2.341), we have to solve a linear Neumann problem to obtain y^n from y^{n-1}. There is no particular difficulty in solving such a problem.

2.11.5.2 Optimality conditions for the discrete control problem (2.334)

We suppose that the discrete control space $\mathcal{U}^{\Delta t}$ is equipped with the scalar product $(\cdot,\cdot)_{\Delta t}$ defined by

$$(\mathbf{v}, \mathbf{w})_{\Delta t} = \Delta t \sum_{n=1}^{N} \mathbf{v}^n \cdot \mathbf{w}^n = \Delta t \sum_{n=1}^{N} \sum_{m=1}^{M} v_m^n w_m^n, \tag{2.342}$$

and the corresponding Euclidean norm. Let

$$\mathbf{u}^{\Delta t} = \{\mathbf{u}^n\}_{n=1}^N = \{\{u_m^n\}_{m=1}^M\}_{n=1}^N$$

be the solution of problem (2.334); the vector $\mathbf{u}^{\Delta t}$ is characterized by the *optimality condition*

$$\nabla J^{\Delta t}(\mathbf{u}^{\Delta t}) = \mathbf{0}. \tag{2.343}$$

In order to make relation (2.343) more explicit, we shall discuss first the calculation of $\nabla J^{\Delta t}(\mathbf{v})$, where $\mathbf{v} = \{\mathbf{v}^n\}_{n=1}^N$ is an arbitrary element of $\mathcal{U}^{\Delta t}$. We proceed as in the continuous case (see Section 2.11.4) by observing that

$$\delta J^{\Delta t}(\mathbf{v}) = (\nabla J^{\Delta t}(\mathbf{v}), \delta \mathbf{v})_{\Delta t}. \tag{2.344}$$

It follows from (2.337)–(2.341) that we also have

$$\delta J^{\Delta t}(\mathbf{v}) = \Delta t \sum_{n=1}^{N} \mathbf{v}^n \cdot \delta \mathbf{v}^n + k_1 \Delta t \sum_{n=1}^{N} \int_\Omega y^n \delta y^n \, dx + k_2 \int_\Omega y^N \delta y^N \, dx,$$

with

$$\delta y^0 = 0, \tag{2.345}$$

and for $n = 1, \ldots, N$,

$$\frac{\delta y^n - \delta y^{n-1}}{\Delta t} - \nu \Delta \delta y^n + \mathbf{V} \cdot \nabla \delta y^{n-1} - C \delta y^{n-1} = 0 \quad \text{in } \Omega, \tag{2.346}$$

$$\nu \frac{\partial}{\partial n} \delta y^n = 0 \quad \text{on } \Gamma \backslash \gamma, \tag{2.347}$$

$$\forall\, m = 1, \ldots, M, \quad \nu \frac{\partial}{\partial n} \delta y^n = v_m^n \text{ on } \gamma_m, \tag{2.348}$$

respectively. Let us introduce now $\{p^n\}_{n=1}^N$, where the p^n are smooth functions of x; multiplying both sides of (2.346) by p^n, integrating continuously over Ω and discretely over $(0, T)$, we obtain

$$\begin{aligned}
0 &= \Delta t \sum_{n=1}^{N} \int_\Omega p^n \left(\frac{\delta y^n - \delta y^{n-1}}{\Delta t} - \nu \Delta \delta y^n + \mathbf{V} \cdot \nabla \delta y^{n-1} - C \delta y^{n-1} \right) \mathrm{d}x \\
&= \Delta t \sum_{n=1}^{N} \int_\Omega p^n \left(\frac{\delta y^n - \delta y^{n-1}}{\Delta t} \right) \mathrm{d}x - \Delta t\, \nu \sum_{n=1}^{N} \int_\Omega p^n \Delta \delta y^n \, \mathrm{d}x \\
&+ \Delta t \sum_{n=1}^{N} \int_\Omega p^n \mathbf{V} \cdot \nabla \delta y^{n-1} \, \mathrm{d}x - \Delta t\, C \sum_{n=1}^{N} \int_\Omega p^n \, \delta y^{n-1} \, \mathrm{d}x.
\end{aligned} \tag{2.349}$$

Taking the initial condition (2.345) into account, we clearly have

$$\begin{aligned}
\Delta t \sum_{n=1}^{N} \int_\Omega p^n \frac{\delta y^n - \delta y^{n-1}}{\Delta t} \, \mathrm{d}x &= \Delta t \sum_{n=1}^{N} \int_\Omega \frac{p^n - p^{n+1}}{\Delta t} \delta y^n \, \mathrm{d}x \\
&+ \int_\Omega p^{N+1} \delta y^N \, \mathrm{d}x,
\end{aligned} \tag{2.350}$$

with p^{N+1} to be specified later on. We have similarly

$$-\Delta t\, C \sum_{n=1}^{N} \int_\Omega p^n \delta y^{n-1} \, \mathrm{d}x = -\Delta t\, C \sum_{n=1}^{N-1} \int_\Omega p^{n+1} \delta y^n \, \mathrm{d}x. \tag{2.351}$$

Using now Green's formula, we have first

$$\begin{aligned}
-\Delta t\, \nu \sum_{n=1}^{N} \int_\Omega p^n \Delta \delta y^n \, \mathrm{d}x &= -\Delta t\, \nu \sum_{n=1}^{N} \int_\Omega \Delta p^n \delta y^n \, \mathrm{d}x \\
&+ \Delta t\, \nu \sum_{n=1}^{N} \int_\Gamma \left(\frac{\partial p^n}{\partial n} \delta y^n - p^n \frac{\partial}{\partial n} \delta y^n \right) \mathrm{d}\Gamma,
\end{aligned} \tag{2.352}$$

and then, taking into account $\mathbf{\nabla} \cdot \mathbf{V} = 0$ in Ω and $\mathbf{V} \cdot \mathbf{n} = 0$ on Γ , we obtain

$$\Delta t \sum_{n=1}^{N} \int_{\Omega} p^n \mathbf{V} \cdot \mathbf{\nabla} \delta y^{n-1}\, dx = \Delta t \sum_{n=1}^{N-1} \int_{\Omega} p^{n+1} \mathbf{V} \cdot \mathbf{\nabla} \delta y^n\, dx$$
$$= -\Delta t \sum_{n=1}^{N-1} \int_{\Omega} \mathbf{V} \cdot \mathbf{\nabla} p^{n+1}\, \delta y^n\, dx. \tag{2.353}$$

Suppose that $\frac{\partial p^n}{\partial n} = 0$, $\forall n = 1, \ldots, N$; it follows then from (2.347), (2.348), and (2.352) that

$$-\Delta t\, \nu \sum_{n=1}^{N} \int_{\Omega} p^n\, \Delta \delta y^n\, dx = -\Delta t\, \nu \sum_{n=1}^{N} \int_{\Omega} \Delta p^n\, \delta y^n\, dx$$
$$-\Delta t \sum_{n=1}^{N} \sum_{m=1}^{M} \left(\int_{\gamma_m} p^n\, d\Gamma \right) \delta v_m^n. \tag{2.354}$$

By summation of (2.350), (2.351), (2.353), and (2.354), and comparison with (2.349), we obtain that

$$\Delta t \sum_{n=1}^{N-1} \int_{\Omega} \left(\frac{p^n - p^{n+1}}{\Delta t} - \nu \Delta p^n - \mathbf{V} \cdot \mathbf{\nabla} p^{n+1} - C p^{n+1} \right) \delta y^n\, dx$$
$$+ \Delta t \int_{\Omega} \left(\frac{p^N - p^{N+1}}{\Delta t} - \nu \Delta p^N \right) \delta y^N\, dx + \int_{\Omega} p^{N+1} \delta y^N\, dx$$
$$= \Delta t \sum_{n=1}^{N} \sum_{m=1}^{M} \left(\int_{\gamma_m} p^n\, d\Gamma \right) \delta v_m^n. \tag{2.355}$$

In order to have a simple expression for $\mathbf{\nabla} J^{\Delta t}(\mathbf{v})$, we shall impose on $\{p^n\}_{n=1}^{N+1}$ the following conditions:

$$p^{N+1} = k_2\, y^N, \tag{2.356}$$

$$\frac{p^N - p^{N+1}}{\Delta t} - \nu \Delta p^N = k_1\, y^N \quad \text{in } \Omega, \tag{2.357}$$

$$\nu \frac{\partial p^N}{\partial n} = 0 \qquad \text{on } \Gamma, \tag{2.358}$$

and for $n = N - 1, \ldots, 1$,

$$\frac{p^n - p^{n+1}}{\Delta t} - \nu \Delta p^n - \mathbf{V} \cdot \mathbf{\nabla} p^{n+1} - C p^{n+1} = k_1\, y^n \quad \text{in } \Omega, \tag{2.359}$$

$$\nu \frac{\partial p^n}{\partial n} = 0 \qquad \text{on } \Gamma. \tag{2.360}$$

Taking the above relations into account, it follows from (2.349) and (2.355) that

$$\delta J^{\Delta t}(\mathbf{v}) = \Delta t \sum_{n=1}^{N} \sum_{m=1}^{M} \left(v_m^n + \int_{\gamma_m} p^n \, d\Gamma \right) \delta v_m^n. \tag{2.361}$$

Since $\delta \mathbf{v}$ is arbitrary, it follows from (2.361) that

$$\nabla J^{\Delta t}(\mathbf{v}) = \left\{ \left\{ v_m^n + \int_{\gamma_m} p^n \, d\Gamma \right\}_{m=1}^{M} \right\}_{n=1}^{N}, \tag{2.362}$$

or in *variational form*, $\forall\, \mathbf{v}, \mathbf{w} \in \mathcal{U}^{\Delta t}$,

$$\left(\nabla J^{\Delta t}(\mathbf{v}), \mathbf{w} \right)_{\Delta t} = \Delta t \sum_{n=1}^{N} \sum_{m=1}^{M} \left(v_m^n + \int_{\gamma_m} p^n \, d\Gamma \right) w_m^n. \tag{2.363}$$

Suppose now that $\mathbf{u}^{\Delta t}$ is the solution of problem (2.343); we still denote by $\{y^n\}_{n=0}^N$ and $\{p^n\}_{n=1}^{N+1}$ the solution of the systems (2.338)–(2.341) and (2.356)–(2.360) corresponding to $\mathbf{v} = \mathbf{u}^{\Delta t}$. The discrete optimal control $\mathbf{u}^{\Delta t}$ is characterized by the following (optimality) system:

$$u_m^n = - \int_{\gamma_m} p^n \, d\Gamma, \quad \forall\, m = 1, \ldots, M, \ \forall\, n = 1, \ldots, N, \tag{2.364}$$

$$y^0 = y_0, \tag{2.365}$$

for $n = 1, \ldots, N$,

$$\frac{y^n - y^{n-1}}{\Delta t} - \nu\, \Delta y^n + \mathbf{V} \cdot \nabla y^{n-1} - C y^{n-1} = 0 \quad \text{in } \Omega, \tag{2.366}$$

$$\forall\, m = 1, \ldots, M, \quad \nu \frac{\partial y^n}{\partial n} = u_m^n \ \text{ on } \gamma_m, \tag{2.367}$$

$$\nu \frac{\partial y^n}{\partial n} = 0 \quad \text{on } \Gamma \setminus \gamma, \tag{2.368}$$

$$p^{N+1} = k_2\, y^N, \tag{2.369}$$

$$\frac{p^N - p^{N+1}}{\Delta t} - \nu\, \Delta p^N = k_1\, y^N \quad \text{in } \Omega, \tag{2.370}$$

$$\nu \frac{\partial p^N}{\partial n} = 0 \qquad \text{on } \Gamma, \tag{2.371}$$

and for $n = N-1, \ldots, 1$,

$$\frac{p^n - p^{n+1}}{\Delta t} - \nu \Delta p^n - \mathbf{V} \cdot \nabla p^{n+1} - C p^{n+1} = k_1\, y^n \quad \text{in } \Omega, \tag{2.372}$$

$$\nu \frac{\partial p^n}{\partial n} = 0 \qquad \text{on } \Gamma. \tag{2.373}$$

Remark 2.24 According to, for example, Glowinski (1984, Appendix 1), *the elliptic* problems (2.366)–(2.368), (2.370), (2.371), and (2.372), (2.373) enjoy the following *variational formulations*:

(i) $y^n \in H^1(\Omega)$; $\forall z \in H^1(\Omega)$ we have

$$\int_\Omega \frac{y^n - y^{n-1}}{\Delta t} z \, dx + \nu \int_\Omega \nabla y^n \cdot \nabla z \, dx$$
$$= C \int_\Omega y^{n-1} z \, dx - \int_\Omega \mathbf{V} \cdot \nabla y^{n-1} z \, dx + \sum_{m=1}^{M} u_m^n \int_{\gamma_m} z \, d\Gamma. \quad (2.374)$$

(ii) $p^N \in H^1(\Omega)$; $\forall z \in H^1(\Omega)$ we have

$$\int_\Omega \frac{p^N - p^{N-1}}{\Delta t} z \, dx + \nu \int_\Omega \nabla p^N \cdot \nabla z \, dx = k_1 \int_\Omega y^N z \, dx. \quad (2.375)$$

(iii) $p^n \in H^1(\Omega)$; $\forall z \in H^1(\Omega)$ we have

$$\int_\Omega \frac{p^n - p^{n+1}}{\Delta t} z \, dx + \nu \int_\Omega \nabla p^n \cdot \nabla z \, dx$$
$$= C \int_\Omega p^{n+1} z \, dx + \int_\Omega \mathbf{V} \cdot \nabla p^{n+1} z \, dx + k_1 \int_\Omega y^n z \, dx. \quad (2.376)$$

We recall that $H^1(\Omega) = \{z \,|\, z \in L^2(\Omega), \frac{\partial z}{\partial x_i} \in L^2(\Omega), \forall i = 1, \ldots, d\}$, the derivatives being in the sense of distributions; see, for example, Glowinski (1984, Appendix 1) and the references therein for more details. In the following section, we shall rely on the above variational formulations to *fully discretize* the control problem (2.248) using a finite-element approximation closely related to the one we employed in Section 2.4.3.

2.11.6 Full discretization of problem (2.248)

2.11.6.1 Generalities

We suppose from now on that Ω is a *polygonal* domain of $\mathbb{R}^2$. We introduce then a *finite-element triangulation* $\mathcal{T}_h$ of Ω as in Section 2.4.3, with h the largest length of the edges of the triangles of $\mathcal{T}_h$. Next, we approximate $H^1(\Omega)$ and $L^2(\Omega)$ by the following finite-dimensional space:

$$V_h = \{z \,|\, z \in C^0(\overline{\Omega}), z|_K \in P_1, \forall K \in \mathcal{T}_h\}, \quad (2.377)$$

where, in (2.377), P_1 is the space of the polynomials in two variables of degree ≤ 1.

2.11.6.2 *Formulation of the fully discrete control problem*

With Δt, $\mathcal{U}^{\Delta t}$, and $\mathbf{v}$ as in Section 2.11.5.1, we approximate the control problem (2.248) (and (2.334)) by

$$\min_{\mathbf{v}\in\mathcal{U}^{\Delta t}} J_h^{\Delta t}(\mathbf{v}), \tag{2.378}$$

with

$$J_h^{\Delta t}(\mathbf{v}) = \frac{\Delta t}{2}\sum_{n=1}^{N}|\mathbf{v}^n|^2 + \frac{k_1\Delta t}{2}\sum_{n=1}^{N}\int_\Omega |y_h^n|^2\,dx + \frac{k_2}{2}\int_\Omega |y_h^N|^2\,dx, \tag{2.379}$$

where $\{y_h^n\}_{n=1}^N$ is defined from the solution of the following fully discrete parabolic problem:

$$y_h^0 = y_{0h}, \tag{2.380}$$

and for $n = 1,\ldots,N$,

$$\begin{cases} y_h^n \in V_h, \\ \displaystyle\int_\Omega \frac{y_h^n - y_h^{n-1}}{\Delta t} z\,dx + \nu\int_\Omega \nabla y_h^n\cdot\nabla z\,dx = C\int_\Omega y_h^{n-1} z\,dx \\ \displaystyle\quad -\int_\Omega \mathbf{V}\cdot\nabla y_h^{n-1} z\,dx + \sum_{m=1}^{M} v_m^n \int_{\gamma_m} z\,d\Gamma, \quad \forall z\in V_h; \end{cases} \tag{2.381}$$

y_h^0 in (2.380) verifies

$$y_{0h}\in V_h,\ \forall h, \quad \text{and} \quad \lim_{h\to 0} y_{0h} = y_0 \text{ in } L^2(\Omega). \tag{2.382}$$

The fully discrete control problem (2.378) has a unique solution $\mathbf{u}_h^{\Delta t}$, which is characterized by

$$\nabla J_h^{\Delta t}(\mathbf{u}_h^{\Delta t}) = \mathbf{0}. \tag{2.383}$$

In the following section, we will return on relation (2.383) in order to make it more practical for computational purposes; this will be achieved through the introduction of an appropriate adjoint system.

2.11.6.3 *Optimality conditions for the fully discrete control problem*

Let $\mathbf{u}_h^{\Delta t} = \{\mathbf{u}^n\}_{n=1}^N \in \mathcal{U}^{\Delta t}$ be the unique solution of the discrete control problem (2.378). Since the optimal control $\mathbf{u}_h^{\Delta t}$ is characterized by relation (2.383), it is imperative to explicit first $\nabla J_h^{\Delta t}$. Proceeding essentially as in the semidiscrete case discussed in Section 2.11.5, we can show that

$$\nabla J_h^{\Delta t}(\mathbf{v}) = \left\{\left\{v_m^n + \int_{\gamma_m} p_h^n\,d\Gamma\right\}_{n=1}^N\right\}_{m=1}^M, \tag{2.384}$$

where p_h^n is obtained from $\mathbf{v}$ via the solution of (2.380), (2.381), and then from the solution of the following discrete adjoint system:

$$p_h^{N+1} = k_2 y_h^N, \tag{2.385}$$

$$\begin{cases} p_h^N \in V_h, \quad \forall z \in V_h, \\ \displaystyle\int_\Omega \frac{p_h^N - p_h^{N+1}}{\Delta t} z \, dx + \nu \int_\Omega \nabla p_h^N \cdot \nabla z \, dx = k_1 \int_\Omega y_h^N z \, dx, \end{cases} \tag{2.386}$$

and for $n = N-1, \ldots, 1$,

$$\begin{cases} p_h^n \in V_h, \\ \displaystyle\int_\Omega \frac{p_h^n - p_h^{n+1}}{\Delta t} z \, dx + \nu \int_\Omega \nabla p_h^n \cdot \nabla z \, dx = C \int_\Omega p_h^{n+1} z \, dx \\ \qquad \displaystyle + \int_\Omega \mathbf{V} \cdot \nabla p_h^{n+1} z \, dx + k_1 \int_\Omega y_h^n z \, dx, \quad \forall z \in V_h. \end{cases} \tag{2.387}$$

From the above relations, the optimality condition (2.383) takes the following form:

$$u_m^n = -\int_{\gamma_m} p_h^n \, d\Gamma, \quad \forall m = 1, \ldots, M, \text{ and } \forall n = 1, \ldots, N, \tag{2.388}$$

where $\{p_h^n\}_{n=1}^{N+1}$ is obtained from $\{y_h^n\}_{n=0}^N$ via (2.385)–(2.387), while the functions y_h^n are obtained from (2.380), (2.381) with $v_m^n = u_m^n$, $\forall m = 1, \ldots, M$, and $\forall n = 1, \ldots, N$.

2.11.6.4 Conjugate gradient solution of the discrete control problem (2.378)

Generalities Proceeding as in Section 2.11.4.4, we can easily prove that the solution $\mathbf{u}_h^{\Delta t} \in \mathcal{U}^{\Delta t}$ of the discrete control problem (2.378) is also a solution of the following linear equation in $\mathcal{U}^{\Delta t}$:

$$\mathbf{A}_h^{\Delta t} \mathbf{u}_h^{\Delta t} = \boldsymbol{\beta}_h^{\Delta t}, \tag{2.389}$$

where $\boldsymbol{\beta}_h^{\Delta t} \in \mathcal{U}^{\Delta t}$, and where the linear operator $\mathbf{A}_h^{\Delta t}$ is symmetric and positive definite over $\mathcal{U}^{\Delta t}$ for the scalar product defined by (2.342). From these properties of the operator $\mathbf{A}_h^{\Delta t}$, a finite-dimensional approximation of the operator $\mathbf{A}$ defined in Section 2.11.4.4, problem (2.378), (2.389) can be solved by a *conjugate gradient* algorithm operating in $\mathcal{U}^{\Delta t}$; this algorithm will be described below.

Description of the conjugate gradient algorithm The fully discrete control problem (2.378) can be solved by the following variant of algorithm (2.311)–(2.333) (see Section 2.11.4.5), where most of the h subscripts have been dropped:

$$\mathbf{u}_0 \text{ is given in } \mathcal{U}^{\Delta t}. \tag{2.390}$$

Solve first

$$y_0^0 = y_{0h}, \tag{2.391}$$

and for $n = 1, \ldots, N$,

$$\begin{cases} y_0^n \in V_h, \\ \displaystyle\int_\Omega \frac{y_0^n - y_0^{n-1}}{\Delta t} z \, dx + \nu \int_\Omega \nabla y_0^n \cdot \nabla z \, dx = C \int_\Omega y_0^{n-1} z \, dx \\ \qquad \displaystyle - \int_\Omega \mathbf{V} \cdot \nabla y_0^{n-1} z \, dx + \sum_{m=1}^{M} u_{0,m}^n \int_{\gamma_m} z \, d\Gamma, \quad \forall z \in V_h. \end{cases} \tag{2.392}$$

Solve now

$$p_0^{N+1} = k_2 y_0^N, \tag{2.393}$$

$$\begin{cases} p_0^N \in V_h, \quad \forall z \in V_h, \\ \displaystyle\int_\Omega \frac{p_0^N - p_0^{N+1}}{\Delta t} z \, dx + \nu \int_\Omega \nabla p_0^N \cdot \nabla z \, dx = k_1 \int_\Omega y_0^N z \, dx, \end{cases} \tag{2.394}$$

and for $n = N - 1, \ldots, 1$,

$$\begin{cases} p_0^n \in V_h, \\ \displaystyle\int_\Omega \frac{p_0^n - p_0^{n+1}}{\Delta t} z \, dx + \nu \int_\Omega \nabla p_0^n \cdot \nabla z \, dx = C \int_\Omega p_0^{n+1} z \, dx \\ \qquad \displaystyle + \int_\Omega \mathbf{V} \cdot \nabla p_0^{n+1} z \, dx + k_1 \int_\Omega y_0^n z \, dx, \quad \forall z \in V_h. \end{cases} \tag{2.395}$$

Compute

$$\mathbf{g}_0 = \left\{ \left\{ u_{0,m}^n + \int_{\gamma_m} p_0^n \, d\Gamma \right\}_{n=1}^{N} \right\}_{m=1}^{M}, \tag{2.396}$$

and set

$$\mathbf{w}_0 = \mathbf{g}_0. \tag{2.397}$$

Then, for $k \geq 0$, *assuming that* $\mathbf{u}_k$, $\mathbf{g}_k$, *and* $\mathbf{w}_k$ *are known, the last two different from* $\mathbf{0}$, *compute* $\mathbf{u}_{k+1}$, $\mathbf{g}_{k+1}$, *and, if necessary,* $\mathbf{w}_{k+1}$ *as follows:*
Solve

$$\bar{y}_k^0 = 0, \tag{2.398}$$

and for $n = 1, \ldots, N$,

$$\begin{cases} \bar{y}_k^n \in V_h, \\ \displaystyle\int_\Omega \frac{\bar{y}_k^n - \bar{y}_k^{n-1}}{\Delta t} z \, dx + \nu \int_\Omega \nabla \bar{y}_k^n \cdot \nabla z \, dx = C \int_\Omega \bar{y}_k^{n-1} z \, dx \\ \qquad \displaystyle - \int_\Omega \mathbf{V} \cdot \nabla \bar{y}_k^{n-1} z \, dx + \sum_{m=1}^{M} w_{k,m}^n \int_{\gamma_m} z \, d\Gamma, \quad \forall z \in V_h. \end{cases} \tag{2.399}$$

Solve now

$$\bar{p}_k^{N+1} = k_2 \bar{y}_k^N, \tag{2.400}$$

$$\begin{cases} \bar{p}_k^N \in V_h, \quad \forall z \in V_h, \\ \displaystyle\int_\Omega \frac{\bar{p}_k^N - \bar{p}_k^{N+1}}{\Delta t} z \,dx + \nu \int_\Omega \nabla \bar{p}_k^N \cdot \nabla z \,dx = k_1 \int_\Omega \bar{y}_k^N z \,dx, \end{cases} \tag{2.401}$$

and for $n = N-1, \ldots, 1,$

$$\begin{cases} \bar{p}_k^n \in V_h, \\ \displaystyle\int_\Omega \frac{\bar{p}_k^n - \bar{p}_k^{n+1}}{\Delta t} z \,dx + \nu \int_\Omega \nabla \bar{p}_k^n \cdot \nabla z \,dx = C \int_\Omega \bar{p}_k^{n+1} z \,dx \\ \displaystyle\quad + \int_\Omega \mathbf{V} \cdot \nabla \bar{p}_k^{n+1} z \,dx + k_1 \int_\Omega \bar{y}_k^n z \,dx, \quad \forall z \in V_h. \end{cases} \tag{2.402}$$

Compute

$$\bar{\mathbf{g}}_k = \left\{ \left\{ w_{k,m}^n + \int_{\gamma_m} \bar{p}_k^n \,d\Gamma \right\}_{n=1}^N \right\}_{m=1}^M, \tag{2.403}$$

$$\rho_k = \frac{(\mathbf{g}_k, \mathbf{g}_k)_{\Delta t}}{(\bar{\mathbf{g}}_k, \mathbf{w}_k)_{\Delta t}}, \tag{2.404}$$

and then

$$\mathbf{u}_{k+1} = \mathbf{u}_k - \rho_k \mathbf{w}_k, \tag{2.405}$$

$$\mathbf{g}_{k+1} = \mathbf{g}_k - \rho_k \bar{\mathbf{g}}_k. \tag{2.406}$$

If $\frac{(\mathbf{g}_{k+1}, \mathbf{g}_{k+1})_{\Delta t}}{(\mathbf{g}_0, \mathbf{g}_0)_{\Delta t}} \leq \varepsilon^2$, *take* $\mathbf{u}_h^{\Delta t} = \mathbf{u}_{k+1}$; *else, compute*

$$\gamma_k = \frac{(\mathbf{g}_{k+1}, \mathbf{g}_{k+1})_{\Delta t}}{(\mathbf{g}_k, \mathbf{g}_k)_{\Delta t}}, \tag{2.407}$$

and then

$$\mathbf{w}_{k+1} = \mathbf{g}_{k+1} + \gamma_k \mathbf{w}_k. \tag{2.408}$$

Do $k = k+1$ *and return to* (2.398).

Remark 2.25 Despite its apparent complexity, algorithm (2.390)–(2.408) is quite easy to implement. Actually, the main difficulty arising with the above algorithm seems to be storage related; indeed, to integrate the discrete adjoint equations (2.393)–(2.395) and (2.400)–(2.402), it seems that we need to store the corresponding solutions of (2.391), (2.392), and (2.398), (2.399), for $n = 1, \ldots, N$. For large multidimensional problems, such requirements are prohibitive; fortunately, these storage requirements can be substantially reduced if one uses the memory saving method discussed in Chapter 1, Section 1.12 (see also the Section 6.5 of He and Glowinski, 1998).

2.11.7 Numerical experiments: controllability enhancement by chattering control

2.11.7.1 Generalities

For the test problems considered in this section we take (the notation being as in Section 2.11.2)

$$\Omega = (0,1) \times (0,1),$$
$$\gamma = \left\{ x \,|\, x = \{x_1, x_2\},\ x_1 = 0,\ |x_2 - 1/2| < \sqrt{3}/32 \right\},$$
$$\nu = 1, \quad \lambda = 1, \quad T = 1;$$

see Figure 2.37. The *initial data* y_0 is defined by interpolating linearly values generated *randomly* and *uniformly* on the interval $[0, 2]$ and associated with the points belonging to $\overline{\Omega}$ and to a regular lattice of stepsize $\frac{1}{8}$; see Figure 2.38 for the graph of y_0. Concerning the *space discretization*, we will use a uniform triangulation like the one shown in Figure 2.1 of Section 2.6.

Remark 2.26 We have taken y_0 so that $\int_\Omega y_0 \, dx \neq 0$, in order to "excite" the *constant* (in space) *mode* of y which is the *most unstable* in the uncontrolled case; see Remark 2.27 below.

Remark 2.27 In the particular case where $\mathbf{V} = \mathbf{0}$, we can use a simple *spectral decomposition* method to predict the controllability properties of the Neumann controls considered here. More precisely, we observe that the eigenfunctions of the operator $-\Delta$ for the homogeneous Neumann boundary condition are given by

$$w_{mn}(x) = \alpha_{mn} \cos(m\pi x_1) \cos(n\pi x_2),$$

with m and n nonnegative integers and $\alpha_{mn} = 2$ if $m > 0$ and $n > 0$, $\alpha_{m0} = \sqrt{2}$ if $m > 0$, $\alpha_{0n} = \sqrt{2}$ if $n > 0$, and $\alpha_{00} = 1$; the corresponding eigenvalues λ_{mn} are

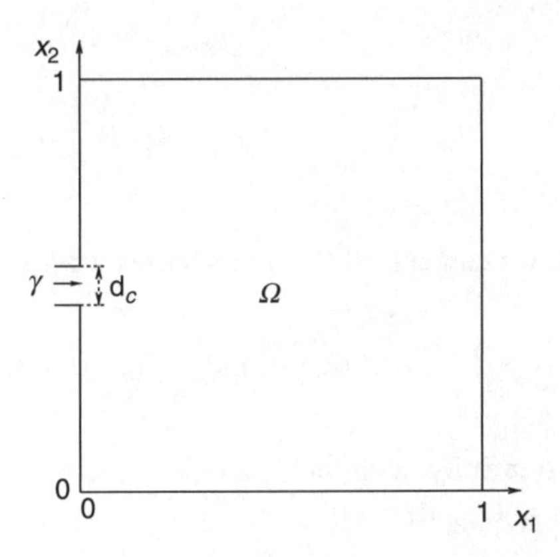

Fig. 2.37. Computational domain and control boundary.

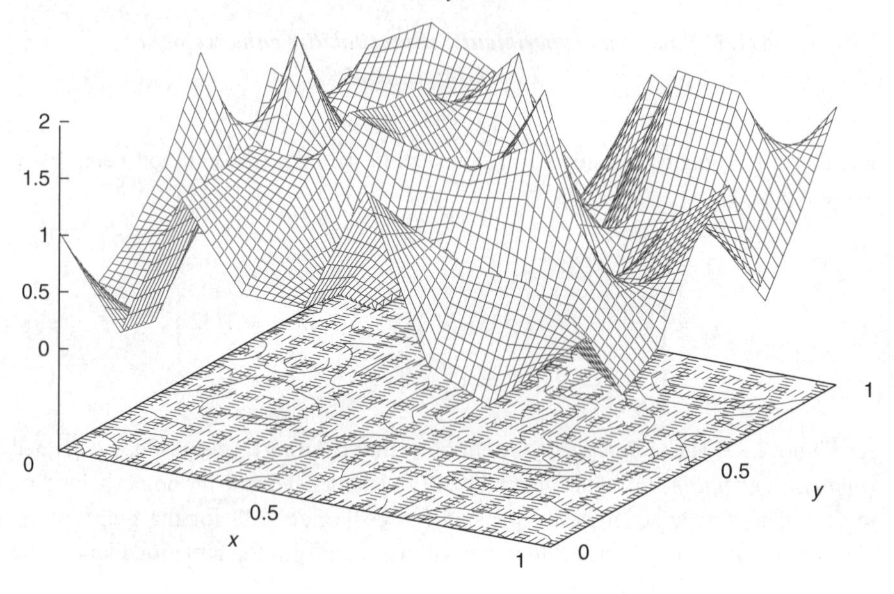

Fig. 2.38. Graph of the initial data y_0.

given by

$$\lambda_{mn} = \left(m^2 + n^2\right)\pi^2.$$

By spectral decomposition of y on the orthonormal basis $\{w_{mn}\}_{m,n\geq 0}$, that is,

$$y(t) = \sum_{m,\, n\geq 0} y_{mn}(t)\, w_{mn},$$

we can diagonalize the system (2.252)–(2.255) and obtain (since $\nu = 1$) the following infinite system of ordinary differential equations (with $m, n \geq 0$):

$$\dot{y}_{mn} + \left[(m^2 + n^2)\pi^2 - C\right] y_{mn} = v \int_\gamma w_{mn}\, d\Gamma, \tag{2.409}$$

$$y_{mn}(0) = \int_\Omega y_0 w_{mn}\, dx. \tag{2.410}$$

From relation (2.409), we can expect that in the cases where

$$C > \left(m^2 + n^2\right)\pi^2 \quad \text{and} \quad y_{mn}(0) \neq 0, \tag{2.411}$$

$y_{mn}(t)$ will converge to infinity, with the sign of $y_{mn}(0)$, as $t \to +\infty$, if either $v = 0$ (uncontrolled case) or $\int_\gamma w_{mn}\, d\Gamma = 0$.

2.11.7.2 *Test problems with* $\mathbf{V} = \mathbf{0}$

The test problems in this paragraph are all related to situations where $\mathbf{V} = \mathbf{0}$ (reaction-diffusion models). Numerical experiments for those situations where $\mathbf{V} \neq \mathbf{0}$ will be reported in Section 2.11.7.3.

First test problem We take $\mathbf{V} = \mathbf{0}$ and $C = 2$ in (2.252)–(2.255). It follows from Remark 2.27 that, without control, the only unstable mode is y_{00}, since $\lambda_{00} = 0$. All the other modes are stable since they satisfy

$$\lambda_{mn} \geq \pi^2 > 2 = C.$$

Without control, $\|y_h(t)\|_{L^2(\Omega)}$ increases exponentially as shown in Figure 2.39, where y has been obtained by the space–time discretization of (2.252)–(2.255), with $v = 0$ on γ, using the finite element method discussed in Section 2.11.6 and the time-discretization scheme (2.380)–(2.382), with (the notation being obvious)

$$\Delta x_1 = \Delta x_2 = \frac{h}{\sqrt{2}} = \frac{1}{64} \quad \text{and} \quad \Delta t = \frac{1}{200}.$$

We observe from Figure 2.39 that $\|y_h(t)\|_{L^2(\Omega)}$ behaves like e^{2t} (modulo a multiplicative constant), which is exactly what we can expect from Remark 2.27.

If we use now a boundary control supported by γ and function of t only, and if we take $k_1 = 10^2, 10^4, 10^6$ and $k_2 = 10^2, 10^4, 10^6$ in (2.250) and (2.379), and use the above space–time discretization parameters, we obtain the results shown in Figure 2.40, where we have visualized the variation of the function $t \to \|y_h(t)\|_{L^2(\Omega)}$. These results show that we have been able to stabilize an intrinsically unstable

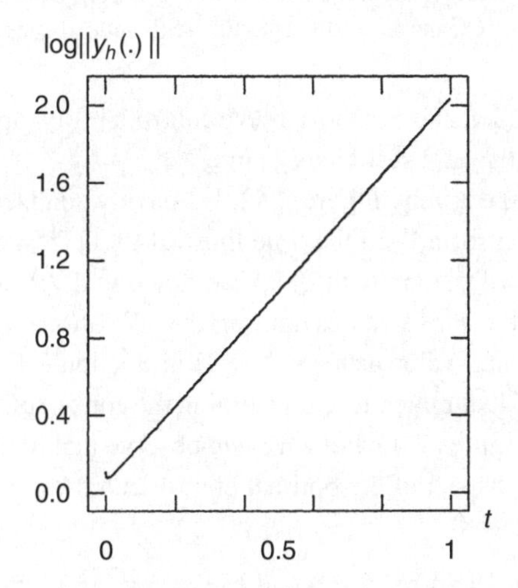

Fig. 2.39. First test problem: Graph of the function $\log \|y_h(.)\|_{L^2(\Omega)}$ for $C = 2$ and $\mathbf{V} = \mathbf{0}$ (no control).

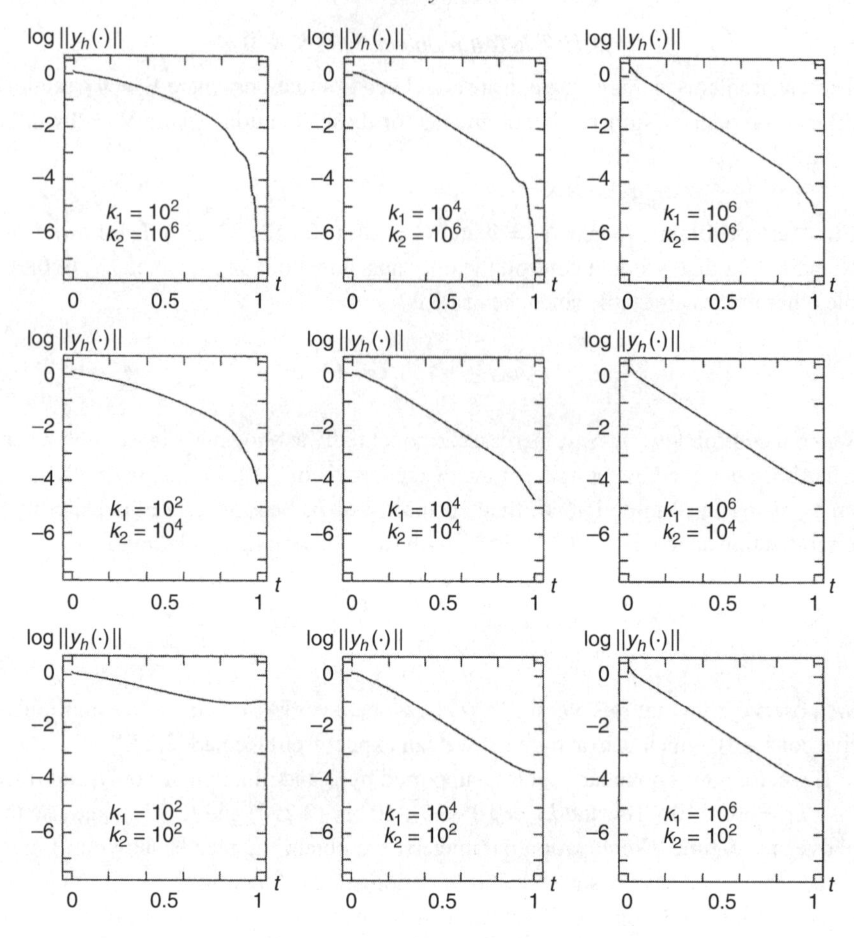

Fig. 2.40. First test problem: Graph of the function $\log \|y_h(.)\|_{L^2(\Omega)}$ for $C = 2$, $\mathbf{V} = \mathbf{0}$, $k_1 = 10^2$, 10^4, 10^6, and $k_2 = 10^2, 10^4, 10^6$ (with optimal control).

reaction-diffusion model, and even to achieve controllability (approximately, at least) at $t = T = 1$ if k_1 and k_2 are sufficiently large.

In Figure 2.41 (respectively, Figure 2.42, we have visualized the variation of the computed optimal control $\mathbf{u}_h^{\Delta t}$ on the time interval [0, 1] (respectively, the variation versus k of the norm of the gradient $\mathbf{g}_k$ of the functional $J_h^{\Delta t}$ at step k of algorithm (2.390)–(2.408)) for the above values of k_1 and k_2. We observe from Figure 2.41 that $\mathbf{u}_h^{\Delta t}$ converges to a small value as $t \to T = 1$; in Section 2.11.9, we shall return to this phenomenon and its implications concerning the control of the original nonlinear system. Also, from Figures 2.41 and 2.42, we observe again that k_1 is the dominant factor concerning the aspect of the optimal control and the convergence behavior of algorithm (2.390)–(2.408).

Second test problem We take $\mathbf{V} = \mathbf{0}$ and $C = 1.5\,\pi^2$ in (2.252)–(2.255). By direct inspection of the system (2.409), (2.410), we can anticipate that, without control, the modes y_{00}, y_{10}, and y_{01} will be unstable, all the other modes being stable. If we use

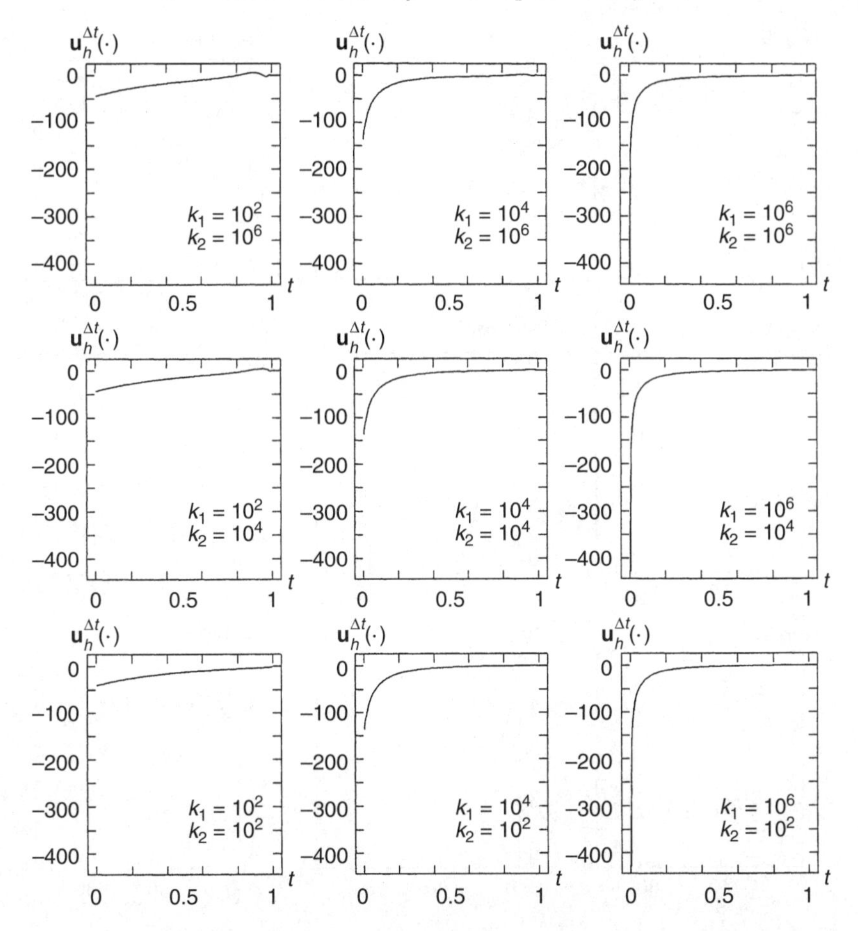

Fig. 2.41. First test problem: Graph of the optimal control $\mathbf{u}_h^{\Delta t}$ for $C = 2$, $\mathbf{V} = \mathbf{0}$, $k_1 = 10^2, 10^4, 10^6$, and $k_2 = 10^2, 10^4, 10^6$.

now the same type of control as for the *first test problem*, we observe that

$$\int_\gamma w_{01} \, d\Gamma = 0,$$

implying that the corresponding mode is not controllable. All these predictions are confirmed by the results displayed in Figures 2.43 and 2.44(a), obtained with $k_1 = k_2 = 10^4$ and $\Delta x_1 = \Delta x_2$, Δt as for the *first test problem*. Figure 2.43 (respectively, Figure 2.44(a)) shows the variation of $t \to \|y_h(t)\|_{L^2(\Omega)}$ over the time interval $[0, 1]$ in the noncontrolled case (respectively, the noncompletely controllable case). We observe that, in the noncontrolled case, $\|y_h(t)\|_{L^2(\Omega)}$ behaves like $\exp(1.5\pi^2 t)$ (modulo a multiplicative constant), while in the second case, and after some oscillations, $\|y_h(t)\|_{L^2(\Omega)}$ behaves like $\exp(\frac{1}{2}\pi^2 t)$, which corresponds to an instability due

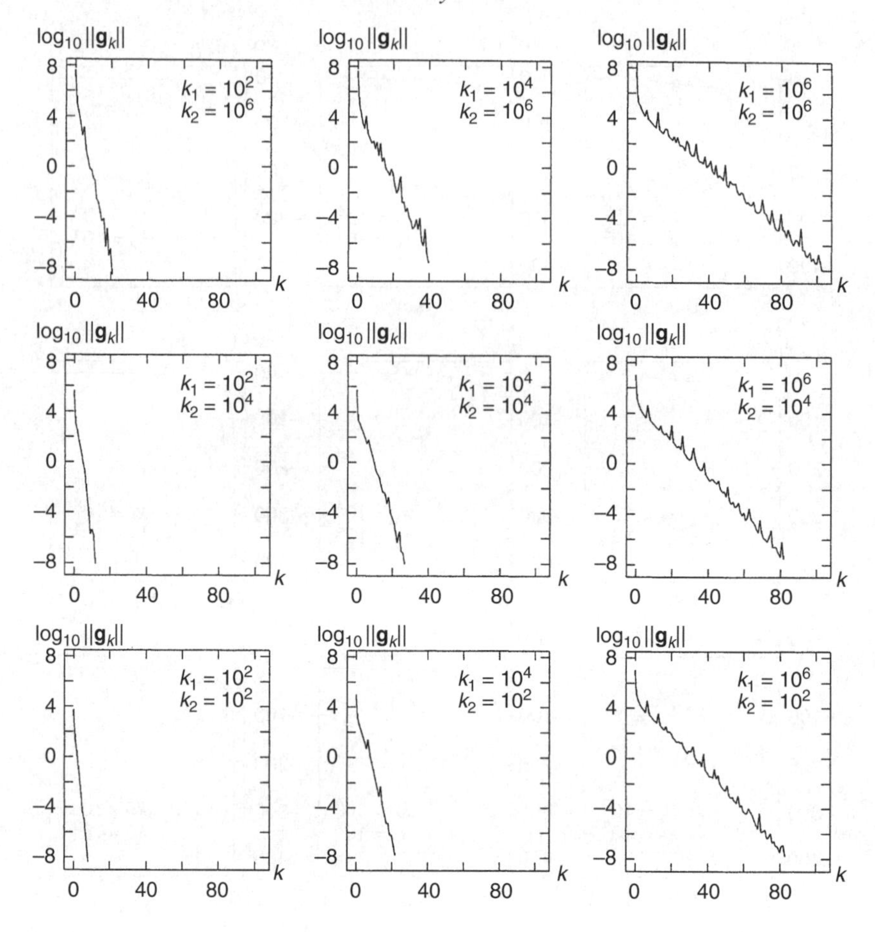

Fig. 2.42. First test problem: Variation of $\|\mathbf{g}_k\|$ versus k for $C = 2$, $\mathbf{V} = \mathbf{0}$, $k_1 = 10^2, 10^4, 10^6$, and $k_2 = 10^2, 10^4, 10^6$ (log scale).

to the mode y_{01}, for which

$$\left(m^2 + n^2\right)\pi^2 - C = -\frac{1}{2}\pi^2$$

precisely. For completeness, we have shown in Figure 2.44(b) (respectively, Figure 2.44(c)), the graph of the computed optimal control (respectively, the variation versus k of the norm of the gradient $\mathbf{g}_k$ in algorithm (2.390)–(2.408)).

We observe once again that the optimal control converges quickly to zero as $t \to T = 1$. We insist on the fact that the optimal control is unable to completely stabilize the system (2.252)–(2.255).

In order to achieve controllability on all unstable modes, we can think of changing the location and/or the length of γ (without increasing the number of connected components of γ); however, since $\lambda_{10} = \lambda_{01}$, it can be shown easily that these approaches cannot provide controllability. To achieve controllability on all modes, we shall apply a *chattering control* technique generalizing to boundary control a

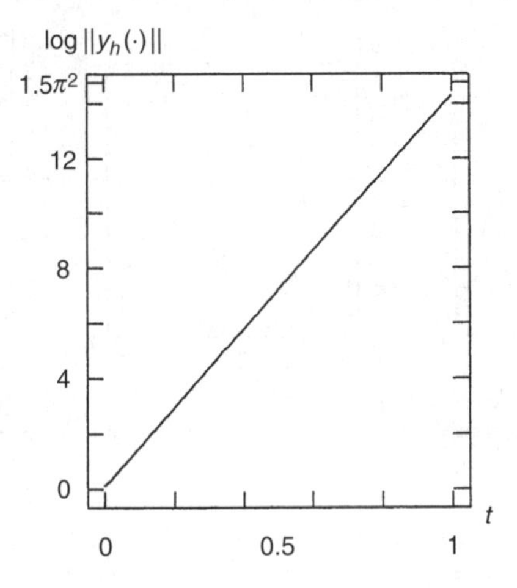

Fig. 2.43. Second test problem: Graph of the function $\log \|y_h(.)\|_{L^2(\Omega)}$ for $C = 1.5\,\pi^2$ and $\mathbf{V} = \mathbf{0}$ (no control).

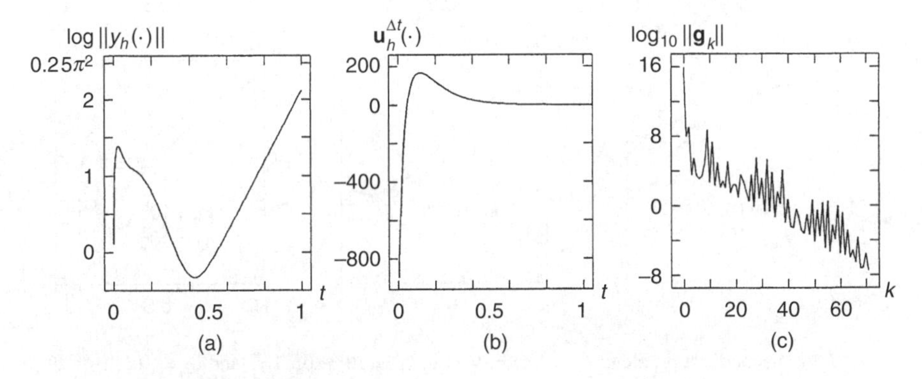

Fig. 2.44. Second test problem: $C = 1.5\,\pi^2$, $\mathbf{V} = \mathbf{0}$, and $k_1 = k_2 = 10^4$ (control with fixed γ). (a) Graph of the function $\log \|y_h(.)\|_{L^2(\Omega)}$ corresponding to the optimal control. (b) Graph of the computed optimal control $\mathbf{u}_h^{\Delta t}$. (c) Variation of $\|\mathbf{g}_k\|$ versus k (log scale).

method advocated and completely analyzed in J.L. Lions (1997b) for the *pointwise control* of linear parabolic equations in one space dimension; see also Berggren (1995) and Barck-Holst (1996) for computer implementations of this methodology. Our chattering control is obtained by giving to γ a vertical sinusoidal motion preserving its length and such that the center c of γ obeys the following motion on the Ox_2-axis:

$$c(t) = \frac{1}{2} + a\sin(2\pi ft). \tag{2.412}$$

There is no basic difficulty *at solving* variant of the control problem (2.248) obtained with moving γ when the motion is known *a priori*. Indeed, the methodology used for the fixed-γ situation is easily generalizable to the moving-γ situation.

Using the above chattering control technique with $a = \frac{1}{16}$ and $f = 20$ in (2.412), we have been able to obtain complete controllability as shown in Figure 2.45 where

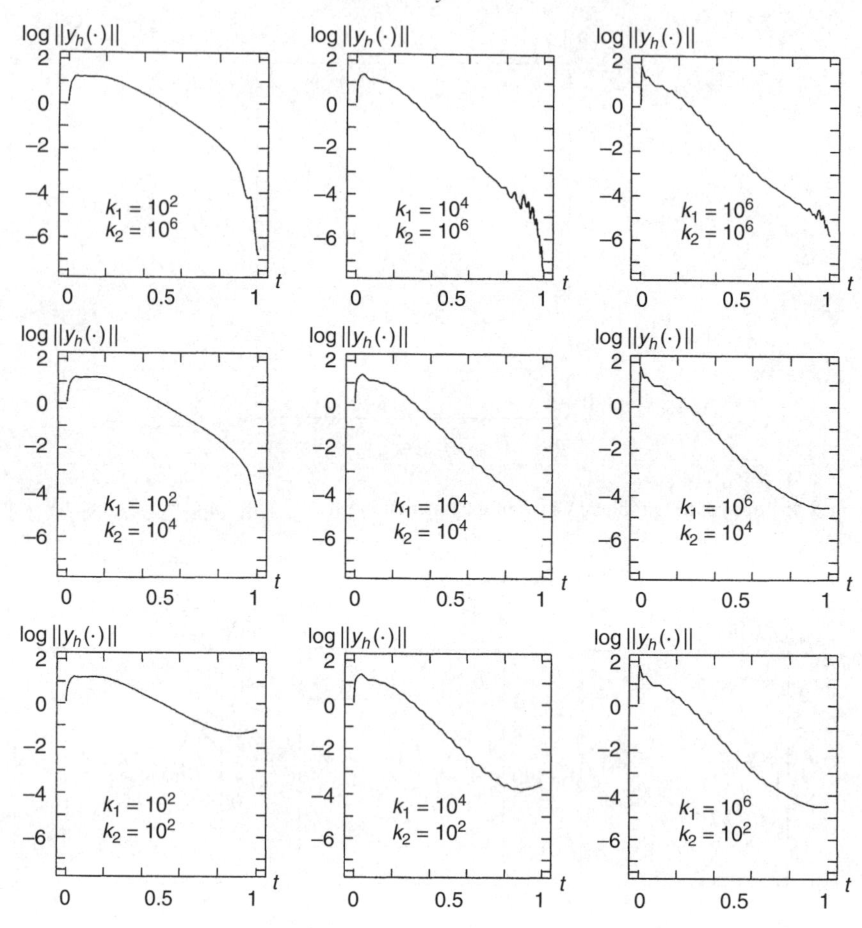

Fig. 2.45. Second test problem: $C = 1.5\pi^2$, $\mathbf{V} = \mathbf{0}$, $k_1 = 10^2, 10^4, 10^6$, and $k_2 = 10^2, 10^4, 10^6$ (chattering control). Graph of the function $\log \|y_h(.)\|_{L^2(\Omega)}$ corresponding to the optimal control.

we have displayed the variation of the function $t \to \log \|y_h(t)\|_{L^2(\Omega)}$ on $[0, 1]$, for $k_1 = 10^2, 10^4, 10^6$ and $k_2 = 10^2, 10^4, 10^6$, the discretization parameters being the same. We observe once again that, for k_1 and/or k_2 large enough, we have a fast convergence of $\|y_h(t)\|_{L^2_{(\Omega)}}$ to a small value as $t \to T = 1$ and also that k_1 seems to be the dominant parameter concerning the behavior of the function $t \to \|y_h(t)\|_{L^2(\Omega)}$. Similar comments apply to the optimal control (which has been visualized in Figure 2.46) and to the variation versus k of the norm of the gradient $\mathbf{g}_k$ in algorithm (2.390)–(2.408) (see Figure 2.47).

2.11.7.3 Test problems with advection ($\mathbf{V} \neq \mathbf{0}$)

For the test problems discussed in this section, we have defined $\mathbf{V} = \{V_1, V_2\}$ by

$$\begin{cases} V_1(x) = -l\pi \sin(l\pi x_1)\cos(l\pi x_2), \\ V_2(x) = \quad l\pi \cos(l\pi x_1)\sin(l\pi x_2), \end{cases} \tag{2.413}$$

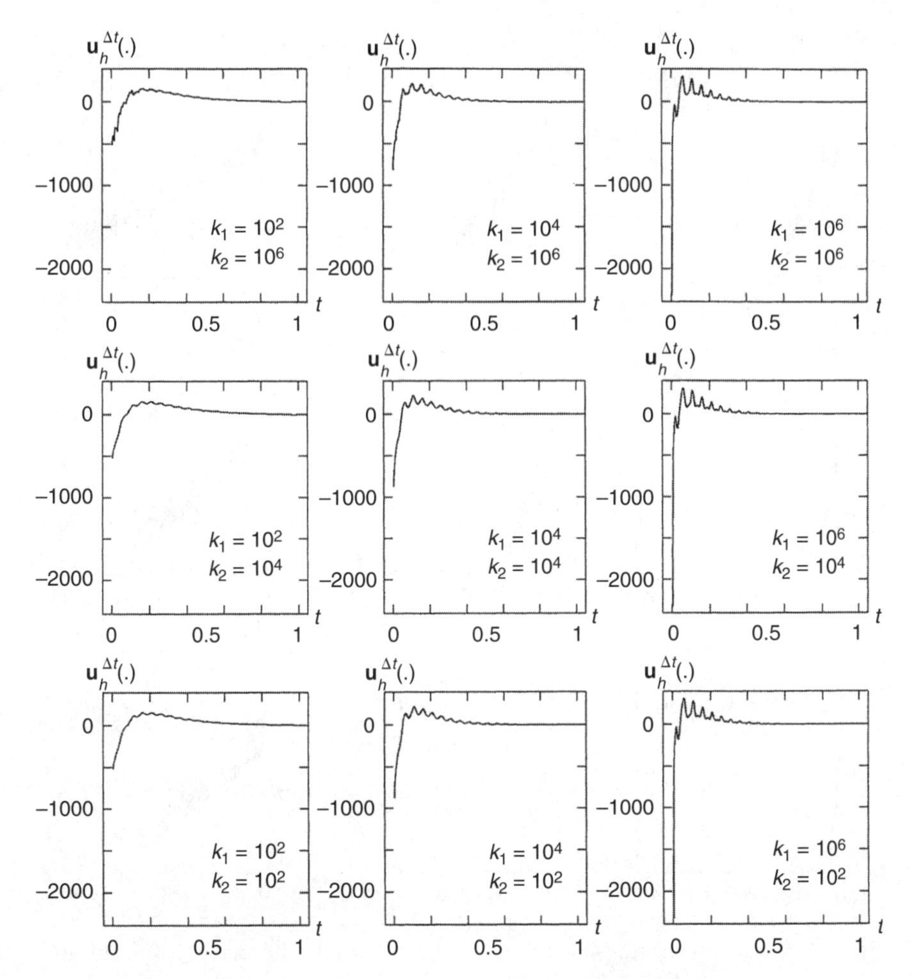

Fig. 2.46. Second test problem: $C = 1.5\pi^2$, $\mathbf{V} = \mathbf{0}$, $k_1 = 10^2, 10^4, 10^6$, and $k_2 = 10^2, 10^4, 10^6$ (chattering control). Graph of the computed optimal control $\mathbf{u}_h^{\Delta t}$.

with l a positive integer. We observe that $\mathbf{V}$ is *divergence-free* and that it satisfies $\mathbf{V} \cdot \mathbf{n} = 0$ on Γ. We have visualized on Figure 2.48 the streamlines of $\mathbf{V}$ corresponding to $l = 4$.

Third test problem We take $C = 2$ in (2.252)–(2.255) and $l = 4$ in (2.413), with the initial data y_0 and the discretization parameters h and Δt as in Section 2.11.7.2. In the *uncontrolled case* ($\mathbf{v} = \mathbf{0}$ in (2.252)–(2.255)), the system is unstable as shown in Figure 2.49, where we have visualized the function $t \to \|y_h(t)\|_{L^2(\Omega)}$, which clearly shows exponential increase. If the control process is the one described in Section 2.11.7.1 with $k_1 = k_2 = 10^4$, we obtain the results described in Figure 2.50, namely the variation of the function $t \to \|y_h(t)\|_{L^2(\Omega)}$ (Figure 2.50(a)), the graph of the optimal control (Figure 2.50(b)), and the variation versus k of the norm of $\mathbf{g}_k$ in algorithm (2.390)–(2.408) (Figure 2.50(c)). In this particular case, there is little difference between the behaviors of $t \to \|y_h(t)\|_{L^2(\Omega)}$ and $t \to \mathbf{u}_h^{\Delta t}(t)$ with or without

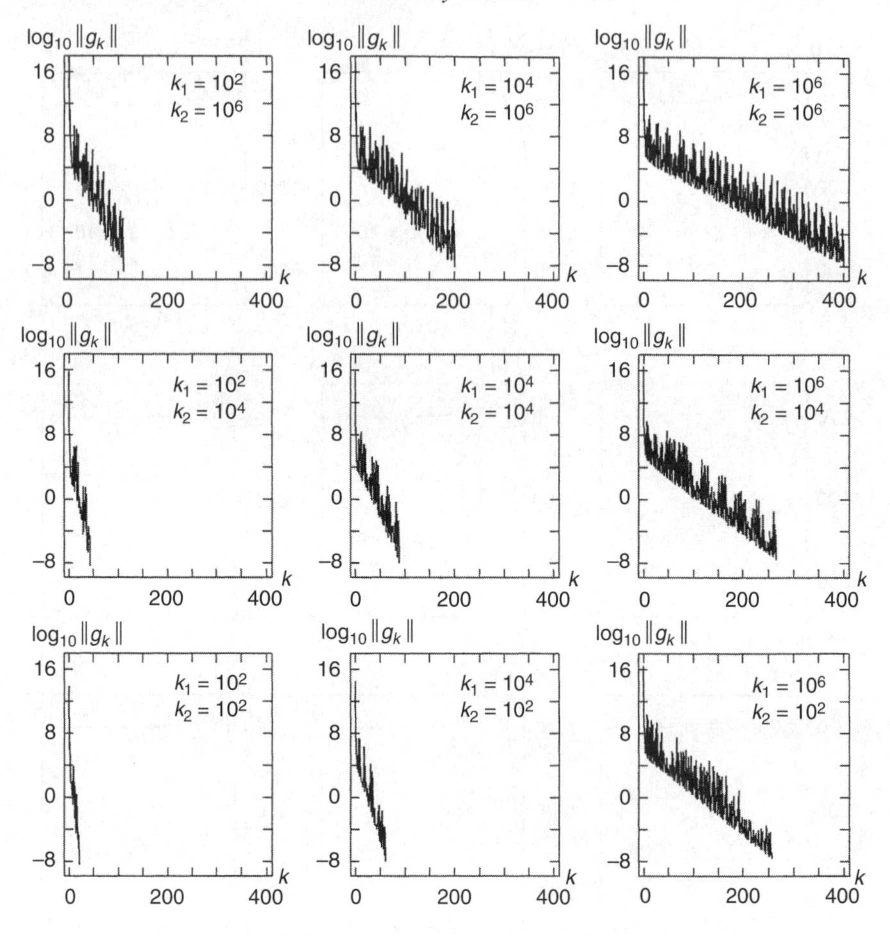

Fig. 2.47. Second test problem: $C = 1.5\pi^2$, $\mathbf{V} = \mathbf{0}$, $k_1 = 10^2, 10^4, 10^6$, and $k_2 = 10^2, 10^4, 10^6$ (chattering control). Variation of $\|\mathbf{g}_k\|$ versus k (log scale).

advection; however, the convergence of algorithm (2.390)–(2.408) is twice slower than in the case without advection.

Fourth test problem We take $C = 1.5\pi^2$ in (2.252)–(2.255), with y_0, k_1, k_2, and the discretization parameters as for the *Third Test Problem*, that is, $k_1 = k_2 = 10^4$, $\Delta x_1 = \Delta x_2 = \frac{1}{64}$, $\Delta t = \frac{1}{200}$. Concerning the parameter l, we shall give it the values $l = 1, 2, 3, 4$. The first series of numerical experiments concerns control with γ and v as in the Third Test Problem (namely, γ fixed and v dependent on t only). For $l = 1, 2, 3, 4$ we have visualized in Figure 2.51 the variation of $t \to \|y_h(t)\|_{L^2(\Omega)}$, the graph of the computed optimal control and the convergence behavior of the conjugate gradient algorithm (2.390)–(2.408); we observe that for $l = 1$ (a one-vortex situation), we have controllability, which was not the case with $\mathbf{V} = \mathbf{0}$; see the Second Test Problem in Section 2.11.7.2. For $l = 2, 3, 4$, we do not have controllability; however, $\|y_h(t)\|_{L^2(\Omega)}$ increases more slowly as $t \to T = 1$, compared to the case $\mathbf{V} = \mathbf{0}$. The same behavior would hold if we replace the coefficient $l\pi$ by 1 in (2.413). For

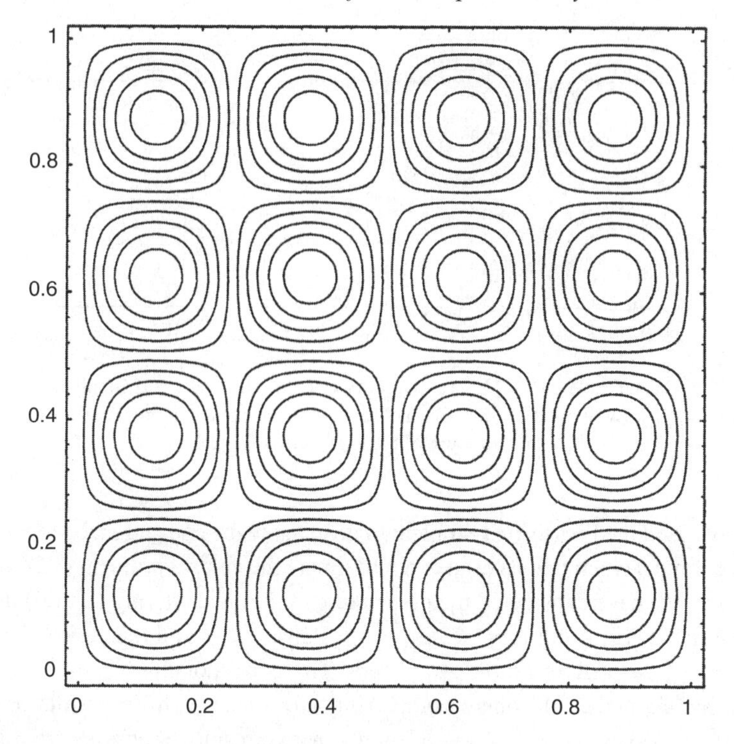

Fig. 2.48. Streamlines of **V** defined by (2.413) with $l = 4$.

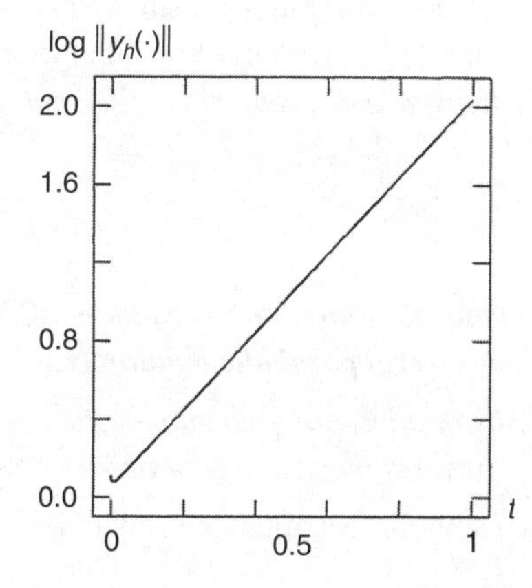

Fig. 2.49. Third test problem: $C = 2, l = 4$. Graph of the function log $\|y_h(.)\|_{L^2(\Omega)}$ (no control).

this particular test problem, advection seems to improve controllability or to delay blowing-up.

In order to improve the stabilization or controllability properties of the boundary control process, we revert to the chattering control approach described for the Second Test Problem in Section 2.11.7.2 with the same values for a and f in (2.412). The corresponding results are displayed in Figure 2.52; they clearly show that, once again, chattering control dramatically improves stability.

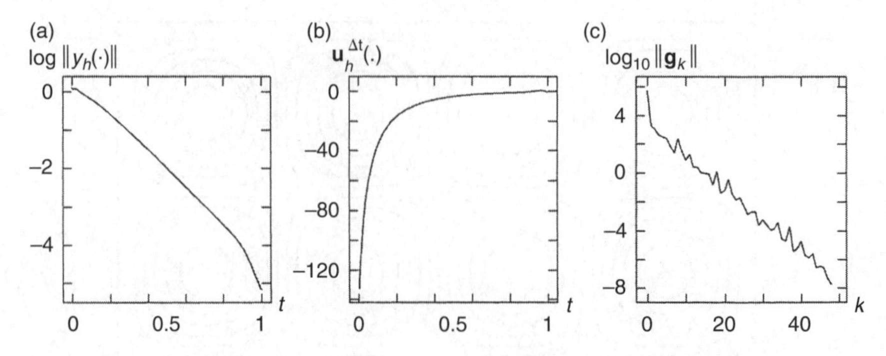

Fig. 2.50. Third test problem: $C = 2, l = 4, k_1 = k_2 = 10^4$ with fixed γ. (a) Graph of the function $\log \|y_h(.)\|_{L^2(\Omega)}$ (with optimal control). (b) Graph of the computed optimal control $\mathbf{u}_h^{\Delta t}$. (c) Variation of $\log \|\mathbf{g}_k\|$ versus k.

We have used the Fourth Test Problem to investigate the convergence properties of the space–time discretizations used to approximate the control problem (2.248). As a benchmark, we have considered the particular case (with chattering control) defined by $l = 4$ in (2.413) using $\Delta x_1 = \Delta x_2 = \frac{1}{128}$, $\Delta t = \frac{1}{200}$, and then $\Delta x_1 = \Delta x_2 = \frac{1}{64}$, $\Delta t = \frac{1}{400}$ as discretization parameters. The corresponding results are shown in Figures 2.53 and 2.54, where comparisons are made with the results obtained with $\Delta x_1 = \Delta x_2 = \frac{1}{64}$, $\Delta t = \frac{1}{200}$. From the above results, it appears from a very close inspection that the main source of approximation errors seems to be the time discretization (albeit all the computed solutions seem pretty close to each other). This is not surprising, since the *time-discretization scheme* that we have employed is *first-order accurate* only, while the *space discretization* is *second-order accurate*, for the $L^2(\Omega)$-norm at least.

2.11.8 Boundary control of advection–reaction–diffusion systems of several parabolic equations

2.11.8.1 A family of advection–reaction–diffusion systems of several parabolic equations

The systems of nonlinear parabolic equations considered in this section are of the following type:

$$\frac{\partial \phi_1}{\partial t} - d_1 \Delta \phi_1 + \mathbf{V} \cdot \nabla \phi_1 - \alpha e^{\phi_1} + \beta e^{\phi_2} = 0 \quad \text{in } \Omega \times (0, T), \tag{2.414}$$

$$\frac{\partial \phi_2}{\partial t} - d_2 \Delta \phi_2 + \mathbf{V} \cdot \nabla \phi_2 + \alpha e^{\phi_1} - \beta e^{\phi_2} = 0 \quad \text{in } \Omega \times (0, T), \tag{2.415}$$

completed by the initial conditions

$$\phi_1(0) = \phi_{10}, \quad \phi_2(0) = \phi_{20}, \tag{2.416}$$

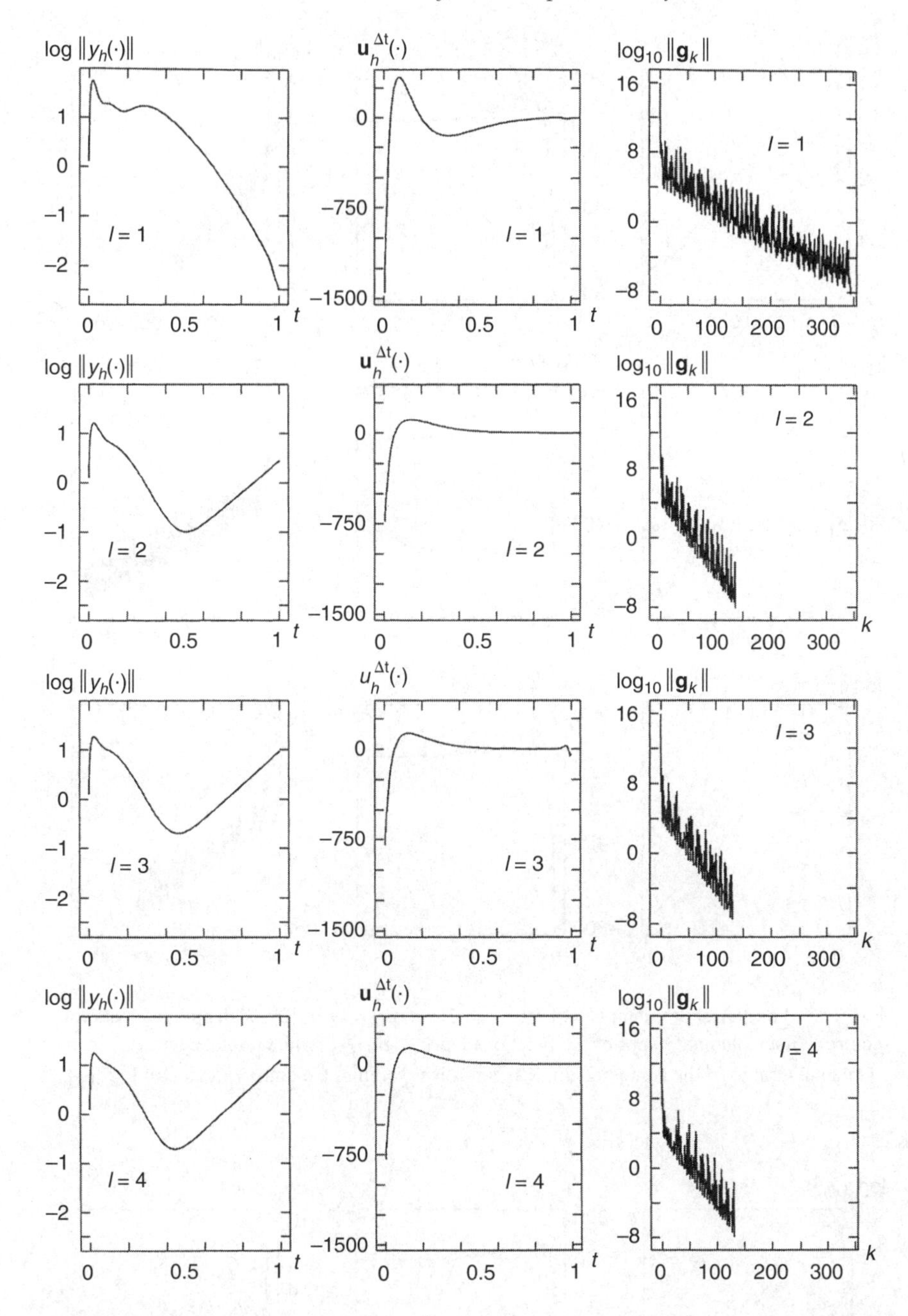

Fig. 2.51. Fourth test problem: $C = 1.5\pi^2$, $l = 1, 2, 3, 4$, $k_1 = k_2 = 10^4$, with fixed γ. First Column: Graphs of the functions $\log \|y_h(.)\|_{L^2(\Omega)}$ (with optimal control). Second Column: Graphs of the computed optimal controls $\mathbf{u}_h^{\Delta t}$. Third Column: Variation of $\log \|\mathbf{g}_k\|$ versus k.

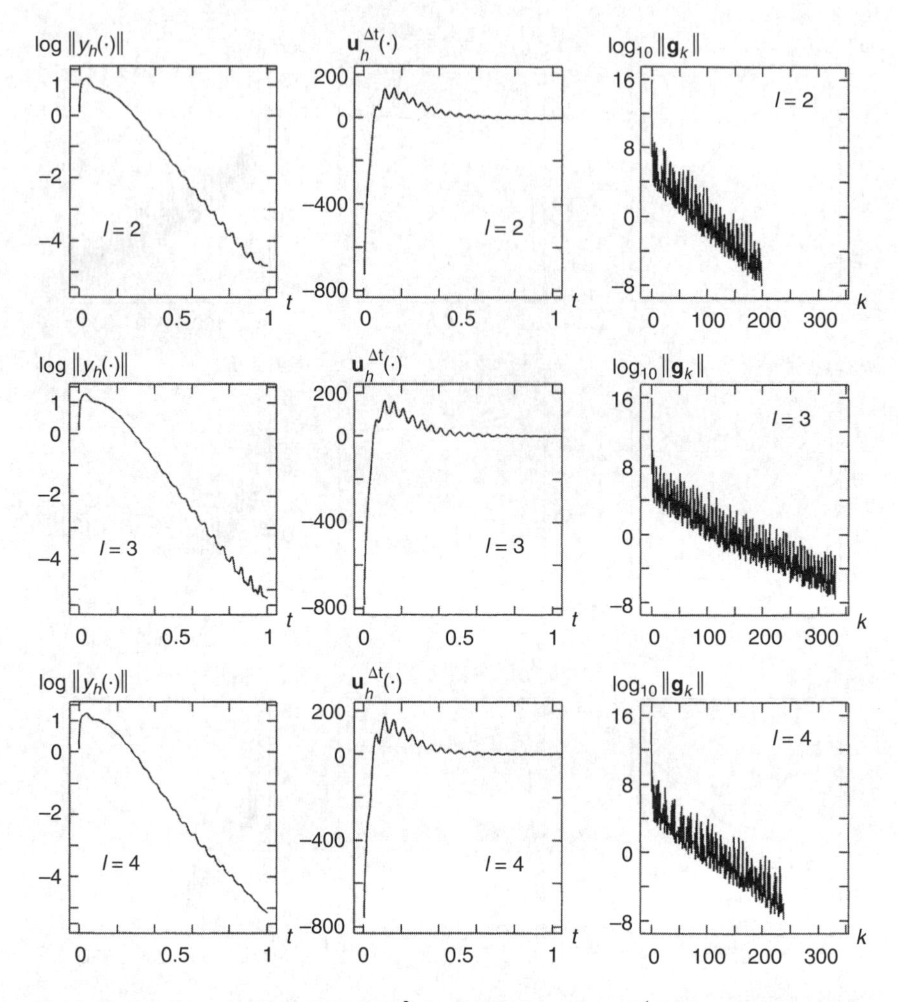

Fig. 2.52. Fourth test problem: $C = 1.5\pi^2$, $l = 2, 3, 4$, $k_1 = k_2 = 10^4$, with optimal chattering control. First Column: Graphs of the functions $\log \|y_h(.)\|_{L^2(\Omega)}$ (with optimal control). Second Column: Graphs of the computed optimal controls $\mathbf{u}_h^{\Delta t}$. Third Column: Variation of $\log \|\mathbf{g}_k\|$ versus k.

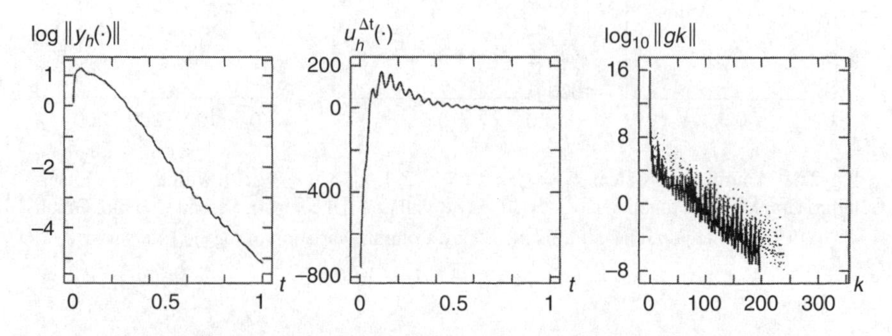

Fig. 2.53. Fourth test problem: $C = 1.5\pi^2$, $l = 4$, $k_1 = k_2 = 10^4$, with optimal chattering control: (—) $\Delta x_1 = \Delta x_2 = \frac{1}{128}$, $\Delta t = \frac{1}{200}$; $(\cdots)$ $\Delta x_1 = \Delta x_2 = \frac{1}{64}$, $\Delta t = \frac{1}{200}$.

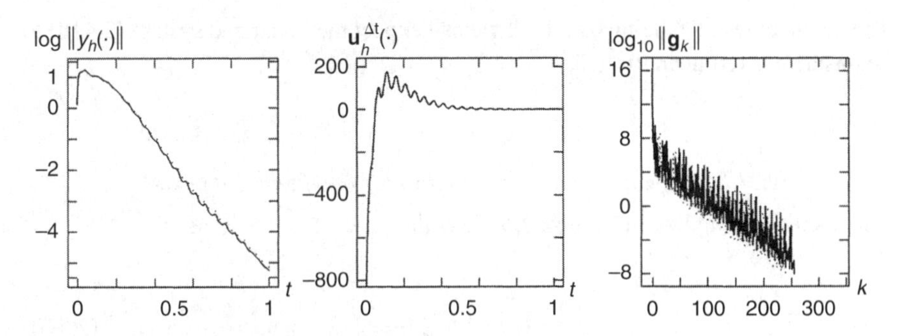

Fig. 2.54. Fourth test problem: $C = 1.5\pi^2$, $l = 4$, $k_1 = k_2 = 10^4$, with optimal chattering control: (—) $\Delta x_1 = \Delta x_2 = \frac{1}{64}$, $\Delta t = \frac{1}{400}$; $(\cdots)$ $\Delta x_1 = \Delta x_2 = \frac{1}{64}$, $\Delta t = \frac{1}{200}$.

and by the boundary conditions

$$\frac{\partial \phi_1}{\partial n} = \frac{\partial \phi_2}{\partial n} = 0 \quad \text{on } \Gamma \times (0, T); \tag{2.417}$$

in (2.414) and (2.415), we have $d_i > 0$, $\mathbf{V}$ as in Section 2.11.2.1, and α, β both positive. We easily verify that the pair $\{a, b\}$, where a and b are two constants, is a steady state solution of the system (2.414), (2.415) if the following condition is verified:

$$a - b = \ln \beta - \ln \alpha. \tag{2.418}$$

Let us show that $\{a, b\}$ verifying (2.418) is an unstable solution of (2.414), (2.415); in order to do that, let us consider perturbations $\delta\phi_1(0)$ and $\delta\phi_2(0)$ of a and b, constant in space. After dropping the perturbation terms of order larger than one, we obtain the following system of linear equations for the perturbation $\{\delta\phi_1, \delta\phi_2\}$ of $\{a, b\}$:

$$\frac{\mathrm{d}\delta\phi_1}{\mathrm{d}t} - \alpha e^a \delta\phi_1 + \beta e^b \delta\phi_2 = 0,$$

$$\frac{\mathrm{d}\delta\phi_2}{\mathrm{d}t} + \alpha e^a \delta\phi_1 - \beta e^b \delta\phi_2 = 0.$$

It follows from (2.418) that

$$\alpha e^a = \beta e^b;$$

let us denote by c this positive common value and by $\mathbf{z}$ the (column) vector $(\delta\phi_1, \delta\phi_2)^t$. The vector $\mathbf{z}$ is the solution of

$$\frac{\mathrm{d}\mathbf{z}}{\mathrm{d}t} + c\,\mathbf{B}\mathbf{z} = \mathbf{0}, \quad \mathbf{z}(0) = (\delta\phi_1(0), \delta\phi_2(0))^t, \tag{2.419}$$

with the matrix $\mathbf{B}$ defined by

$$\mathbf{B} = \begin{bmatrix} -1 & 1 \\ 1 & -1 \end{bmatrix}.$$

The eigenvalues of $\mathbf{B}$ being 0 and -2 we can expect the linearized system (2.419) to be unstable if left uncontrolled.

2.11.8.2 Neumann boundary control for the linearized models

The notation being like in Section 2.11.3, with

$$\gamma^i = \bigcup_{m=1}^{M_i} \gamma_m^i, \quad \forall i = 1, 2, \tag{2.420}$$

we consider the following control problem:

$$\min_{\mathbf{v}\in\mathcal{U}} J(\mathbf{v}), \tag{2.421}$$

where

$$\mathbf{v} = \{\mathbf{v}^1, \mathbf{v}^2\}, \quad \text{with } \mathbf{v}^i = \{v_m^i\}_{m=1}^{M_i}, \ \forall i = 1, 2,$$

and

$$\mathcal{U} = L^2(0, T; \mathbb{R}^{M_1}) \times L^2(0, T; \mathbb{R}^{M_2}), \tag{2.422}$$

$$J(\mathbf{v}) = \frac{1}{2}\sum_{i=1}^{2}\int_0^T |\mathbf{v}^i|^2\,dt + \frac{k_1}{2}\sum_{i=1}^{2}\int_Q |y_i|^2\,dx\,dt + \frac{k_2}{2}\sum_{i=1}^{2}\int_\Omega |y_i(T)|^2\,dx, \tag{2.423}$$

with $k_1 \geq 0, k_2 \geq 0, k_1 + k_2 > 0$, $|\mathbf{v}^i| = \left(\sum_{m=1}^{M_i} |v_m^i|^2\right)^{\frac{1}{2}}$, and $\{y_1, y_2\}$ obtained from $\mathbf{v}$ via the solution of the following linear parabolic system:

$$\frac{\partial y_1}{\partial t} - d_1\Delta y_1 + \mathbf{V}\cdot\nabla y_1 - c(y_1 - y_2) = 0 \quad \text{in } \Omega\times(0, T), \tag{2.424}$$

$$\frac{\partial y_2}{\partial t} - d_2\Delta y_2 + \mathbf{V}\cdot\nabla y_2 + c(y_1 - y_2) = 0 \quad \text{in } \Omega\times(0, T), \tag{2.425}$$

$$y_1(0) = y_{10} = \delta\phi_1(0), \tag{2.426}$$

$$y_2(0) = y_{20} = \delta\phi_2(0), \tag{2.427}$$

$$\forall m = 1, \ldots, M_i, \ \forall i = 1, 2, \quad d_i\frac{\partial y_i}{\partial n} = v_m^i \quad \text{on } \gamma_m^i \times (0, T), \tag{2.428}$$

$$\forall i = 1, 2, \quad d_i\frac{\partial y_i}{\partial n} = 0 \quad \text{on } \Gamma\setminus\gamma^i \times (0, T). \tag{2.429}$$

Remarks 2.17 and 2.18 from Section 2.11.3 still apply to the control problem (2.421).

2.11.8.3 Optimality conditions for problem (2.421)

Proceeding as in Section 2.11.4, we can easily show that the unique solution to the control problem (2.421) is characterized by the following (optimality) system:

$$u_m^i = -\int_{\gamma_m^i} p_i \, d\Gamma, \quad \forall m = 1, \ldots, M_i, \quad \forall i = 1, 2, \tag{2.430}$$

$$\frac{\partial y_1}{\partial t} - d_1 \Delta y_1 + \mathbf{V} \cdot \nabla y_1 - c(y_1 - y_2) = 0 \quad \text{in } \Omega \times (0, T), \tag{2.431}$$

$$\frac{\partial y_2}{\partial t} - d_2 \Delta y_2 + \mathbf{V} \cdot \nabla y_2 + c(y_1 - y_2) = 0 \quad \text{in } \Omega \times (0, T), \tag{2.432}$$

$$y_1(0) = y_{10} = \delta\phi_1(0), \tag{2.433}$$

$$y_2(0) = y_{20} = \delta\phi_2(0), \tag{2.434}$$

$$\forall m = 1, \ldots, M_i, \quad \forall i = 1, 2, \quad d_i \frac{\partial y_i}{\partial n} = u_m^i \text{ on } \gamma_m^i \times (0, T), \tag{2.435}$$

$$\forall i = 1, 2, \quad d_i \frac{\partial y_i}{\partial n} = 0 \quad \text{on } \Gamma \setminus \gamma^i \times (0, T), \tag{2.436}$$

$$-\frac{\partial p_1}{\partial t} - d_1 \Delta p_1 - \mathbf{V} \cdot \nabla p_1 - c(p_1 - p_2) = k_1 y_1 \quad \text{in } \Omega \times (0, T), \tag{2.437}$$

$$-\frac{\partial p_2}{\partial t} - d_2 \Delta p_2 - \mathbf{V} \cdot \nabla p_2 + c(p_1 - p_2) = k_1 y_2 \quad \text{in } \Omega \times (0, T), \tag{2.438}$$

$$\forall i = 1, 2, \quad p_i(T) = k_2 y_i(T), \tag{2.439}$$

$$\forall i = 1, 2, \quad d_i \frac{\partial p_i}{\partial n} = 0 \qquad \text{on } \Gamma \times (0, T). \tag{2.440}$$

From the optimality conditions (2.430)–(2.440), we can derive a conjugate gradient algorithm operating in the space $\mathcal{U}$ to solve the control problem (2.421). This algorithm is a straightforward generalization of algorithm (2.311)–(2.333) in Section 2.11.4.5. Similarly, there is no practical difficulty to extend to problem (2.421) the space–time-discretization techniques used in Sections 2.11.5 and 2.11.6 to approximate the control problem (2.248).

2.11.8.4 Numerical experiments

Generalities As in Section 2.11.7, we take $\Omega = (0, 1)^2$ and $T = 1$. Next, we define the control support sets γ^1 and γ^2 by

$$\gamma^1 = \{x \,|\, x = \{x_1, x_2\},\, x_1 = 0,\, |x_2 - 1/2| < \sqrt{3}/32\}, \tag{2.441}$$

$$\gamma^2 = \{x \,|\, x = \{x_1, x_2\},\, x_1 = 1,\, |x_2 - 1/2| < \sqrt{3}/32\}, \tag{2.442}$$

respectively. Concerning the various parameters in (2.414), (2.415), we have taken $d_1 = 1, d_2 = \frac{1}{4}, \beta = \pi^2, a = 0$, and $\mathbf{V}$ defined by (2.413) with $l = 4$. The initial data y_{10} (respectively, y_{20}) is defined by interpolating linearly values generated randomly on $[0, 2]$ (respectively, $[-1.5, 0]$) and associated to the points belonging to $\overline{\Omega}$ and to a regular lattice of stepsize $\frac{1}{8}$. Concerning the space–time discretization, we use a regular triangulation like the one in Figure 2.1 of Section 2.6, with $\Delta x_1 = \Delta x_2 = \frac{1}{64}$, $\Delta t = \frac{1}{200}$.

Fifth test problem We take $\alpha = 0.3\pi^2$, which implies that

$$b = a - \ln\frac{\beta}{\alpha} = \ln(0.3) < 0, \quad c = \alpha e^a = 0.3\pi^2.$$

Without control, $\|y_{ih}(t)\|_{L^2(\Omega)}$ increases exponentially, $\forall i = 1, 2$. If we use boundary controls constant in space on γ^1 and γ^2 we obtain, with $k_1 = k_2 = 10^4$ in (2.423), the results shown in Figure 2.55, namely, (a) the variation on $(0, 1)$ of $\|y_{1h}(t)\|_{L^2(\Omega)}$ and $\|y_{2h}(t)\|_{L^2(\Omega)}$, (b) the variation on $(0, 1)$ of $u^1(t)$ and $u^2(t)$ (in fact, of their discrete analogues), and finally, (c) the variation versus k of the norm of g_k in the conjugate gradient algorithm used to solve the discrete control problem (algorithm which generalizes algorithm (2.390)–(2.408)). We observe from Figure 2.55(a) that, for $c = 0.3\pi^2$, we have essentially approximate controllability.

Sixth test problem We take $\alpha = 0.5\pi^2$, which implies that

$$b = a - \ln\frac{\beta}{\alpha} = \ln(0.5) < 0, \quad c = \alpha e^a = 0.5\pi^2.$$

We take again $k_1 = k_2 = 10^4$. With fixed γ^1 and γ^2, still defined by (2.441) and (2.442), we have no controllability as shown in Figure 2.56(a), which represents the variation over $(0, 1)$ of $\|y_{1h}(t)\|_{L^2(\Omega)}$ and $\|y_{2h}(t)\|_{L^2(\Omega)}$. The corresponding controls are shown in Figure 2.56(b); we observe their fast convergence to zero as $t \to 1$. The

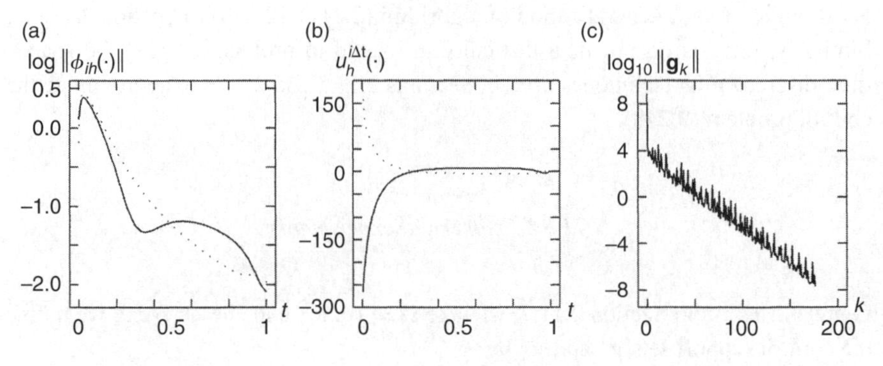

Fig. 2.55. Fifth test problem: $c = 0.3\pi^2, l = 4, k_1 = k_2 = 10^4$; optimal control with fixed γ and $\Delta x_1 = \Delta x_2 = \frac{1}{64}, \Delta t = \frac{1}{200}$: (a) Graph of the function $\log \|y_{1h}(t)\|_{L^2(\Omega)}$ and $\log \|y_{2h}(t)\|_{L^2(\Omega)}$ (with optimal control). (b) Graph of the computed optimal control $u_h^{1\Delta t}(t)$ and $u_h^{2\Delta t}(t)$. (c) Variation of $\log \|\mathbf{g}_k\|$ versus k.

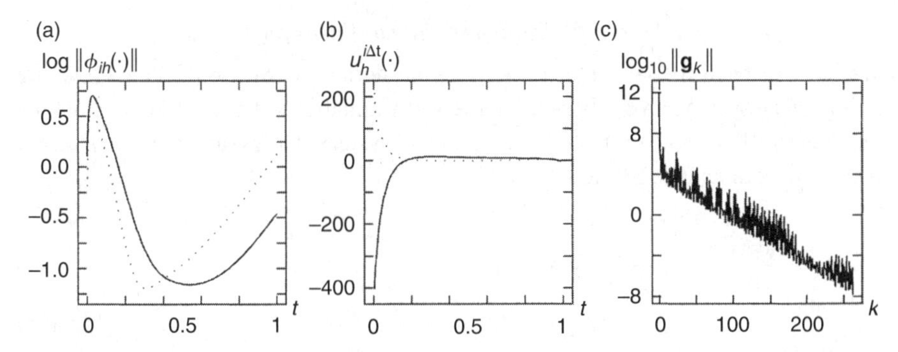

Fig. 2.56. Fifth test problem: $c = 0.5\pi^2$, $l = 4$, $k_1 = k_2 = 10^4$; optimal control with fixed γ and $\Delta x_1 = \Delta x_2 = \frac{1}{64}$, $\Delta t = \frac{1}{200}$: (a) Graph of the function $\log \|y_{1h}(t)\|_{L^2(\Omega)}$ and $\log \|y_{2h}(t)\|_{L^2(\Omega)}$ (with optimal control). (b) Graph of the computed optimal control $u_h^{1\Delta t}(t)$ and $u_h^{2\Delta t}(t)$. (c) Variation of $\log \|\mathbf{g}_k\|$ versus k.

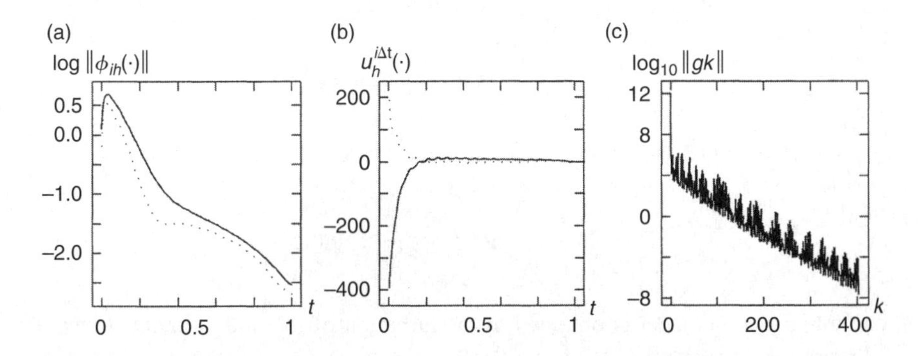

Fig. 2.57. Fifth test problem: $c = 0.5\pi^2$, $l = 4$, $k_1 = k_2 = 10^4$; optimal chattering control and $\Delta x_1 = \Delta x_2 = \frac{1}{64}$, $\Delta t = \frac{1}{200}$: (a) Graph of the function $\log \|y_{1h}(t)\|_{L^2(\Omega)}$ and $\log \|y_{2h}(t)\|_{L^2(\Omega)}$ (with optimal control). (b) Graph of the computed optimal control $u_h^{1\Delta t}(t)$ and $u_h^{2\Delta t}(t)$. (c) Variation of $\log \|\mathbf{g}_k\|$ versus k.

variation versus k of the norm of g_k in the conjugate gradient algorithm is shown in Figure 2.56(c).

In order to enhance controllability, we shall rely again on chattering control; proceeding as for the fourth test problem (see Section 2.11.7.3), we give, for $i = 1, 2$, to the center c^i of γ^i a vertical sinusoidal motion preserving the length of γ^i and defined by

$$c^i(t) = \frac{1}{2} + \frac{1}{16}\sin(2\pi f t). \tag{2.443}$$

Using the above chattering control with $f = 20$ in (2.443) enhances indeed controllability as shown in Figure 2.57(a); the corresponding controls (respectively, the variation versus k of the norm of g_k in the conjugate gradient algorithm) are shown in Figure 2.57(b) (respectively, Figure 2.57(c)).

2.11.9 Stabilization of the nonlinear models

We have reached the last step of the program defined in Section 2.11.2.1; we are going thus to verify, on the basis of numerical simulations, if the controls computed via linearization stabilize the original nonlinear system. The system to be considered is the one defined (see Section 2.11.2) by

$$\frac{\partial \phi}{\partial t} - \nu \Delta \phi + \mathbf{V} \cdot \nabla \phi = \lambda e^{\phi} - C \quad \text{in } \Omega \times (0, +\infty), \tag{2.444}$$

$$\phi(0) = \phi_0, \tag{2.445}$$

$$\nu \frac{\partial \phi}{\partial n} = 0 \quad \text{on } (\Gamma \setminus \gamma) \times (0, +\infty), \tag{2.446}$$

$$\nu \frac{\partial \phi}{\partial n} = v \quad \text{on } \gamma \times (0, +\infty), \tag{2.447}$$

with

$$\nu = 1, \quad \lambda = 1, \quad C = 1.5\pi^2$$

and $\mathbf{V}$ defined by (2.413) with $l = 4$. The steady state solution to stabilize is

$$\phi_s = \ln \frac{C}{\lambda} = \ln(1.5\pi^2) = 2.69492 \cdots$$

The initial condition ϕ_0 is obtained by perturbation of ϕ_s, and is parametrized as follows:

$$\phi_0 = \phi_s + \xi \phi_P, \tag{2.448}$$

where ϕ_P is a given function of the order of 1 and where ξ is a small positive or negative number; we have taken for ϕ_P the function visualized in Figure 2.38 of Section 2.11.7.1, where it was denoted by y_0.

Taking Δx_1, Δx_2, Δt as for the First Test Problem (see Section 2.11.7.2), we have computed the solution ϕ of problem (2.444)–(2.447) in the uncontrolled case ($v = 0$ in (2.447)). In Figure 2.58(a) (respectively, 2.58(b)) we have visualized the function $t \to \|\phi_h(t)\|_{L^2(\Omega)}$ for $\xi = 1, 10^{-1}, 10^{-2}, 10^{-3}, 10^{-4}$ (respectively, $\xi = -1, -10^{-1}, -10^{-2}, -10^{-3}, -10^{-4}$) in (2.448). The results confirm the analysis done in Section 2.11.2 concerning the instability of the model (2.232)–(2.234).

Next, we apply to the system the chattering optimal control computed in Section 2.11.7.3 for the fourth test problem (after multiplication by ξ); in Figure 2.59(a) (respectively, Figure 2.59(b)) , we have shown the variation of the function $t \to \|\phi_h(t)\|_{L^2(\Omega)}$ for $\xi = 1, 10^{-1}, 10^{-2}, 10^{-3}, 10^{-4}$ (respectively, $\xi = -1, -10^{-1}, -10^{-2}, -10^{-3}, -10^{-4}$). We observe that control delays explosion, but does not prevent it. Such a behavior could have been predicted, since Figures 2.52–2.54 show that the control obtained from the linearized model is quite small for t sufficiently large (say $t > 0.6$ for this example); the small values of the control between $t = 0.6$

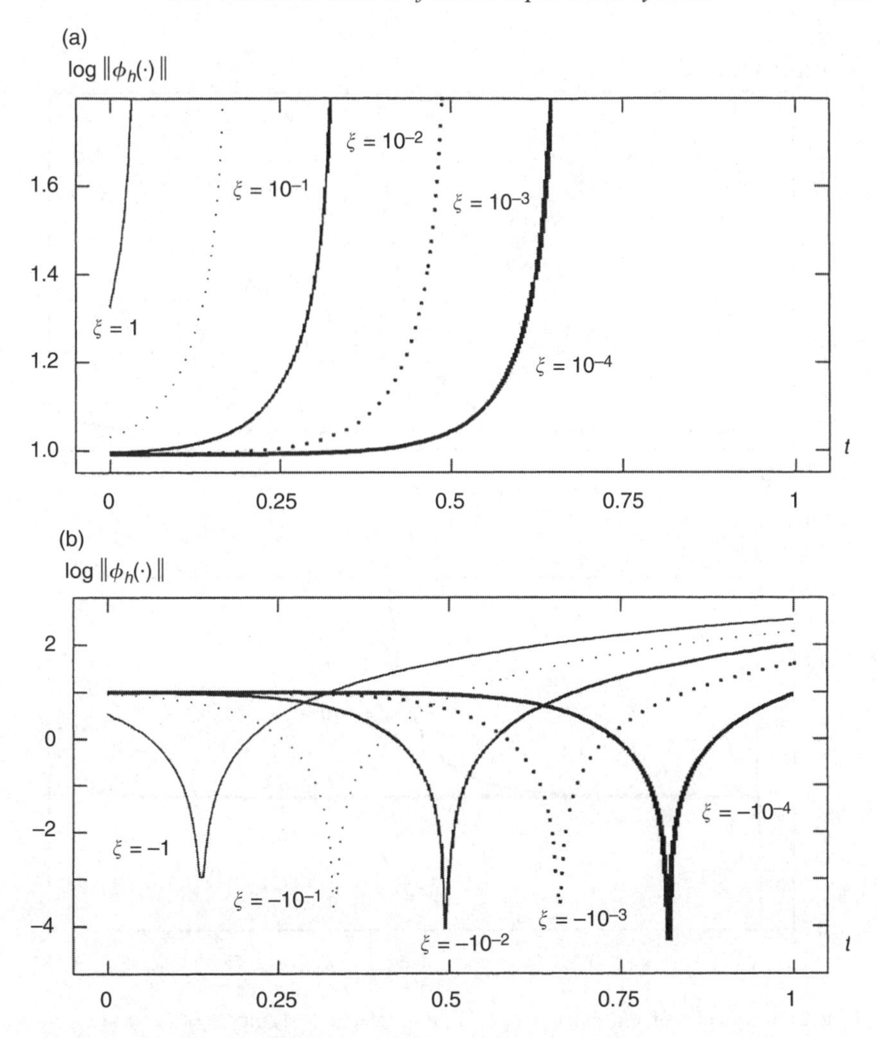

Fig. 2.58. Variation of $\log \|\phi_h(t)\|_{L^2(\Omega)}$ on the time interval (0, 1) for $C = 1.5\pi^2$ and $l = 4$ (no control): (a) with $\xi = 1, 10^{-1}, 10^{-2}, 10^{-3}, 10^{-4}$; (b) with $\xi = -1, -10^{-1}, -10^{-2}, -10^{-3}, -10^{-4}$.

and $t = 1$ are sufficient to stabilize the linearized system but are unable to do the same for the original nonlinear system.

In order to overcome this difficulty, we have divided the time interval under consideration into subintervals of smaller length $\Delta T = T/Q$, and we denote $q\Delta T$ by T_q for $q = 1, \ldots, Q$; we proceed then as follows:

(i) For $q = 0$, we denote by y_0 the difference $\phi_0 - \phi_s$, and we solve the linearized control problem (2.248) on $[0, T_1]$; let us denote by u_1 the corresponding control. This control is injected in (2.444)–(2.447), and we denote by y_1 the difference $\phi(T_1) - \phi_s$.

(ii) For $q > 0$, we denote by y_q the difference $\phi(T_q) - \phi_s$; we solve the linearized control problem (2.248) on $[T_q, T_{q+1}]$, with y_0 replaced by y_q, and we denote by

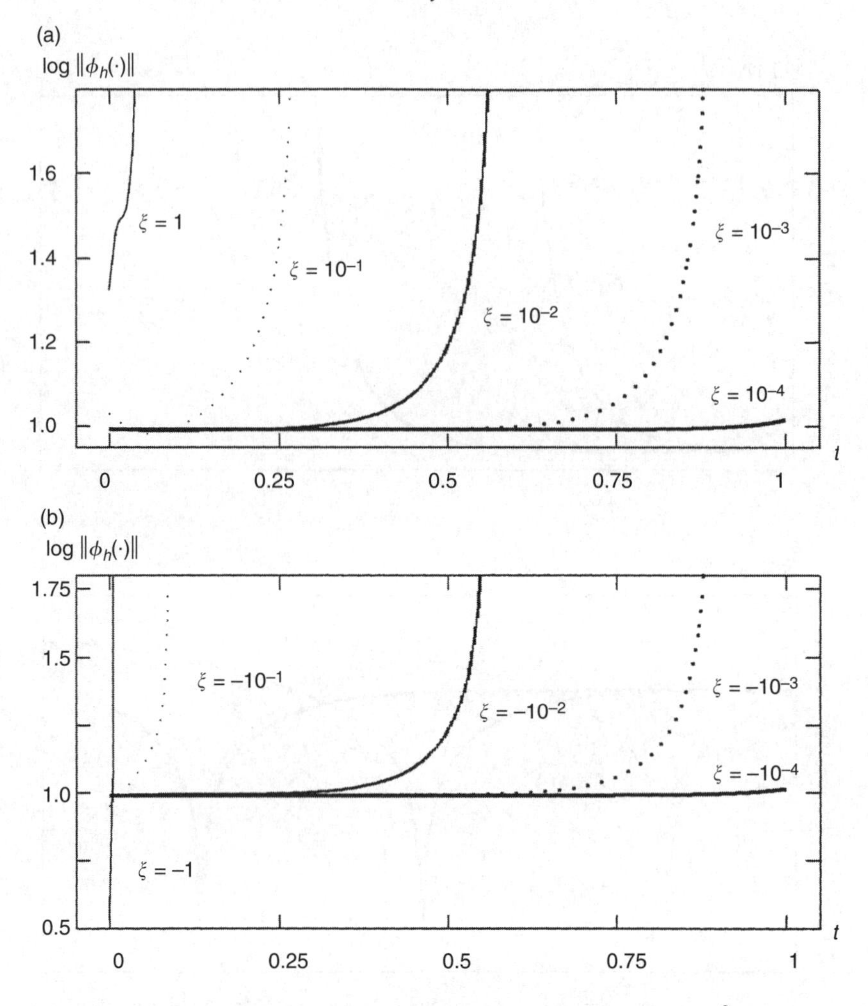

Fig. 2.59. Variation of $\log \|\phi_h(t)\|_{L^2(\Omega)}$ on the time interval (0, 1) for $C = 1.5\pi^2$ and $l = 4$ by injecting the chattering optimal control, computed by solving the linearized control problems: (a) with $\xi = 1, 10^{-1}, 10^{-2}, 10^{-3}, 10^{-4}$; (b) with $\xi = -1, -10^{-1}, -10^{-2}, -10^{-3}, -10^{-4}$.

u_{q+1} the corresponding optimal control. The control u_{q+1} is injected in (2.448) and we denote by y_{q+1} the difference $\phi(T_{q+1}) - \phi_s$.

(iii) We do $q = q + 1$ and we repeat the process.

The above time partitioning method has been applied to the system (2.444)–(2.447), with ϕ_0 defined by (2.448) with $\xi = 10^{-3}$, the time interval under consideration being $[0, 3]$; we have used $\Delta T = 0.25$. After $t = 3$, we have taken $v = 0$ in (2.444)–(2.447) to observe the evolution of the suddenly uncontrolled nonlinear system. The results are reported in Figures 2.60–2.62. We observe that the system is practically stabilized for $1 \leq t \leq 3$, but if one stops controlling, the small residual perturbations of the system at $t = 3$, are sufficient to destabilize the system and induce blowing-up in finite time.

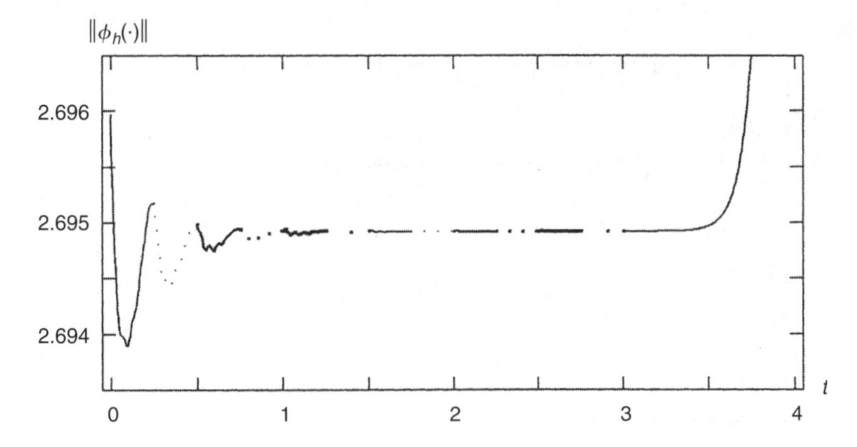

Fig. 2.60. Variation of $\|\phi_h(t)\|_{L^2(\Omega)}$ on the time interval $[0, 4]$ for $C = 1.5\pi^2$ and $l = 4$ and $\xi = 10^{-3}$ (on the time interval $[0, 3]$ we have injected the chattering optimal control computed by solving the sequence of linearized control problems associated with the time partitioning method with $\Delta T = 0.25$).

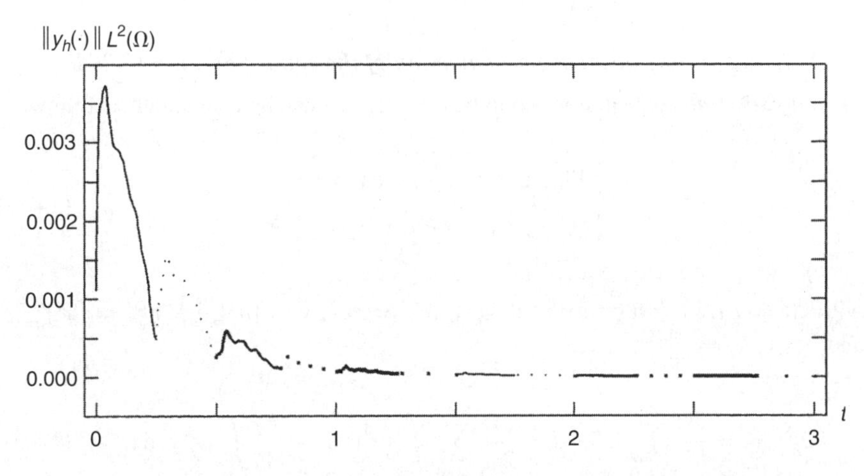

Fig. 2.61. Variation of $\|y_h(t)\|_{L^2(\Omega)}$ on the time interval $[0, 3]$ for $C = 1.5\pi^2$ and $l = 4$ and $\xi = 10^{-3}$ (chattering optimal control computed by solving the sequence of linearized control problems associated with the time partitioning method with $\Delta T = 0.25$).

2.12 Closed-loop Neumann control of unstable parabolic systems via the Riccati equation approach

2.12.1 Generalities

In order to speed up the *Neumann* control of the parabolic systems considered in Section 2.11, make this control *recursive* and be able to express it as a function of the state variable, we are going to investigate a *Riccati equation* approach (which follows closely Section 2.5 of He, Glowinski, Metcalfe, and Périaux (1998); see also Section 1.13 in Chapter 1 of the present book for an introduction to Riccati equation based control methods). The control problem to be investigated via a Riccati equation based method is thus (2.248) (see Section 2.11.3 for its precise definition).

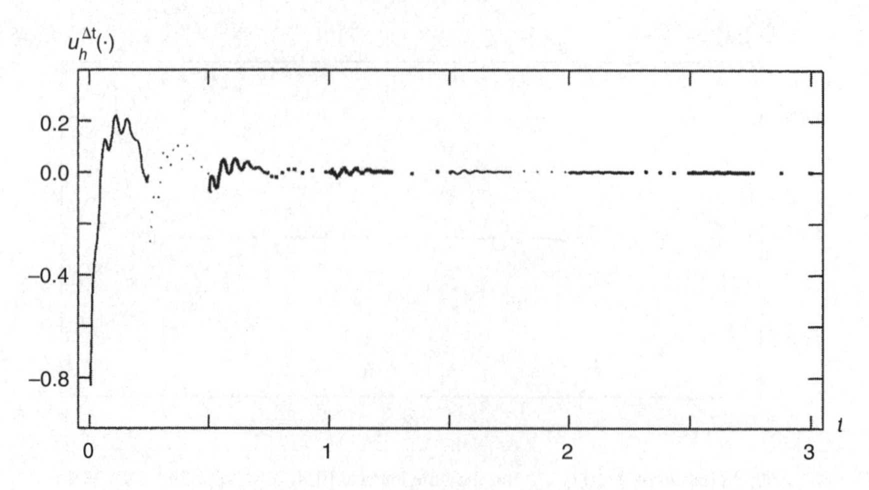

Fig. 2.62. Chattering optimal control $u_h^{\Delta t}(t)$ on the time interval $[0, 3]$ for $C = 1.5\pi^2$ and $l = 4$ and $\xi = 10^{-3}$ (computed by solving the sequence of linearized control problems associated with the time partitioning method with $\Delta T = 0.25$).

2.12.2 A fully discrete approximation of the control problem (2.248)

The *fully discrete* control problem to be investigated can be formulated as follows:

$$\begin{cases} \text{Find } \mathbf{u}_h^{\Delta t} \in \mathcal{U}^{\Delta t}, \text{ such that} \\ J_h^{\Delta t}(\mathbf{u}_h^{\Delta t}) \leq J_h^{\Delta t}(\mathbf{v}), \quad \mathbf{v} \in \mathcal{U}^{\Delta t}, \end{cases} \tag{2.449}$$

with $\Delta t = T/N$ (N: a positive integer), $\mathcal{U}^{\Delta t} = \mathbb{R}^N$, $\mathbf{v} = \{v^n\}_{n=0}^{N-1}$, $\mathbf{u}_h^{\Delta t} = \{u^n\}_{n=0}^{N-1}$, and

$$J_h^{\Delta t}(\mathbf{v}) = \frac{\Delta t}{2} \sum_{n=0}^{N-1} |v^n|^2 + \frac{k_1\, \Delta t}{2} \sum_{n=0}^{N-1} \int_\Omega |y_h^n|^2\, dx + \frac{k_2}{2} \int_\Omega |y_h^N|^2\, dx, \tag{2.450}$$

with $\{y_h^n\}_{n=0}^N$ obtained from $\mathbf{v}$ as follows:

$$y_h^0 = y_{0h} \in V_h (y_{0h} \approx y_0); \tag{2.451}$$

for $n = 0, \ldots, N-1$,

$$\begin{cases} y_h^{n+1} \in V_h, \\ \displaystyle\int_\Omega \frac{y_h^{n+1} - y_h^n}{\Delta t} z\, dx + \nu \int_\Omega \nabla y_h^n \cdot \nabla z\, dx + \int_\Omega \mathbf{V} \cdot \nabla y_h^n\, z\, dx \\ \displaystyle\qquad - C \int_\Omega y_h^n\, z\, dx = v^n \int_\gamma z\, d\Gamma, \quad \forall z \in V_h. \end{cases} \tag{2.452}$$

In (2.451) and (2.452), V_h is the *finite-dimensional subspace* of $H^1(\Omega)$ defined in Section 2.11.6.1; that is, it is obtained via a classical *finite element method*, based on *piecewise linear approximations*.

Remark 2.28 We have chosen, for simplicity, a *one step forward Euler scheme* for the *time discretization*. As done in Section 2.11, we can use time-discretization schemes which are more sophisticated than forward Euler's, leading to better accuracy and stability properties; however, these schemes will introduce highly nontrivial complications when deriving and solving the corresponding discrete *Riccati equations*.

In the following, we are going to use a *matrix formalism* in order to apply results of standard Control Theory for *Linear-Quadratic systems* (a relevant reference in that direction is Bittanti, Laub, and Willems, 1991). Let $\mathcal{B}_h = \{w_i\}_{i=1}^{NV}$ be a vector basis of the space V_h (that is, $NV = \dim(V_h)$). Expanding y_h^n over $\mathcal{B}_h$ we obtain thus

$$y_h^n = \sum_{i=1}^{NV} y_i^n \, w_i. \tag{2.453}$$

The cost function $J_h^{\Delta t}$ and the linear discrete state-equation can be expressed in matrix form as follows:

$$J_h^{\Delta t}(\mathbf{v}) = \frac{1}{2} \sum_{n=0}^{N-1} \left(\Delta t \, |v^n|^2 + (\mathbf{y}^n)^t \, \mathbf{Q} \, \mathbf{y}^n \right) + \frac{1}{2} \, (\mathbf{y}^N)^t \, \mathbf{P}^N \, \mathbf{y}^N, \tag{2.454}$$

and

$$\mathbf{y}^{n+1} = \mathbf{A} \, \mathbf{y}^n + \mathbf{B} \, v^n, \quad \forall \, n = 0, 1, \ldots, N-1, \tag{2.455}$$

respectively, with

$$\mathbf{y}^n = \left[y_1^n, y_2^n, \ldots, y_{NV}^n\right]^t, \quad \mathbf{Q} = k_1 \, \Delta t \, \mathbf{M}_h, \quad \mathbf{P}^N = k_2 \, \mathbf{M}_h,$$
$$\mathbf{A} = \mathbf{I} - \Delta t \, \mathbf{M}_h^{-1} \, \mathbf{L}_h, \quad \mathbf{B} = \Delta t \, \mathbf{M}_h^{-1} \, \mathbf{F}_h,$$

where

$$\mathbf{L}_h = (l_{ij}), \quad \mathbf{M}_h = (m_{ij}), \quad \mathbf{F}_h = [f_1, f_2, \ldots, f_{NV}]^t,$$

and

$$l_{ij} = \nu \int_\Omega \nabla w_i \cdot \nabla w_j \, dx + \int_\Omega \mathbf{V} \cdot \nabla w_j \, w_i \, dx - C \int_\Omega w_i \, w_j \, dx,$$
$$m_{ij} = \int_\Omega w_i \, w_j \, dx, \quad f_i = \int_\gamma w_i \, d\Gamma.$$

Remark 2.29 Using the *trapezoidal rule* to compute the m_{ij} yields a *diagonal matrix* for $\mathbf{M}_h$. This simple observation makes scheme (2.451), (2.452) *fully explicit* and greatly facilitates actual computations.

2.12.3 Optimality conditions for the discrete control problem

The discrete optimal control (that is, the solution of problem (2.449)) is characterized by the following relations:

$$\begin{cases} \mathbf{y}^0 = \mathbf{y}_0, \\ \mathbf{y}^{n+1} = \mathbf{A}\,\mathbf{y}^n + \mathbf{B}\,u^n, \quad \forall n = 0, 1, \ldots, N-1, \end{cases} \tag{2.456}$$

$$\begin{cases} \mathbf{p}^N = \mathbf{P}^N\,\mathbf{y}^N, \\ \mathbf{p}^n = \mathbf{A}^t\,\mathbf{p}^{n+1} + \mathbf{Q}\,\mathbf{y}^n, \quad \forall n = N-1, \ldots, 1, \end{cases} \tag{2.457}$$

$$u^n = -\frac{1}{\Delta t}\mathbf{B}^t\,\mathbf{p}^{n+1}, \quad \forall n = 0, 1, \ldots, N-1. \tag{2.458}$$

2.12.4 Discrete Riccati equations

Let assume that, $\forall n = N, N-1, \ldots, 1$, there exists a $NV \times NV$ matrix $\mathbf{P}^n$ so that

$$\mathbf{p}^n = \mathbf{P}^n\,\mathbf{y}^n. \tag{2.459}$$

Via algebraic manipulations (which are indeed classical) we can show that $\mathbf{P}^n$ has to satisfy the following *backward discrete Riccati equation*

$$\mathbf{P}^n = \left(\mathbf{A} - \mathbf{B}\mathbf{K}^n\right)^t \mathbf{P}^{n+1} \left(\mathbf{A} - \mathbf{B}\mathbf{K}^n\right) + \Delta t \left(\mathbf{K}^n\right)^t \mathbf{K}^n + \mathbf{Q}, \tag{2.460}$$

with the *Kalman gain* $\mathbf{K}^n$ defined by

$$\mathbf{K}^n = \left(\Delta t + \mathbf{B}^t\,\mathbf{P}^{n+1}\,\mathbf{B}\right)^{-1} \mathbf{B}^t\,\mathbf{P}^{n+1}\,\mathbf{A}. \tag{2.461}$$

Relations (2.460) and (2.461) have to be completed by the "initial" condition

$$\mathbf{P}^N = k_2\,\mathbf{M}_h. \tag{2.462}$$

Observe that $\mathbf{B}^t\,\mathbf{P}^{n+1}\,\mathbf{B}$ in (2.461) is a real number. Relations (2.460), (2.461) correspond to the so-called *Joseph stabilized version* of the Riccati equation; it has very good numerical properties, concerning particularly the *monotonicity* and *positivity* of matrices $\mathbf{P}^n$ (see Bittanti, Laub, and Willems, 1991 for further details).

From the above relations the optimal control u^n is obtained from $\mathbf{y}^n$ via the following *feedback law* (forward in time)

$$u^n = -\mathbf{K}^n\,\mathbf{y}^n, \tag{2.463}$$

the corresponding "trajectory" being defined, from (2.456), by

$$\mathbf{y}^{n+1} = \mathbf{A}\,\mathbf{y}^n + \mathbf{B}\,u^n; \tag{2.464}$$

both $\mathbf{y}^n$ and u^n can be computed on-line.

Remark 2.30 We observe that in practice matrices $\mathbf{P}^n$ and $\mathbf{K}^n$ can be computed *off-line* using relations (2.460) and (2.461); from a practical point of view, we only have to store the gain sequence, namely, $\{\mathbf{K}^n\}_{n=0}^{N-1}$. The Riccati matrices $\mathbf{P}^n$ do not have to be stored.

Remark 2.31 It is important to observe that when applying the above method, the computational requirements are dependent of the ways (2.460) and (2.461) are implemented. By exploiting the *symmetry* of matrices $\mathbf{P}^n$ (obvious from relations (2.460) and (2.462)), relations (2.460)–(2.462) can be implemented *off-line* as follows:

$$\mathbf{P} \Leftarrow \mathbf{P}^N$$

and for $n = N-1, \ldots, 0$

$$\begin{aligned}
\mathbf{w} &\Leftarrow \mathbf{B}^t\mathbf{P}, \\
d &\Leftarrow \left(\Delta t + \mathbf{B}^t\mathbf{w}^t\right)^{-1}, \\
\mathbf{K}^{nt} &\Leftarrow \mathbf{A}^t\mathbf{w}^t, \\
\mathbf{K}^{nt} &\Leftarrow \mathbf{K}^{nt}d, \\
\mathbf{Z} &\Leftarrow \mathbf{A}^t\mathbf{P}, \\
\mathbf{Z} &\Leftarrow \mathbf{Z} - \mathbf{K}^{nt}\mathbf{w}, \\
\mathbf{Z} &\Leftarrow \mathbf{Z}^t, \\
\mathbf{w} &\Leftarrow \mathbf{B}^t\mathbf{Z}, \\
\mathbf{P} &\Leftarrow \mathbf{A}^t\mathbf{Z}, \\
\mathbf{w} &\Leftarrow \mathbf{w} - \Delta t\mathbf{K}^n, \\
\mathbf{P} &\Leftarrow \mathbf{P} - \mathbf{K}^{nt}\mathbf{w}, \\
\mathbf{P} &\Leftarrow \mathbf{P} + \mathbf{Q}.
\end{aligned}$$

By exploiting the *sparsity* of the matrices $\mathbf{A}$ and $\mathbf{B}$, the above algorithm requires $O(N \times NV \times NV)$ multiplications; for large problems, this may be quite memory demanding.

2.12.5 Application to the boundary control of a reaction–diffusion model

The test problem that we consider is the first problem discussed in Section 2.11.7.2; it is, therefore, defined as follows:

$$\Omega = (0,1)^2 \ (\subset \mathbb{R}^2), \quad T = 1, \quad \nu = 1, \quad \mathbf{V} = \mathbf{0}, \quad C = 2,$$

and

$$\gamma = \left\{ x \,|\, x = \{x_1, x_2\},\, x_1 = 0,\, |x_2 - 1/2| < \sqrt{3}/32 \right\}.$$

The geometrical aspects of the test problem under consideration have been visualized in Section 2.11.7.1, Figure 2.37. The finite element triangulation used to define the space V_h introduced in Section 2.12.2 is similar to the one shown in Section 2.6, Figure 2.1, with $h = \Delta x_1 = \Delta x_2 = \frac{1}{32}$. The initial data y_0 is the randomly generated one already visualized in Section 2.11.7.1, Figure 2.38.

As shown in Section 2.11.7.2, without control the function $t \to \|y_h(t)\|_{L^2(\Omega)}$ behaves like $\to \|y_0\|_{L^2(\Omega)}\, e^{2t}$ (a behavior explained in Sections 2.11.7.1 and 2.11.7.2).

Remark 2.32 As already mentioned, the open loop control of the current test problem has been discussed in Section 2.11.7.2, the discussion including variants where the advective field $\mathbf{V}$ is different from $\mathbf{0}$.

To solve the current test problem via a feedback control à la Riccati, we have chosen $\Delta t = \frac{1}{5000}$, which is small enough to preserve the positivity and monotonicity of the Riccati matrices $\mathbf{P}^n$, properties which are crucial for the *stability* of the *backward recursion algorithm*. Actually, in order to overcome the memory requirements associated with $N = T/\Delta t = 5000$, we have divided $[0, T]$ in J subintervals of smaller length ($J = \frac{1}{100}$, here) and we have denoted by T_j, for $j = 1, \ldots, J$, the discrete time $j\Delta T$, where $T = T/J$. We proceed then as follows:

(i) We solve the Riccati equation *backward* on the time interval $[0, T_1]$, using the matrix $k_2\, \mathbf{M}_h$ as starting value. The optimal gains $\mathbf{K}^n$ are also computed and stored, $\forall\, n = 0, \ldots, \Delta T/\Delta t = N/J$.

(ii) For $q = 0$, starting from the state vector $\mathbf{y}_0$ (that is, the one associated with y_0), we use the gains $\mathbf{K}^n$ to compute the optimal control and the corresponding state trajectory on $[0, T_1]$, via the feedback relations (2.463) and (2.264), and we denote by $\mathbf{y}_1$ the computed state vector at $t = T_1$.

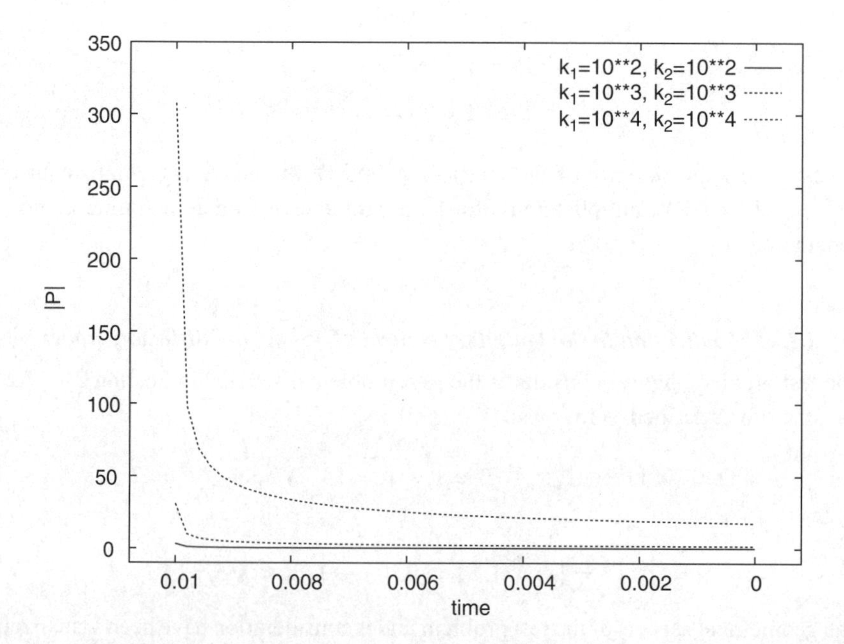

Fig. 2.63. Variation versus $n\Delta t$ of the Frobenius norm of matrix $\mathbf{P}^n$.

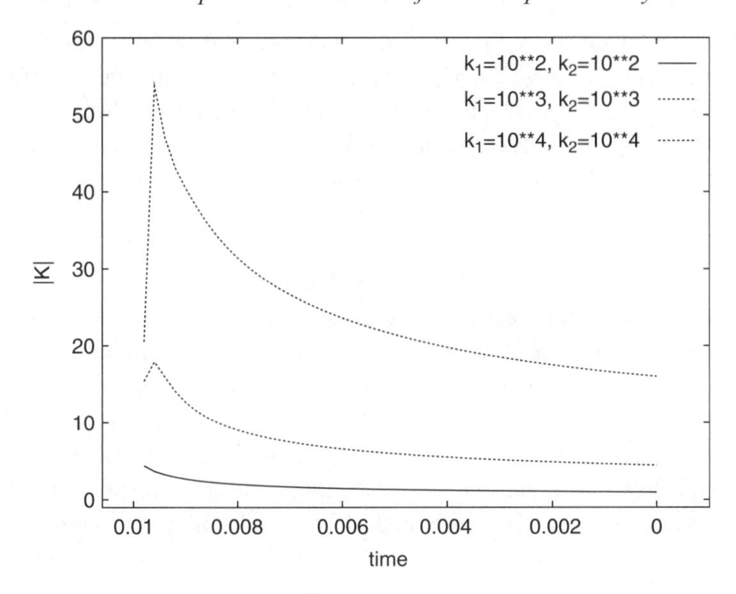

Fig. 2.64. Variation versus $n\Delta t$ of the Euclidean norm of vector $\mathbf{K}^n$.

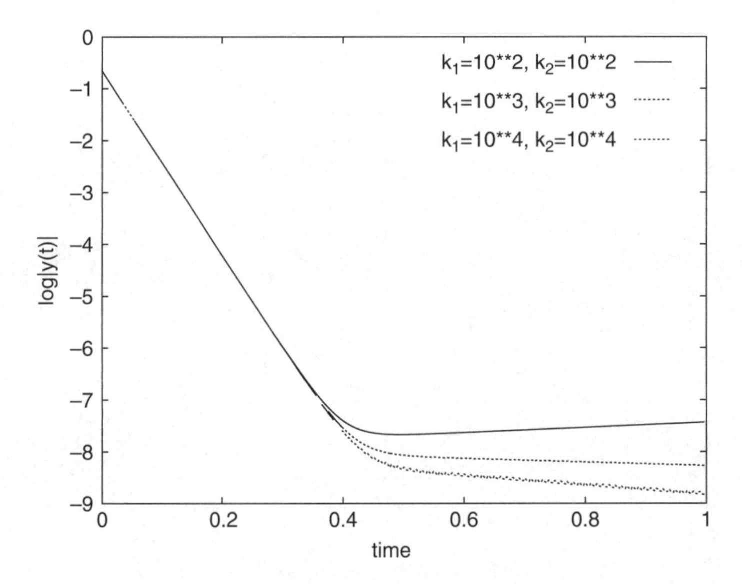

Fig. 2.65. Variation versus $n\Delta t$ of $\|y^n\|_{L^2(\Omega)}$.

(iii) For $q > 0$, we proceed as in (ii) with $[0, T_1]$ (respectively, $\mathbf{y}_0$) replaced by $[T_q, T_{q+1}]$ (respectively, $\mathbf{y}_q$). We denote by $\mathbf{y}_{q+1}$ the state vector at $t = T_{q+1}$.
(iv) We do $q \Leftarrow q + 1$ and repeat the process.

For various values of the penalty parameters k_1 and k_2 we have shown on Figure 2.63 (respectively Figure 2.64) the time variation of the Frobenius norm (respectively, Euclidean norm) of matrix $\mathbf{P}^n$ (respectively, vector $\mathbf{K}^n$). To respect the backward in time character of (some parts of) the computational process, we have reversed the time orientation in the above figures (that is, t decreases from left to right). Finally, we have represented on Figure 2.65 (using a log-scale) the variation of the function $t \rightarrow \|y_h(t)\|_{L^2(\Omega)}$; the closed-loop control method that we have just discussed, clearly drives $y_h(t)$ to 0 exponentially on the time interval $[0, 0.4]$ (beyond $t = 0.4$, we suspect that round-off errors limit the decay of $\|y_h(t)\|_{L^2(\Omega)}$).

2.12.6 Further comments

We have discussed in this section the closed-loop control of systems modeled by reaction–diffusion equations. When implementable, the closed-loop approach is robust, stable and fast. The memory requirement associated with the Riccati equation approach can be reduced via *dimension reduction techniques* like those provided by the *Proper Orthogonal Decomposition* (POD) methods (see, for example, Holmes, Lumley, and Berkooz, 1996; and Antoulas, 2005, and the references therein, for dimension reduction via POD related methods and applications to Control).

3

Control of the Stokes system

3.1 Generalities. Synopsis

In the original text (that is, *Acta Numerica 1995*), the content of this chapter was considered as a preliminary step to a more ambitious goal, namely, the *control of systems governed by the Navier–Stokes equations modeling incompressible viscous flow*. Indeed, substantial progress concerning this objective took place in the late 1990s (some of them to be reported in Part III of this book), making – in some sense – this chapter obsolete. We decided to keep it since it addresses some important issues that will not be considered in Part III (and also because it reflects some of the J.L. Lions scientific concerns at the time).

Back to the original text, let us say that the control problems and methods which have been discussed so far in this book have been mostly concerned with systems governed by *linear diffusion equations of the parabolic type*, associated with *second-order elliptic operators*. Indeed, these methods have been applied in, for example, Berggren (1992) and Berggren, Glowinski, and J.L. Lions (1996b), to the solution of *approximate boundary controllability* problems for systems governed by *strongly* advection dominated *linear advection–diffusion equations*. These methods can also be applied to *systems* of linear advection–diffusion equations and to *higher-order parabolic equations* (or *systems* of such equations). Motivated by the solution of controllability problems for the *Navier–Stokes equations* modeling *incompressible* viscous flow, we will discuss now controllability issues for a system of partial differential equations *which is not of the Cauchy–Kowalewska type*, namely, the classical *Stokes system*.

3.2 Formulation of the Stokes system. A fundamental controllability result

In the following, we equip the Euclidean space $\mathbb{R}^d$ ($d \geq 2$) with its classical scalar product and with the corresponding norm, that is,

$$\mathbf{a} \cdot \mathbf{b} = \sum_{i=1}^{d} a_i\, b_i, \quad \forall\, \mathbf{a} = \{a_i\}_{i=1}^{d},\ \mathbf{b} = \{b_i\}_{i=1}^{d} \in \mathbb{R}^d;$$

$$|\mathbf{a}| = (\mathbf{a} \cdot \mathbf{a})^{\frac{1}{2}}, \qquad \forall\, \mathbf{a} \in \mathbb{R}^d.$$

We suppose from now on that the control $\mathbf{v}$ is distributed over Ω, with its support in $\overline{\mathcal{O}} \subset \Omega$ (as in Chapter 1, Sections 1.1 to 1.8, whose notation is kept). The *state equation* is given by

$$\begin{cases} \dfrac{\partial \mathbf{y}}{\partial t} - \Delta \mathbf{y} = \mathbf{v}\,\chi_{\mathcal{O}} - \nabla \pi & \text{in } Q, \\ \nabla \cdot \mathbf{y} = 0 & \text{in } Q, \end{cases} \tag{3.1}$$

subject to the following *initial* and *boundary conditions*:

$$\mathbf{y}(0) = \mathbf{0} \quad \text{and} \quad \mathbf{y} = \mathbf{0} \ \text{ on } \Sigma. \tag{3.2}$$

In (3.1), we shall assume that

$$\mathbf{v} \in \mathcal{V} = \text{ a closed subspace of } (L^2(\mathcal{O} \times (0, T)))^d. \tag{3.3}$$

To fix ideas, we shall take $d = 3$, and consider the following cases for $\mathcal{V}$:

$$\mathcal{V} = (L^2(\mathcal{O} \times (0, T)))^3, \tag{3.4}$$

$$\mathcal{V} = \{\mathbf{v} \,|\, \mathbf{v} = \{v_1, v_2, 0\},\ \{v_1, v_2\} \in (L^2(\mathcal{O} \times (0, T)))^2\}, \tag{3.5}$$

$$\mathcal{V} = \{\mathbf{v} \,|\, \mathbf{v} = \{v_1, 0, 0\},\ v_1 \in L^2(\mathcal{O} \times (0, T))\}. \tag{3.6}$$

Problem (3.1), (3.2) has a unique solution, such that (in particular)

$$\begin{cases} \mathbf{y}(\mathbf{v}) \in L^2(0, T; (H_0^1(\Omega))^3), \quad \nabla \cdot \mathbf{y}(\mathbf{v}) = 0, \\ \dfrac{\partial}{\partial t}\mathbf{y}(\mathbf{v}) \in L^2(0, T; V'), \end{cases} \tag{3.7}$$

where V' is the *dual space* of

$$V = \{\boldsymbol{\varphi} \,|\, \boldsymbol{\varphi} \in (H_0^1(\Omega))^3,\ \nabla \cdot \boldsymbol{\varphi} = 0\}. \tag{3.8}$$

It follows from (3.7) that

$$t \to \mathbf{y}(t; \mathbf{v}) \quad \text{belongs to} \quad C^0([0, T]; H), \tag{3.9}$$

where

$$\begin{aligned} H &= \text{ closure of } V \text{ in } (L^2(\Omega))^3 \\ &= \{\boldsymbol{\varphi} \,|\, \boldsymbol{\varphi} \in (L^2(\Omega))^3,\ \nabla \cdot \boldsymbol{\varphi} = 0,\ \boldsymbol{\varphi} \cdot \mathbf{n} = 0 \text{ on } \Gamma\} \end{aligned} \tag{3.10}$$

($\mathbf{n}$ denotes the unit outward normal vector at Γ).

We are now going to prove the following:

Proposition 3.1 *If V is defined by either* (3.4) *or* (3.5), *then the space spanned by* $\mathbf{y}(T;\mathbf{v})$ *is dense in* H.

Proof. It suffices to prove the above results for the case where V is defined by (3.5). Let us therefore consider $\mathbf{f} \in H$ such that,

$$\int_\Omega \mathbf{y}(T;\mathbf{v}) \cdot \mathbf{f}\, dx = 0, \quad \forall \mathbf{v} \in \mathcal{V}. \tag{3.11}$$

With $\mathbf{f}$ we associate the solution $\boldsymbol{\psi}$ of the following *backward* Stokes problem:

$$\begin{cases} -\dfrac{\partial \boldsymbol{\psi}}{\partial t} - \Delta \boldsymbol{\psi} = -\nabla\sigma & \text{in } Q, \\ \nabla \cdot \boldsymbol{\psi} = 0 & \text{in } Q, \end{cases} \tag{3.12}$$

$$\boldsymbol{\psi}(T) = \mathbf{f}, \quad \boldsymbol{\psi} = \mathbf{0} \quad \text{on } \Sigma. \tag{3.13}$$

Multiplying the first equation in (3.12) by $\mathbf{y} = \mathbf{y}(\mathbf{v})$ and integrating by parts we obtain

$$\iint_{\Omega\times(0,T)} \boldsymbol{\psi} \cdot \mathbf{v}\, dx\, dt = 0, \quad \forall \mathbf{v} \in \mathcal{V}. \tag{3.14}$$

Suppose that $\boldsymbol{\psi} = \{\psi_1, \psi_2, \psi_3\}$; it follows then from (3.14) that

$$\psi_2 = \psi_2 = 0 \quad \text{in } \mathcal{O} \times (0,T). \tag{3.15}$$

Since $\boldsymbol{\psi}$ is (among other things) *continuous* in t and *real analytic* in x in $Q(= \Omega \times (0,T))$, it follows from (3.15) that

$$\psi_1 = \psi_2 = 0 \quad \text{in} \quad \Omega \times (0,T). \tag{3.16}$$

Since (from (3.12)) $\nabla \cdot \boldsymbol{\psi} = 0$, it follows from (3.16) that $\frac{\partial \psi_3}{\partial x_3} = 0$ in $\Omega \times (0,T)$, which combined with the boundary condition $\psi_3 = 0$ on Σ implies that $\psi_3 = 0$ in $\Omega \times (0,T)$; the t-continuity of $\boldsymbol{\psi}$ implies that $\boldsymbol{\psi}(T) = \mathbf{0}$, that is, $\mathbf{f} = \mathbf{0}$ (from (3.13)), which completes the proof. □

Remark 3.2 The above density result does not always hold if $\mathcal{V}$ is defined by (3.6), as shown in Diaz and Fursikov (1997).

Remark 3.3 Proposition 3.1 was proved in the lectures of the second author (J.L. Lions) at *Collège de France* in 1990–91. Other results along these lines are due to Fursikov (1992).

The density result in Proposition 3.1 implies (at least) *approximate controllability*. Thus, we shall formulate and discuss, in the following sections, two approximate controllability problems.

3.3 Two approximate controllability problems

The *first problem* is defined by

$$\min_{\mathbf{v}\in\mathcal{U}_f} \frac{1}{2} \iint_{\mathcal{O}\times(0,T)} |\mathbf{v}|^2 \, dx \, dt, \tag{3.17}$$

where

$$\mathcal{U}_f = \{\mathbf{v} | \mathbf{v} \in \mathcal{V}, \{\mathbf{v}, \mathbf{y}\} \text{ verifies (3.1), (3.2), and } \mathbf{y}(T) \in \mathbf{y}_T + \beta B_H\}; \tag{3.18}$$

in (3.18), $\mathbf{y}_T$ is given in H, β is a positive number arbitrarily small, B_H is the unit ball of H and – to fix ideas – the control space $\mathcal{V}$ is defined by (3.5).

The *second problem* is obtained by *penalization* of the final condition $\mathbf{y}(T) = \mathbf{y}_T$; we have then

$$\min_{\mathbf{v}\in\mathcal{V}} \left\{ \frac{1}{2} \iint_{\mathcal{O}\times(0,T)} |\mathbf{v}|^2 \, dx \, dt + \frac{1}{2} k \int_\Omega |\mathbf{y}(T) - \mathbf{y}_T|^2 \, dx \right\}, \tag{3.19}$$

where, in (3.19), k is an arbitrarily large positive number, $\mathbf{y}$ is obtained from $\mathbf{v}$ via (3.1), (3.2), and $\mathcal{V}$ is as above.

It follows from Proposition 3.1 that both control problems (3.17) and (3.19) *have a unique solution.*

3.4 Optimality conditions and dual problems

We start with problem (3.19), since it is (by far) simpler than problem (3.17). If we denote by J_k the cost functional in (3.19), we have

$$\begin{aligned} (J_k'(\mathbf{v}), \mathbf{w}) &= \lim_{\substack{\theta\to 0\\ \theta\neq 0}} \frac{J_k(\mathbf{v} + \theta\mathbf{w}) - J_k(\mathbf{v})}{\theta} \\ &= \iint_{\mathcal{O}\times(0,T)} (\mathbf{v} - \mathbf{p}) \cdot \mathbf{w} \, dx \, dt, \quad \forall \mathbf{v}, \mathbf{w} \in \mathcal{V}, \end{aligned} \tag{3.20}$$

where, in (3.20), the *adjoint velocity field* $\mathbf{p}$ is solution of the following *backward Stokes problem*:

$$\begin{cases} -\dfrac{\partial \mathbf{p}}{\partial t} - \Delta\mathbf{p} + \nabla\sigma = \mathbf{0} & \text{in } Q, \\ \nabla \cdot \mathbf{p} = 0 & \text{in } Q, \end{cases} \tag{3.21}$$

$$\mathbf{p} = \mathbf{0} \text{ on } \Sigma, \quad \mathbf{p}(T) = k\,(\mathbf{y}_T - \mathbf{y}(T)). \tag{3.22}$$

Suppose now that $\mathbf{u}$ is the unique solution of problem (3.19); it is *characterized* by

$$\begin{cases} \mathbf{u} \in \mathcal{V}, \\ (J_k'(\mathbf{u}), \mathbf{w}) = 0, \quad \forall \mathbf{w} \in \mathcal{V}, \end{cases} \tag{3.23}$$

which implies in turn that the *optimal triple* $\{\mathbf{u}, \mathbf{y}, \mathbf{p}\}$ is characterized by

$$u_1 = p_1|_{\mathcal{O}}, \quad u_2 = p_2|_{\mathcal{O}}, \quad u_3 = 0, \tag{3.24}$$

$$\begin{cases} \dfrac{\partial \mathbf{y}}{\partial t} - \Delta \mathbf{y} + \nabla \pi = \mathbf{u}\chi_{\mathcal{O}} & \text{in } Q, \\ \nabla \cdot \mathbf{y} = 0 & \text{in } Q, \end{cases} \tag{3.25}$$

$$\mathbf{y}(0) = \mathbf{0}, \quad \mathbf{y} = \mathbf{0} \text{ on } \Sigma, \tag{3.26}$$

to be completed by (3.21), (3.22).

To obtain the *dual problem* of (3.19) from the above optimality conditions, we proceed as in Chapters 1 and 2 by introducing an operator $\mathbf{\Lambda} \in \mathcal{L}(H, H)$ defined as follows:

$$\mathbf{\Lambda}\,\mathbf{g} = \boldsymbol{\varphi}_g(T), \quad \forall \mathbf{g} \in H, \tag{3.27}$$

where to obtain $\boldsymbol{\varphi}_g(T)$ we solve first

$$\begin{cases} -\dfrac{\partial \boldsymbol{\psi}_g}{\partial t} - \Delta \boldsymbol{\psi}_g + \nabla \sigma_g = \mathbf{0} & \text{in } Q, \\ \nabla \cdot \boldsymbol{\psi}_g = 0 & \text{in } Q, \end{cases} \tag{3.28}$$

$$\boldsymbol{\psi}_g(T) = \mathbf{g}, \quad \boldsymbol{\psi}_g = \mathbf{0} \text{ on } \Sigma, \tag{3.29}$$

and then (with obvious notation)

$$\begin{cases} \dfrac{\partial \boldsymbol{\phi}_g}{\partial t} - \Delta \boldsymbol{\phi}_g + \nabla \pi_g = \{\psi_{1g}, \psi_{2g}, 0\}\,\chi_{\mathcal{O}} & \text{in } Q, \\ \nabla \cdot \boldsymbol{\varphi}_g = 0 & \text{in } Q, \end{cases} \tag{3.30}$$

$$\boldsymbol{\varphi}_g(0) = \mathbf{0}, \quad \boldsymbol{\varphi}_g = \mathbf{0} \text{ on } \Sigma \tag{3.31}$$

(the two above Stokes problems are well posed).

Integrating by parts in time and using Green's formula, we can show (again with obvious notation) that

$$\int_\Omega (\mathbf{\Lambda}\mathbf{g}) \cdot \mathbf{g}' \, dx = \iint_{\mathcal{O}\times(0,T)} (\psi_1 \psi_1' + \psi_2 \psi_2') \, dx \, dt, \quad \forall \mathbf{g}, \mathbf{g}' \in H. \tag{3.32}$$

It follows from relation (3.32) that the operator $\mathbf{\Lambda}$ is *symmetric* and *positive semi definite* over H; indeed, using the approach taken in Section 3.2 to prove Proposition 3.1, we can show that the operator $\mathbf{\Lambda}$ is *positive definite* over H. Back to the *optimality conditions*, let us denote by $\mathbf{f}$ the function $\mathbf{p}(T)$; it follows then from (3.22) and from the definition of $\mathbf{\Lambda}$ that $\mathbf{f}$ satisfies

$$k^{-1}\,\mathbf{f} + \mathbf{\Lambda}\,\mathbf{f} = \mathbf{y}_T \tag{3.33}$$

which is precisely the dual problem of (3.19). From the *symmetry* and *positivity* of $\mathbf{\Lambda}$, the dual problem (3.33) can be solved by a *conjugate gradient algorithm* operating in the space H.

We consider now the control problem (3.17); applying, as done previously, the *Fenchel–Rockafellar duality theory* it can be shown that the unique solution $\mathbf{u}$ of problem (3.17) can be obtained via

$$u_1 = p_1 \chi_{\mathcal{O}}, \quad u_2 = p_2 \chi_{\mathcal{O}}, \quad u_3 = 0, \tag{3.34}$$

where, in (3.34), $\mathbf{p}$ is the solution of the following backward Stokes problem:

$$\begin{cases} -\dfrac{\partial \mathbf{p}}{\partial t} - \Delta \mathbf{p} + \nabla \sigma = \mathbf{0} \quad \text{in } Q, \\ \nabla \cdot \mathbf{p} = 0 \quad \text{in } Q, \end{cases} \tag{3.35}$$

$$\mathbf{p}(T) = \mathbf{f}, \quad \mathbf{p} = \mathbf{0} \text{ on } \Sigma, \tag{3.36}$$

where, in (3.36), $\mathbf{f}$ is *the* solution of the following *variational inequality*:

$$\begin{cases} \mathbf{f} \in H, \quad \forall \mathbf{g} \in H, \\ \displaystyle\int_\Omega (\mathbf{\Lambda f}) \cdot (\mathbf{g} - \mathbf{f})\, dx + \beta \|\mathbf{g}\|_H - \beta \|\mathbf{f}\|_H \geq \int_\Omega \mathbf{y}_T \cdot (\mathbf{g} - \mathbf{f})\, dx, \end{cases} \tag{3.37}$$

with $\|\mathbf{g}\|_H = \left(\int_\Omega |\mathbf{g}|^2\, dx\right)^{1/2}$. Problem (3.37) can be viewed as the *dual* of problem (3.17).

3.5 Iterative solution of the control problem (3.19)

The various *primal* and *dual* control problems considered in Sections 3.3 and 3.4 can be solved by variants of the algorithms which have been used to solve their scalar diffusion analogues; these algorithms have been described in Chapter 1, Section 1.8. Here, we shall focus on the *direct solution* of the control problem (3.19) by a *conjugate gradient algorithm*, since we will use this approach to solve the test problems considered in Section 3.7. The unique solution $\mathbf{u}$ of the control problem (3.19) is *characterized* as being also the unique solution of the *linear variational* problem (3.23). From the properties of the functional J_k, this problem is a particular case of problem (1.123) in Chapter 1, Section 1.8.2; applying thus the conjugate gradient algorithm (1.124)–(1.131) to problem (3.23) we obtain

$$\mathbf{u}^0 \text{ chosen in } \mathcal{V}; \tag{3.38}$$

solve

$$\begin{cases} \dfrac{\partial \mathbf{y}^0}{\partial t} - \Delta \mathbf{y}^0 + \nabla \pi^0 = \mathbf{u}^0 \chi_{\mathcal{O}} \quad \text{in } Q, \\ \nabla \cdot \mathbf{y}^0 = 0 \quad \text{in } Q, \end{cases} \tag{3.39}$$

$$\mathbf{y}^0(0) = \mathbf{0}, \quad \mathbf{y}^0 = \mathbf{0} \text{ on } \Sigma, \tag{3.40}$$

and then

$$\begin{cases} -\dfrac{\partial \mathbf{p}^0}{\partial t} - \Delta \mathbf{p}^0 + \nabla \sigma^0 = \mathbf{0} & \text{in } Q, \\ \nabla \cdot \mathbf{p}^0 = 0 & \text{in } Q, \end{cases} \tag{3.41}$$

$$\mathbf{p}^0(T) = k\,(\mathbf{y}_T - \mathbf{y}^0(T)), \quad \mathbf{p}^0 = \mathbf{0} \text{ on } \Sigma. \tag{3.42}$$

Solve now

$$\begin{cases} \mathbf{g}^0 \in \mathcal{V}, \\ \displaystyle\iint_{\mathcal{O}\times(0,T)} \mathbf{g}^0 \cdot \mathbf{v}\,\mathrm{d}x\,\mathrm{d}t = \iint_{\mathcal{O}\times(0,T)} (\mathbf{u}^0 - \mathbf{p}^0) \cdot \mathbf{v}\,\mathrm{d}x\,\mathrm{d}t, \quad \forall \mathbf{v} \in \mathcal{V}, \end{cases} \tag{3.43}$$

and set

$$\mathbf{w}^0 = \mathbf{g}^0. \tag{3.44}$$

Then for $n \geq 0$, assuming that $\mathbf{u}^n$, $\mathbf{g}^n$, and $\mathbf{w}^n$, are known, the last two different from $\mathbf{0}$*, we obtain $\mathbf{u}^{n+1}$, $\mathbf{g}^{n+1}$, and $\mathbf{w}^{n+1}$ as follows:*
Solve

$$\begin{cases} \dfrac{\partial \bar{\mathbf{y}}^n}{\partial t} - \Delta \bar{\mathbf{y}}^n + \nabla \bar{\pi}^n = \mathbf{w}^n\,\chi_{\mathcal{O}} & \text{in } Q, \\ \nabla \cdot \bar{\mathbf{y}}^n = 0 & \text{in } Q, \end{cases} \tag{3.45}$$

$$\bar{\mathbf{y}}^n(0) = \mathbf{0}, \quad \bar{\mathbf{y}}^n = \mathbf{0} \text{ on } \Sigma, \tag{3.46}$$

and then

$$\begin{cases} -\dfrac{\partial \bar{\mathbf{p}}^n}{\partial t} - \Delta \bar{\mathbf{p}}^n + \nabla \bar{\sigma}^n = \mathbf{0} & \text{in } Q, \\ \nabla \cdot \bar{\mathbf{p}}^n = 0 & \text{in } Q, \end{cases} \tag{3.47}$$

$$\bar{\mathbf{p}}^n(T) = -k\,\bar{\mathbf{y}}^n(T), \quad \bar{\mathbf{p}}^n = \mathbf{0} \text{ on } \Sigma. \tag{3.48}$$

Solve now

$$\begin{cases} \bar{\mathbf{g}}^n \in \mathcal{V}, \quad \forall \mathbf{v} \in \mathcal{V}, \\ \displaystyle\iint_{\mathcal{O}\times(0,T)} \bar{\mathbf{g}}^n \cdot \mathbf{v}\,\mathrm{d}x\,\mathrm{d}t = \iint_{\mathcal{O}\times(0,T)} (\mathbf{w}^n - \bar{\mathbf{p}}^n) \cdot \mathbf{v}\,\mathrm{d}x\,\mathrm{d}t, \end{cases} \tag{3.49}$$

and compute

$$\rho_n = \iint_{\mathcal{O}\times(0,T)} |\mathbf{g}^n|^2\,\mathrm{d}x\,\mathrm{d}t \Big/ \iint_{\mathcal{O}\times(0,T)} \bar{\mathbf{g}}^n \cdot \mathbf{w}^n\,\mathrm{d}x\,\mathrm{d}t, \tag{3.50}$$

$$\mathbf{u}^{n+1} = \mathbf{u}^n - \rho_n \mathbf{w}^n, \tag{3.51}$$

$$\mathbf{g}^{n+1} = \mathbf{g}^n - \rho_n \bar{\mathbf{g}}^n. \tag{3.52}$$

If $\|\mathbf{g}^{n+1}\|_{(L^2(\mathcal{O}\times(0,T)))^d} / \|\mathbf{g}^0\|_{(L^2(\mathcal{O}\times(0,T)))^d} \leq \epsilon$, *take* $\mathbf{u} = \mathbf{u}^{n+1}$*; else compute*

$$\gamma_n = \iint_{\mathcal{O}\times(0,T)} |\mathbf{g}^{n+1}|^2 \,dx\,dt \Big/ \iint_{\mathcal{O}\times(0,T)} |\mathbf{g}^n|^2 \,dx\,dt, \tag{3.53}$$

and update $\mathbf{w}^n$ *by*

$$\mathbf{w}^{n+1} = \mathbf{g}^{n+1} + \gamma_n \mathbf{w}^n. \tag{3.54}$$

Do $n = n + 1$ *and go to* (3.45).

Remark 3.4 For a given value of ϵ the number of iterations necessary to obtain the convergence of algorithm (3.38)–(3.54) varies like $k^{1/2}$ (as before for closely related algorithms).

Remark 3.5 The implementation of algorithm (3.38)–(3.54) requires *efficient Stokes solvers*, for solving problems (3.39)–(3.40), (3.41)–(3.42), (3.45)–(3.46) and (3.47)–(3.48). Such solvers can be found in, for example, Glowinski and Le Tallec (1989), Glowinski (1991, 1992a, and 2003, Chapter 5); actually, this issue is fully addressed in the related article by Berggren, Glowinski, and Lions (1996b) (and in Glowinski, 2003, Chapter 5) for *more general* boundary conditions than Dirichlet.

3.6 Time discretization of the control problem (3.19)

The practical implementation of algorithm (3.38)–(3.54) requires space–time approximation of the control problem (3.19). Focusing on the *time discretization* only (the space discretization is addressed in Berggren, Glowinski, and Lions, 1996b) we introduce a *time-discretization step* $\Delta t = T/N$ (with N a *positive* integer), denote by $\mathbf{v}$ the vector $\{\mathbf{v}^n\}_{n=1}^N$ and approximate problem (3.19) by

$$\min_{\mathbf{v}\in\mathcal{V}^{\Delta t}} \left\{ \frac{1}{2}\Delta t \sum_{n=1}^{N} \int_{\mathcal{O}} |\mathbf{v}^n|^2 \,dx + \frac{1}{2}k \int_{\Omega} |\mathbf{y}^N - \mathbf{y}_T|^2 \,dx \right\}, \tag{3.55}$$

where, by analogy with (3.4)–(3.6), $\mathcal{V}^{\Delta t}$ is defined by either

$$\mathcal{V}^{\Delta t} = \left\{ \{\mathbf{v}^n\}_{n=1}^N \mid \mathbf{v}^n = \{v_1^n, v_2^n, v_3^n\} \in (L^2(\mathcal{O}))^3, \forall n = 1, \ldots, N \right\},$$

or

$$\mathcal{V}^{\Delta t} = \left\{ \{\mathbf{v}^n\}_{n=1}^N \mid \mathbf{v}^n = \{v_1^n, v_2^n, 0\}, \{v_1^n, v_2^n\} \in (L^2(\mathcal{O}))^2, \forall n = 1, \ldots, N \right\},$$
$$\mathcal{V}^{\Delta t} = \left\{ \{\mathbf{v}^n\}_{n=1}^N \mid \mathbf{v}^n = \{v_1^n, 0, 0\}, v_1^n \in L^2(\mathcal{O}), \forall n = 1, \ldots, N \right\},$$

and where $\mathbf{y}^n$ is obtained from $\mathbf{v}$ via

$$\mathbf{y}^0 = \mathbf{0}; \tag{3.56}$$

for $n = 1, \ldots, N$, we obtain $\{\mathbf{y}^n, \pi^n\}$ from $\mathbf{y}^{n-1}$ by solving the following Stokes type problem:

$$\begin{cases} \dfrac{\mathbf{y}^n - \mathbf{y}^{n-1}}{\Delta t} - \Delta \mathbf{y}^n + \nabla \pi^n = \mathbf{v}^n \chi_{\mathcal{O}} & \text{in } \Omega, \\ \nabla \cdot \mathbf{y}^n = 0 & \text{in } \Omega, \end{cases} \tag{3.57}$$

$$\mathbf{y}^n = \mathbf{0} \quad \text{on } \Gamma. \tag{3.58}$$

The above scheme is nothing but a *backward Euler discretization* of problems (3.1) and (3.2). Efficient algorithms for solving problems (3.57) and (3.58) (and finite element approximations of it) can be found in, for example, Glowinski and Le Tallec (1989), Glowinski (1991, 1992a) (see also Berggren, Glowinski, and Lions, 1996b; Glowinski, 2003, Chapter 5). The discrete control problem (3.55) has a unique solution; for the optimality conditions and a discrete analogue of of the conjugate gradient algorithm (3.38)–(3.54) see Berggren, Glowinski, and Lions (1996b) (see also the above reference for a *full discretization* of problem (3.19) and solution methods for the fully discrete problem).

3.7 Numerical experiments

Following Berggren, Glowinski, and Lions (1996b), we (briefly) consider the practical solution of the following two-dimensional variant of problem (3.19):

$$\min_{\mathbf{v} \in \mathcal{V}} \left\{ \frac{1}{2} \iint_{\mathcal{O} \times (0,T)} |\mathbf{v}|^2 \, dx \, dt + \frac{1}{2} k \int_\Omega |\mathbf{y}(T) - \mathbf{y}_T|^2 \, dx \right\}, \tag{3.59}$$

where, in (3.59), $\mathcal{O} \subset \Omega \subset \mathbb{R}^2$, $\mathbf{v} = \{v_1, 0\}$, $\mathcal{V} = \big\{ \mathbf{v} \,|\, \mathbf{v} = \{v_1, 0\},\ v_1 \in L^2(\mathcal{O} \times (0,T)) \big\}$, where $\mathbf{y}(T)$ is obtained from $\mathbf{v}$ via the solution of the following *Stokes problem*:

$$\begin{cases} \dfrac{\partial \mathbf{y}}{\partial t} - \nu \Delta \mathbf{y} + \nabla \pi = \mathbf{v}\, \chi_{\mathcal{O}} & \text{in } Q, \\ \nabla \cdot \mathbf{y} = 0 & \text{in } Q, \end{cases} \tag{3.60}$$

$$\begin{cases} \mathbf{y}(0) = \mathbf{y}_0, \ \text{with} \ \mathbf{y}_0 \in (L^2(\Omega))^2, \ \nabla \cdot \mathbf{y}_0 = 0, \\ \mathbf{y}_0 \cdot \mathbf{n} = 0 \ \text{on } \Sigma_0 \ (= \Gamma_0 \times (0,T)), \end{cases} \tag{3.61}$$

$$\mathbf{y} = \mathbf{g}_0 \quad \text{on } \Sigma_0, \tag{3.62}$$

$$\nu \frac{\partial \mathbf{y}}{\partial n} - \mathbf{n}\pi = \mathbf{g}_1 \quad \text{on } \Sigma_1 \times (0,T), \tag{3.63}$$

and where the target function $\mathbf{y}_T$ is given in $(L^2(\Omega))^2$. In (3.60)–(3.63), $\nu(> 0)$ is a viscosity parameter and $\Gamma_0 \cap \Gamma_1 = \emptyset$ while the closure of $\Gamma_0 \cup \Gamma_1 = \Gamma$. Actually, the boundary condition (3.63) is not particularly physical, but it can be used to implement downstream boundary conditions for flow in unbounded regions. The test problem that we consider is the particular problem (3.59) where:

(i) $\Omega = (0,2) \times (0,1)$, $\mathcal{O} = (1/2,3/2) \times (1/4,3/4)$, $T = 1$.
(ii) $\Gamma_0 = \left\{ \{x_i\}_{i=1}^2 \,|\, x_2 \in \{0,1\},\ 0 < x_1 < 2 \right\}$,
$\Gamma_1 = \left\{ \{x_i\}_{i=1}^2 \,|\, x_1 \in \{0,2\},\ 0 < x_2 < 1 \right\}$.
(iii) $\mathbf{g}_0 = \mathbf{0}$, $\mathbf{g}_1 = 0$.
(iv) $\mathbf{y}_T = \mathbf{0}$, $k = 20$.
(v) $\nu = 5 \times 10^{-2}$.
(vi) $\mathbf{y}_0$ corresponds to a plane *Poiseuille flow* of maximum velocity equal to 1, that is,

$$\mathbf{y}_0(\mathbf{x}) = \big(4x_2(1 - x_2), 0\big), \quad \forall \mathbf{x} \in \Omega.$$

Integrating equations (3.60)–(3.63) with $\mathbf{v} = \mathbf{0}$ will lead to a solution that decays in time with a rate determined by the size of the viscosity parameter ν. Here, the problem is to find – via (3.59) – a control that will speed up this decay as much as possible at time T. The *time discretization* has been obtained through a variant of scheme (3.56)–(3.58), using $\Delta t = \frac{1}{50}$; the space discretization was achieved using a *finite element* approximation associated with a 32×16 (respectively, 16×8) regular grid for the *velocity* (respectively, the *pressure*) (see Berggren, Glowinski, and Lions, 1996b, for details). A *fully discrete* variant of the conjugate gradient algorithm (3.38)–(3.54) was used to compute the *approximate optimal control* $\mathbf{u}_h^{\Delta t}$ and the associated velocity field $\mathbf{y}_h^{\Delta t}$. The results displayed in Figures 3.1–3.3, below, were obtained at the 70th iteration of the above algorithm.

On Figure 3.1 we compared the decays between $t = 0$ and $t = T = 1$ of the uncontrolled flow velocity ($\cdots$) and of the controlled one (—) (we have shown the

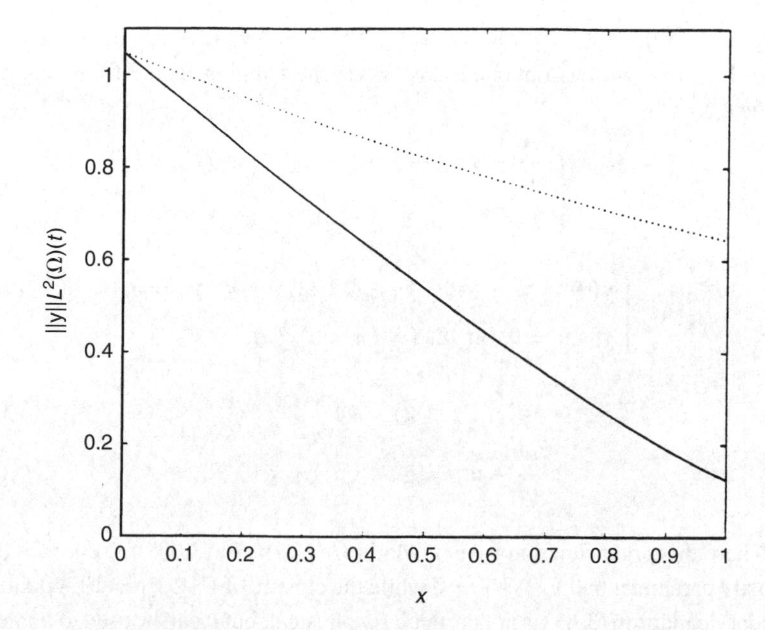

Fig. 3.1. Variation of $\|\mathbf{y}_h^{\Delta t}(t)\|_{(L^2(\Omega))^2}$ with (—) and without ($\cdots$) control.

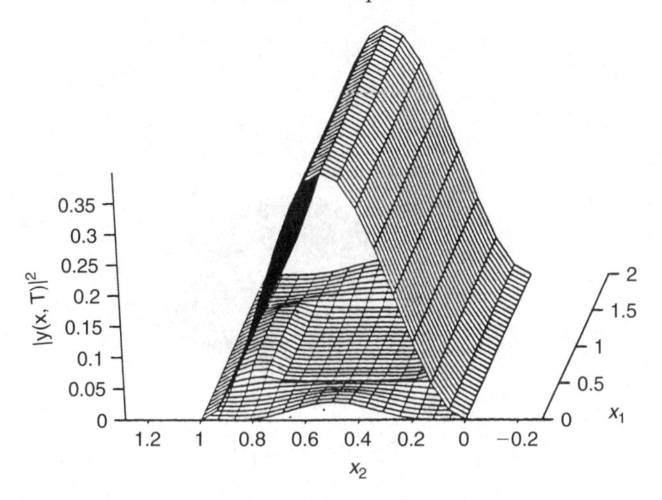

Fig. 3.2. Kinetic energy distribution at time $t = T\ (= 1)$ of the controlled flow (lower graph) and uncontrolled flow (upper graph).

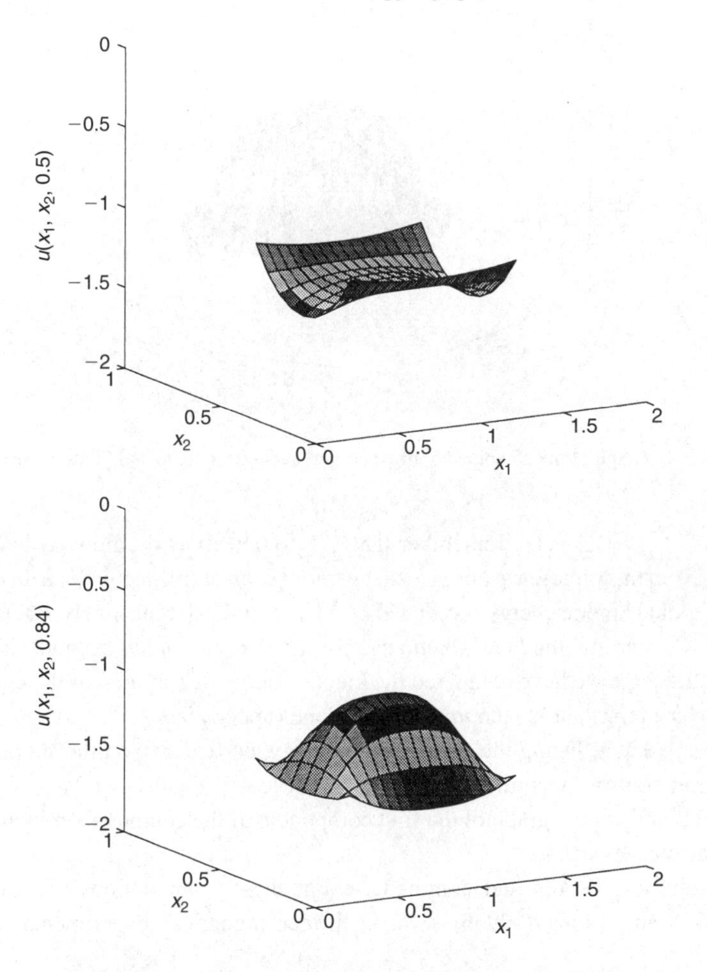

Fig. 3.3. Graph of the computed optimal control at (top) $t = 0.5$ and (bottom) $t = 0.84$.

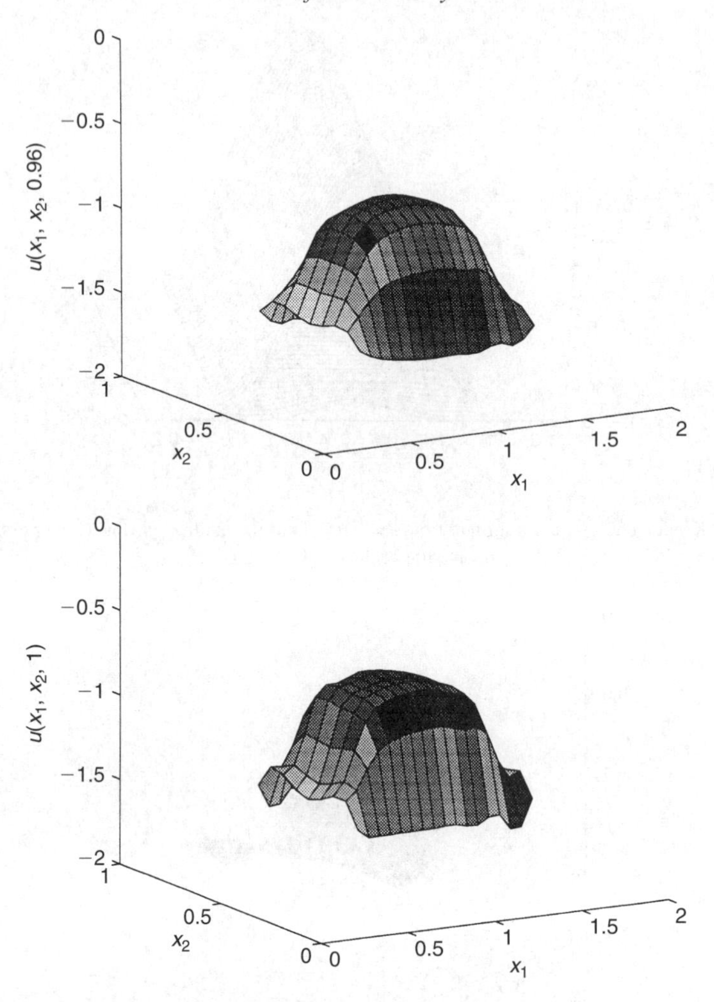

Fig. 3.4. Graph of the computed optimal control at (top) $t = 0.96$ and (bottom) $t = 1$.

values of $\left(\int_\Omega |\mathbf{y}(t)|^2\, dx\right)^{\frac{1}{2}}$; remember that $\frac{1}{2}\int_\Omega |\mathbf{y}(t)|^2\, dx$ is the flow *global kinetic energy*); from this picture we observe that *without control* (respectively, *with control*) the flow global kinetic energy is reduced by a factor of 2.4 (respectively, 66) (approximately). Concerning the *local kinetic energy* at $t = T = 1$, it has been visualized on Figure 3.2, where we have compared the kinetic energy distribution of the *controlled* flow (lower graph) and of the *uncontrolled* one (upper graph). Clearly, control has been effective at reducing the flow kinetic energy, particularly on the support $\mathcal{O}$ of the optimal control (according to Figure 3.2, at least). Finally, we have shown on Figures 3.3 and 3.4 the graph of the first component of the computed optimal control $\mathbf{u}_h^{\Delta t}$ at various values of t.

For additional details and comments about these computations see Berggren, Glowinski, and Lions (1996b), where further numerical experiments are also discussed.

4

Control of nonlinear diffusion systems

4.1 Generalities. Synopsis

The various *controllability problems* which have been discussed so far have all been associated with systems governed by linear diffusion equations (or by linearized diffusion equations, as in Chapter 2, Section 2.11). In this chapter we briefly address the *nonlinear* situation and would like to show that *nonlinearity* may bring *noncontrollability*, as seen in Section 4.2. In Section 4.3 we will discuss the solution of *pointwise control problems* for the *viscous Burgers equation*, and then, in Section 4.4 the control of the *Kuramoto–Sivashinsky* (K.S.) equation, a notably unstable model.

The control of nonlinear diffusion systems has motivated such a large number of publications that it has become practically impossible to give a comprehensive list of related references. Let us mention, however, the pioneering works of J.L. Lions (1991a) and Neittaanmäki and Tiba (1994) (additional references will be given in the following paragraphs). The *control of the Navier–Stokes equations* is, in principle, relevant to this chapter, but considering the importance of this last topic, we have dedicated to it the Part III of this volume.

4.2 Example of a noncontrollable nonlinear system

In this section, we want to emphasize that *approximate controllability may be very unstable under "small" nonlinear perturbations.*

Let us consider again the state equation

$$\frac{\partial y}{\partial t} - \Delta y = v\,\chi_{\mathcal{O}\times(0,T)} \text{ in } Q, \quad y(0) = 0, \quad y = 0 \text{ on } \Sigma, \tag{4.1}$$

which is the same equation as in Section 1.1, but where we take $\mathcal{A} = -\Delta$ to make things as simple as possible. We consider now the *nonlinear* partial differential equation

$$\frac{\partial y}{\partial t} - \Delta y + \alpha y^3 = v\,\chi_{\mathcal{O}\times(0,T)} \text{ in } Q, \quad y(0) = 0, \quad y = 0 \text{ on } \Sigma, \tag{4.2}$$

where α is a *positive* constant, otherwise arbitrary small. Problem (4.2) has a *unique* solution (see, for example, J.L. Lions, 1969). Contrary to what happens with (4.1), the set described by $y(T; v)$ (here, $y(v)$ is the solution of (4.2)) when v spans $L^2(\mathcal{O} \times (0, T))$ *is far from being dense in* $L^2(\Omega)$. There are several proofs of this result, some of them based on *maximum principles*. The following one is due to A. Bamberger (Private Communication, 1977) and is reported in the PhD dissertation of J. Henry (1978). It is based on a simple *energy estimate*. One multiplies (4.2) by my, where $m \in C^1(\overline{\Omega})$, $m \geq 0$, and $m|_{\mathcal{O}} = 0$. Then

$$\frac{1}{2}\frac{\mathrm{d}}{\mathrm{d}t}\int_\Omega my^2\,\mathrm{d}x + \int_\Omega m\,|\nabla y|^2\,\mathrm{d}x + \int_\Omega y\nabla y \cdot \nabla m\,\mathrm{d}x + \alpha\int_\Omega my^4\,\mathrm{d}x = 0. \quad (4.3)$$

Let us write

$$\int_\Omega y\nabla y \cdot \nabla m\,\mathrm{d}x = \int_\Omega m^{\frac{1}{4}}y\left(m^{\frac{1}{2}}\nabla y\right) \cdot \left(m^{-\frac{3}{4}}\nabla m\right)\mathrm{d}x,$$

and observe that, from the above relation we have

$$\left|\int_\Omega y\nabla y \cdot \nabla m\,\mathrm{d}x\right| \leq \alpha\int_\Omega my^4\,\mathrm{d}x + \int_\Omega m\,|\nabla y|^2\,\mathrm{d}x + \frac{\alpha^{-1}}{64}\int_\Omega m^{-3}|\nabla m|^4\,\mathrm{d}x. \quad (4.4)$$

Combining (4.3) and (4.4) gives

$$\frac{\mathrm{d}}{\mathrm{d}t}\int_\Omega my^2\,\mathrm{d}x \leq \frac{\alpha^{-1}}{32}\int_\Omega m^{-3}|\nabla m|^4\,\mathrm{d}x,$$

so that

$$\int_\Omega m(x)|y(x, T; v)|^2\,\mathrm{d}x \leq \frac{\alpha^{-1}}{32}T\int_\Omega m^{-3}|\nabla m|^4\,\mathrm{d}x, \quad (4.5)$$

no matter how v *is chosen*, since the right-hand side of (4.5) does not depend on v. Of course, the above calculation assumes that we can choose m as above and such that $\int_\Omega m^{-3}|\nabla m|^4\,\mathrm{d}x < +\infty$; such functions m are easy to construct. Actually, suppose that $\Omega = (a, b)$ with

$$-\infty < a < -1 < 1 < b < +\infty,$$

and that $\mathcal{O} = (-1, 1)$. For this one-dimensional situation, the function m defined by

$$m(x) = 0 \text{ if } x \in (-1, 1), \quad m(x) = \left(|x| - 1\right)^s \text{ if } x \in [a, b] \setminus (-1, 1)$$

verifies the above required assumptions if $s > 1$.

Remark 4.1 The inequality (4.5) shows that the larger α (that is, the more nonlinear is model (4.2)), the stronger is the *"obstruction" property* resulting from the nonlinearity. Indeed, relation (4.5) quantifies this property with respect to α.

Remark 4.2 Examples and counter-examples of controllability for nonlinear diffusion-type equations are given in Diaz (1991, 1994, 1996) and in Diaz and Ramos (1997) (see also the references therein).

4.3 Pointwise control of the viscous Burgers equation

4.3.1 Motivation

The *inviscid* and *viscous Burgers equations* have, for many years, attracted the attention of many investigators, from both the theoretical and numerical points of view. There are several reasons for this "popularity", one of them being certainly the fact that the Burgers equations provide not too unrealistic simplifications of the *Euler* and *Navier–Stokes* equations from *Fluid Dynamics*. Among the features that the Burgers equations have in common with these more complicated equations, the fact that it involves a *nonlinear advection operator* is certainly the most important single one. It is not surprising, therefore, that the Burgers equations have also attracted the attention of the *Control Community* (see, for example, Burns and Kang, 1991a,b; Burns and Marrekchi, 1993; Ramos, Glowinski, and Périaux, 2002). The present section is another contribution in that direction: we shall address here the solution of controllability problems for the *viscous Burgers equation* via *pointwise controls*; from that point of view this section can be seen as a generalization of Section 1.10 where we addressed the pointwise control of linear diffusion systems (the viscous Burgers equation considered here belongs to the class of the *nonlinear advection–diffusion* systems whose most celebrated representative is the *Navier–Stokes equations* system (whose some control aspects will be discussed in Part III, for incompressible flows)).

4.3.2 Formulation of the control problems

As in Berggren, Glowinski, and Lions (1996a) (see also Berggren, 1992; Dean and Gubernatis, 1991) we can consider the *following pointwise control problem for the viscous Burgers equation*:

$$\min_{\mathbf{v}\in\mathcal{U}} \left\{ \frac{1}{2}\,\|\mathbf{v}\|_{\mathcal{U}}^2 + \frac{1}{2}\,k\,\|y(T)-y_T\|_{L^2(0,1)}^2 \right\}, \tag{4.6}$$

where, in (4.6), we have

(i) $\mathbf{v} = \{v_m\}_{m=1}^M$, $\mathcal{U} = L^2(0,T;\mathbb{R}^M)$, $\|\mathbf{v}\|_{\mathcal{U}} = \left(\sum_{m=1}^M \int_0^T |v_m|^2\,\mathrm{d}t\right)^{\frac{1}{2}}$.
(ii) $k > 0$, arbitrary large.

(iii) $y_T \in L^2(0,1)$ and $y(T)$ is obtained from $\mathbf{v}$ via the solution of the *viscous Burgers equation*, below

$$\frac{\partial y}{\partial t} - \nu \frac{\partial^2 y}{\partial x^2} + y\frac{\partial y}{\partial x} = f + \sum_{m=1}^{M} v_m \delta(x - a_m) \text{ in } Q, \tag{4.7}$$

$$\frac{\partial y}{\partial x}(0,t) = 0 \text{ and } y(1,t) = 0 \quad \text{a.e. on } (0,T), \tag{4.8}$$

$$y(0) = y_0 \ (\in L^2(0,1)); \tag{4.9}$$

in (4.7), $Q = (0,1) \times (0,T)$, ν (> 0) is a *viscosity parameter*, f is a *forcing term*, $a_m \in (0,1)$, $\forall m = 1,\ldots,M$ and $x \to \delta(x - a_m)$ denotes the *Dirac measure* at a_m.

Let us denote by V the (Sobolev) space defined by

$$V = \{ z \,|\, z \in H^1(0,1),\ z(1) = 0 \}, \tag{4.10}$$

and suppose that $f \in L^2(0,T;V')$ (V': dual space of V); it follows then from J.L. Lions (1969) that for $\mathbf{v}$ given in $\mathcal{U}$ the Burgers system (4.7)–(4.9) has a unique solution in $L^2(0,T;V') \cap C^0([0,T];L^2(0,1))$. From this result, we can show that the control problem (4.6) *has a solution (not necessarily unique*, from the *non convexity* of the functional $J : \mathcal{U} \to \mathbb{R}$, where J is the *cost function* in (4.6)).

Remark 4.3 In Berggren, Glowinski, and Lions (1996a), we have discussed the solution of a variant of problem (4.6) where the location of the a_ms on (0, 1) is unknown (a_m: "support" of the mth pointwise control). The solution methods described in the following can be easily modified to accommodate this more complicated situation (see the above reference for details and numerical results).

4.3.3 Optimality conditions for problem (4.6)

To compute a control $\mathbf{u}$ solution of problem (4.6) we shall derive first *necessary optimality conditions* and take advantage of them (in Section 4.3.4) through a *conjugate gradient algorithm* to obtain the above solution.

The derivative $J'(\mathbf{v})$ of J at $\mathbf{v}$ can be obtained from

$$(J'(\mathbf{v}), \mathbf{w})_{\mathcal{U}} = \lim_{\substack{\theta \to 0 \\ \theta \neq 0}} \frac{J(\mathbf{v} + \theta \mathbf{w}) - J(\mathbf{v})}{\theta}. \tag{4.11}$$

Actually, instead of (4.11), we shall use a (formal) perturbation analysis to obtain $J'(\mathbf{v})$:

First, we have

$$\delta J(\mathbf{v}) = (J'(\mathbf{v}), \delta\mathbf{v})_{\mathcal{U}}$$
$$= \sum_{m=1}^{M} \int_0^T v_m\, \delta v_m \,\mathrm{d}t + k \int_0^1 (y(T) - y_T)\delta y(T)\,\mathrm{d}x \tag{4.12}$$

where (from (4.7) to (4.9)) $\delta y(T)$ is obtained from $\delta\mathbf{v}$ via the solution of the following variational problem:

$$\begin{cases} \delta y(t) \in V \text{ a.e. on } (0,T); \quad \forall z \in V \text{ we have} \\ \left\langle \dfrac{\partial}{\partial t}\delta y, z \right\rangle + \nu \displaystyle\int_0^1 \frac{\partial}{\partial x}\delta y \,\frac{\partial z}{\partial x}\,\mathrm{d}x + \int_0^1 \delta y\,\frac{\partial y}{\partial x}\, z\,\mathrm{d}x \\ \qquad + \displaystyle\int_0^1 y\frac{\partial}{\partial x}\delta y\, z\,\mathrm{d}x = \sum_{m=1}^{M} \delta v_m\, z(a_m) \quad \text{a.e. on } (0,T), \end{cases} \tag{4.13}$$

$$\delta y(0) = 0; \tag{4.14}$$

in (4.13), $\langle \cdot, \cdot \rangle$ denotes the *duality pairing* between V' and V.

Consider now $p \in L^2(0,T;V) \cap C^0([0,T];L^2(0,1))$ such that $\frac{\partial p}{\partial t} \in V'$; taking $z = p(t)$ in (4.13) we obtain

$$\int_0^T \left\langle \frac{\partial}{\partial t}\delta y, p \right\rangle \mathrm{d}t + \nu \int_0^T \int_0^1 \frac{\partial}{\partial x}\delta y\,\frac{\partial p}{\partial x}\,\mathrm{d}x\,\mathrm{d}t$$
$$+ \int_0^T \int_0^1 \left(\delta y \frac{\partial y}{\partial x} + y\frac{\partial}{\partial x}\delta y \right) p\,\mathrm{d}x\,\mathrm{d}t = \sum_{m=1}^{M} \int_0^T p(a_m,t)\delta v_m\,\mathrm{d}t. \tag{4.15}$$

Integrating by parts over $(0,T)$ it follows from (4.14), (4.15) that

$$\int_0^1 p(T)\,\delta y(T)\,\mathrm{d}x - \int_0^T \left\langle \frac{\partial p}{\partial t}, \delta y \right\rangle \mathrm{d}t + \nu \int_0^T \int_0^1 \frac{\partial}{\partial x}\delta y\,\frac{\partial p}{\partial x}\,\mathrm{d}x\,\mathrm{d}t$$
$$+ \int_0^T \int_0^1 \left(\delta y \frac{\partial y}{\partial x} + y\frac{\partial}{\partial x}\delta y \right) p\,\mathrm{d}x\,\mathrm{d}t = \sum_{m=1}^{M} \int_0^T p(a_m,t)\,\delta v_m\,\mathrm{d}t. \tag{4.16}$$

Suppose now that p satisfies also

$$\begin{cases} -\left\langle \dfrac{\partial p}{\partial t}, z \right\rangle + \nu \displaystyle\int_0^1 \frac{\partial p}{\partial x}\frac{\partial z}{\partial x}\,\mathrm{d}x + \int_0^1 p \left(\frac{\partial y}{\partial x} z + y\frac{\partial z}{\partial x} \right) \mathrm{d}x = 0, \\ \qquad\qquad \forall z \in V, \quad \text{a.e. on } (0,T), \end{cases} \tag{4.17}$$

and

$$p(T) = k\,(y_T - y(T)); \tag{4.18}$$

it follows then from (4.16) that

$$k \int_0^1 (y(T) - y_T)\,\delta y(T)\,\mathrm{d}x = -\sum_{m=1}^{M} \int_0^T p(a_m, t)\,\delta v_m\,\mathrm{d}t,$$

which combined with (4.12) implies in turn that

$$(J'(\mathbf{v}), \delta\mathbf{v})_{\mathcal{U}} = \sum_{m=1}^{M} \int_0^T (v_m(t) - p(a_m, t))\,\delta v_m(t)\,\mathrm{d}t.$$

We have thus "proved" that, $\forall \mathbf{v}, \mathbf{w} \in \mathcal{U}$

$$(J'(\mathbf{v}), \mathbf{w})_{\mathcal{U}} = \sum_{m=1}^{M} \int_0^T (v_m(t) - p(a_m, t))\,w_m(t)\,\mathrm{d}t. \tag{4.19}$$

Remark 4.4 A rigorous proof of (4.19) can be derived from (4.11).

Suppose now that $\mathbf{u}$ is a solution of problem (4.6); we have then $J'(\mathbf{u}) = \mathbf{0}$ which provides the following *optimality system*:

$$u_m(t) = p(a_m, t) \text{ on } (0, T), \quad \forall\, m = 1, \ldots, M, \tag{4.20}$$

$$\frac{\partial y}{\partial t} - \nu\,\frac{\partial^2 y}{\partial x^2} + y\,\frac{\partial y}{\partial x} = f + \sum_{m=1}^{M} u_m\,\delta(x - a_m) \text{ in } Q, \tag{4.21}$$

$$\frac{\partial y}{\partial x}(0, t) = 0 \text{ and } y(1, t) = 0 \quad \text{on } (0, T), \tag{4.22}$$

$$y(0) = y_0, \tag{4.23}$$

$$-\frac{\partial p}{\partial t} - \nu\,\frac{\partial^2 p}{\partial x^2} - y\,\frac{\partial p}{\partial x} = 0 \text{ in } Q, \tag{4.24}$$

$$\nu\,\frac{\partial p}{\partial x}(0, t) + y(0, t)\,p(0, t) = 0 \text{ and } p(1, t) = 0 \quad \text{on } (0, T), \tag{4.25}$$

$$p(T) = k\,(y_T - y(T)). \tag{4.26}$$

4.3.4 Iterative solution of the control problem (4.6)

Conjugate gradient algorithms are particularly attractive for large-scale nonlinear problems since their application requires only – in principle – *first derivative information* (see, for example, Daniel, 1970; Polack, 1971; Nocedal, 1992; Hiriart-Urruty and Lemarechal, 1993 for further comments and convergence proofs, and Glowinski, 2003 for large scale applications). Problem (4.6) is a particular case of the minimization problem

$$\begin{cases} u \in H, \\ j(u) \le j(v), \quad \forall v \in H, \end{cases} \tag{4.27}$$

where H is a real Hilbert space for the scalar product $(\cdot, \cdot)$ and the associated norm $\|\cdot\|$, and where the functional $j : H \to \mathbb{R}$ is differentiable; we denote by $j'(v)$ ($\in H'$; H' being the dual space of H) the differential of j at v.

A *conjugate gradient* algorithm for the solution of (4.27) is defined as follows:

$$u^0 \text{ is given in } H; \tag{4.28}$$

solve

$$\begin{cases} g^0 \in H, \\ (g^0, v) = \langle j'(u^0), v\rangle, \quad \forall v \in H, \end{cases} \tag{4.29}$$

and set

$$w^0 = g^0. \tag{4.30}$$

For $n \ge 0$, assuming that u^n, g^n, and w^n are known, compute u^{n+1}, g^{n+1}, and w^{n+1} by

$$\begin{cases} \text{Find } \rho_n \in \mathbb{R} \text{ such that} \\ j(u^n - \rho_n w^n) \le j(u^n - \rho w^n), \quad \forall \rho \in \mathbb{R}, \end{cases} \tag{4.31}$$

set

$$u^{n+1} = u^n - \rho_n w^n, \tag{4.32}$$

and solve

$$\begin{cases} g^{n+1} \in H, \\ (g^{n+1}, v) = \langle j'(u^{n+1}), v\rangle, \quad \forall v \in H. \end{cases} \tag{4.33}$$

If $\|g^{n+1}\|/\|g^0\| \le \epsilon$, take $u = u^{n+1}$; else compute either

$$\gamma_n = \frac{\|g^{n+1}\|^2}{\|g^n\|^2} \quad \text{(Fletcher–Reeves update)} \tag{4.34a}$$

or

$$\gamma_n = \frac{(g^{n+1}, g^{n+1} - g^n)}{\|g^n\|^2} \quad \text{(Polack–Ribière update)} \tag{4.34b}$$

and then

$$w^{n+1} = g^{n+1} + \gamma_n w^n. \tag{4.35}$$

Do $n = n + 1$ and go to (4.31).

We observe that each iteration of the conjugate gradient algorithm (4.28)–(4.35) requires the solution of a *linear variational problem* ((4.29) for $n = 0$, (4.33) for $n \ge 1$) and the *line search* (4.31). The popular belief is that in most applications the Polack–Ribière variant of algorithm (4.28)–(4.35) is faster than the Fletcher–Reeves' one (see, for example, Powell, 1976 for an explanation of this fact); actually, we know of several instances where it is the other way around (these include the boundary controllability problem for the viscous Burgers equation under consideration; other examples are given in Foss (2006) (they concern the least-squares solution of nonlinear eigenvalue problems)). Updates other than Fletcher–Reeves' and Polack–Ribière's can be found in the literature (see, for example, Hiriart-Urruty and Lemarechal, 1993; and Glowinski, 2003); personally, we never found an application were these updates perform better than Fletcher–Reeves' or Pollack–Ribière's (a numerical-experiment-based systematic comparison can be found in Foss, 2006).

Application to problem (4.6) Problem (4.6) is a particular case of (4.27) where $H = \mathcal{U} = L^2(0, T; \mathbb{R}^M)$; combining (4.19) and (4.28)–(4.35) we obtain the following solution method for problem (4.6):

$$\mathbf{u}^0 \textit{ is given in } \mathcal{U}; \tag{4.36}$$

solve

$$\begin{cases} \dfrac{\partial y^0}{\partial t} - \nu \dfrac{\partial^2 y^0}{\partial x^2} + y^0 \dfrac{\partial y^0}{\partial x} = f + \displaystyle\sum_{m=1}^{M} u_m^0 \, \delta(x - a_m) \;\; \textit{in } Q, \\ \dfrac{\partial y^0}{\partial x}(0, t) = 0 \textit{ and } y^0(1, t) = 0 \;\; \textit{on } (0, T), \\ y^0(0) = y_0, \end{cases} \tag{4.37}$$

and

$$\begin{cases} -\dfrac{\partial p^0}{\partial t} - \nu \dfrac{\partial^2 p^0}{\partial x^2} - y^0 \dfrac{\partial p^0}{\partial x} = 0 \;\; \textit{in } Q, \\ \nu \dfrac{\partial p^0}{\partial x}(0, t) + y^0(0, t) p^0(0, t) = 0 \;\; \textit{and} \;\; p^0(1, t) = 0 \;\; \textit{on } (0, T), \\ p^0(T) = k\,(y_T - y^0(T)). \end{cases} \tag{4.38}$$

Solve then

$$\begin{cases} \mathbf{g}^0 \in \mathcal{U}, \quad \forall \mathbf{v} (= \{v_m\}_{m=1}^M) \in \mathcal{U}, \\ \displaystyle\int_0^T \mathbf{g}^0 \cdot \mathbf{v} \, dt = \sum_{m=1}^{M} \int_0^T (u_m^0(t) - p^0(a_m, t)) \, v_m(t) \, dt, \end{cases} \tag{4.39}$$

and set

$$\mathbf{w}^0 = \mathbf{g}^0. \tag{4.40}$$

Then for $n \geq 0$, assuming that $\mathbf{u}^n$, $\mathbf{g}^n$, and $\mathbf{w}^n$ are known compute $\mathbf{u}^{n+1}$, $\mathbf{g}^{n+1}$, and $\mathbf{w}^{n+1}$ as follows:

Solve the following one-dimensional minimization problem:

$$\begin{cases} \rho_n \in \mathbb{R}, \\ J(\mathbf{u}^n - \rho_n \mathbf{w}^n) \leq J(\mathbf{u}^n - \rho \mathbf{w}^n), \quad \forall \rho \in \mathbb{R}, \end{cases} \tag{4.41}$$

and update $\mathbf{u}^n$ *by*

$$\mathbf{u}^{n+1} = \mathbf{u}^n - \rho_n \mathbf{w}^n. \tag{4.42}$$

Next, solve

$$\begin{cases} \dfrac{\partial y^{n+1}}{\partial t} - \nu \dfrac{\partial^2 y^{n+1}}{\partial x^2} + y^{n+1} \dfrac{\partial y^{n+1}}{\partial x} = f + \displaystyle\sum_{m=1}^{M} u_m^{n+1} \delta(x - a_m) \quad \textit{in } Q, \\ \dfrac{\partial y^{n+1}}{\partial x}(0,t) = 0 \ \textit{ and } \ y^{n+1}(1,t) = 0 \ \textit{ on } \ (0,T), \\ y^{n+1}(0) = y_0, \end{cases} \tag{4.43}$$

and

$$\begin{cases} -\dfrac{\partial p^{n+1}}{\partial t} - \nu \dfrac{\partial^2 p^{n+1}}{\partial x^2} - y^{n+1} \dfrac{\partial p^{n+1}}{\partial x} = 0 \ \textit{ in } Q, \\ \nu \dfrac{\partial p^{n+1}}{\partial x}(0,t) + y^{n+1}(0,t)\, p^{n+1}(0,t) = 0 \quad \textit{on } \ (0,T), \\ \qquad\qquad p^{n+1}(1,t) = 0 \quad \textit{on } \ (0,T), \\ p^{n+1}(T) = k\left(y_T - y^{n+1}(T)\right). \end{cases} \tag{4.44}$$

Solve then

$$\begin{cases} \mathbf{g}^{n+1} \in \mathcal{U}, \quad \forall \mathbf{v} \in \mathcal{U}, \\ \displaystyle\int_0^T \mathbf{g}^{n+1} \cdot \mathbf{v}\, dt = \sum_{m=1}^{M} \int_0^T (u_m^{n+1}(t) - p^{n+1}(a_m, t))\, v_m(t)\, dt. \end{cases} \tag{4.45}$$

If $\|\mathbf{g}^{n+1}\|_{\mathcal{U}} / \|\mathbf{g}^0\|_{\mathcal{U}} \leq \epsilon$, *take* $\mathbf{u} = \mathbf{u}^{n+1}$; *else compute either*

$$\gamma_n = \int_0^T |\mathbf{g}^{n+1}|^2\, dt \Big/ \int_0^T |\mathbf{g}^n|^2\, dt \quad \textit{(Fletcher–Reeves)}, \tag{4.46a}$$

or

$$\gamma_n = \int_0^T \mathbf{g}^{n+1} \cdot (\mathbf{g}^{n+1} - \mathbf{g}^n)\, dt \Big/ \int_0^T |\mathbf{g}^n|^2\, dt \quad \textit{(Polack–Ribière)}, \tag{4.46b}$$

and update $\mathbf{w}^n$ *by*

$$\mathbf{w}^{n+1} = \mathbf{g}^{n+1} + \gamma_n \mathbf{w}^n. \tag{4.47}$$

Do $n = n + 1$ *and go to (4.41).*

The practical implementation of algorithm (4.36)–(4.47) will rely on the *numerical integration* of the parabolic problems (4.37), (4.38), (4.43), and (4.44) (to be discussed in Section 4.3.5) and on the accuracy and efficiency of the line search in (4.41); actually, to solve the one-dimensional minimization problem (4.41) we have employed the *cubic backtracking* strategy advocated in Dennis and Schnabel (1983, Chapter 6, 1996, Chapter 6).

4.3.5 Space–time discretization of the control problem (4.6). Optimality conditions

In order to achieve the *space–time discretization* of problem (4.6), we shall employ a combination of *finite element* and *finite difference* methods; for simplicity we shall use *uniform meshes* for both discretizations. We consider therefore two positive integers I and N (to be "large" in practice) and define the *discretization steps* h and Δt by $h = 1/I$ and $\Delta t = T/N$. Next, we define $x_i = ih$, $i = 0, 1, \ldots, I$ and approximate $L^2(0,1)$ and $H^1(0,1)$ by

$$H_h^1 = \left\{ z_h \,|\, z_h \in C^0[0,1],\ z_h|_{e_i} \in P_1,\ \forall i = 1, \ldots, I \right\},$$

where P_1 denotes the space of the polynomials in one variable of degree ≤ 1 and $e_i = [x_{i-1}, x_i]$. The space V in (4.10) is approximated by

$$V_h = \left\{ z_h \,|\, z_h \in H_h^1,\ z_h(1) = 0 \right\} (= V \cap H_h^1),$$

while the *control space* $\mathcal{U}$ $(= L^2(0,T;\mathbb{R}^M))$ in (4.6) is approximated by

$$\mathcal{U}^{\Delta t} = (\mathbb{R}^M)^N = \left\{ \mathbf{v} \,|\, \mathbf{v} = \{\{v_m^n\}_{m=1}^M\}_{n=1}^N \right\}, \tag{4.48}$$

to be equipped with the following scalar–product:

$$(\mathbf{v}, \mathbf{w})_{\Delta t} = \Delta t \sum_{n=1}^{N} \sum_{m=1}^{M} v_m^n\, w_m^n, \quad \forall \mathbf{v}, \mathbf{w} \in \mathcal{U}^{\Delta t}.$$

We approximate then the control problem (4.6) by

$$\min_{\mathbf{v} \in \mathcal{U}^{\Delta t}} J_h^{\Delta t}(\mathbf{v}), \tag{4.49}$$

where the functional $J_h^{\Delta t}(\mathbf{v}) : \mathcal{U}^{\Delta t} \to \mathbb{R}$ is defined by

$$J_h^{\Delta t}(\mathbf{v}) = \frac{1}{2}\,(\mathbf{v}, \mathbf{v})_{\Delta t} + \frac{1}{2}\,k\,\|y^N - y_T\|_{L^2(0,1)}^2, \tag{4.50}$$

with y^N defined from $\mathbf{v}$ via the solution of the following *discrete Burgers equation*:

$$y^0 = y_{0h} \in H_h^1 \quad \text{with} \quad \lim_{h\to 0} \|y_{0h} - y_0\|_{L^2(0,1)} = 0; \tag{4.51}$$

for $n = 1, \ldots, N$ we obtain y^n from y^{n-1} via the solution of the following discrete linear (elliptic) variational problem:

$$\begin{cases} y^n \in V_h; \quad \forall z \in V_h \quad \text{we have} \\ \displaystyle\int_0^1 \frac{y^n - y^{n-1}}{\Delta t} z \, \mathrm{d}x + \nu \int_0^1 \frac{\mathrm{d}y^n}{\mathrm{d}x} \frac{\mathrm{d}z}{\mathrm{d}x} \, \mathrm{d}x + \int_0^1 y^{n-1} \frac{\mathrm{d}y^{n-1}}{\mathrm{d}x} z \, \mathrm{d}x \\ \displaystyle\qquad = \int_0^1 f \, z \, \mathrm{d}x + \sum_{m=1}^{M} v_m^n z(a_m). \end{cases} \tag{4.52}$$

Scheme (4.51), (4.52) is *semi-implicit* since the nonlinear term $y \frac{\mathrm{d}y}{\mathrm{d}x}$ is treated *explicitely*; we can expect therefore that Δt has to satisfy a *stability condition*. It is easily verified that obtaining y^n from y^{n-1} is equivalent to solving a linear system for a matrix (the discrete analogue of the elliptic operator $\frac{I}{\Delta t} - \nu \frac{\mathrm{d}^2}{\mathrm{d}x^2}$) which is *tridiagonal*, symmetric, and *positive definite*. If Δt is constant over the time interval $(0, T)$ this matrix being independent of n can be *Cholesky* factored once for all.

The *approximate control problem* (4.49) *has at least one solution*

$$\mathbf{u}_h^{\Delta t} = \left\{ \{u_m^n\}_{m=1}^M \right\}_{n=1}^N.$$

Any solution of problem (4.49) satisfies the *(necessary) optimality condition*

$$\nabla J_h^{\Delta t}(\mathbf{u}_h^{\Delta t}) = \mathbf{0}, \tag{4.53}$$

where $\nabla J_h^{\Delta t}$ is the gradient of the functional $J_h^{\Delta t}$. Following the approach taken in Section 4.3.3 for the continuous problem (4.6) we can show that

$$\left(\nabla J_h^{\Delta t}(\mathbf{v}), \mathbf{w}\right)_{\Delta t} = \Delta t \sum_{n=1}^{N} \sum_{m=1}^{M} \left(v_m^n - p(a_m^n)\right) w_m^n, \quad \forall \mathbf{v}, \mathbf{w} \in \mathcal{U}^{\Delta t}, \tag{4.54}$$

where $\{p^n\}_{n=1}^N$ is obtained from $\mathbf{v}$ via the solution of the discrete Burgers equation (4.51), (4.52), followed by the solution of the *discrete adjoint* equation below
Compute

$$\begin{cases} p^{N+1} \in V_h, \\ \displaystyle\int_0^1 p^{N+1} z \, \mathrm{d}x = k \int_0^1 (y_T - y^N) z \, \mathrm{d}x, \quad \forall z \in V_h, \end{cases} \tag{4.55}$$

and then for $n = N, N-1, \dots, 1$, p^n is obtained from p^{n+1} via the solution of the following discrete elliptic problems:

$$\begin{cases} p^N \in V_h, \\ \displaystyle\int_0^1 \frac{p^N - p^{N+1}}{\Delta t} z \, dx + \nu \int_0^1 \frac{dp^N}{dx} \frac{dz}{dx} \, dx = 0, \quad \forall z \in V_h, \end{cases} \tag{4.56a}$$

and, for $n = N-1, \dots, 1$,

$$\begin{cases} p^n \in V_h, \\ \displaystyle\int_0^1 \frac{p^n - p^{n-1}}{\Delta t} z \, dx + \nu \int_0^1 \frac{dp^n}{dx} \frac{dz}{dx} \, dx \\ \displaystyle\quad + \int_0^1 p^{n+1} \left(y^n \frac{dz}{dx} + \frac{dy^n}{dx} z \right) dx = 0, \quad \forall z \in V_h. \end{cases} \tag{4.56b}$$

The comments concerning the calculation of y^n from y^{n-1} still apply here; actually, the linear systems to be solved at each time step to obtain p^n from p^{n+1} have the same matrix than those encountered in the calculation of y^n from y^{n-1}.

From (4.54), we can derive a fully discrete variant of algorithm (4.36)–(4.47) to solve the approximate control problem (4.49) via the optimality conditions (4.53); such an algorithm is discussed in Berggren, Glowinski, and Lions (1996a).

4.3.6 Numerical experiments

Following Berggren, Glowinski, and Lions (1996a) (see also Dean and Gubernatis, 1991; Glowinski, 1991) we consider particular cases of problem (4.6) which have in common

$$T = 1, \quad \nu = \frac{1}{100}, \quad k = 8, \quad y_0 = 0,$$

$$f(x,t) = \begin{cases} 1 & \text{if } \{x,t\} \in (0, 0.5) \times (0, T), \\ 2(1-x) & \text{if } \{x,t\} \in (0.5, 1) \times (0, T), \end{cases}$$

$$y_T(x) = 1 - x^2, \quad \forall x \in (0,1).$$

To discretize the corresponding control problems, we have used the methods described in Section 4.3.5 with $h = 1/128$ and $\Delta t = 1/256$. The discrete control problems (4.49) have been solved by the fully discrete variant of algorithm (4.36)–(4.47) mentioned in Section 4.3.5, using $\mathbf{u}^0 = \mathbf{0}$ as *initial guess* and $\epsilon = 10^{-5}$ as the *stopping criterion*.

First, several experiments have been performed with a *single* control point ($M = 1$) for different values of a $(= a_1)$. In Table 4.1 we have summarized some of the numerical results concerning the *computed optimal control* $u_h^{\Delta t}$ and the corresponding

Table 4.1. *Summary of numerical results.*

a	$\frac{1}{5}$	$\frac{2}{3}$
Nit	89	47
$\frac{\|y_h^{\Delta t}(T)-y_T\|_{L^2(0,1)}}{\|y_T\|_{L^2(0,T)}}$	2×10^{-1}	9×10^{-2}
$\|u_h^{\Delta t}\|_{L^2(0,T)}$	0.125	0.115

discrete state function $y_h^{\Delta t}$ (in Table 4.1, *Nit* denotes the number of iterations necessary to achieve convergence).

For $a = \frac{1}{5}$ (respectively, $\frac{2}{3}$) we have visualized on Figure 4.1(a) (respectively, Figure 4.2(a)) the *computed optimal control* $u_h^{\Delta t}$ while on Figure 4.1(b) (respectively, Figure 4.2(b)) we have compared the *target function* y_T ($\cdots$) with the *computed approximation* $y_h^{\Delta t}(T)$ (—) of $y(T)$.

For $a = \frac{2}{3}$ a good fit can be observed *downstream* from the control point, while the solution seems to be *close to noncontrollable upstream*. The positive sign of the solution implies that the *convection* is directed toward the increasing values of x, which is why it seems reasonable to assume that the system is at least locally controllable in that direction. The only way for controlling the system upstream is through the *diffusion* term, which is small here ($\nu = \frac{1}{100}$) compared with the convection term. If $a = \frac{1}{5}$, there are clearly problems with controllability far downstream of the control point (recall that there is a distributed, uncontrolled forcing term, namely, f, which affects the solution).

Figure 4.3 shows the target and the final state when *two* control points are used, namely, $a_1 = \frac{1}{5}$ and $a_2 = \frac{3}{5}$; concerning controllability, the results are significantly better. Actually, the results become "very good" (as shown in Figure 4.4) if one uses the five control points $a_1 = 0.1$, $a_2 = 0.3$, $a_3 = 0.5$, $a_4 = 0.7$, and $a_5 = 0.9$; in that case we are "close" to a control distributed over the whole interval $(0, 1)$. Some further results are summarized in Table 4.2.

Remark 4.5 Concerning the *convergence* of the *conjugate gradient* algorithm used to solve the approximate control problems (4.49) let us mention that

(i) The *Fletcher–Reeves*' variant seems to have here a faster convergence than the *Polak–Ribière's* one.

(ii) The computational time does not depend too much on the number M of control points. For example, the CPU time (user time on a SUN Workstation SPARC10) was about 22s for the case with *one* control point at $a = \frac{1}{5}$, to be compared with 27s for the *five* control points test problem. Thus, the time-consuming part is the solution of the discrete state and adjoint state equations and not the manipulations of the control vectors (see Berggren, Glowinski, and Lions, 1996a for further details).

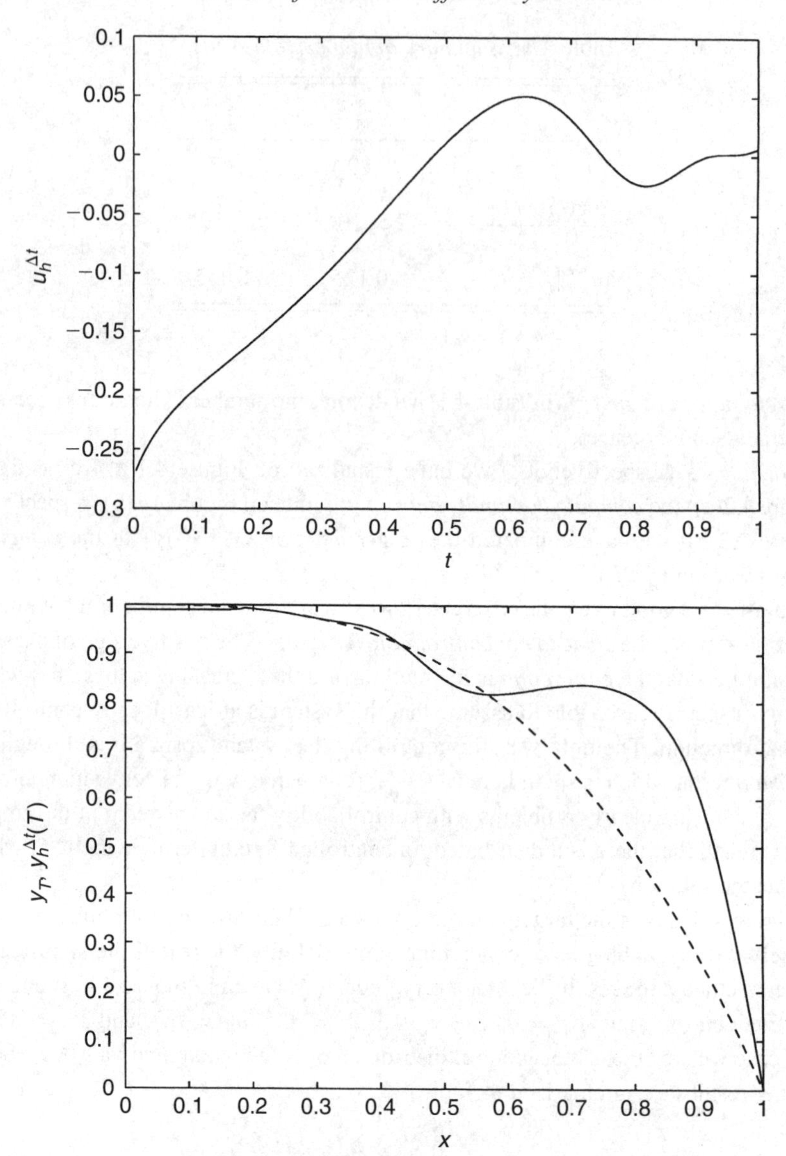

Fig. 4.1. (a) Graph of the computed optimal control $u_h^{\Delta t}$ $(a = \frac{1}{5})$ (top). (b) Comparison between y_T $(\cdots)$ and $y_h^{\Delta t}(T)$ (—) $(a = \frac{1}{5})$ (bottom).

Remark 4.6 In Berggren, Glowinski, and Lions (1996a) we have also addressed and solved the more complicated problem where the control **u** and the location **a** of the control points are unknown; this new problem can also be solved by a *conjugate gradient algorithm* operating in $L^2(0, T; \mathbb{R}^M) \times \mathbb{R}^M$; compared with the case where **a** is fixed the convergence of the new algorithm is significantly (about four times) slower (see Berggren, Glowinski, and Lions, 1996a for the computational aspects and for numerical results).

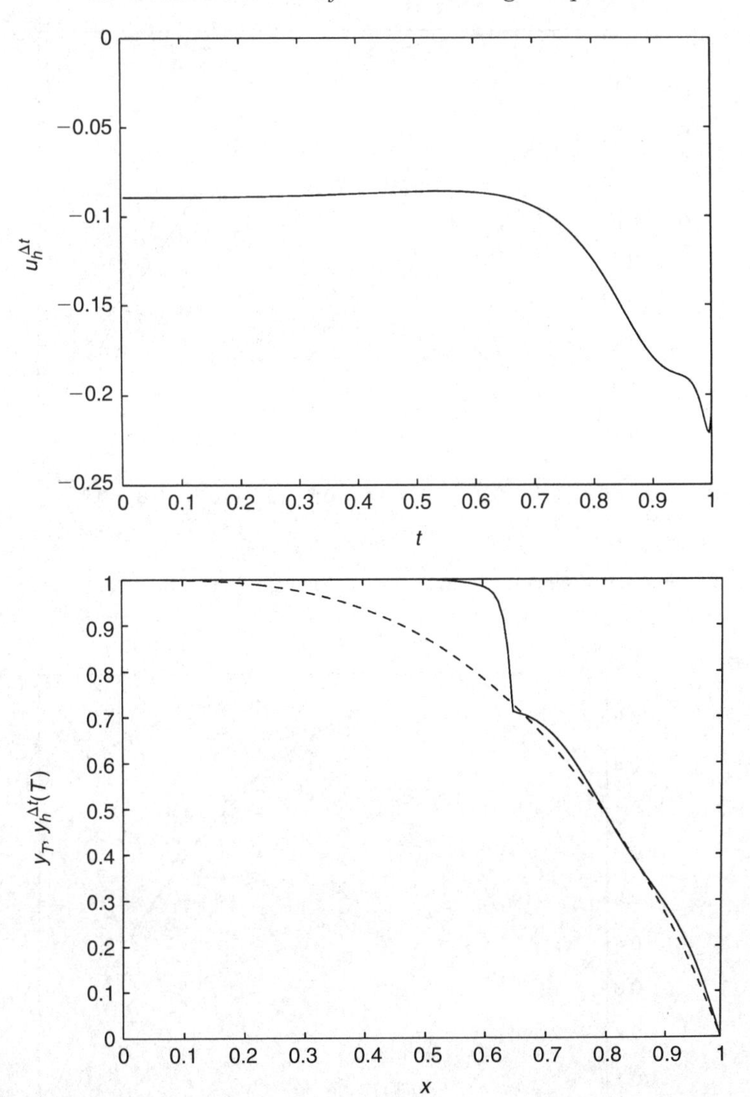

Fig. 4.2. (a) Graph of the computed optimal control $u_h^{\Delta t}$ $(a = \frac{2}{3})$ (top). (b) Comparison between y_T $(\cdots)$ and $y_h^{\Delta t}(T)$ (—) $(a = \frac{2}{3})$ (bottom).

4.3.7 Controllability and the Navier–Stokes equations

In Part II of the *Acta Numerica* article (Glowinski and Lions, 1995; J.L. Lions, 2003) at the origin of the present publication we wrote:

> *Flow control* is an important part of Engineering Sciences and from that point of view has been around for quite many years. However the corresponding *mathematical problems* are quite difficult and most of them are still open; it is therefore our opinion that a survey on the numerical aspects of these problems is still premature.
>
> It is nevertheless worth mentioning that a most important issue in that direction is the *control of turbulence* motivated, for example, by *drag reduction* (see, for example,

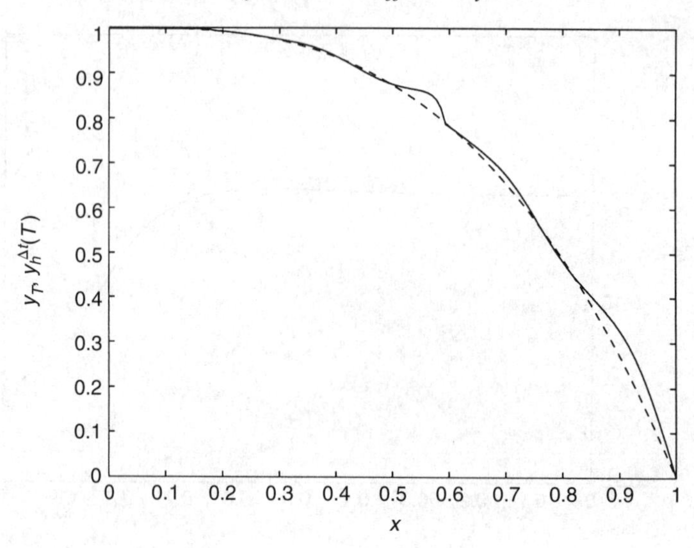

Fig. 4.3. Comparison between y_T $(\cdots)$ and $y_h^{\Delta t}(T)$ (—) ($\mathbf{a} = \{\frac{1}{5}, \frac{3}{5}\}$).

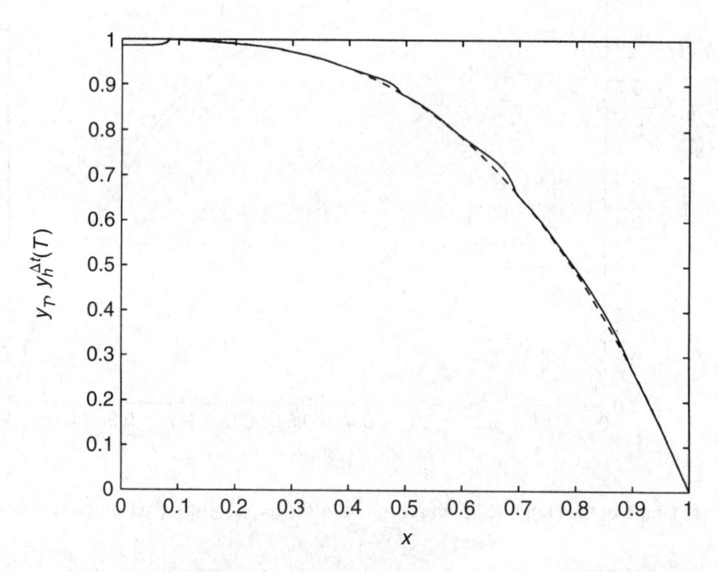

Fig. 4.4. Comparison between y_T $(\cdots)$ and $y_h^{\Delta t}(T)$ (—) ($\mathbf{a} = \{0.1, 0.3, 0.5, 0.7, 0.9\}$).

Table 4.2. *Summary of numerical results.*

a	$\left\{\frac{1}{5}, \frac{3}{5}\right\}$	$\{0.1, 0.3, 0.5, 0.7, 0.9\}$
Nit	86	82
$\frac{\Vert y_h^{\Delta t}(T) - y_T \Vert_{L^2(0,1)}}{\Vert y_T \Vert_{L^2(0,1)}}$	2.5×10^{-2}	8.5×10^{-3}

Buschnell and Hefner, 1990 and Sellin and Moses, 1989). Another important issue concerns the control of *turbulent combustion* as discussed in, for example, McManus, Poinsot, and Candel (1993) and Samaniego et al. (1993).

Despite the lack of theoretical results there is an enormous amount of literature on flow control topics (see, for example, the above four publications and the references therein). Focusing on recent work in the spirit of the present article, let us mention Abergel and Temam (1990), J.L. Lions (1991a), Glowinski (1991), this list being far from complete. In the following we shall give further references; they concern the application of *Dynamic Programming* to the control of systems governed by the Navier–Stokes equations.

Clearly, a lot has changed over the last ten years concerning the *mathematics* and *numerics* of flow control, making the above quotation largely obsolete. Indeed, the changes on these topics (particularly for *incompressible viscous* flow modeled by the *Navier–Stokes equations*) have been so dramatic that we felt compelled to add to Parts I and II a third one totally dedicated to flow control, mostly from the *numerical point of view*. However, this Part III will provide also up to date references relevant to the *mathematical aspects* of flow control.

4.4 On the controllability and the stabilization of the K.S. equation in one space dimension

4.4.1 Motivation. Synopsis

The main objective of this section (which follows closely He et al. 1998) is the numerical investigation of the *controllability* and *stabilization* properties of systems modeled by the K.S. equation. It will be shown that under some circumstances the *controllability* is enhanced when the positive coefficient of the biharmonic term in the K.S. equation gets smaller, a result conjectured by J.L. Lions in 1997, on the basis of the null-controllability property for the heat equation proved by G. Lebeau and L. Robbiano (cf. Lebeau and Robbiano, 1995) and reported in Chapter 1, Section 1.2, Remark 1.7. The control methods described in this section will be also applied to the stabilization of the above systems in the neighborhood of *unstable steady-state solutions*.

4.4.2 Generalities

The K.S. equation is one of simplest models used for describing the *thermo-diffusive instability* in the evolution of a *flame front*. It is derived, by using high activation energy asymptotics, from the so-called *constant density model of combustion*, where reaction–diffusion and hydrodynamics are decoupled in order to describe flame phenomena; for the derivation of the K.S. equation see, for example, Sivashinsky (1977). Indeed, the K.S. equation is essentially valid for those situations where hydrodynamical effects play a secondary role compared to reactive and diffusive ones, in other words, when the gas flow is not far from being uniform. The K.S. model has been validated by the experiments carried out at University of Houston by M. Gorman and his collaborators (see Gorman, El-Hamdi, and Robbins, 1994). Despite its relative simplicity, the K.S. model has played an important role in the understanding of flame propagation since it retains many of the qualitative features of flame propagation phenomena, including cellular flame instabilities such as bifurcation to hexagonal cellular patterns.

In an appropriate system of physical units, the K.S. equation can be written as

$$\frac{\partial \varphi}{\partial t} + \nabla^2 \varphi + \frac{1}{2}|\nabla \varphi|^2 + \nu \nabla^4 \varphi = 0 \quad \text{in } \Omega \times (0, T), \tag{4.57}$$

where φ is a small perturbation of the height of the plane flame front, $\nu > 0$, and Ω ($\subset \mathbb{R}^2$) is the cross section (parallel to the plane x_1Ox_2) of the tube where the combustion is taking pace. Equation (4.57) has to be completed by appropriate *boundary conditions* and by an *initial condition* such as

$$\varphi(0) = \varphi_0; \tag{4.58}$$

in (4.58) (and in the sequel) we denote by $\varphi(t)$ the function $x \to \varphi(x, t)$.

Remark 4.7 One may find in the literature a more complicated K.S. equation than the one given by (4.57); this other equation reads as follows:

$$\frac{\partial \varphi}{\partial t} + \epsilon \varphi + (I + \nabla^2)^2 \varphi + \frac{1}{2}|\nabla \varphi|^2 = 0 \quad \text{in } \Omega \times (0, T), \tag{4.59}$$

with $\epsilon \in \mathbb{R}$. Actually, the simulation, control and stabilization techniques for (4.57), (4.58) still apply to (4.58), (4.59).

The K.S. equation has become a prototypical model to study the *chaotic behavior of nonlinear physical systems*, its solutions being *very sensitive* to the variation of the data φ_0 and coefficient ν and displaying a quite complicated dynamics. This complex dynamical behavior follows from the combination of the *negative diffusion* term $\nabla^2\varphi$ with the *nonlinear* one $\frac{1}{2}|\nabla\varphi|^2$, the term $\nu\nabla^4\varphi$ keeping the number of unstable modes *finite* (this number increases with $\frac{1}{\nu}$).

Denote $\nabla\varphi$ by $\mathbf{p}$; we have then for $\mathbf{p}$ the following (*mixed*) formulation:

$$\frac{\partial \mathbf{p}}{\partial t} + \nabla^2 \mathbf{p} + (\mathbf{p} \cdot \nabla)\,\mathbf{p} + \nu\, \nabla^4 \mathbf{p} = \mathbf{0} \quad \text{in } \Omega \times (0, T), \tag{4.60}$$

$$\mathbf{p} = \nabla \varphi, \tag{4.61}$$

relation (4.61) being *equivalent* to

$$\nabla \times \mathbf{p} = \mathbf{0}. \tag{4.62}$$

The system (4.60)–(4.62) has definitely a *Navier–Stokes "flavor"; actually, the following system of equations*

$$\begin{cases} \dfrac{\partial \mathbf{u}}{\partial t} + \nabla^2 \mathbf{u} + (\mathbf{u} \cdot \nabla)\mathbf{u} + \nu \nabla^4 \mathbf{u} + \nabla p = \mathbf{0}, \\ \nabla \cdot \mathbf{u} = 0, \end{cases} \tag{4.63}$$

is known as the Kuramoto–Sivashinsky–Navier–Stokes equations and is used as a model for some turbulence phenomena. The numerical solution, by spectral methods, of (4.63) completed by periodic boundary conditions and an initial condition is discussed in Gama, Frisch, and Scholl (1991).

4.4.3 On the numerical solution of the K.S. equation

Solving *numerically* the K.S. equation is not an easy task; surprisingly, the numerical difficulties are not coming from $\nabla^2\varphi + \frac{1}{2}|\nabla\varphi|^2$ but from the *linear* operator

$$\varphi \to \frac{\partial\varphi}{\partial t} + \nu\nabla^4\varphi.$$

In order to understand the reasons of these difficulties, let us assume that the *backward* Euler scheme is used to *time discretize* the K.S. equation; we obtain then

$$\frac{\varphi^n - \varphi^{n-1}}{\Delta t} + \nabla^2\varphi^n + \frac{1}{2}|\nabla^2\varphi^n|^2 + \nu\nabla^4\varphi^n = 0 \quad \text{in } \Omega. \tag{4.64}$$

Suppose that $\Omega \subset \mathbb{R}$; if we *space discretize* (4.64) by *finite differences*, with h as space-discretization step, we obtain with obvious notation,

$$\left(\frac{\partial\varphi}{\partial t} + \varphi_{xx} + \frac{1}{2}|\varphi_x|^2 + \nu\,\varphi_{xxxx}\right)(ih, n\Delta t)$$

$$\approx \frac{\varphi_i^n - \varphi_i^{n-1}}{\Delta t} + \frac{1}{h^2}\left(\varphi_{i+1}^n + \varphi_{i-1}^n - 2\varphi_i^n\right) + \frac{1}{4h^2}\left(\varphi_{i+1}^n - \varphi_i^n\right)^2$$

$$+ \frac{1}{4h^2}\left(\varphi_{i-1}^n - \varphi_1^n\right)^2 + \frac{\nu}{h^4}\left(\varphi_{i+2}^n - 4\varphi_{i+1}^n + 6\varphi_i^n - 4\varphi_{i-1}^n + \varphi_{i-2}^n\right) = 0.$$

Suppose that, for example, $\nu = 0.1$ and $\Delta t = h = 10^{-2}$; the coefficient of φ_i^n is then equal to

$$6 \times 10^7 - 2 \times 10^4 + 10^2,$$

which shows that the contribution of $\frac{\partial\varphi}{\partial t}$ (10^2, here) is essentially lost in the sum of the two other terms. The above observation holds for the spectral approximations of the K.S. equations.

An obvious (and costly) cure is to take a very small Δt (everything else staying the same), as done for example in Hyman and Nicolaenko (1986). Another (and less costly) way to overcome the above difficulty is to observe that

$$I + \nu\,\Delta t\,\nabla^4 = \left(I - \sqrt{\nu\Delta t}\,\nabla^2\right)^2 + 2\sqrt{\nu\Delta t}\,\nabla^2.$$

The above *approximate factorization method* (due, to our knowledge, to the first author) has been applied to the solution of the one-dimensional K.S. equation in Nicolas-Carrizosa (1991), and also in Dean, Glowinski, and Trevas (1996) to the solution of the two-dimensional *Cahn–Hilliard equation* modeling the *spinodal decomposition of some binary alloys* (the Cahn-Hilliard equation involves also the linear operator $\varphi \to \frac{\partial\varphi}{\partial t} + \nu\nabla^4\varphi$); actually, for space dimensions ≥ 2 one used a *mixed formulation* to implement the above approximate factorization approach (see Dean, Glowinski, and Trevas, 1996 for details). It is worth mentioning that the *Neumann boundary control* of the Cahn–Hilliard equation is discussed in Glowinski and Ramos (2002).

4.4.4 On the modeling of some combustion phenomena: the Burgers–Kuramoto–Sivashint (B.K.S.) equation

For combustion phenomena taking place in tubes whose cross sections are narrow-rings (as the one shown in Figure 4.5) we have $0 < (R_2 - R_1)/(R_2 + R_1) << 1$ and the K.S. equation can be approximated by

$$\frac{\partial \varphi}{\partial t} + \frac{1}{R^2}\frac{\partial^2 \varphi}{\partial \theta^2} + \frac{1}{2R^2}\left|\frac{\partial \varphi}{\partial \theta}\right|^2 + \frac{\nu}{R^4}\frac{\partial^4 \varphi}{\partial \theta^4} = 0$$

where $R = \frac{1}{2}(R_1 + R_2)$ and θ is a *polar angle*. Taking R as unit length, rescaling t and ν if necessary, and denoting θ by x, we obtain the following K.S. system:

$$\begin{cases} \dfrac{\partial \varphi}{\partial t} + \dfrac{\partial^2 \varphi}{\partial x^2} + \dfrac{1}{2}\left|\dfrac{\partial \varphi}{\partial x}\right|^2 + \nu \dfrac{\partial^4 \varphi}{\partial x^4} = 0 \quad \text{in } (0, 2\pi) \times (0, T), \\ \dfrac{\partial^k \varphi}{\partial x^k}(0, t) = \dfrac{\partial^k \varphi}{\partial x^k}(2\pi, t) \quad \text{in } (0, T), \quad \forall k \in \{0, 1, 2, 3\}, \\ \varphi(0) = \varphi_0. \end{cases} \tag{4.65}$$

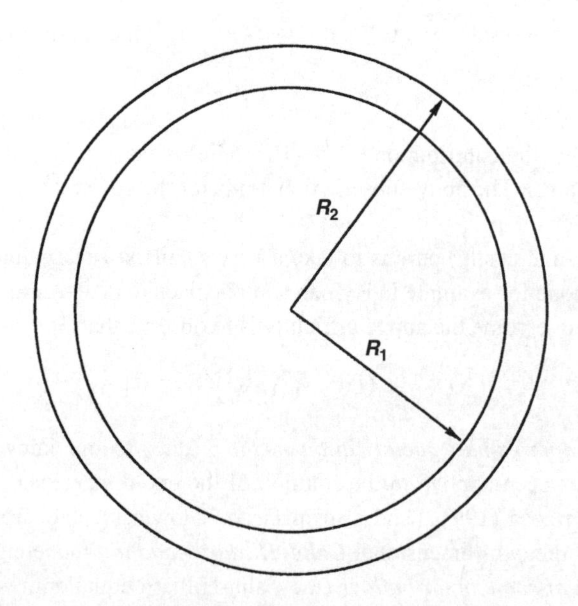

Fig. 4.5. A ring-shaped domain Ω.

Denoting $\frac{\partial\varphi}{\partial x}$ by y, we can easily show that y is a solution of the following variant of the viscous *Burgers equation*

$$\begin{cases} \dfrac{\partial y}{\partial t} + \dfrac{\partial^2 y}{\partial x^2} + y\,\dfrac{\partial y}{\partial x} + \nu\,\dfrac{\partial^4 y}{\partial x^4} = 0 \quad \text{in } (0, 2\pi) \times (0, T), \\ \dfrac{\partial^k y}{\partial x^k}(0, t) = \dfrac{\partial^k y}{\partial x^k}(2\pi, t) \quad \text{in } (0, T), \quad \forall k \in \{0, 1, 2, 3\}, \\ y(0) = y_0 \left(= \dfrac{\mathrm{d}\varphi_0}{\mathrm{d}x} \right). \end{cases} \tag{4.66}$$

Remark 4.8 The combustion experiments carried out in the laboratory of Professor M. Gorman at University of Houston validate the one-dimensional model (4.65) in the case of tubes with narrow-ring cross sections.

In the following paragraphs of this section we are going to discuss the controllability and stabilization properties of the systems modeled by (4.66) that we shall call the *Burgers–Kuramoto–Sivashinsky* (B.K.S.) equation.

In Section 4.4.5, we shall briefly discuss the numerical solution of the B.K.S. equation, then address its *controllability* properties in Section 4.4.6 and, in Section 4.4.7, apply the results of Section 4.4.6 to the *stabilization* of the B.K.S. equation in the neighborhood of *unstable steady-state solutions*.

4.4.5 Numerical solution of the B.K.S. equation: Numerical results

In order to solve the B.K.S. system (4.66) we will use

(i) A *spectral-Galerkin space-discretization* method based on the Fourier modes $\{\cos qx\}_{q\geq 1}$ and $\{\sin qx\}_{q\geq 1}$ which provide *high accuracy* and are well suited to *periodic boundary conditions* (since $y = \frac{\partial\varphi}{\partial x}$, with φ the solution of (4.65), the constant mode, which corresponds to $q = 0$, has been left out).
(ii) The following *second-order accurate time-discretization* scheme

$$y^0 = y_0, \tag{4.67}$$

$$\frac{y^1 - y^0}{\Delta t} + y^1_{xx} + y^0\, y^0_x + \nu y^1_{xxxx} = 0 \quad \text{in } (0, 2\pi), \tag{4.68}$$

and for $n \geq 2$,

$$\frac{3y^n - 4y^{n-1} + y^{n-2}}{2\Delta t} + y^n_{xx} + (2y^{n-1} - y^{n-2})\,(2y^{n-1} - y^{n-2})_x + \nu y^n_{xxxx} = 0 \quad \text{in } (0, 2\pi). \tag{4.69}$$

The above discretization techniques have been combined to solve the B.K.S. system (4.66). In order to convince those readers not very familiar with the K.S. equations of the complicated dynamics that their solution can develop, we have visualized on Figures 4.6–4.8 and then Figures 4.9–4.11, for $\nu = 0.2$ and 0.1, respectively, the

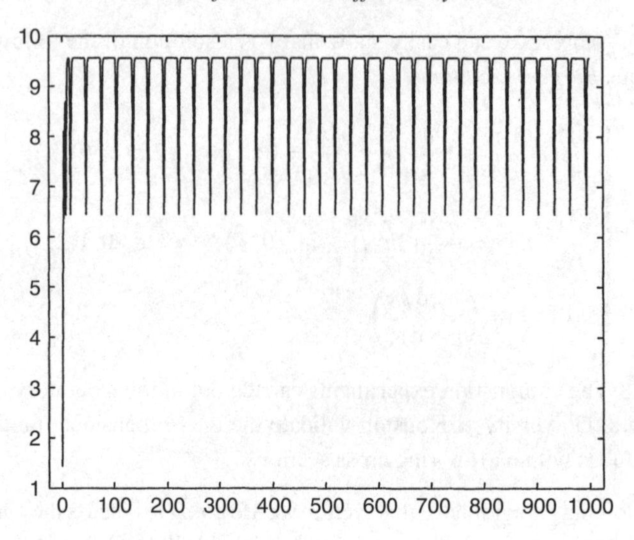

Fig. 4.6. Graph of the function $t \to \|y(t)\|_{L^2(0,2\pi)}$ for $\nu = 0.2$.

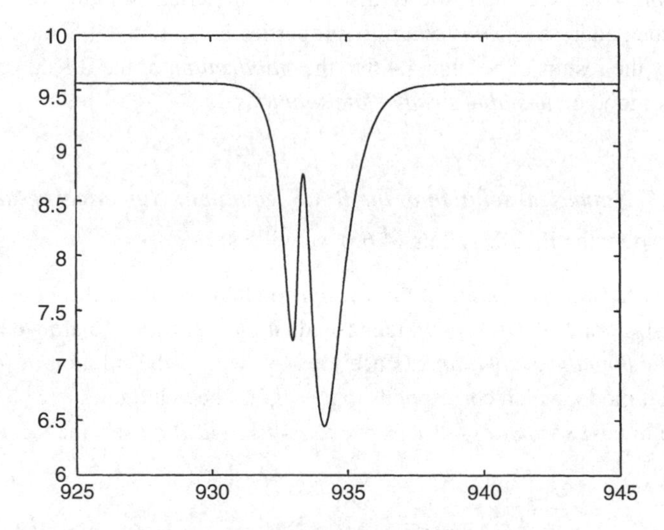

Fig. 4.7. Details of the graph of the function $t \to \|y(t)\|_{L^2(0,2\pi)}$ for $\nu = 0.2$.

graph of the function $t \to \|y(t)\|_{L^2(0,2\pi)}$ (Figures 4.6 and 4.9) on the time interval (0, 1000), a "zoom" of the above graphs in the neighborhood of $t = 940$ (Figures 4.7 and 4.10) and, finally, a phase plane representation of the solution showing the trajectory of the point $\{\|y(t)\|_{L^2(0,2\pi)}, \frac{\mathrm{d}}{\mathrm{d}t}\|y(t)\|_{L^2(0,2\pi)}\}$ as t varies from 0 to 1000 (Figures 4.8 and 4.11). These results have been obtained using the spectral approximation mentioned in (i), with $q = 1, \ldots, 30$, and $\Delta t = 10^{-2}$ in scheme (4.67)–(4.69). For $\nu = 0.2$, the solution evolves to a *periodic homoclinic* burst as $t \to +\infty$, while, for $\nu = 0.1$, the solution evolves to a *chaotic homoclinic* burst. The asymptotic behavior of the solution of the K.S. equations is discussed in, for example, Temam (1988) and

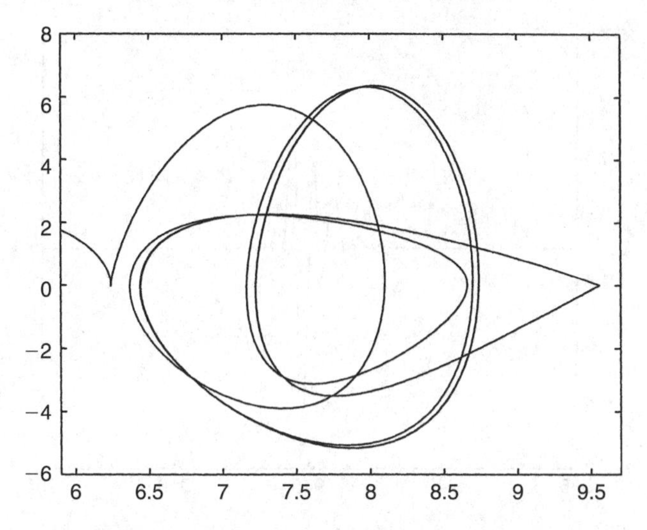

Fig. 4.8. Phase plane representation for $\nu = 0.2$.

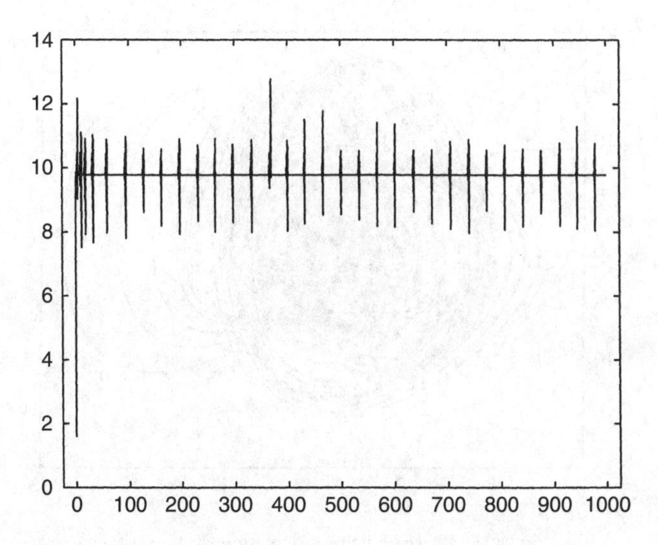

Fig. 4.9. Graph of the function $t \to \|y(t)\|_{L^2(0,2\pi)}$ for $\nu = 0.1$.

Armbruster, Guckenheimer, and Holmes (1989). We recall that the dynamics complexity "increase" follows from the fact that the number of unstable modes increases as $\nu \to 0_+$.

4.4.6 On the controllability of the K.S. equation

We are going to discuss now the *control* of the K.S. equations, via the B.K.S. system (4.66) whose numerical solution has been briefly discussed in Section 4.4.5. To the best of our knowledge, He et al. (1998) (that we follow closely, here) was one of the very first publications to invest computationally the controllability properties of

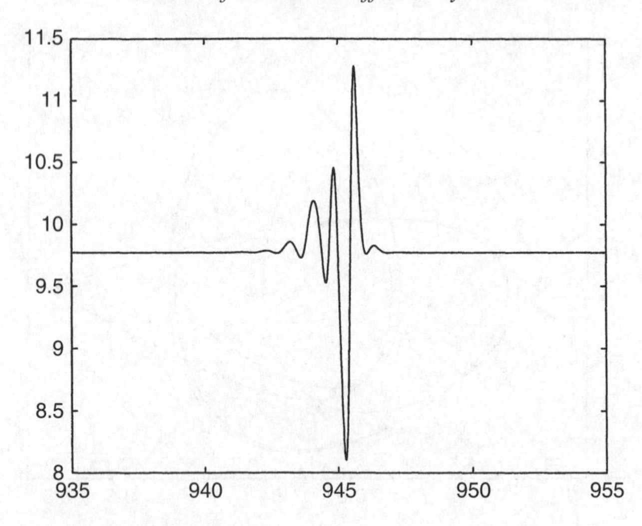

Fig. 4.10. Details of the graph of the function $t \to \|y(t)\|_{L^2(0,2\pi)}$ for $\nu = 0.1$.

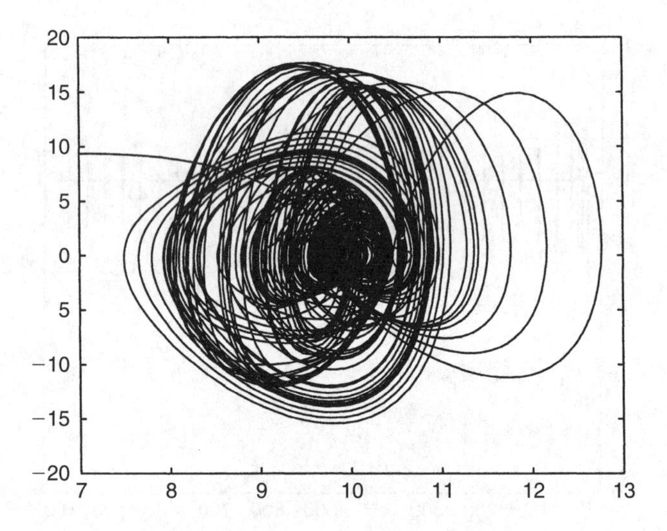

Fig. 4.11. Phase plane representation for $\nu = 0.1$.

the K.S. equations (see also Glowinski and He, 1998 for a related discussion and preliminary results).

The control problem to be considered can be formulated as follows:

$$\begin{cases} \text{Find } u \in \mathcal{U} \text{ so that} \\ J(u) \leq J(v), \quad \forall v \in \mathcal{U}, \end{cases} \tag{4.70}$$

with

$$\mathcal{U} = L^2(\omega \times (0, T)), \tag{4.71}$$

ω being an open subset of $(0, 2\pi)$ and $0 < T < +\infty$, and

$$J(v) = \frac{1}{2}\epsilon \int_{\omega\times(0,T)} |v|^2 \, dx \, dt + \frac{1}{2}\int_0^{2\pi} |y(T) - y_T|^2 \, dx, \tag{4.72}$$

where, in (4.72), ϵ (>0) is a *regularization parameter* (ϵ^{-1} can be viewed as a *penalty parameter* forcing the closeness of $y(T)$ to the target function y_T), y_T is a given function of $L^2(0, 2\pi)$, and y is a *nonlinear* function of the control v via the solution of the following B.K.S. system:

$$\begin{cases} \dfrac{\partial y}{\partial t} + \dfrac{\partial^2 y}{\partial x^2} + y\,\dfrac{\partial y}{\partial x} + \nu\,\dfrac{\partial^4 y}{\partial x^4} = v\,\chi_\omega \quad \text{in } Q\ (= (0, 2\pi) \times (0, T)), \\[2ex] \dfrac{\partial^k y}{\partial x^k}(0, t) = \dfrac{\partial^k y}{\partial x^k}(2\pi, t) \quad \text{in } (0, T), \quad \forall\, k \in \{0, 1, 2, 3\}, \\[2ex] y(0) = y_0 \ (\in L^2(0, 2\pi)); \end{cases} \tag{4.73}$$

in (4.73), χ_ω denotes the *characteristic function* of the set ω (that is, $\chi_\omega(x) = 1$ if $x \in \omega$, $\chi_\omega(x) = 0$ elsewhere).

Remark 4.9 We can also take as control space $\mathcal{U}$ a closed strict subspace of $L^2(\omega \times (0, T))$.

If u is a solution of problem (4.70), it verifies, *necessarily*,

$$\int_{\omega\times(0,T)} J'(u)\, v \, dx \, dt = 0, \quad \forall\, v \in \mathcal{U}, \tag{4.74}$$

with

$$J'(v) = \epsilon\, v + p\,\chi_\omega, \tag{4.75}$$

where, in (4.75), p is the solution of the following *adjoint system* of equations:

$$\begin{cases} -\dfrac{\partial p}{\partial t} + \dfrac{\partial^2 p}{\partial x^2} - y\,\dfrac{\partial p}{\partial x} + \nu\,\dfrac{\partial^4 p}{\partial x^4} = 0 \quad \text{in } Q, \\[2ex] \dfrac{\partial^k p}{\partial x^k}(0, t) = \dfrac{\partial^k p}{\partial x^k}(2\pi, t) \quad \text{in } (0, T), \quad \forall\, k \in \{0, 1, 2, 3\}, \\[2ex] p(T) = y(T) - y_T. \end{cases} \tag{4.76}$$

After *space–time discretization*, we can derive from relations (4.72)–(4.76) minimization algorithms of the *limited memory* BFGS (LMBFGS) type, to achieve the solution of the discrete variants of the control problem (4.70) (see, for example, Liu and Nocedal, 1989; Nocedal, 1992 for details and further references on LMBFGS algorithms; see also Part III of this book).

Actually, besides applications to *combustion*, one of the main motivations for investigating the controllability properties of the K.S. equation is the verification of the following conjecture of the second author (made in 1997):

$$\begin{cases} \textit{Starting from } y_0 = 0 \textit{ we have approximate controllability} \\ \textit{at } t = T, \textit{for any target } y_T \in L^2(0, 2\pi); \\ \textit{moreover reaching } y_T \textit{ is easier as } \nu \textit{ decreases to } 0_+. \end{cases} \tag{4.77}$$

The conjecture (4.77) is based on the following facts:

(i) For $\nu = 0$, the system (4.73) reduces to a nonlinear variant of the *backward heat equation* (change t in $T - t$).
(ii) The result from Lebeau and Robbiano (1995), mentioned in Remark 1.7 of Chapter 1, Section 1.2. According to this result the *forward heat equation* is *exactly null-controllable*. This means that for *any* initial data $y_0 \in L^2(0, 2\pi)$ and $\forall T \in (0, +\infty)$, there exists a distributed control $v \in L^2(\omega \times (0, T))$ so that $y(T) = 0$.

By changing t in $T-t$ it follows from (ii) that there exists controls v in $L^2(\omega\times(0, T))$ and solutions of

$$\begin{cases} \dfrac{\partial y}{\partial t} + \dfrac{\partial^2 y}{\partial x^2} = v\,\chi_\omega \quad \text{in } (0, 2\pi) \times (0, T), \\ y(0) = 0, \end{cases}$$

(completed by reasonable boundary conditions) so that $y(T) = y_T$, $\forall y_T \in L^2(0, 2\pi)$; we have thus exact controllability, $\forall T > 0$, assuming that the initial value y_0 is zero.

Combining (i) and (ii) suggests that if approximate controllability holds for the B.K.S. equation initialized by $y(0) = 0$, we can expect that reaching y_T will be easier as ν gets closer to 0_+. Suppose that the above conjecture is true, then, taking into account the fact that the number of *unstable* modes in the B.K.S. equation goes to infinity as $\nu \to 0_+$, we shall have found another instance where *instability enhances controllability* in the sense of Ott, Sauer, and Yorke (1994). For a discussion of the controllability properties of *semilinear heat equations* see Zuazua (1997) and the references therein.

In order to check computationally conjecture (4.77), the following numerical experiments have been carried out:

We have solved the controllability problem (4.70) using the computational methods described in Section 4.4.5, combined with a LMBFGS minimization algorithm. The control v in (4.70), (4.72)–(4.75) is approximated by $\mathbf{v} = \{v^n\}_{n=1}^N$, with $N = T/\Delta t$ and

$$v^n = \chi_\omega \sum_{i=1}^{I} \left(v_{1i}^n \cos ix + v_{2i}^n \sin ix\right), \quad \forall n = 1, \ldots, N; \tag{4.78}$$

in (4.78), $\{v_{1i}^n, v_{2i}^n\} \in \mathbb{R}^2$. Similarly, we suppose that for $n = 1, \ldots, N$, we have

$$y^n = \sum_{k=1}^{K} (y_{1k}^n \cos kx + y_{2k}^n \sin kx), \quad \forall n = 1, \ldots, N, \tag{4.79}$$

with $\{y_{1k}^n, y_{2k}^n\} \in \mathbb{R}^2$; in (4.79), $y^n \approx y(n\Delta t)$. We take $y^0 = 0$ and define y_T by

$$y_T = \sum_{j=1}^{J} (y_{1j}^T \cos jx + y_{2j}^T \sin jx), \tag{4.80}$$

with $\{y_{1j}^T, y_{2j}^T\} \in \mathbb{R}^2$; the Fourier coefficients of y_T have been generated *randomly* on $(-\frac{1}{2}, \frac{1}{2})$.

For the numerical experiments described below we have taken $T = 1$, $\epsilon = 10^{-6}$, $\Delta t = 10^{-2}$, $\omega = (0, 2\pi)$, $I = J = 5$, and $K = 10$ and 20. On Figure 4.12, we have visualized the spectrum of the *target function* y_T, that is, its Fourier coefficients as a function of the integer j, with the positive (respectively, negative) values of j corresponding to the $\sin jx$ (respectively, $\cos jx$) modes. On Figure 4.13, we have visualized the variation of $\|u^*\|_{L^2((0,2\pi)\times(0,T))}$ (u^* being the computed optimal control) as a function of ν for $K = 10$ and 20; the two graphs are practically identical. On Table 4.3 (respectively, Table 4.4) we have shown the influence of the parameter ν, for $K = 10$, (respectively, of the time-discretization step Δt, for $\nu = 0.2$) on $\|u^*\|_{L^2((0,2\pi)\times(0,T))}$ and on the discrete analogues of $\|y(T) - y_T\|_{L^2(0,2\pi)}$ and $J(u)$ (denoted $J_*(u^*)$, here). We observe (not surprisingly) that accurate results require small values of Δt.

Remark 4.10 From the above numerical results, we observe that as $\nu \to 0_+$ controlling is cheaper and we are getting closer to the target function. This shows

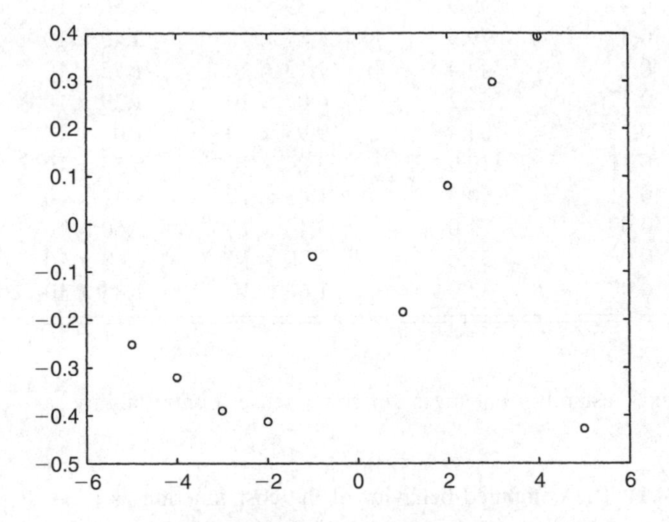

Fig. 4.12. Spectrum of the target function y_T.

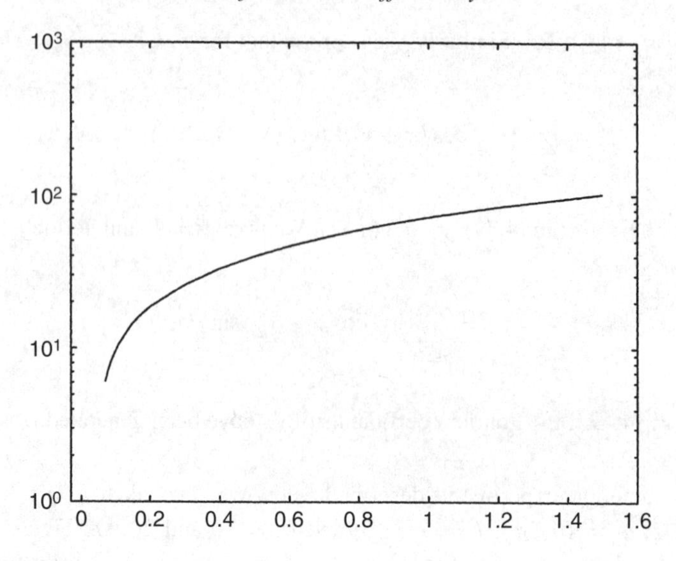

Fig. 4.13. Variation of $\|u^*\|_{L^2((0,2\pi)\times(0,T))}$ versus ν ($\Delta t = 10^{-2}, I = J = 5, K = 10$ and 20).

Table 4.3. *Influence of* ν *(*$K = 10$ *and* $\Delta t = 10^{-2}$*).*

ν	$\|u^*\|_{L^2((0,2\pi)\times(0,T))}$	$\|y^N - y_T\|_{L^2(0,2\pi)}$	$J_*(u^*)$
1.5	10454.2	9.37×10^{-5}	5.27×10^{-3}
1.2	7074.0	4.24×10^{-5}	3.56×10^{-3}
1.0	5158.1	2.21×10^{-5}	2.59×10^{-3}
0.9	4302.3	1.57×10^{-5}	2.16×10^{-3}
0.8	3517.6	1.02×10^{-5}	1.76×10^{-3}
0.7	2804.1	6.38×10^{-6}	1.41×10^{-3}
0.6	2161.3	3.72×10^{-6}	1.08×10^{-3}
0.5	1591.8	2.14×10^{-6}	7.97×10^{-4}
0.4	1099.9	9.89×10^{-7}	5.50×10^{-4}
0.3	679.5	3.81×10^{-7}	3.40×10^{-4}
0.2	344.4	1.13×10^{-7}	1.72×10^{-4}
0.17	258.2	6.07×10^{-7}	1.29×10^{-4}
0.15	204.4	9.93×10^{-8}	1.03×10^{-4}
0.11	110.9	2.24×10^{-8}	5.55×10^{-5}
0.1	90.4	1.43×10^{-8}	4.52×10^{-5}
0.09	72.0	3.40×10^{-9}	3.60×10^{-5}
0.08	53.6	3.30×10^{-9}	2.68×10^{-5}
0.07	37.0	1.66×10^{-9}	1.85×10^{-5}

that, indeed, instability enhances, in some sense, controllability, as conjectured in (4.77).

Remark 4.11 The computed behavior of the cost function as $\nu \to 0_+$ is in full agreement with the mathematical results obtained by J.L. Lions and Zuazua (1997).

Table 4.4. *Influence of* Δt *(*$K = 10$ *and* $\nu = 0.2$*).*

Δt	$\|u^*\|_{L^2((0,2\pi)\times(0,T))}$	$\|y^N - y_T\|_{L^2(0,2\pi)}$	$J_*(u^*)$
1/100	344.4	1.13×10^{-7}	1.72×10^{-4}
1/200	287.2	4.47×10^{-7}	1.44×10^{-4}
1/400	255.8	3.96×10^{-8}	1.28×10^{-4}
1/800	240.2	6.44×10^{-8}	1.20×10^{-4}
1/1600	232.0	3.93×10^{-8}	1.16×10^{-4}
1/3200	227.8	4.60×10^{-8}	1.14×10^{-4}
1/6400	225.8	3.68×10^{-8}	1.13×10^{-4}

4.4.7 On the stabilization of the K.S. equation

Suppose that we want to stabilize the solution of the B.K.S. equation in the neighborhood of a function y_d, where y_d can be, for example, an *unstable steady-state solution* of the B.K.S. equation. Suppose that we want to achieve this stabilization on the finite interval $(0, T)$; a systematic way to achieve this goal is to solve the following variant of the control problem (4.70), (4.71):

$$\begin{cases} \text{Find } u \in \mathcal{U} \text{ so that} \\ J(u) \leq J(v), \quad \forall\, v \in \mathcal{U}, \end{cases} \tag{4.81}$$

with

$$\mathcal{U} = L^2(\omega \times (0, T)), \tag{4.82}$$

ω being as in Section 4.4.6 and, this time, $J(\cdot)$ defined by

$$J(v) = \frac{1}{2}\epsilon \int_{\omega\times(0,T)} |v|^2\, dx\, dt + \frac{1}{2}\int_{(0,2\pi)\times(0,T)} |y - y_d|^2\, dx\, dt; \tag{4.83}$$

in (4.83), y is still the solution of problem (4.73).

Any solution of the control problem (4.81) verifies

$$\int_{\omega\times(0,T)} J'(u)\, v\, dx\, dt = 0 \quad \forall\, v \in \mathcal{U}, \tag{4.84}$$

with

$$J'(v) = \epsilon\, v + p\, \chi_\omega, \tag{4.85}$$

p being the solution of the following *adjoint system*:

$$\begin{cases} -\dfrac{\partial p}{\partial t} + \dfrac{\partial^2 p}{\partial x^2} - y\dfrac{\partial p}{\partial x} + \nu\dfrac{\partial^4 p}{\partial x^4} = y - y_d \quad \text{in } Q, \\ \dfrac{\partial^k p}{\partial x^k}(0,t) = \dfrac{\partial^k p}{\partial x^k}(2\pi,t) \quad \text{in } (0,T), \quad \forall k \in \{0,1,2,3\}, \\ p(T) = 0. \end{cases} \tag{4.86}$$

Suppose that y_d is a *steady-state solution*, possibly *unstable*, of the B.K.S. equation. The complex dynamic of the solutions of the time-dependent B.K.S. equations is associated with an equally complex structure of the set of the steady-state solutions (we assume here that ν is sufficiently small) as shown by the *bifurcation diagrams* shown in, for example, Krevedis, Nicolaenko, and Scovel (1990), which visualize the locus of the steady-state solutions as ν varies. Several authors have shown, computationally, that if $\nu \in S_1 = (0.307602\ldots, 1)$ the B.K.S. equation has a global unimodal fixed point attractor. Denote such an attractor by $x \to Y(x;\nu)$, with $\nu \in S_1$. For any constants ρ and c, the B.K.S. equation has the following two-parameter family of solutions:

$$y(x,t) = \rho\, Y(\rho(x - ct); \rho^2\nu) + c. \tag{4.87}$$

For stationary waves, we have $c = 0$, while periodicity implies that ρ is an integer. Relation (4.87) can be used to generate steady-states solutions of the B.K.S. equation, the parameter ν varying like $\nu_0 k^{-2}$ with $\nu_0 \in S_1$ and $k = 1, 2, \ldots$. If, for example, the 2π-periodic steady-state is known for $\nu = 0.8$, then we can construct other steady-state solutions, via (4.87), by taking $\nu = 0.8/2^2, 0.8/3^2$, and so on. The solution for $\nu = 0.8/2^2$ is bimodal and its amplitude is twice the one associated with $\nu = 0.8$. This folding and scaling process can be repeated *ad infinitum* to obtain k-modal steady-states solutions of the B.K.S. equation, for any integer k. On Figure 4.14 we

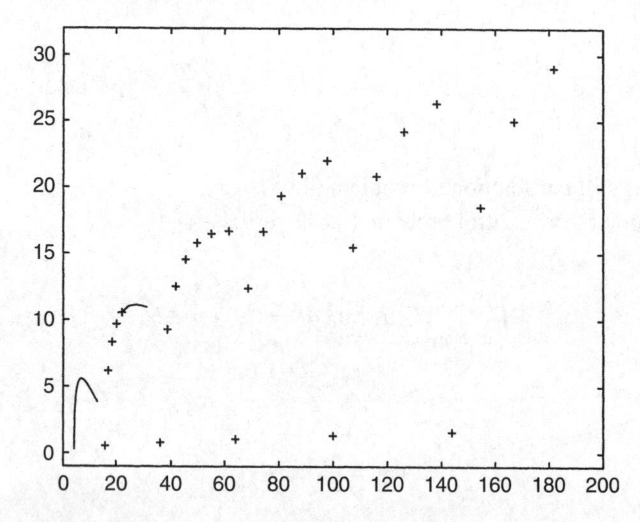

Fig. 4.14. Variation of the L^2-norm of the k-modal steady-state solution versus $4/\nu$.

have shown the variation of the L^2-norm of the k-fold replication of a 2π-periodic steady state as a function of $4/\nu$. For small values of ν the amplitude of such k-fold replications is of the order of $\nu^{-\frac{1}{2}}$, which is also the order of the L^2-norm of these steady-state solutions.

In practice, stabilization should take place over $(0, +\infty)$ (*infinite horizon control*). Solving the control problem, even for large finite values of T, will be extremely costly for the following reasons:

(i) The fast dynamics of the B.K.S. solutions requires a small time discretization step Δt implying that the number of time steps, that is, $T/\Delta t$, will be extremely large.
(ii) For small values of ν, as nonlinear effects develop, a very large number of spectral modes is required for the space approximation in order to have an accurate representation of the solutions.
(iii) If the optimal control is computed via the adjoint equation approach, we need to store the values of the discrete analogue of the state function y at every time step, in order to integrate from T to 0 the discrete analogue of the adjoint equation (4.86). Of course, we can significantly reduce this storage requirement by using the method described in Chapter 1, Section 1.12, but for very large values of T, the memory requirement will still be prohibitive.

In order to overcome the above difficulties, we are going to compromise between optimality and storage by using the following *piecewise optimal control strategy* (already employed in Chapter 2, Section 2.11):

First, we divide $[0, +\infty)$ in sub intervals of equal length ΔT and we denote $l\Delta T$ by T_l, with $l = 1, 2, \ldots$. We proceed then as follows:

(a) For $l = 0$, starting from y_0, we solve on $(0, T_1)$ a control problem such as (4.81) and we denote by y_1 the value of the optimal state at $t = T_1$.
(b) For $l > 0$, we proceed as in (a), with $(0, T_1)$ (respectively, y_0) replaced by (T_l, T_{l+1}) (respectively, y_l). We denote by y_{l+1} the optimal state at $t = T_{l+1}$.
(c) We do $l = l + 1$ and repeat the process.

Let us discuss now the results of numerical experiments. The function y_d being a steady-state solution, we define y_0 as

$$y_0 = y_d + \delta y_d,$$

with

$$\delta y_d = \sum_{j=1}^{J} \left(\delta y_{1j}^d \cos jx + \delta y_{2j}^d \sin jx\right),$$

the Fourier coefficients of δy_d being generated randomly on $\left[-\frac{1}{2}, \frac{1}{2}\right]$.

If $\nu > 1$, the only steady-state solution of the B.K.S. equation is $y = 0$, which is stable. There is a bifurcation at $\nu = 1$ and the steady-state solution $y = 0$ becomes unstable if $\nu < 1$. The *first series of numerical experiments* corresponds to $y_d = 0$ in

Table 4.5. *First series of numerical experiments: Influence of* ν *on the stabilization.*

ν	$J(u)$	$\|u\|_{L^2(Q)}$	$\|y(T) - y_d\|_{L^2(0,2\pi)}$
0.9	4.75×10^{-11}	0.97	1.37×10^{-9}
0.8	5.23×10^{-11}	1.02	1.76×10^{-9}
0.7	5.74×10^{-11}	1.07	1.54×10^{-9}
0.6	6.30×10^{-11}	1.12	3.99×10^{-9}
0.5	6.92×10^{-11}	1.18	2.03×10^{-9}
0.4	7.77×10^{-11}	1.25	1.95×10^{-9}
0.3	1.01×10^{-10}	1.42	7.79×10^{-10}
0.2	1.95×10^{-10}	1.97	1.86×10^{-9}
0.15	3.48×10^{-10}	2.64	1.11×10^{-8}
0.1	1.23×10^{-8}	15.68	2.77×10^{-6}
0.09	2.37×10^{-8}	21.71	1.42×10^{-5}
0.08	6.25×10^{-8}	27.57	2.21×10^{-4}

the cost functional and $T = \Delta T = 1$. We have taken $\epsilon = 10^{-8}$ in (4.83), $\Delta t = 10^{-2}$, $\omega = (0, 2\pi)$, $I = 15$, $J = K = 30$, and ν varying on the interval [0.08, 0.9]. We have displayed on Table 4.5 the computed values of $J(u)$, $\|u\|_{L^2(Q)}$ and $\|y(T) - y_d\|_{L^2(0,2\pi)}$. We observe that, despite the fact that $y_d = 0$ is an unstable solution of the B.K.S. equation, this solution can be tracked and we can stay "very" close to it, at a fast increasing cost of the control, however, as ν decreases.

For the *second series of numerical experiments* we have taken for y_d an unstable steady-state solution of the B.K.S. equation obtained, via (4.87), by a 6-fold replication of a stable solution of the B.K.S. equation. The target function y_d has been visualized on Figure 4.15.

We have taken $\Delta T = 2.5 \times 10^{-2}$, $\epsilon = 10^{-6}$, $\Delta t = 0.5 \times 10^{-4}$, $\omega = (0, 2\pi)$, $I = 30$, and $J = K = 60$. We have applied to this stabilization problem the *piecewise optimal control* strategy described above. We have visualized on Tables 4.6 (for $\nu = 0.02585$) and 4.7 (for $\nu = 0.022008$) the computed values of $J(u)$, $\|u\|_{L^2(\omega \times (T_{l-1}, T_l))}$ and $\|y(T_l) - y_d\|_{L^2(0,2\pi)}$ for each intermediate control problem. These results show the good stabilization properties of the piecewise optimal control approach: for $l = 3$, we are already close to the target at low cost for the control. These results show also that the cost per sub interval decreases very quickly with l.

Remark 4.12 For those readers wondering what is the interest of stabilizing a dynamical system in the neighborhood of an unstable steady-state solution, we would like to mention, focusing, for example, on combustion phenomena, that there are many instances where the best compromise between energy production and pollution emissions corresponds to a regime which is unstable. Similar situations are commonly encountered in aero-elasticity, chemical and nuclear engineerings, and so on.

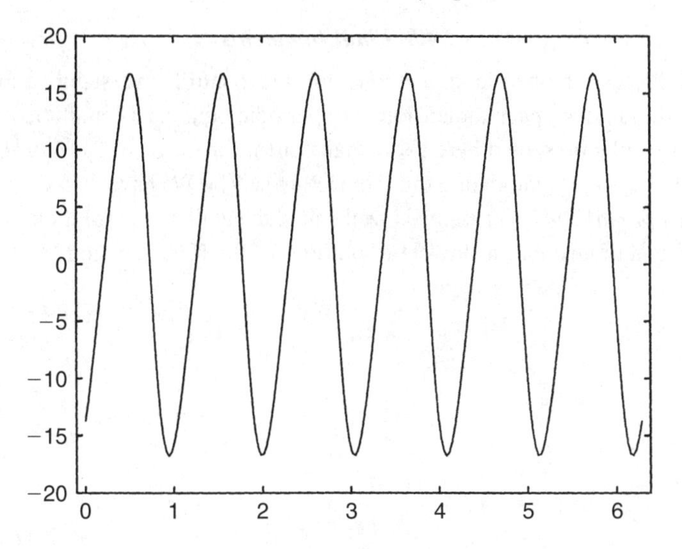

Fig. 4.15. Second series of numerical experiments: Graph of the target function y_d for $\nu = 0.022008$.

Table 4.6. *Second series of numerical experiments: Piecewise optimal control history ($\nu = 0.02585$).*

l	$J(u)$	$\|u\|_{L^2(\omega\times(T_{l-1},T_l))}$	$\|y(T_l) - y_d\|_{L^2(0,2\pi)}$
1	2.53×10^{-4}	20.93	8.26×10^{-3}
2	9.00×10^{-10}	4.08×10^{-2}	1.17×10^{-5}
3	5.93×10^{-16}	3.25×10^{-5}	1.13×10^{-8}
4	1.32×10^{-21}	4.91×10^{-8}	1.56×10^{-11}
5	1.81×10^{-27}	5.65×10^{-11}	2.08×10^{-14}

Table 4.7. *Second series of numerical experiments: Piecewise optimal control history ($\nu = 0.022008$).*

l	$J(u)$	$\|u\|_{L^2(\omega\times(T_{l-1},T_l))}$	$\|y(T_l) - y_d\|_{L^2(0,2\pi)}$
1	6.85×10^{-3}	32.35	1.13×10^{-1}
2	1.39×10^{-6}	1.10	1.25×10^{-3}
3	1.45×10^{-11}	3.61×10^{-3}	4.00×10^{-6}
4	7.31×10^{-16}	1.95×10^{-5}	3.29×10^{-8}
5	4.22×10^{-20}	1.08×10^{-7}	2.69×10^{-10}

4.4.8 Final comments

We have discussed open-loop methods for the control and stabilization of the K.S. equation in one-space dimension with periodic boundary conditions. The computational results presented here show that, starting from $y_0 = 0$, *controllability is enhanced as* $\nu \to 0_+$, validating thus conjecture (4.77). We have also shown that the *piecewise optimal control strategy* is well suited to stabilize the solution, even in the neighborhood of unstable steady-state solutions of the K.S. equation.

5

Dynamic programming for linear diffusion equations

5.1 Introduction. Synopsis

We address in this section the "*real-time*" aspect of the controllability problems. We proceed in a largely formal fashion. The content of this chapter is based on J.L. Lions (1991b).

We consider again the *state equation*

$$\frac{\partial y}{\partial t} + Ay = v\,\chi_{\mathcal{O}}, \tag{5.1}$$

now in the time interval $(s, T]$, $0 \leq s \leq T$; the "*initial*" condition is

$$y(s) = h, \tag{5.2}$$

where h is an *arbitrary* function in $L^2(\Omega)$; the *boundary* condition is

$$y = 0 \quad \text{on } \Sigma_s = \Gamma \times (s, T). \tag{5.3}$$

Consider now the following *control problem*

$$\inf \left\{ \frac{1}{2} \iint_{\mathcal{O}\times(s,T)} v^2 \, dx \, dt, \quad \forall v \in L^2(\mathcal{O} \times (s, T)) \right. \\ \left. \text{so that} \quad y(T; v) \in y_T + \beta B \right\}, \tag{5.4}$$

where in (5.4), $\beta > 0$, B is the closed unit ball of $L^2(\Omega)$ centered at 0, $y_T \in L^2(\Omega)$ and $t \to y(t; v)$ is the solution of (5.1)–(5.3).

The *minimum* in (5.4) is now a function of h and s, we define $\phi(h, s)$ by

$$\phi(h, s) \text{ is the minimal value of the cost function in (5.4).} \tag{5.5}$$

In the following section, we are going to derive the *Hamilton–Jacobi–Bellman* (H.J.B) *equation* satisfied by ϕ on $L^2(\Omega) \times (0, T)$.

5.2 Derivation of the Hamilton–Jacobi–Bellman equation

As said above, we shall proceed in a largely formal fashion. We take

$$v(x,t) = w(x) \quad \text{in } (s, s+\varepsilon), \qquad \text{with } \varepsilon > 0 \text{ and “very small.”} \tag{5.6}$$

With this choice of v, the state function $y(t)$ moves on the time interval $(s, s+\varepsilon)$ from h to an element of $L^2(\Omega)$ “very close” to

$$h_\varepsilon = h - \varepsilon\, A\, h + \varepsilon\, w\, \chi_{\mathcal{O}}, \tag{5.7}$$

assuming that $h \in H^2(\Omega) \cap H_0^1(\Omega)$ and that the coefficients of operator A are smooth enough (we observe that h_ε is obtained from h by the *explicit Euler scheme*).

On the time interval $(s+\varepsilon, T)$ we consider the whole process starting from h_ε at time $s+\varepsilon$. The *optimality principle* leads to

$$\phi(h,s) = \inf_w \left\{ \frac{1}{2}\varepsilon \int_{\mathcal{O}} w^2 \, dx + \phi(h_\varepsilon, s+\varepsilon) \right\} + \text{“negligible terms.”} \tag{5.8}$$

Taking now the ε-Taylor expansion of the function $\phi(h_\varepsilon, s+\varepsilon)$ we obtain from (5.7) that

$$\begin{aligned} \phi(h_\varepsilon, s+\varepsilon) &= \phi(h - \varepsilon A h + \varepsilon w \chi_{\mathcal{O}}, s+\varepsilon), \\ &= \phi(h,x) - \varepsilon \left(\frac{\partial \phi}{\partial h}(h,s), Ah \right) + \varepsilon \left(\frac{\partial \phi}{\partial h}(h,s), w\chi_{\mathcal{O}} \right) \\ &\quad + \varepsilon \frac{\partial \phi}{\partial s}(h,s) + \text{ higher-order terms}, \end{aligned} \tag{5.9}$$

where

$$\left(\frac{\partial \phi}{\partial h}(h,s), g \right) = \lim_{\lambda \to 0} \frac{d}{d\lambda} \phi(h + \lambda g, s),$$

with h and g belonging to $H^2(\Omega) \cap H_0^1(\Omega)$ $(\subset L^2(\Omega))$.

Combining (5.8) with (5.9) and dividing by ε, we obtain at the limit when $\varepsilon \to 0_+$

$$\begin{aligned} \inf_w \Bigg\{ \frac{1}{2} \int_{\mathcal{O}} w^2 \, dx &- \left(\frac{\partial \phi}{\partial h}(h,s), Ah \right) \\ &+ \left(\frac{\partial \phi}{\partial h}(h,s), w\chi_{\mathcal{O}} \right) + \frac{\partial \phi}{\partial s}(h,s) \Bigg\} = 0; \end{aligned} \tag{5.10}$$

the unique minimizer of problem (5.10) is $-\frac{\partial \phi}{\partial h}(h,s)\chi_{\mathcal{O}}$ which implies in turn that

$$-\frac{\partial \phi}{\partial s}(h,s) + \left(\frac{\partial \phi}{\partial h}(h,s), Ah \right) + \frac{1}{2} \int_{\mathcal{O}} \left(\frac{\partial \phi}{\partial h}(h,s) \right)^2 dx = 0. \tag{5.11}$$

The functional equation (5.11) is the H.J.B. equation. It is a *partial differential equation in infinite dimensions* since $h \in L^2(\Omega)$ (in fact, $h \in H^2(\Omega) \cap H_0^1(\Omega)$), and where $s \in (0,T)$. We have to add an *“initial” condition*, here for $t = T$, since we integrate (5.11) *backward in time*. When $s \to T$, we have less and less time to "correct" the trajectory, implying (this is again formal but it can be proved rigorously) that

$$\phi(h, T) = \begin{cases} 0 & \text{if } h \in y_T + \beta B, \\ +\infty & \text{otherwise.} \end{cases} \tag{5.12}$$

5.3 Some remarks

Remark 5.1 The "solution" of the system (5.11), (5.12) should be defined in the framework of the *viscosity solutions* of M. Crandall and P.L. Lions (1985, 1986a,b, 1990, 1991), which was generalized by those authors to the infinite-dimensional case, which is the present situation.

Remark 5.2 Let h be given in $L^2(\Omega)$ and let y_h be the solution of

$$\frac{\partial y_h}{\partial t} + A y_h = 0 \text{ in } \Omega \times (s, T), \quad y_h(s) = h, \quad y_h = 0 \text{ on } \Sigma_s \tag{5.13}$$

(that is, we choose $v = 0$ in (5.1)–(5.3)). Let us denote by E_s the set of those functions h in (5.13) such that

$$y_h(T) \in y_T + \beta B. \tag{5.14}$$

If relation (5.14) holds, we clearly have (from (5.11) and (5.12)):

$$\phi(h, s) = 0 \quad \text{if } h \in E_s. \tag{5.15}$$

Collecting the above results we have "visualized" in Figure 5.1, the behavior of the function ϕ in the set $L^2(\Omega) \times (0, T)$.

Remark 5.3 As usual in the *dynamic programming* approach, the *best decision* at time s corresponds to the element w in $L^2(\mathcal{O})$ which achieves the minimum in (5.10), namely,

$$u(s) = -\frac{\partial \phi}{\partial h}(h, s)\chi_{\mathcal{O}}. \tag{5.16}$$

This is the "real time" optimal policy, provided we know how to compute $\frac{\partial \phi}{\partial h}(h, s)$*, a most formidable task indeed*!

Remark 5.4 The *duality* formulas of Chapter 1, Section 1.4 can of course be applied. We would obtain then

$$\phi(h, s) = -\inf_{g \in L^2(\Omega)} \left\{ \frac{1}{2} \iint_{\mathcal{O} \times (s,T)} \psi_g^2 \, dx \, dt \right.$$
$$\left. - \int_\Omega g \left(y_T - y_h(T)\right) dx + \beta \, \|g\|_{L^2(\Omega)} \right\}, \tag{5.17}$$

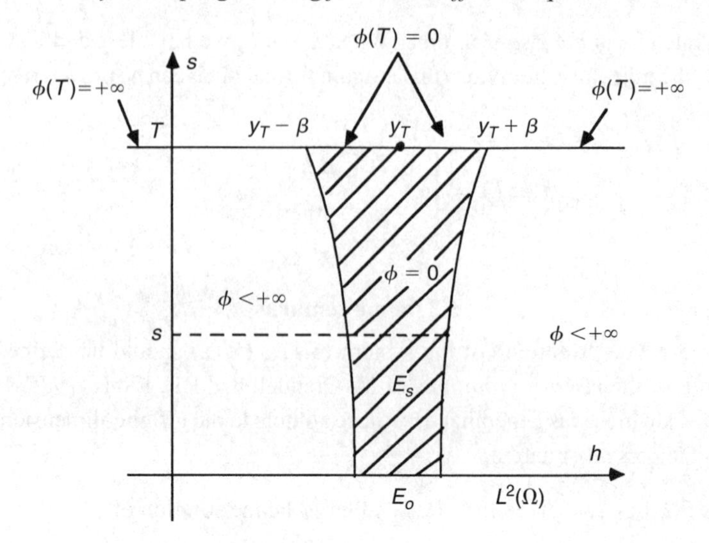

Fig. 5.1. Distribution of ϕ in the set $L^2(\Omega) \times (0, T)$.

where y_h is defined by (5.13) and where ψ_g is defined by

$$-\frac{\partial \psi_g}{\partial t} + A^* \psi_g = 0 \text{ in } \Omega \times (s, T), \quad \psi_g(T) = g, \quad \psi_g = 0 \text{ on } \Sigma_s. \tag{5.18}$$

Remark 5.5 *Dynamic programming* has been applied to the *closed loop* control of the *Navier–Stokes equations* modeling *incompressible viscous flow* in Sritharan (1991a,b).

Part II

Wave Models

6

Wave equations

6.1 Wave equations: Dirichlet boundary control

Let Ω be a *bounded* domain in $\mathbb{R}^d$, with a smooth boundary Γ. In $Q = \Omega \times (0, T)$, we consider the *wave equation*

$$\frac{\partial^2 y}{\partial t^2} + \mathcal{A}y = 0, \tag{6.1}$$

where $\mathcal{A}$ is a *second-order elliptic operator*, with *smooth coefficients*, and such that

$$\mathcal{A} = \mathcal{A}^*. \tag{6.2}$$

A classical case is

$$\mathcal{A} = -\Delta \quad \left(= -\nabla^2 = -\sum_{i=1}^{d} \frac{\partial^2}{\partial x_i^2} \right). \tag{6.3}$$

We assume that

$$y(0) = 0, \quad \frac{\partial y}{\partial t}(0) = 0, \tag{6.4}$$

and we suppose that the *control is applied on a part of the boundary*. More precisely, let Γ_0 be a "smooth" subset of Γ. Then

$$y = \begin{cases} v & \text{on } \Sigma_0 = \Gamma_0 \times (0, T), \\ 0 & \text{on } \Sigma \backslash \Sigma_0, \quad \Sigma = \Gamma \times (0, T). \end{cases} \tag{6.5}$$

We denote by $y(v) : t \to y(t; v)$ the solution of the wave problem (6.1), (6.4), (6.5), assuming that the control v satisfies "some" further properties. Indeed, we shall assume that

$$v \in L^2(\Sigma_0), \tag{6.6}$$

since this is – as far as the control itself is concerned! – certainly the simplest possible choice. However, a few preliminary remarks are necessary here.

Remark 6.1 Even assuming that Γ_0, Γ and the coefficients of operator $\mathcal{A}$ are very smooth, once the choice (6.6) has been made, one *has* to deal with *weak solutions* of

(6.1), (6.4), and (6.5). In fact (cf. J.L. Lions, 1988a and 1988b, Vol. 1) the (unique) solution $y(v)$ of (6.1), (6.4), and (6.5) satisfies the following properties:

$$y(v) \in C^0([0,T];L^2(\Omega)), \tag{6.7}$$

$$y_t(v) \in C^0([0,T];H^{-1}(\Omega)), \tag{6.8}$$

where, in (6.8) and in the following, we have set:

$$\varphi_t = \frac{\partial \varphi}{\partial t}, \quad \varphi_{tt} = \frac{\partial^2 \varphi}{\partial t^2}.$$

The solution $y = y(v)$ is defined by *transposition* as in J.L. Lions and Magenes (1968). If we consider the *adjoint equation*

$$\begin{cases} \varphi_{tt} + \mathcal{A}\varphi = f \quad \text{in } Q, \\ \varphi(T) = \varphi_t(T) = 0, \quad \varphi = 0 \quad \text{on } \Sigma, \end{cases} \tag{6.9}$$

where $f \in L^1(0,T;L^2(\Omega))$, then y is defined by

$$\int_Q yf \,\mathrm{d}x\,\mathrm{d}t = -\int_{\Sigma_0} \frac{\partial \varphi}{\partial n_{\mathcal{A}}} v \,\mathrm{d}\Sigma, \quad \forall f \in L^1(0,T;L^2(\Omega)), \tag{6.10}$$

where $\frac{\partial}{\partial n_{\mathcal{A}}}$ denotes the *normal derivative* associated with $\mathcal{A}$ (it is the usual normal derivative if $\mathcal{A} = -\Delta$). The linear functional

$$f \to -\int_{\Sigma_0} \frac{\partial \varphi}{\partial n_{\mathcal{A}}} v \,\mathrm{d}\Sigma$$

is *continuous* over $L^1(0,T;L^2(\Omega))$; this is the *key point* since we can show that

$$\left\| \frac{\partial \varphi}{\partial n_{\mathcal{A}}} \right\|_{L^2(\Sigma)} \le C \|f\|_{L^1(0,T;L^2(\Omega))}. \tag{6.11}$$

One uses then the restriction of $\frac{\partial \varphi}{\partial n_{\mathcal{A}}}$ to Σ_0 and therefore

$$y \in L^\infty(0,T;L^2(\Omega)).$$

One proceeds then to obtain (6.7), (6.8).

Remark 6.2 The original proof (J.L. Lions, 1983) assumes that Γ is *smooth*. Strangely enough, it took ten years – and a nontrivial technical proof – to generalize (6.11) to *Lipschitz continuous* boundaries (in the sense of Nečas, 1967); this was done by Chiara (1993).

We want now to study the *controllability* for systems modeled by (6.1), (6.4), and (6.5), that is,

$$\begin{cases} \text{given } T \ (0 < T < +\infty) \text{ and } \{z^0, z^1\} \in L^2(\Omega) \times H^{-1}(\Omega) \\ \text{can we find } v \text{ such that} \\ y(T; v) = z^0 \quad \textit{or} \quad y(T; v) \textit{ ``very close'' to } z^0, \\ y_t(T; v) = z^1 \quad \textit{or} \quad y_t(T; v) \textit{ ``very close'' to } z^1 \ ? \end{cases} \tag{6.12}$$

There is a *fundamental difference* between the present situation and those discussed in Part I for *diffusion* equations, due to the *finite propagation speed* of the waves (or singularities) the solution is made of, whereas this speed is *infinite* for *diffusion* equations (and for *Petrowsky*'s type equations as well). It follows from this property that

$$\textit{Conditions (6.12) may be possible only if } T \textit{ is sufficiently large.} \tag{6.13}$$

This will be made precise in the following sections.

6.2 Approximate controllability

For technical reasons we shall always consider the mapping

$$L : v \to \{-y_t(T; v), y(T; v)\}, \tag{6.14}$$

(which is *linear* and *continuous* from $L^2(\Sigma_0)$ into $H^{-1}(\Omega) \times L^2(\Omega)$), instead of the mapping

$$v \to \{y(T; v), y_t(T; v)\}$$

(but this does not go beyond simplifying – we hope – some formulae).

Let us discuss first the range $R(L)$ of operator L; we consider thus $\mathbf{f} = \{f^0, f^1\}$ such that

$$\mathbf{f} \in H_0^1(\Omega) \times L^2(\Omega), \tag{6.15}$$

and

$$\langle Lv, \mathbf{f} \rangle = 0, \quad \forall v \in L^2(\Sigma_0), \tag{6.16}$$

that is,

$$-\langle y_t(T; v), f^0 \rangle + \int_\Omega y(T; v) f^1 \, dx = 0, \quad \forall v \in L^2(\Sigma_0);$$

in (6.16) (respectively, in the above relation), $\langle \cdot, \cdot \rangle$ denotes the *duality pairing* between $H^{-1}(\Omega) \times L^2(\Omega)$ and $H_0^1(\Omega) \times L^2(\Omega)$ (respectively, $H^{-1}(\Omega)$ and $H_0^1(\Omega)$).

We introduce ψ solution of

$$\psi_{tt} + \mathcal{A}\psi = 0 \ \text{ in } Q, \quad \psi(T) = f^0, \quad \psi_t(T) = f^1, \quad \psi = 0 \ \text{ on } \Sigma. \tag{6.17}$$

It is a smooth function, which satisfies in particular

$$\frac{\partial \psi}{\partial n_{\mathcal{A}}} \in L^2(\Sigma), \quad \left\| \frac{\partial \psi}{\partial n_{\mathcal{A}}} \right\|_{L^2(\Sigma)} \leq C \left(\|f^0\|_{H_0^1(\Omega)} + \|f^1\|_{L^2(\Omega)} \right). \tag{6.18}$$

Multiplying the first equation in (6.17) by y and integrating by parts, we obtain

$$\langle Lv, \mathbf{f} \rangle = \int_{\Sigma_0} \frac{\partial \psi}{\partial n_{\mathcal{A}}} v \, d\Sigma. \tag{6.19}$$

Thus (6.16) is equivalent to

$$\frac{\partial \psi}{\partial n_{\mathcal{A}}} = 0 \quad \text{on } \Sigma_0. \tag{6.20}$$

Therefore, the *Cauchy data* are zero for ψ on Σ_0. According to the *Holmgren's Uniqueness Theorem* (cf. Hörmander, 1976) it follows that

$$\begin{gathered} \text{If } T > 2\,\text{diam}(\Omega), \text{ then the vector-valued function} \\ \{y(T; v), y_t(T; v)\} \\ \text{describes a dense subspace of } L^2(\Omega) \times H^{-1}(\Omega) \end{gathered} \tag{6.21}$$

(in (6.21), diam(Ω) denotes the *diameter* of Ω related to the *geodetic distance* associated with operator $\mathcal{A}$; it is the usual Euclidean distance if $\mathcal{A} = -\Delta$). According to *Holmgren's Theorem*, we have $\psi = 0$ in Q, which implies in turn that $\mathbf{f}$ $(= \{\psi(T), \psi_t(T)\}) = \mathbf{0}$ (see, for example, J.L. Lions, 1988b, Vol. 1).

Remark 6.3 Holmgren's theorem applies with the conditions

$$\psi = 0 \text{ and } \frac{\partial \psi}{\partial n_{\mathcal{A}}} = 0 \quad \text{on } \Sigma_0,$$

without having necessarily $\psi = 0$ on $\Sigma \backslash \Sigma_0$. The fact that in the present situation we have $\psi = 0$ on the whole Σ provides some additional flexibility to obtain uniqueness results. We shall return to this later on.

6.3 Formulation of the approximate controllability problem

We shall make the following hypothesis:

$$\Sigma_0 \textit{ allows the application of the Holmgren's Uniqueness Theorem.} \tag{6.22}$$

Then, $\{z^0, z^1\}$ being given in $L^2(\Omega) \times H^{-1}(\Omega)$, there exist always controls v (actually, an infinite number of them) such that

$$y(T; v) \in z^0 + \beta_0 B, \quad y_t(T; v) \in z^1 + \beta_1 B_{-1}, \tag{6.23}$$

where B (respectively B_{-1}) denotes the unit ball of $L^2(\Omega)$ (respectively $H^{-1}(\Omega)$), and where β_0, β_1 are given positive numbers, arbitrarily small.

The *optimal control* problem that we consider is

$$\inf_v \left\{ \frac{1}{2} \int_{\Sigma_0} v^2 \, d\Sigma, \quad \text{with } v \in L^2(\Sigma_0) \text{ and satisfying (6.23)} \right\}. \tag{6.24}$$

Remark 6.4 *Exact controllability* corresponds to $\beta_0 = \beta_1 = 0$.

6.4 Dual problems

We proceed essentially as in Chapter 1, Section 1.4. We introduce therefore

$$F_1(v) = \frac{1}{2} \int_{\Sigma_0} v^2 \, d\Sigma, \quad \forall v \in L^2(\Sigma_0), \tag{6.25}$$

and then $F_2 : H^{-1}(\Omega) \times L^2(\Omega) \to \mathbb{R} \cup \{+\infty\}$ by

$$F_2(\mathbf{g}) = F_2(g^0, g^1) = \begin{cases} 0 & \text{if } \; g^0 \in -z^1 + \beta_1 B_{-1} \\ & \text{and } g^1 \in z^0 + \beta_0 B, \\ +\infty & \text{otherwise.} \end{cases} \tag{6.26}$$

With this notation, the control problem (6.20) can be formulated as

$$\inf_{v \in L^2(\Sigma_0)} \{F_1(v) + F_2(Lv)\}. \tag{6.27}$$

with operator L defined by (6.14). Using, as in Chapter 1, Section 1.4, *Duality Theory* we obtain

$$\inf_{v \in L^2(\Sigma_0)} \{F_1(v) + F_2(Lv)\} + \inf_{\mathbf{g} \in H_0^1(\Omega) \times L^2(\Omega)} \{F_1^*(L^*\mathbf{g}) + F_2^*(-\mathbf{g})\} = 0, \tag{6.28}$$

where

$$F_1^*(v) = \frac{1}{2} \int_{\Sigma_0} v^2 \, d\Sigma, \quad \forall v \in L^2(\Sigma_0), \tag{6.29a}$$

and

$$F_2^*(\mathbf{g}) = -\langle z^1, g^0 \rangle + \int_\Omega z^0 g^1 \, dx + \beta_1 \|g^0\|_{H_0^1(\Omega)} + \beta_0 \|g^1\|_{L^2(\Omega)},$$
$$\forall \, \mathbf{g} = \{g^0, g^1\} \in H_0^1(\Omega) \times L^2(\Omega). \tag{6.29b}$$

Using (6.19) we have

$$L^*\mathbf{g} = \frac{\partial \psi_{\mathbf{g}}}{\partial n_A} \quad \text{on } \Sigma_0, \tag{6.30}$$

where $\psi_{\mathbf{g}}$ is given by (6.17) (with $\mathbf{f} = \mathbf{g}$). We have therefore the following.

Theorem 6.5 *We suppose that* (6.22) *holds true. For β_0 and β_1 given arbitrarily small, problem* (6.24) *has a unique solution and*

$$\inf_v \frac{1}{2}\int_{\Sigma_0} v^2 \, d\Sigma = -\inf_{\mathbf{g}} \left\{ \frac{1}{2}\int_{\Sigma_0} \left(\frac{\partial \psi_{\mathbf{g}}}{\partial n_{\mathcal{A}}}\right)^2 d\Sigma + \langle z^1, g^0 \rangle \right.$$

$$\left. - \int_\Omega z^0 g^1 \, dx + \beta_1 \|g^0\|_{H_0^1(\Omega)} + \beta_0 \|g^1\|_{L^2(\Omega)} \right\}, \quad (6.31)$$

where, in (6.31), *$v \in L^2(\Sigma_0)$ and verifies* (6.23), *$\mathbf{g} \in H_0^1(\Omega) \times L^2(\Omega)$, and where $\psi_{\mathbf{g}}$ is given by* (6.17), *with $\mathbf{f} = \mathbf{g}$.*

The *dual problem* is the minimization problem in the right-hand side of (6.31). If $\mathbf{f}$ is the solution of the dual problem and if ψ is the corresponding solution of (6.17) then the *optimal control*, that is. the solution u of problem (6.24) is given by

$$u = \frac{\partial \psi}{\partial n_{\mathcal{A}}} \quad \text{on } \Sigma_0. \quad (6.32)$$

6.5 Direct solution of the dual problem

One can formulate the dual problem in an equivalent fashion which will be useful when β_0 and β_1 converge to zero, and also for numerical calculations. To this effect, we introduce the following operator $\mathbf{\Lambda}$:

Given $\mathbf{g} = \{g^0, g^1\} \in H_0^1(\Omega) \times L^2(\Omega)$, we define $\psi_{\mathbf{g}}$ and $y_{\mathbf{g}}$ as the solutions of the two wave problems below

$$\begin{cases} \dfrac{\partial^2 \psi_{\mathbf{g}}}{\partial t^2} + \mathcal{A}\psi_{\mathbf{g}} = 0 \quad \text{in } Q, \\ \\ \psi_{\mathbf{g}}(T) = g^0, \quad \dfrac{\partial \psi_{\mathbf{g}}}{\partial t}(T) = g^1, \quad \psi_{\mathbf{g}} = 0 \text{ on } \Sigma, \end{cases} \quad (6.33a)$$

$$\begin{cases} \dfrac{\partial^2 y_{\mathbf{g}}}{\partial t^2} + \mathcal{A} y_{\mathbf{g}} = 0 \quad \text{in } Q, \\ \\ y_{\mathbf{g}}(0) = \dfrac{\partial y_{\mathbf{g}}}{\partial t}(0) = 0, \quad y_{\mathbf{g}} = \dfrac{\partial \psi_{\mathbf{g}}}{\partial n_{\mathcal{A}}} \text{ on } \Sigma_0, \quad y_{\mathbf{g}} = 0 \text{ on } \Sigma \backslash \Sigma_0, \end{cases} \quad (6.33b)$$

and we set

$$\mathbf{\Lambda g} = \left\{ -\frac{\partial y_{\mathbf{g}}}{\partial t}(T), y_{\mathbf{g}}(T) \right\}. \quad (6.34)$$

We define in this way an operator $\mathbf{\Lambda}$ such that

$$\mathbf{\Lambda} \in \mathcal{L}(H_0^1(\Omega) \times L^2(\Omega), H^{-1}(\Omega) \times L^2(\Omega)). \quad (6.35)$$

If we multiply both sides of the first equation in (6.33b) by $\psi_{\mathbf{g}^*}$ (which corresponds to $\mathbf{g}^* \in H_0^1(\Omega) \times L^2(\Omega)$) and if we integrate by parts we obtain (with obvious notation):

$$\langle \mathbf{\Lambda g}, \mathbf{g}^* \rangle = \int_{\Sigma_0} \frac{\partial \psi_{\mathbf{g}}}{\partial n_A} \frac{\partial \psi_{\mathbf{g}^*}}{\partial n_A} \, d\Sigma. \tag{6.36}$$

It follows from (6.36) that the operator $\mathbf{\Lambda}$ is *self-adjoint* and *positive semidefinite*.

The *dual problem* is then *equivalent* to

$$\inf_{\mathbf{g}} \left\{ \frac{1}{2} \langle \mathbf{\Lambda g}, \mathbf{g} \rangle + \langle z^1, g^0 \rangle - \int_\Omega z^0 g^1 \, dx + \beta_1 \|g^0\|_{H_0^1(\Omega)} + \beta_0 \|g^1\|_{L^2(\Omega)} \right\}, \tag{6.37}$$

where, in (6.37), $\mathbf{g} \in H_0^1(\Omega) \times L^2(\Omega)$. Assuming that the condition (6.22) holds true, problem (6.37) has a *unique* solution for β_0 and $\beta_1 > 0$, arbitrarily small. If we denote by $\mathbf{f}$ the solution of problem (6.37), it is also the solution of the following *variational inequality*

$$\begin{cases} \mathbf{f} \in H_0^1(\Omega) \times L^2(\Omega); \quad \forall \mathbf{g} \in H_0^1(\Omega) \times L^2(\Omega), \\ \langle \mathbf{\Lambda f}, \mathbf{g} - \mathbf{f} \rangle + \langle z^1, g^0 - f^0 \rangle - \displaystyle\int_\Omega z^0 (g^1 - f^1) \, dx \\ \quad + \beta_1 \big(\|g^0\|_{H_0^1(\Omega)} - \|f^0\|_{H_0^1(\Omega)} \big) + \beta_0 \big(\|g^1\|_{L^2(\Omega)} - \|f^1\|_{L^2(\Omega)} \big) \geq 0. \end{cases} \tag{6.38}$$

Remark 6.6 Problems (6.37) and (6.38) are equivalent to the minimization problem in the right-hand side of (6.31), but they are better suited for the solution of the dual problem.

Remark 6.7 The operator $\mathbf{\Lambda}$ (defined by (6.33), (6.34)) is the same as the one introduced in the *Hilbert Uniqueness Method* (HUM). This is made more precise in the following section (see also J.L. Lions, 1986, 1988a,b).

Remark 6.8 Relation (6.36) makes sense, because there exists a constant C such that

$$\int_{\Sigma_0} \left| \frac{\partial \psi_{\mathbf{g}}}{\partial n_A} \right|^2 d\Sigma \leq C \big(\|g^0\|^2_{H_0^1(\Omega)} + \|g^1\|^2_{L^2(\Omega)} \big),$$
$$\forall \mathbf{g} \; (= \{g^0, g^1\}) \in H_0^1(\Omega) \times L^2(\Omega), \tag{6.39}$$

where, in (6.39), $\psi_{\mathbf{g}}$ and $\mathbf{g}$ are related by (6.33).

6.6 Exact controllability and new functional spaces

Let us consider now problem (6.37), (6.38) with the idea of letting β_0 and β_1 converge to zero. We introduce on $H_0^1(\Omega) \times L^2(\Omega)$ the following new functional:

$$[\mathbf{g}] = \big(\langle \mathbf{\Lambda g}, \mathbf{g} \rangle \big)^{1/2}. \tag{6.40}$$

Since we assume that the condition (6.22) holds true, the functional $[\cdot]$ is in fact a *norm*, of a *pre-Hilbertian* nature. We introduce then

$$E = \text{Completion of } H_0^1(\Omega) \times L^2(\Omega) \text{ for the norm } [\cdot]; \tag{6.41}$$

with this notation we can state that

$$\mathbf{\Lambda} \text{ is an isomorphism from } E \text{ onto } E'. \tag{6.42}$$

If $\beta_0 = \beta_1 = 0$, problem (6.37), (6.38) is *equivalent* to

$$\inf_{\substack{\mathbf{g}=\{g^0,g^1\}\\ \in H_0^1(\Omega)\times L^2(\Omega)}} \left\{ \frac{1}{2}[\mathbf{g}]^2 + \langle z^1, g^0\rangle - \int_\Omega z^0 g^1 \, dx \right\}, \tag{6.43}$$

Problem (6.43) *has a unique solution if and only if*

$$\{-z^1, z^0\} \in E'. \tag{6.44}$$

If we denote by $\mathbf{f}_\beta = \{f_\beta^0, f_\beta^1\}$ the solution of problem (6.37), (6.38), where $\boldsymbol{\beta} = \{\beta_0, \beta_1\}$, then

$$\lim_{\boldsymbol{\beta}\to\mathbf{0}} \mathbf{f}_\beta = \mathbf{f}_0 = \text{ the solution of (6.43)} \tag{6.45}$$

if and only if condition (6.44) holds true.

Remark 6.9 The solution method that we have just presented is what is called HUM (cf. J.L. Lions, 1986, 1988a,b) since the key element is the introduction of a new Hilbert space E based on a *uniqueness* property.

Remark 6.10 Problems (6.38) and (6.43) give a *constructive* approach to exact or approximate controllability; we shall make this more precise in the next sections.

Remark 6.11 Condition (6.44) means that *exact controllability is possible if and only if z^0 and z^1 belong to well chosen Hilbert spaces.*

Remark 6.12 The approach taken in the present section is closely related to the one followed in Chapter 1, Section 1.5. With the notation of Chapter 1, Section 1.5, Remark 1.17, we would have

$$E = \widehat{H_0^1(\Omega) \times L^2}(\Omega).$$

There is, however, a very important technical difference between the two situations, since for the *diffusion* problems discussed in Chapter 1, Section 1, the space $\widehat{L^2(\Omega)}$ is *never* a "simple" *distribution space* (except for the case $\mathcal{O} = \Omega$, that is the control is distributed over the whole domain Ω). For the *wave equation* the situation is quite different, as we shall see in the next section.

6.7 On the structure of space E

Following Bardos, Lebeau, and Rauch (1988) here we shall say that Σ_0 enjoys the *geometrical control condition* if any ray, originating from any point of Ω at $t = 0$, reaches eventually (after geometrical reflexions on Γ) the set Γ_0 before time $t = T$.

The main result is then

$$\textit{If } \Sigma_0 \textit{ satisfies the geometrical control condition, then } E = H_0^1(\Omega) \times L^2(\Omega). \tag{6.46}$$

Actually, the geometrical control condition is also *necessary* in order for (6.46) to be true. The inequality corresponding to (6.46) is the *reverse* of inequality (6.39): there exists a constant $C_1 > 0$ such that

$$\int_{\Sigma_0} \left| \frac{\partial \psi_g}{\partial n_A} \right|^2 d\Sigma \geq C_1 \left(\|g^0\|^2_{H_0^1(\Omega)} + \|g^1\|^2_{L^2(\Omega)} \right), \tag{6.47}$$

if and only if Σ_0 satisfies the geometrical control condition.

Remark 6.13 We refer to J.L. Lions (1988b) for the various contributions, by several authors, which led to the fundamental inequality (6.47).

Remark 6.14 If Σ_0 *does not* satisfy the geometrical control condition, but *does* satisfy the condition for the Holmgren's Uniqueness Theorem to apply, then

$$[g] = \left(\int_{\Sigma_0} \left| \frac{\partial \psi_g}{\partial n_A} \right|^2 d\Sigma \right)^{1/2}$$

is a norm, *strictly weaker* than the $H_0^1(\Omega) \times L^2(\Omega)$ norm. In that case, the space E is a *new* Hilbert space such that

$$H_0^1(\Omega) \times L^2(\Omega) \subset E, \quad \text{strictly}, \tag{6.48}$$

the exact structure of E is far from being simple, since the space E *can contain elements which are not distributions on* Ω.

6.8 Numerical methods for the Dirichlet boundary controllability of the wave equation

6.8.1 Generalities. Synopsis

In this section which is largely inspired by Dean, Glowinski, and Li (1989), Glowinski, Li and J.L. Lions (1990), Glowinski and Li (1990), and Glowinski (1992a) we shall discuss the *numerical solution* of the *exact* and *approximate Dirichlet boundary controllability problems* considered in the preceding sections.

To make things simpler we shall assume that Σ_0 satisfies the *geometrical control condition* (see the above section), so that

$$E = H_0^1(\Omega) \times L^2(\Omega), \tag{6.49}$$

and the operator $\mathbf{\Lambda}$ is an *isomorphism* from E onto E' $(= H^{-1}(\Omega) \times L^2(\Omega))$. The properties of $\mathbf{\Lambda}$ (*symmetry* and *strong ellipticity*) will make the solution of the exact controllability problem possible by a *conjugate gradient algorithm* operating in the space E; this algorithm will be described in Section 6.8.3. Next, we shall describe the *time* and *space discretizations* of the exact controllability problem by a combination of *finite difference* (FD) and *finite element* (FE) methods and then discuss the iterative solution of the corresponding approximate problem. Finally, we shall describe solution methods for the *approximate* boundary controllability problem (6.24).

Both exact and approximate controllability problems will be solved using their *dual formulation* since the corresponding control problems are *easier* to solve than their primal counterparts.

The results of numerical experiments testing the methods discussed in the present section will be reported in Section 6.9. In Section 6.10, we will briefly review some very recent results, by various authors, concerning the solution of boundary controllability problems closely related to those discussed here.

Remark 6.15 A *spectral* method – still based on HUM – for solving *directly* (that is noniteratively) the *exact* Dirichlet boundary controllability problem is discussed in Bourquin (1993), where numerical results are also presented.

6.8.2 Dual formulation of the exact controllability problem. Further properties of $\mathbf{\Lambda}$

To obtain the *dual problem* corresponding to *exact controllability* it suffices to take $\beta_0 = \beta_1 = 0$ in formulations (6.37), (6.38); we obtain then

$$\mathbf{\Lambda f} = \{-z^1, z^0\}. \tag{6.50}$$

Since we supposed (see Section 6.8.1) that the *geometrical control condition holds*, we know (from Section 6.6) that

$$\mathbf{\Lambda} \text{ is an isomorphism from } E \text{ onto } E', \tag{6.51}$$

with $E = H_0^1(\Omega) \times L^2(\Omega)$, and $E' = H^{-1}(\Omega) \times L^2(\Omega)$. Problem (6.50) has therefore a *unique* solution, $\forall \{z^0, z^1\} \in L^2(\Omega) \times H^{-1}(\Omega)$. The solution $\mathbf{f}$ of (6.50) is also *the* solution of the following *linear variational* problem

$$\begin{cases} \mathbf{f} \in E; \quad \forall \mathbf{g} = \{g^0, g^1\} \in E \text{ we have} \\ \langle \mathbf{\Lambda f}, \mathbf{g} \rangle = -\langle z^1, g^0 \rangle + \displaystyle\int_\Omega z^0 g^1 \, dx. \end{cases} \tag{6.52}$$

Since (from Sections 6.5–6.7) $\mathbf{\Lambda}$ is *continuous, self-adjoint* and *strongly elliptic* (in the sense that there exists $\alpha > 0$ such that

$$\langle \mathbf{\Lambda g}, \mathbf{g} \rangle \geq \alpha \, \|\mathbf{g}\|_E^2, \quad \forall \mathbf{g} \in E)$$

the *bilinear functional*

$$\{\mathbf{g}, \mathbf{g}'\} \mapsto \langle \mathbf{\Lambda g}, \mathbf{g}' \rangle : \ E \times E \to \mathbb{R}$$

is *continuous, symmetric* and *E-elliptic* over $E \times E$. On the other hand, the *linear* functional in the right-hand side of (6.52) is clearly *continuous* over E, implying (cf. Chapter 1, Section 1.8.2) that problem (6.50), (6.52) can be solved by a *conjugate gradient algorithm* operating in the space E. such an algorithm will be described in the following section.

Remark 6.16 We suppose here that $\Gamma_0 = \Gamma$ and that $\mathcal{A} = -\Delta$; we suppose also that there exists $x_0 \in \Omega$ and $C > 0$ such that

$$\overrightarrow{x_0 M} \cdot \mathbf{n} = C, \quad \forall M \in \Gamma, \tag{6.53}$$

with $\mathbf{n}$ the unit vector of the outward normal at Γ, at M. Domains satisfying the condition (6.53) are easy to characterize geometrically, simple cases being disks and squares. Now let us denote by $\mathbf{\Lambda}_T$ the operator $\mathbf{\Lambda}$ associated with T. It has been proved by J.L. Lions (unpublished result) and Bensoussan (1990) that

$$\lim_{T \to +\infty} \frac{\mathbf{\Lambda}_T}{T} = \frac{1}{C} \begin{bmatrix} -\Delta & 0 \\ 0 & I \end{bmatrix}. \tag{6.54}$$

Result (6.54) is quite important for, among other things, the validation of the numerical methods described hereafter; indeed, (6.54) easily provides

$$\lim_{T \to +\infty} T \, \mathbf{f}_T = \{\chi^0, \chi^1\}, \tag{6.55}$$

where

$$\Delta \chi^0 = C z^1 \ \text{ in } \Omega, \quad \chi^0 = 0 \ \text{ on } \Gamma, \tag{6.56}$$

and

$$\chi^1 = C z^0. \tag{6.57}$$

6.8.3 Conjugate gradient solution of problem (6.50), (6.52)

Assuming that the *geometrical control condition* holds, it follows from Section 6.8.2 that we can apply the general *conjugate gradient* algorithm (1.124)–(1.131) (see Chapter 1, Section 1.8.2) to the solution of problem (6.50), (6.52); indeed, it suffices to take

$$V = E, \quad a(\cdot, \cdot) = \langle \mathbf{\Lambda} \cdot, \cdot \rangle, \quad L : \mathbf{g} \to -\langle z^1, g^0 \rangle + \int_\Omega z^0 g^1 \, dx.$$

On E, we will use as scalar product

$$\{\mathbf{v}, \mathbf{w}\} \rightarrow \int_\Omega \left(\nabla v^0 \cdot \nabla w^0 + v^1 w^1\right) \mathrm{d}x, \quad \forall \mathbf{v}, \mathbf{w} \in E,$$

other choices being possible. We obtain then the following algorithm:

Step 0: Initialization.

$$f_0^0 \in H_0^1(\Omega) \text{ and } f_0^1 \in L^2(\Omega) \text{ are given;} \tag{6.58}$$

solve then

$$\begin{cases} \dfrac{\partial^2 \psi_0}{\partial t^2} + A\psi_0 = 0 \text{ in } Q, \quad \psi_0 = 0 \text{ on } \Sigma, \\ \psi_0(T) = f_0^0, \quad \dfrac{\partial \psi_0}{\partial t}(T) = f_0^1, \end{cases} \tag{6.59}$$

and

$$\begin{cases} \dfrac{\partial^2 \varphi_0}{\partial t^2} + A\varphi_0 = 0 \text{ in } Q, \quad \varphi_0 = \dfrac{\partial \psi_0}{\partial n_A} \text{ on } \Sigma_0, \quad \varphi_0 = 0 \text{ on } \Sigma \backslash \Sigma_0, \\ \varphi_0(0) = 0, \quad \dfrac{\partial \varphi_0}{\partial t}(0) = 0. \end{cases} \tag{6.60}$$

Compute $\mathbf{g}_0 = \{g_0^0, g_0^1\} \in E$ *by*

$$-\Delta g_0^0 = z^1 - \frac{\partial \varphi_0}{\partial t}(T) \text{ in } \Omega, \quad g_0^0 = 0 \text{ on } \Gamma, \tag{6.61}$$

$$g_0^1 = \varphi_0(T) - z^0, \tag{6.62}$$

respectively. Set then

$$\mathbf{w}_0 = \mathbf{g}_0. \tag{6.63}$$

Now, for $n \geq 0$, *assuming that* $\mathbf{f}_n$, $\mathbf{g}_n$, *and* $\mathbf{w}_n$ *are known, compute* $\mathbf{f}_{n+1}$, $\mathbf{g}_{n+1}$, *and* $\mathbf{w}_{n+1}$ *as follows.*

Step 1: Descent.
Solve

$$\begin{cases} \dfrac{\partial^2 \bar{\psi}_n}{\partial t^2} + A\bar{\psi}_n = 0 \text{ in } Q, \quad \bar{\psi}_n = 0 \text{ on } \Sigma, \\ \bar{\psi}_n(T) = w_n^0, \quad \dfrac{\partial \bar{\psi}_n}{\partial t}(T) = w_n^1, \end{cases} \tag{6.64}$$

$$\begin{cases} \dfrac{\partial^2 \bar{\varphi}_n}{\partial t^2} + A\bar{\varphi}_n = 0 \text{ in } Q, \ \bar{\varphi}_n = \dfrac{\partial \bar{\psi}^n}{\partial n_A} \text{ on } \Sigma_0, \ \bar{\varphi}_n = 0 \text{ on } \Sigma \backslash \Sigma_0, \\ \bar{\varphi}_n(0) = 0, \quad \dfrac{\partial \bar{\varphi}_n}{\partial t}(0) = 0, \end{cases} \tag{6.65}$$

$$\Delta \bar{g}_n^0 = \frac{\partial \bar{\varphi}_n}{\partial t}(T) \text{ in } \Omega, \quad \bar{g}_n^0 = 0 \text{ on } \Gamma, \tag{6.66}$$

and set

$$\bar{g}_n^1 = \bar{\varphi}_n(T). \tag{6.67}$$

Compute now

$$\rho_n = \int_\Omega \left(|\nabla g_n^0|^2 + |g_n^1|^2\right) \mathrm{d}x \bigg/ \int_\Omega \left(\nabla \bar{g}_n^0 \cdot \nabla w_n^0 + \bar{g}_n^1 w_n^1\right) \mathrm{d}x, \tag{6.68}$$

$$\mathbf{f}_{n+1} = \mathbf{f}_n - \rho_n \mathbf{w}_n, \tag{6.69}$$

$$\mathbf{g}_{n+1} = \mathbf{g}_n - \rho_n \bar{\mathbf{g}}_n. \tag{6.70}$$

Step 2: Test of the convergence and construction of the new descent direction. If $\mathbf{g}_{n+1} = \mathbf{0}$, *or is sufficiently small, that is*

$$\int_\Omega \left(|\nabla g_{n+1}^0|^2 + |g_{n+1}^1|^2\right) \mathrm{d}x \bigg/ \int_\Omega \left(|\nabla g_0^0|^2 + |g_0^1|^2\right) \mathrm{d}x \le \epsilon^2), \tag{6.71}$$

take $\mathbf{f} = \mathbf{f}_{n+1}$*; else, compute*

$$\gamma_n = \int_\Omega \left(|\nabla g_{n+1}^0|^2 + |g_{n+1}^1|^2\right) \mathrm{d}x \bigg/ \int_\Omega \left(|\nabla g_n^0|^2 + |g_n^1|^2\right) \mathrm{d}x, \tag{6.72}$$

and set

$$\mathbf{w}_{n+1} = \mathbf{g}_{n+1} + \gamma_n \mathbf{w}_n. \tag{6.73}$$

Do $n = n + 1$ *and go to* (6.64).

Remark 6.17 It appears at first glance that algorithm (6.58)–(6.73) is quite memory demanding since it seems to require the storage of $\frac{\partial \bar{\psi}_n}{\partial n_A}$ on Σ_0 (in practice the storage of $\frac{\partial \bar{\psi}_n}{\partial n_A}$ over a discrete – but still large – subset of Σ_0). In fact, we can avoid this storage problem by observing that since the wave equation in (6.64) is *reversible* we can integrate *simultaneously*, from 0 to T, the wave equations (6.65) and

$$\begin{cases} \dfrac{\partial^2 \bar{\psi}_n}{\partial t^2} + A\bar{\psi}_n = 0 \text{ in } Q, \quad \bar{\psi}_n = 0 \text{ on } \Sigma, \\ \bar{\psi}_n(0) \text{ and } \dfrac{\partial \bar{\psi}_n}{\partial t}(0) \text{ being known from the integration} \\ \text{of (6.64) from } T \text{ to } 0. \end{cases} \tag{6.74}$$

In the particular case where an *explicit scheme* is used for solving the wave equations (6.64), (6.65), and (6.74), the extra cost associated with the solution of (6.74) is *negligible* compared with the saving due to not having to store $\frac{\partial \bar{\psi}_n}{\partial n_A}$ on Σ_0.

Remark 6.18 Once the solution $\mathbf{f}$ of the dual problem (6.50), (6.52) is known, it suffices to integrate the wave equation (6.33a) with $\mathbf{g} = \mathbf{f}$ to obtain ψ. The optimal control u, solution of the exact controllability problem is given then by

$$u = \left. \frac{\partial \psi}{\partial n_A} \right|_{\Sigma_0}. \tag{6.75}$$

6.8.4 Finite difference approximation of the dual problem (6.50), (6.52)

6.8.4.1 Generalities

A finite element/finite difference approximation of problem (6.50), (6.52) will be discussed in Section 6.8.7 (see also Glowinski, Li, and J.L. Lions, 1990; Glowinski, 1992a, and the references therein). At the present moment, we shall focus on the case where

$$\Omega = (0,1)^2, \quad \mathcal{A} = -\Delta, \quad \Gamma_0 = \Gamma,$$

and where finite difference methods are used both for the space–time discretizations. Indeed, these approximations can also be obtained via space discretizations associated with finite element grids like the one shown in Figure 2.1 of Chapter 2, Section 2.6 (we should use, as shown in Glowinski, Li, and J.L. Lions, 1990, piecewise linear space approximations and numerical integration by the trapezoidal rule). Let I and N be positive integers; we define h (*space discretization step*) and Δt (*time-discretization step*) by

$$h = \frac{1}{I+1}, \quad \Delta t = \frac{T}{N}, \tag{6.76}$$

respectively, and then denote by M_{ij} the point $\{ih, jh\}$.

6.8.4.2 Approximation of the wave equation (6.33a)

Let us discuss first the discretization of the following wave problem:

$$\begin{cases} \psi_{tt} - \Delta\psi = 0 \text{ in } Q, \quad \psi = 0 \text{ on } \Sigma, \\ \psi(T) = f^0, \quad \psi_t(T) = f^1. \end{cases} \tag{6.77}$$

With ψ_{ij}^n an approximation of $\psi(M_{ij}, n\Delta t)$, we approximate the wave problem (6.77) by the following *explicit* finite difference scheme:

$$\frac{\psi_{ij}^{n-1} + \psi_{ij}^{n+1} - 2\psi_{ij}^n}{|\Delta t|^2} - \frac{\psi_{i+1j}^n + \psi_{i-1j}^n + \psi_{ij+1}^n + \psi_{ij-1}^n - 4\psi_{ij}^n}{h^2} = 0,$$

$$1 \le i,j \le I, \quad 0 \le n \le N, \tag{6.78a}$$

$$\psi_{kl}^n = 0 \quad \text{if } M_{kl} \in \Gamma, \tag{6.78b}$$

$$\psi_{ij}^N = f^0(M_{ij}), \quad \psi_{ij}^{N+1} - \psi_{ij}^{N-1} = 2\Delta t f^1(M_{ij}), \quad 1 \le i,j \le I. \tag{6.78c}$$

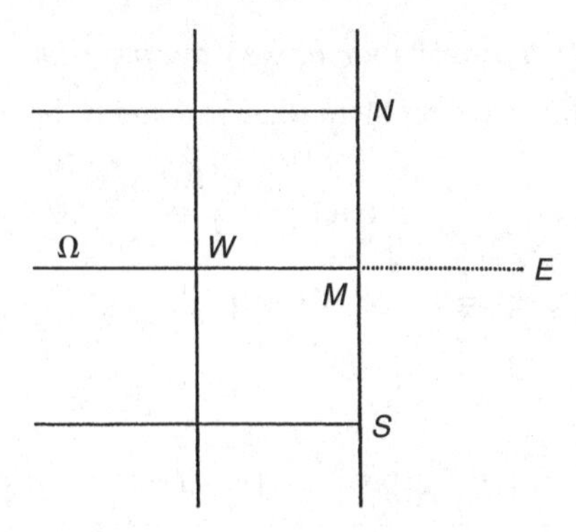

Fig. 6.1. Approximation of $\frac{\partial\psi}{\partial n}(M)$.

To be *stable*, the above scheme has to satisfy the following (*stability*) condition:

$$\Delta t \leq h/\sqrt{2}. \tag{6.79}$$

6.8.4.3 Approximation of $(\partial\psi/\partial n)|_\Sigma$

Suppose that we want to approximate $\partial\psi/\partial n$ at $M \in \Gamma$, as shown in Figure 6.1. Suppose that ψ is known at E; we shall then approximate $\partial\psi/\partial n$ at M by

$$\frac{\partial\psi}{\partial n}(M) \approx \frac{\psi(E) - \psi(W)}{2h}, \tag{6.80}$$

a reasonable choice, indeed.

In fact, $\psi(E)$ is not known since $E \notin \bar{\Omega}$. However – formally at least – $\psi = 0$ on Σ implies $\psi_{tt} = 0$ on Σ, which combined with $\psi_{tt} - \Delta\psi = 0$ implies $\Delta\psi = 0$ on Σ; discretizing this last relation at M yields

$$\frac{\psi(W) + \psi(E) + \psi(N) + \psi(S) - 4\psi(M)}{h^2} = 0. \tag{6.81}$$

Since N, M, S belong to Γ, relation (6.81) reduces to

$$\psi(W) = -\psi(E), \tag{6.82}$$

which, combined with (6.80), implies that

$$\frac{\partial\psi}{\partial n}(M) \approx -\frac{\psi(W)}{h} = \frac{0 - \psi(W)}{h} = \frac{\psi(M) - \psi(W)}{h}. \tag{6.83}$$

In that particular case, the *centered* approximation (6.80) (which is *second-order accurate*) coincides with the *one-sided* one in (6.83) (which is only *first-order accurate* in general). In the sequel, we shall use, therefore, relation (6.83) to approximate $\partial\psi/\partial n$ at M and we shall denote by $\delta_{kl}\psi$ the corresponding approximation of $\partial\psi/\partial n$ at $M_{kl} \in \Gamma$.

6.8.4.4 Approximation of the wave problem (6.33b)

As done with (6.33a), the wave problem (6.33b), namely, here,

$$\begin{cases} \varphi_{tt} - \Delta\varphi = 0 \text{ in } Q, \quad \varphi = \dfrac{\partial \psi}{\partial n} \text{ on } \Sigma, \\ \varphi(0) = 0, \quad \varphi_t(0) = 0 \end{cases} \tag{6.84}$$

will be approximated by

$$\frac{\varphi_{ij}^{n+1} + \varphi_{ij}^{n-1} - 2\varphi_{ij}^n}{|\Delta t|^2} - \frac{\varphi_{i+1j}^n + \varphi_{i-1j}^n + \varphi_{ij+1}^n + \varphi_{ij-1}^n - 4\varphi_{ij}^n}{h^2} = 0,$$

$$1 \leq i,j \leq I, \quad 0 \leq n \leq N, \tag{6.85a}$$

$$\varphi_{kl}^n = \delta_{kl}\psi^n \quad \text{if } M_{kl} \in \Gamma, \tag{6.85b}$$

$$\varphi_{ij}^0 = 0, \quad \frac{\varphi_{ij}^1 - \varphi_{ij}^{-1}}{2\Delta t} = 0, \quad 1 \leq i,j \leq I. \tag{6.85c}$$

6.8.4.5 Approximation of $\mathbf{\Lambda}$

Starting from

$$\mathbf{f}_h = \left\{\{f_{ij}^0, f_{ij}^1\}\right\}_{1 \leq i,j \leq I},$$

and via the solution of the discrete wave equations (6.78), (6.85) we approximate $\mathbf{\Lambda f}$ by

$$\mathbf{\Lambda}_h^{\Delta t}\mathbf{f}_h = \left\{\left\{-\frac{\varphi_{ij}^{N+1} - \varphi_{ij}^{N-1}}{2\Delta t}, \varphi_{ij}^N\right\}\right\}_{1 \leq i,j \leq I}. \tag{6.86}$$

It is proved in Glowinski, Li, and J.L. Lions (1990, pp. 17–19) that we have (with obvious notation)

$$\begin{aligned} \langle \mathbf{\Lambda}_h^{\Delta t}\mathbf{f}_h, \mathbf{g}_h\rangle_{h,\Delta t} &= h^2 \sum_{1 \leq i,j \leq I} \left(\varphi_{ij}^N g_{ij}^1 - \left(\frac{\varphi_{ij}^{N+1} - \varphi_{ij}^{N-1}}{2\Delta t}\right) g_{ij}^0\right) \\ &= h\Delta t \sum_{n=0}^{N} \alpha_n \sum_{M_{kl} \in \Gamma^*} \delta_{kl}\psi^n \delta_{kl}\psi_{\mathbf{g}}^n, \end{aligned} \tag{6.87}$$

where, in (6.87), $\alpha_0 = \alpha_N = \frac{1}{2}$, $\alpha_n = 1$, $\forall n = 1, \ldots, N-1$, and where $\Gamma^* = \Gamma \setminus \left\{\{0,0\},\{0,1\},\{1,0\},\{1,1\}\right\}$. It follows from (6.87) that $\mathbf{\Lambda}_h^{\Delta t}$ is *symmetric* and *positive semidefinite*. Actually, it is proved in Glowinski, Li, and J.L. Lions (1990, Section 6.2) that the operator $\mathbf{\Lambda}_h^{\Delta t}$ is *positive definite* if $T > T_{\min} \approx \Delta t/h$. This property implies that if $T(> 0)$ is given, it suffices to take the ratio $\Delta t/h$ sufficiently small to have exact boundary controllability for the discrete wave equation. This property is in contradiction with the continuous case where the exact boundary controllability property is lost if T is too small ($T < 1$, here). The reasons for this discrepancy have been thoroughly discussed in Zuazua (2005) (see also the references therein and the brief discussion in Section 6.8.6, hereafter).

6.8.4.6 Approximation of the dual problem (6.50), (6.52)

With $\mathbf{z}_h$ a convenient approximation of $\mathbf{z} = \{z^0, z^1\}$ we approximate problem (6.50), (6.52) by

$$\mathbf{\Lambda}_h^{\Delta t}\mathbf{f}_h^{\Delta t} = \sigma \mathbf{z}_h, \tag{6.88}$$

where, in (6.88), σ denotes the matrix $\begin{pmatrix} 0 & -1 \\ 1 & 0 \end{pmatrix}$. In Glowinski, Li, and J.L. Lions (1990) we may find a discrete variant of the conjugate gradient algorithm (6.58)–(6.73) which can be used to solve the approximate problem (6.88).

6.8.5 Numerical solution of a test problem; ill-posedness of the discrete problem (6.88)

Following Glowinski, Li, and J.L. Lions (1990, Section 7), Dean, Glowinski, and Li (1989, Section 2.7), and Glowinski (1992a, Section 2.7) we still consider the case $\Omega = (0,1)^2$, $\Gamma_0 = \Gamma$, $\mathcal{A} = -\Delta$; we take $T = 3.75/\sqrt{2}$ (> 1, so that the exact controllability property holds) and $\mathbf{f} = \{f^0, f^1\}$ defined by

$$f^0(x_1, x_2) = \sin \pi x_1 \sin \pi x_2, \quad f^1 = -\pi\sqrt{2} f^0. \tag{6.89}$$

It is shown in Glowinski, Li, and J.L. Lions (1990, Section 7) that using *separation of variables methods* we can compute a *Fourier series* expansion of $\mathbf{\Lambda f}$. The corresponding functions z^0 and z^1 (both computed by *Fast Fourier Transform*) have been visualized on Figures 6.2 and 6.3, respectively (the graph on Figure 6.3 is the one of the function $-z^1$).

From the above figures, z^0 is a *Lipschitz continuous function* which is not C^1; similarly, z^1 is *bounded* but *discontinuous*. On Figure 6.4, we have shown the graph of the function $t \rightarrow \|\partial\psi/\partial n(t)\|_{L^2(\Gamma)}$ where ψ, given by

$$\psi(x,t) = \sqrt{2} \cos \pi\sqrt{2}\left(t - \frac{7}{2\sqrt{2}}\right) \sin \pi x_1 \sin \pi x_2,$$

is the solution of the wave equation (6.77) when f^0 and f^1 are given by (6.89); we recall that $\partial\psi/\partial n|_\Sigma$ ($= u$) is precisely the optimal Dirichlet control for which we have exact boundary controllability.

The numerical methods described in Sections 6.8.3 and 6.8.4 have been applied to the solution of the above test problem taking $\Delta t = h/\sqrt{2}$. Interestingly enough, the numerical results deteriorate as h and Δt converge to zero; moreover, taking Δt twice smaller (that is, $\Delta t = h/2\sqrt{2}$) does not improve the situation. Also, the number of conjugate gradient iterations necessary to achieve convergence (*Nit*) increases as h and Δt decrease. Results of the numerical experiments are reported on Table 6.1, where f_*^0, f_*^1, and u_*, are the computed approximations of f^0, f^1, and u, respectively.

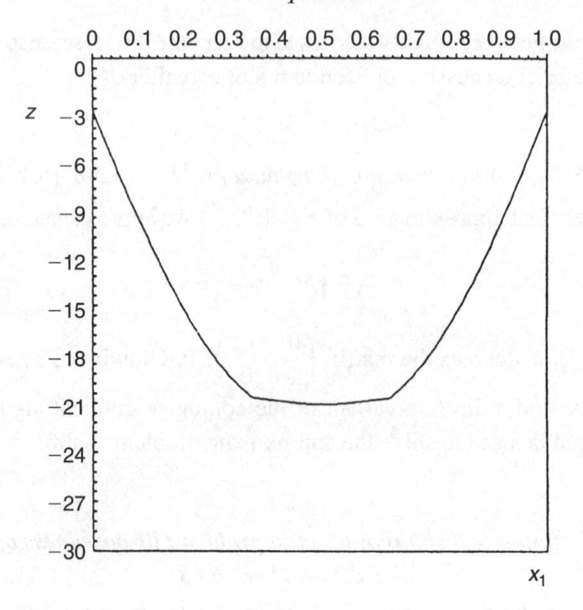

Fig. 6.2. Graph of the function $x_1 \to z^0(x_1, \frac{1}{2})$.

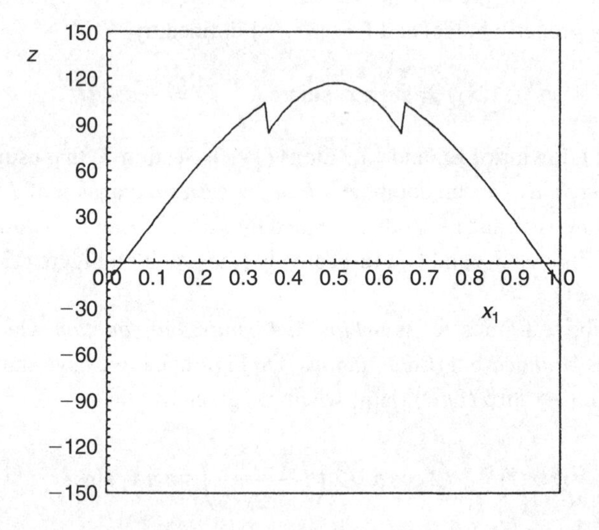

Fig. 6.3. Graph of the function $x_1 \to z^1(x_1, \frac{1}{2})$.

The most striking fact coming from Table 6.1 is the deterioration in the numerical results as h and Δt tend to zero; indeed, for $h = 1/128$, convergence was not achieved after 1000 iterations. To further illustrate this deterioration as h and $\Delta t \to 0$ we have compared, in Figures 6.5–6.8, f^0 and f^1 with their computed approximations f_*^0 and f_*^1, for $h = 1/32$ and $1/64$; we observe that for $h = 1/64$ the variations in f_*^0 and f_*^1 are so large that we have been obliged to use a very large scale to be able to picture them (actually we have plotted $-f^1, -f_*^1$).

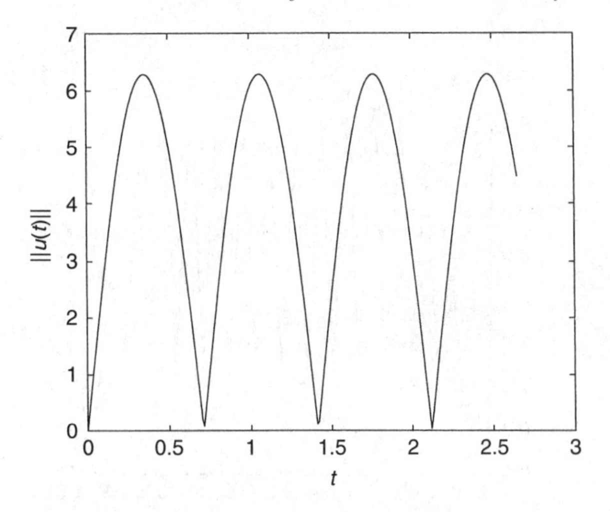

Fig. 6.4. Graph of the function $t \to \|(\partial\psi/\partial n)(t)\|_{L^2(\Gamma)}$.

Table 6.1. *Summary of numerical results (n.c.: no convergence).*

h	$\frac{1}{8}$	$\frac{1}{16}$	$\frac{1}{32}$	$\frac{1}{64}$	$\frac{1}{128}$
Nit	20	38	84	363	n.c.
$\|f^0 - f^0_*\|_{L^2(\Omega)}$	0.42×10^{-1}	0.18×10^{-1}	0.41×10^{-1}	3.89	n.c.
$\|f^0 - f^0_*\|_{H^1_0(\Omega)}$	0.65	0.54	2.54	498.1	n.c.
$\|f^1 - f^1_*\|_{L^2(\Omega)}$	0.20	0.64×10^{-1}	1.18	170.6	n.c.
$\|u - u_*\|_{L^2(\Sigma)}$	0.51	0.24	0.24	1.31	n.c.
$\|u_*\|_{L^2(\Sigma)}$	7.320	7.395	7.456	7.520	n.c.

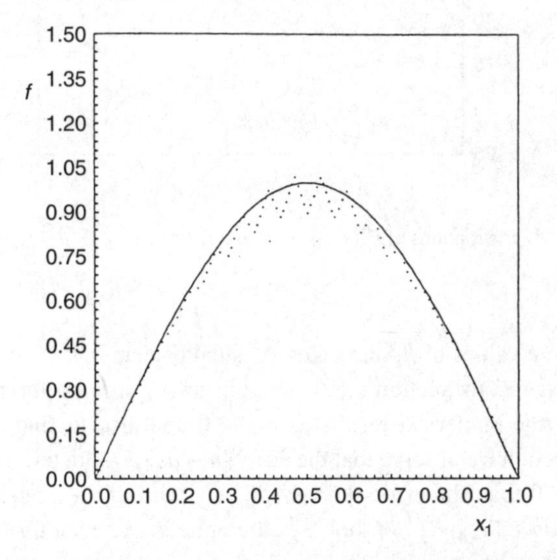

Fig. 6.5. Variations of $f^0(x_1, \frac{1}{2})$ (—) and $f^0_*(x_1, \frac{1}{2})$ ($\cdots$) ($h = 1/32$).

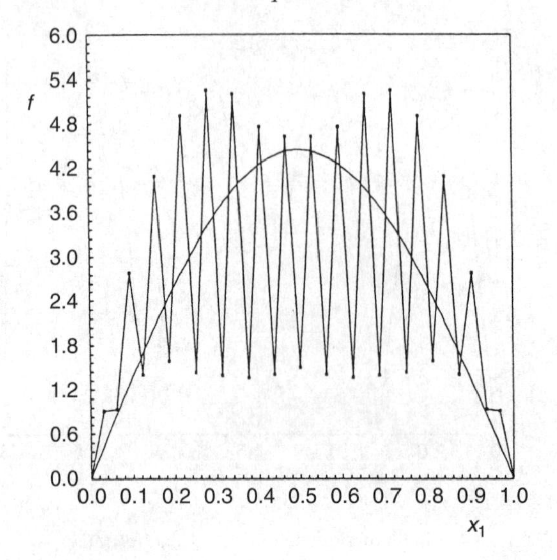

Fig. 6.6. Variations of $f^1(x_1, \frac{1}{2})$ (—) and $f_*^1(x_1, \frac{1}{2})$ $(\cdots)$ $(h = 1/32)$.

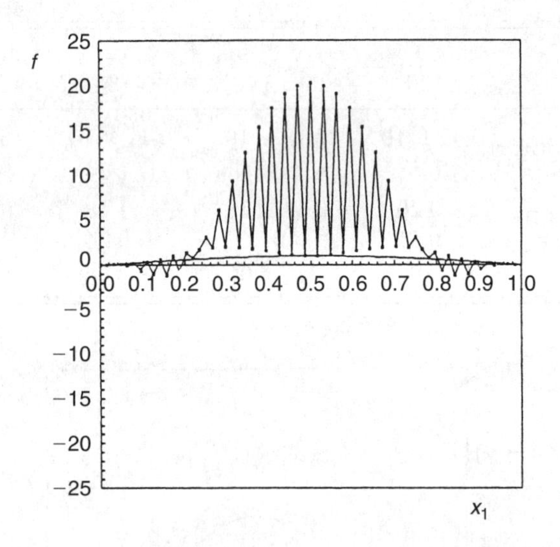

Fig. 6.7. Variations of $f^0(x_1, \frac{1}{2})$ (—) and $f_*^0(x_1, \frac{1}{2})$ $(\cdots)$ $(h = 1/64)$.

If, for the same values of h, one takes Δt smaller than $h/\sqrt{2}$, the results remain practically the same. In Section 6.8.6, we will try to analyze the reasons for this deterioration in the numerical results as $h \to 0$ and also to find cures for it. To conclude this section we observe that the error $\|u - u_*\|_{L^2(\Sigma)}$ deteriorates much more slowly as $h \to 0$ than the errors $\|f^0 - f_*^0\|_{H_0^1(\Omega)}$ and $\|f^1 - f_*^1\|_{L^2(\Omega)}$ in fact, the approximate values $\|u_*\|_{L^2(\Sigma)}$ of $\|u\|_{L^2(\Sigma)}$ are quite good, even for $h = 1/64$ if one realizes that the exact value of $\|u\|_{L^2(\Sigma)}$ is 7.386 68. For further details see Glowinski (1992a, Section 2.7) and the references therein.

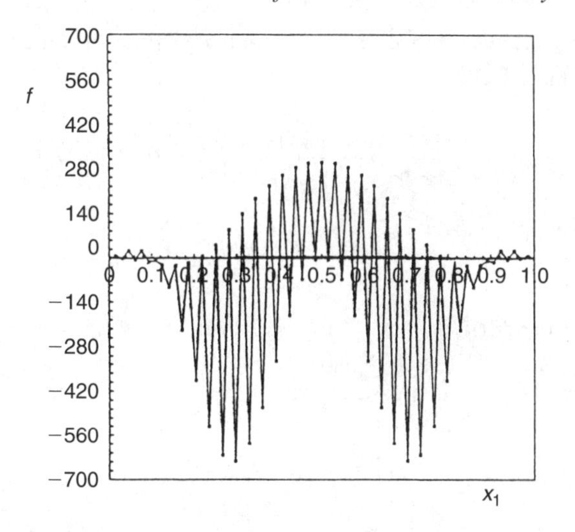

Fig. 6.8. Variations of $f^1(x_1, \frac{1}{2})$ (—) and $f_*^1(x_1, \frac{1}{2})$ $(\cdots)$ $(h = 1/64)$.

6.8.6 Analysis and cures of the ill-posedness of the approximate problem (6.88)

It follows from the numerical results discussed in Section 6.8.5, that when $h \to 0$, the *ill-posedness* of the discrete problem gets worse. From the oscillatory results shown in Figures 6.5–6.8 it is quite clear that the trouble lies with the *high-frequency components* of the discrete solution or, to be more precise, with the way in which the discrete operator $\mathbf{\Lambda}_h^{\Delta t}$ acts on the *short wave length* components of $\mathbf{f}_h$. Before analyzing the mechanism producing these unwanted oscillations let us introduce a vector basis of $\mathbb{R}^{I \times I}$, well suited to the following discussion: This basis $\mathcal{B}_h$ is defined by

$$\mathcal{B}_h = \left\{ w_{pq} \right\}_{1 \le p,q \le I}, \tag{6.90}$$

$$w_{pq} = \left\{ \sin p\pi ih \times \sin q\pi jh \right\}_{1 \le i,j \le I}; \tag{6.91}$$

we recall that $h = 1/(I + 1)$.

From the oscillatory results described in Section 6.8.5 it is reasonable to assume that the discrete operator $\mathbf{\Lambda}_h^{\Delta t}$ damps too strongly those components of $\mathbf{f}_h^{\Delta t}$ with *large wave numbers* p and q; in other words, we can expect that if p and/or q are large then $\mathbf{\Lambda}_h^{\Delta t}\{w_{pq}, 0\}$ or $\mathbf{\Lambda}_h^{\Delta t}\{0, w_{pq}\}$ will be quite small, implying in turn (this is typical of ill-posed problems) that small perturbations of the right-hand side of the discrete problem (6.88) can produce very large variations of the corresponding solution.

Operator $\mathbf{\Lambda}_h^{\Delta t}$ is fairly complicated (see Section 6.8.4 for its precise definition) and we can wonder which stage in it in particular acts as a *low pass filter* (that is, damping selectively the large wave-number components of the discrete solutions). Starting from the observation that the ill-posedness persists if, for a fixed h, we decrease Δt, it is then natural (and much simpler) to consider the *semidiscrete* case, where only the space derivatives have been discretized. In such a case, problem (6.77) is discretized as follows (with $\dot{\psi} = \partial\psi/\partial t$ and $\ddot{\psi} = \partial^2\psi/\partial t^2$) if $\Omega = (0,1)^2$ as in Sections 6.8.4 and 6.8.5:

$$\ddot{\psi}_{ij} - \frac{\psi_{i+1j} + \psi_{i-1j} + \psi_{ij+1} + \psi_{ij-1} - 4\psi_{ij}}{h^2} = 0, \quad 1 \leq i,j \leq I, \tag{6.92a}$$

$$\psi_{kl} = 0 \quad \text{if } \{kh, lh\} \in \Gamma, \tag{6.92b}$$

$$\psi_{ij}(T) = f_h^0(ij,jh), \quad \dot{\psi}_{ij}(T) = f_h^1(ih,jh), \quad 1 \leq i,j \leq I. \tag{6.92c}$$

Consider now the particular case where

$$f_h^0 = w_{pq}, \quad f_h^1 = 0. \tag{6.93}$$

Since the vectors w_{pq} are for $1 \leq p,q \leq I$ the *eigenvectors* of the *discrete Laplace operator* occurring in (6.92a) and that the corresponding eigenvalues $\lambda_{pq}(h)$ are given by

$$\lambda_{pq}(h) = \frac{4}{h^2}\left(\sin^2 p\pi\frac{h}{2} + \sin^2 q\pi\frac{h}{2}\right), \tag{6.94}$$

we can easily prove that the solution of (6.92), (6.93) (a system of *ordinary differential equations*) is given by

$$\psi_{ij}(t) = \sin p\pi ih \sin q\pi jh \cos\left(\sqrt{\lambda_{pq}(h)}(T-t)\right), \ 0 \leq i,j \leq I+1. \tag{6.95}$$

Next, we use (6.83) (see Section 6.8.4.3) to compute, from (6.95), the approximation of $\partial\psi/\partial n$ at the boundary point $M_{0j} = \{0, jh\}$, with $1 \leq j \leq I$; thus at time t, $\partial\varphi/\partial n$ is approximated at M_{0j} by

$$\delta\psi_h(M_{0j}, t) = -\frac{1}{h}\sin p\pi h \sin q\pi jh \cos\left(\sqrt{\lambda_{pq}(h)}(T-t)\right). \tag{6.96}$$

If $1 \leq p \ll I$, the coefficient $K_h(p)$ defined by

$$K_h(p) = \frac{\sin p\pi h}{h} \tag{6.97}$$

is an approximation of $p\pi$ which is second-order accurate (with respect to h); now if $p \approx I/2$ we have $K_h(p) \approx I$ and if $p = I$ we have (since $h = 1/(I+1)$) $K_h(I) \approx \pi$.

Back to the *continuous problem*, it is quite clear that (6.92), (6.93) is in fact a semidiscrete approximation of the wave problem

$$\psi_{tt} - \Delta\psi = 0 \ \text{in } Q, \quad \psi = 0 \ \text{on } \Sigma, \tag{6.98a}$$

$$\psi(x,T) = \sin p\pi x_1 \sin q\pi x_2, \quad \psi_t(x,T) = 0. \tag{6.98b}$$

The solution of (6.98) is given by

$$\psi(x,t) = \sin p\pi x_1 \sin q\pi x_2 \cos\left(\pi\sqrt{p^2+q^2}(T-t)\right). \tag{6.99}$$

Computing $(\partial\psi/\partial n)|_\Sigma$ we obtain

$$\frac{\partial\psi}{\partial n}(M_{0j},t) = -p\pi \sin q\pi jh \cos\left(\pi\sqrt{p^2+q^2}(T-t)\right). \tag{6.100}$$

We observe that if $p \ll I$ and $q \ll I$, then $(\partial\psi/\partial n)(M_{0j},t)$ and $\delta\psi_h(M_{0j},t)$ are close quantities. Now, if the wave-number p is large, then the coefficient $K(p) = \pi p$ in (6.100) is much larger than the corresponding coefficient $K_h(p)$ in (6.97); we have, in fact,

$$\frac{K(I/2)}{K_h(I/2)} \approx \frac{\pi}{2}, \quad \frac{K(I)}{K_h(I)} \approx I.$$

Figure 6.9, (where we have visualized, with an appropriate scaling, the function $p\pi \to p\pi$ and its discrete analogue, namely the function $p\pi \to \sin p\pi h/h$) shows that for $p,\ q > (I+1)/2$, the approximate normal derivative operator introduces a very strong damping. We would have obtained similar results by considering, instead of (6.93), initial conditions such as

$$f_h^0 = 0, \quad f_h^1 = w_{pq}. \tag{6.101}$$

From the above analysis it appears that the approximation of $(\partial\psi/\partial n)|_\Sigma$, which is used to construct operator $\mathbf{\Lambda}_h^{\Delta t}$, introduces a very strong damping of the *large wave-number components* of $\mathbf{f}_h$. Possible cures for the ill-posedness of the discrete problem have been discussed in Glowinski, Li, and J.L. Lions (1990), Dean, Glowinski, and Li (1989), and Glowinski (1992a). The first reference, in particular, contains a detailed

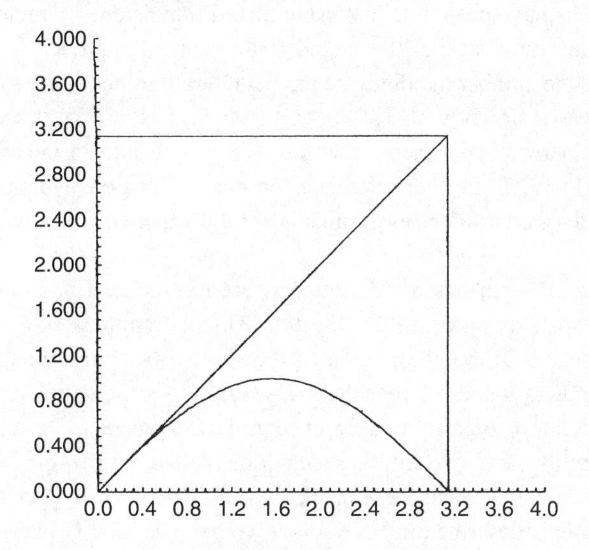

Fig. 6.9. Damping of the discrete normal derivative.

discussion of a *biharmonic Tychonoff regularization procedure*, where problem (6.50) is approximated by a discrete version of

$$\varepsilon \mathbf{M}\mathbf{f}_\varepsilon + \mathbf{\Lambda}\mathbf{f}_\varepsilon = \begin{pmatrix} -z^1 \\ z^0 \end{pmatrix} \quad \text{in } \Omega, \tag{6.102a}$$

$$\Delta f_\varepsilon^0 = f_\varepsilon^0 = f_\varepsilon^1 = 0 \quad \text{on } \Gamma, \tag{6.102b}$$

where, in (6.102), $\varepsilon > 0$, $\mathbf{f}_\varepsilon = \{f_\varepsilon^0, f_\varepsilon^1\}$, and where

$$\mathbf{M} = \begin{pmatrix} \Delta^2 & 0 \\ 0 & -\Delta \end{pmatrix}.$$

Various theoretical and numerical issues associated with (6.102) are discussed in Glowinski, Li, and J.L. Lions (1990), including the choice of ε as a function of h; indeed elementary *boundary layer* considerations show that ε has to be of the order of h^2. The numerical results presented in Glowinski, Li, and J.L. Lions (1990) and Dean, Glowinski, and Li (1989) validate convincingly the above regularization approach. Also, in Glowinski, Li, and J.L. Lions (1990, p. 42), we suggest that *mixed finite element* approximations (see, for example, Roberts and Thomas, 1991; Brezzi and Fortin, 1991 for introductions to mixed finite element methods) may improve the quality of the numerical results; one of the reasons for this potential improvement is that mixed finite element methods are known to provide accurate approximations of derivatives and also that derivative values at selected nodes (including boundary ones) are natural degrees of freedom for these approximations. As shown in Glowinski, Kinton, and Wheeler (1989) and Dupont et al. (1992) this approach substantially reduces the unwanted oscillations, since *without* any regularization good numerical results have been obtained using mixed finite element implementation of HUM. The main drawback of the mixed finite element approach is that (without regularization) the number of conjugate gradient iterations necessary to achieve convergence increases slowly with $1/h$ (in fact, roughly, as $h^{-1/2}$); it seems, also, on the basis of numerical experiments, that the level of unwanted oscillations increases (slowly, again) with T. Further references on mixed finite element approximations of exact boundary controllability problems for the wave equation will be given in Section 6.10.

Another cure for spurious oscillations has been introduced in Glowinski and Li (1990) (see also Glowinski, 1992a, Section 3); this (simple) cure, suggested by Figure 6.9, consists in the elimination of the *short wave length* components of $\mathbf{f}_h$ with wave numbers p and q larger than $(I + 1)/2$; to achieve this radical filtering it suffices to define $\mathbf{f}_h$ on a finite difference grid of step size $\geq 2h$. A *finite element* implementation of the above filtering technique (called the *two-grid method* by E. Zuazua and his collaborators; see Zuazua, 2005) is discussed in Section 6.8.7; also, for the calculations described in Section 6.9 we have defined $\mathbf{f}_h$ over a grid of step size $2h$.

6.8.7 A finite element implementation of the two-grid filtering technique of Section 6.8.6

6.8.7.1 Generalities

We go back to the case where (possibly) $\Omega \neq (0,1)^2$, $\Gamma_0 \neq \Gamma$ and $\mathcal{A} \neq -\Delta$; the most natural way to combine HUM with the two-grid filtering technique discussed in Section 6.8.6 is to use *finite elements* for the space approximation; in fact, as shown in Glowinski, Li, and J.L. Lions (1990, Section 6.2), special triangulations (like the one shown in Figure 2.1 of Section 2.6.1) will give back finite difference approximations closely related to the one discussed in Section 6.8.6. For simplicity, we suppose that Ω is a polygonal domain of $\mathbb{R}^2$; we introduce then a triangulation $\mathcal{T}_h$ of Ω such that $\bar{\Omega} = \bigcup_{T \in \mathcal{T}_h} T$, with h the length of the largest edge (*s*) of $\mathcal{T}_h$. From $\mathcal{T}_h$, we define $\mathcal{T}_{h/2}$ by joining (see Figure 6.10), the midpoints of the edges of the triangles of $\mathcal{T}_h$.

With P_1 the space of the polynomials in two variables of degree ≤ 1, we define the spaces H_h^1 and H_{0h}^1 by

$$\begin{cases} H_h^1 = \{ v \mid v \in C^0(\bar{\Omega}),\, v|_T \in P_1,\, \forall\, T \in \mathcal{T}_h \}, \\ H_{0h}^1 = \{ v \mid v \in H_h^1,\, v|_\Gamma = 0 \}; \end{cases} \tag{6.103}$$

similarly, we define $H_{h/2}^1$ and $H_{0h/2}^1$ by replacing h by $h/2$ in (6.103). We observe that $H_h^1 \subset H_{h/2}^1$, $H_{0h}^1 \subset H_{0h/2}^1$. We then approximate the $L^2(\Omega)$-scalar product over H_h^1 by

$$(v, w)_h = \frac{1}{3} \sum_Q \omega_Q\, v(Q)\, w(Q), \quad \forall\, v, w \in H_h^1, \tag{6.104}$$

where, in (6.104), Q describes the set of the vertices of $\mathcal{T}_h$ and where ω_Q is the area of the polygonal domain, union of those triangles of $\mathcal{T}_h$ which have Q as a common vertex. Similarly, we define $(\cdot,\cdot)_{h/2}$ by substituting $h/2$ to h in (6.104).

Finally, assuming that the points at the interface of Γ_0 and $\Gamma \backslash \Gamma_0$ are vertices of $\mathcal{T}_{h/2}$, we define $V_{0h/2}$ by

$$V_{0h/2} = \{ v \mid v \in H_{h/2}^1,\ v = 0 \ \text{ on } \Gamma_0 \backslash \Gamma \}. \tag{6.105}$$

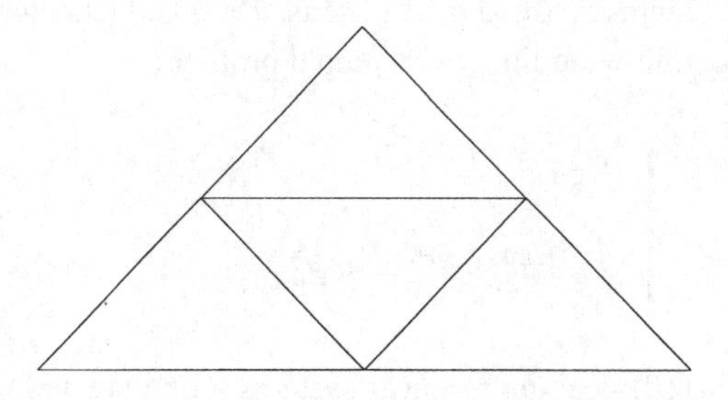

Fig. 6.10. Triangles of $\mathcal{T}_h$ and $\mathcal{T}_{h/2}$.

6.8.7.2 Approximation of problem (6.50)

We approximate the fundamental equation $\mathbf{\Lambda} f = \{-z^1, z^0\}$ by the following linear variational problem in $H^1_{0h} \times H^1_{0h}$:

$$\begin{cases} \mathbf{f}_h^{\Delta t} \in H^1_{0h} \times H^1_{0h}, \quad \forall \mathbf{v} = \{v^0, v^1\} \in H^1_{0h} \times H^1_{0h}, \\ \lambda_h^{\Delta t}(\mathbf{f}_h, \mathbf{v}) = -\langle z^1, v^0 \rangle + \int_\Omega z^0 v^1 \, dx. \end{cases} \tag{6.106}$$

In (6.106), $\langle \cdot, \cdot \rangle$ denotes the duality pairing between $H^{-1}(\Omega)$ and $H^1_0(\Omega)$, and the bilinear functional $\lambda_h^{\Delta t}(\cdot, \cdot)$ is defined as follows:

(i) Take $\mathbf{g}_h = \{g_h^0, g_h^1\} \in H^1_{0h} \times H^1_{0h}$ and solve, for $n = N, \ldots, 0$, the following *discrete variational problem*

$$\begin{cases} \psi_{\mathbf{g}h}^{n-1} \in H^1_{0h/2}, \quad \forall v \in H^1_{0h/2}, \\ \left(\psi_{\mathbf{g}h}^{n-1} + \psi_{\mathbf{g}h}^{n+1} - 2\psi_{\mathbf{g}h}^n, v\right)_{h/2} + |\Delta t|^2 a(\psi_{\mathbf{g}h}^n, v) = 0, \end{cases} \tag{6.107}$$

with the final conditions

$$\psi_{\mathbf{g}h}^N = g_h^0, \quad \psi_{\mathbf{g}h}^{N+1} - \psi_{\mathbf{g}h}^{N-1} = 2\Delta t g_h^1; \tag{6.108}$$

we recall that $a(\cdot, \cdot)$ denotes the bilinear functional defined by

$$a(v, w) = \langle Av, w \rangle, \quad \forall v \in H^1(\Omega), \ w \in H^1_0(\Omega).$$

(ii) To approximate $\partial \psi_{\mathbf{g}} / \partial n_A$ over Σ_0, we introduce first the *complementary* subspace $M_{h/2}$ of $H^1_{0h/2}$ defined by

$$\begin{cases} M_{h/2} \oplus H^1_{0h/2} = V_{0h/2}, \\ v \in M_{h/2} \iff v|_T = 0, \ \forall K \in \mathcal{T}_{h/2} \text{ such that } K \cap \Gamma = \emptyset; \end{cases} \tag{6.109}$$

we observe that $M_{h/2}$ is isomorphic to the space $\gamma V_{0h/2}$ of the traces over Γ_0 of the functions of $V_{0h/2}$. The approximation of $(\partial \psi_{\mathbf{g}} / \partial n_A)|_{\Gamma_0}$ at $t = n\Delta t$ is then defined (cf. Glowinski, Li, and J.L. Lions, 1990 and Chapter 2, Section 2.4.3) by solving the following linear variational problem:

$$\begin{cases} \delta\psi_{\mathbf{g}h}^n \in \gamma V_{0h/2}, \\ \int_{\Gamma_0} \delta\psi_{\mathbf{g}h}^n v \, d\Gamma = a(\psi_{\mathbf{g}h}^n, v), \quad \forall v \in M_{h/2}. \end{cases} \tag{6.110}$$

Variants of (6.110), leading to linear systems with diagonal matrices are given in Glowinski, Li, and J.L. Lions (1990).

(iii) Now, for $n = 0, \ldots, N$, solve the *discrete variational problem*

$$\begin{cases} \varphi_{\mathbf{g}h}^{n+1} \in V_{0h/2},\ \varphi_{\mathbf{g}h}^{n+1} = \delta\psi_{\mathbf{g}h}^{n+1} \text{ on } \Gamma_0; \quad \forall v \in H_{0h/2}^1, \\ \left(\varphi_{\mathbf{g}h}^{n+1} + \varphi_{\mathbf{g}h}^{n-1} - 2\varphi_{\mathbf{g}h}^{n}, v\right)_{h/2} + |\Delta t|^2 a(\varphi_{\mathbf{g}h}^{n}, v) = 0 \end{cases} \tag{6.111}$$

initialized via

$$\varphi_{\mathbf{g}h}^{0} = 0, \quad \varphi_{\mathbf{g}h}^{1} - \varphi_{\mathbf{g}h}^{-1} = 0. \tag{6.112}$$

(iv) Finally, define $\lambda_h^{\Delta t}(\cdot,\cdot)$ by

$$\lambda_h^{\Delta t}(\mathbf{g}_h, \mathbf{v}) = (\lambda_{\mathbf{g}h}^0, v^0)_{h/2} + (\lambda_{\mathbf{g}h}^1, v^1)_{h/2}, \ \forall \mathbf{v} = \{v^0, v^1\} \in H_{0h}^1 \times H_{0h}^1, \tag{6.113}$$

where, in (6.113), $\lambda_{\mathbf{g}h}^0$ and $\lambda_{\mathbf{g}h}^1$ belong both to $H_{0h/2}^1$ and satisfy

$$\begin{cases} \lambda_{\mathbf{g}h}^0(P) = -\dfrac{\varphi_{\mathbf{g}h}^{N+1}(P) - \varphi_{\mathbf{g}h}^{N-1}(P)}{2\Delta t}, \\ \lambda_{\mathbf{g}h}^1(P) = \varphi_{\mathbf{g}h}^{N}(P), \\ \forall P \text{ interior vertex of } \mathcal{T}_{h/2}. \end{cases} \tag{6.114}$$

Following Glowinski, Li, and J.L. Lions (1990, Section 6) we can prove (with obvious notation) that

$$\lambda_h^{\Delta t}(\mathbf{g}_{1h}, \mathbf{g}_{2h}) = \Delta t \sum_{n=0}^{N} \alpha_n \int_{\Gamma_0} \delta\psi_{1h}^n \delta\psi_{2h}^n \, d\Gamma, \ \forall \mathbf{g}_{1h}, \mathbf{g}_{2h} \in H_{0h}^1 \times H_{0h}^1, \tag{6.115}$$

where, in (6.115), $\alpha_0 = \alpha_N = 1/2$, and $\alpha_n = 1$ if $0 < n < N$.

It follows from (6.115) that the bilinear functional $\lambda_h^{\Delta t}(\cdot,\cdot)$ is *symmetric* and *positive semidefinite*. As in Glowinski, Li, and J.L. Lions (1990, Section 6.2), we can prove that $\lambda_h^{\Delta t}(\cdot,\cdot)$ is *positive definite* if $\mathcal{A} = -\Delta$, T is sufficiently large, Ω is a square (or a rectangle), and $\mathcal{T}_h$ and $\mathcal{T}_{h/2}$ *regular* triangulations of Ω. From the above properties of $\lambda_h^{\Delta t}(\cdot,\cdot)$ the linear variational problem (6.106) (which approximates problem (6.50)) can be solved by a *conjugate gradient algorithm* operating in $H_{0h}^1 \times H_{0h}^1$. This algorithm is described in Section 6.8.7.3.

6.8.7.3 Conjugate gradient solution of the approximate problem (6.106)

The following *conjugate gradient algorithm* for solving problem (6.106) is an finite element implementation of algorithm (6.58)–(6.73) (see Section 6.8.3):

Description of the Conjugate Gradient Algorithm

Step 0: Initialization.

$$\mathbf{f}_0 = \{f_0^0, f_0^1\} \in H_{0h}^1 \times H_{0h}^1 \text{ is given;} \tag{6.116}$$

solve then, for $n = N, N-1, \ldots, 0$, *the following discrete linear variational problem*

$$\begin{cases} \psi_0^{n-1} \in H^1_{0h/2}, \\ \left(\psi_0^{n-1} + \psi_0^{n+1} - 2\psi_0^n, v\right)_{h/2} + |\Delta t|^2\, a(\psi_0^n, v) = 0, \quad \forall\, v \in H^1_{0h/2}, \end{cases} \tag{6.117}$$

initialized by

$$\psi_0^N = f_0^0 \text{ and } \psi_0^{N+1} - \psi_0^{N-1} = 2\Delta t f_0^1, \tag{6.118}$$

and store ψ_0^0 and ψ_0^{-1}.

Then for $n = 0, 1, \ldots, N$, *compute* ψ_0^n, $\delta\psi_0^n$ *and* φ_0^{n+1} *by forward (discrete) time integration, as follows*:

(i) *If* $n = 0$, *compute* $\delta\psi_0^0$ *from* ψ_0^0 *using* (6.110).
If $n > 0$, *compute first* ψ_0^n *by solving*

$$\begin{cases} \psi_0^n \in H^1_{0h/2}, \quad \forall\, v \in H^1_{0h/2}, \\ \left(\psi_0^n + \psi_0^{n-2} - 2\psi_0^{n-1}, v\right)_{h/2} + |\Delta t|^2\, a(\psi_0^{n-1}, v) = 0 \end{cases} \tag{6.119}$$

and then $\delta\psi_0^n$ *by using* (6.110).

(ii) *Take* $\varphi_0^n = \delta\psi_0^n$ *on* Γ_0 *and use*

$$\left(\varphi_0^{n+1} + \varphi_0^{n-1} - 2\varphi_0^n, v\right)_{h/2} + |\Delta t|^2 a(\varphi_0^n, v) = 0, \ \forall v \in H^1_{0h/2}, \tag{6.120}$$

to compute the values taken by φ_0^{n+1} $(\in V_{0h/2})$ *at the interior vertices of* $\mathcal{T}_{h/2}$. *These calculations are initialized by*

$$\varphi_0^0(P) = 0, \varphi_0^1(P) - \varphi_0^{-1}(P) = 0, \ \forall P \textit{ interior vertex of } \mathcal{T}_{h/2}. \tag{6.121}$$

Compute then $\mathbf{g}_0 = \{g_0^0, g_0^1\} \in H^1_{0h} \times H^1_{0h}$ *by solving the following discrete Dirichlet problem*

$$\begin{cases} g_0^0 \in H^1_{0h}, \quad \forall v \in H^1_{0h}, \\ \displaystyle\int_\Omega \nabla g_0^0 \cdot \nabla v \, dx = \langle z^1, v\rangle - \left(\frac{\varphi_0^{N+1} - \varphi_0^{N-1}}{2\Delta t}, v\right)_{h/2}, \end{cases} \tag{6.122}$$

and then

$$\begin{cases} g_0^1 \in H^1_{0h}, \\ (g_0^1, v)_h = (\varphi_0^N, v)_{h/2} - \displaystyle\int_\Omega z^0 v \, dx, \ \forall v \in H^1_{0h}. \end{cases} \tag{6.123}$$

If $\mathbf{g}_0 = \mathbf{0}$*, or is "small," take* $\mathbf{f}_h^{\Delta t} = \mathbf{f}_0$*; else, set*

$$\mathbf{w}_0 = \mathbf{g}_0. \tag{6.124}$$

Then for $k \geq 0$*, assuming that* $\mathbf{f}_k$*,* $\mathbf{g}_k$*, and* $\mathbf{w}_k$ *are known (the last two different from* $\mathbf{0}$*), compute* $\mathbf{f}_{k+1}$, $\mathbf{g}_{k+1}$, and, if necessary, $\mathbf{w}_{k+1}$ *as follows:*

Step 1: Descent.
For $n = N, N-1, \ldots, 0$*, solve the following discrete backward wave equation:*

$$\begin{cases} \bar{\psi}_k^{n-1} \in H^1_{0h/2}, \\ \left(\bar{\psi}_k^{n-1} + \bar{\psi}_k^{n+1} - 2\bar{\psi}_k^n, v\right)_{h/2} + |\Delta t|^2 a(\bar{\psi}_k^n, v) = 0, \ \forall v \in H^1_{0h/2}, \end{cases} \tag{6.125}$$

initialized by

$$\bar{\psi}_k^N = w_k^0, \quad \bar{\psi}_k^{N+1} - \bar{\psi}_k^{N-1} = 2\Delta t w_k^1, \tag{6.126}$$

and store $\bar{\psi}_k^0, \bar{\psi}_k^{-1}$.
Then for $n = 0, 1, \ldots, N$*, compute* $\bar{\psi}_k^n$*,* $\delta\bar{\psi}_k^n$ and $\bar{\varphi}_k^{n+1}$ *by forward discrete time integration as follows:*

(i) *If* $n = 0$*, compute* $\delta\bar{\psi}_k^0$ *from* $\bar{\psi}_k^0$ *using* (6.110).
If $n > 0$*, compute first* $\bar{\psi}_k^n$ *by solving*

$$\begin{cases} \bar{\psi}_k^n \in H^1_{0h/2}, \quad \forall v \in H^1_{0h/2}, \\ \left(\bar{\psi}_k^n + \bar{\psi}_k^{n-2} - 2\bar{\psi}_k^{n-1}, v\right)_{h/2} + |\Delta t|^2 a(\bar{\psi}_k^{n-1}, v) = 0, \end{cases} \tag{6.127}$$

and then $\delta\bar{\psi}_k^n$ *by using* (6.110).
(ii) *Take* $\bar{\varphi}_k^n = \delta\bar{\psi}_k^n$ *on* Γ_0 *and use*

$$\left(\bar{\varphi}_k^{n+1} + \bar{\varphi}_k^{n-1} - 2\bar{\varphi}_k^n, v\right)_{h/2} + |\Delta t|^2 a(\bar{\varphi}_k^n, v) = 0, \ \forall v \in H^1_{0h/2}, \tag{6.128}$$

to compute the values taken by $\bar{\varphi}_k^{n+1} (\in V_{0h/2})$ *at the interior vertices of* $\mathcal{T}_{h/2}$*. These calculations are initialized by*

$$\bar{\varphi}_k^1(P) - \bar{\varphi}_k^{-1}(P) = \bar{\varphi}_k^0(P) = 0, \ \forall P \text{ interior vertex of } \mathcal{T}_{h/2}. \tag{6.129}$$

Compute now $\bar{\mathbf{g}}_k (= \{\bar{g}_k^0, \bar{g}_k^1\}) \in H^1_{0h} \times H^1_{0h}$ *by solving*

$$\begin{cases} \bar{g}_k^0 \in H^1_{0h}, \\ \displaystyle\int_\Omega \nabla \bar{g}_k^0 \cdot \nabla v \, dx = -\left(\frac{\bar{\varphi}_k^{N+1} - \bar{\varphi}_k^{N-1}}{2\Delta t}, v\right)_{h/2}, \ \forall v \in H^1_{0h}, \end{cases} \tag{6.130}$$

and

$$\begin{cases} \bar{g}_k^1 \in H_{0h}^1, \\ (\bar{g}_k^1, v)_h = (\bar{\varphi}_k^N, v)_{h/2}, \quad \forall v \in V_{0h}, \end{cases} \tag{6.131}$$

and then ρ_k *by*

$$\rho_k = \frac{\int_\Omega |\nabla g_k^0|^2 \, dx + (g_k^1, g_k^1)_h}{\int_\Omega \nabla \bar{g}_k^0 \cdot \nabla w_k^0 \, dx + (\bar{g}_k^1, w_k^1)_h}. \tag{6.132}$$

Once ρ_k *is known, compute*

$$\mathbf{f}_{k+1} = \mathbf{f}_k - \rho_k \mathbf{w}_k, \tag{6.133}$$

$$\mathbf{g}_{k+1} = \mathbf{g}_k - \rho_k \bar{\mathbf{g}}_k. \tag{6.134}$$

Step 2: Test of the convergence and construction of the new descent direction.
If $\left(\int_\Omega |\nabla g_{k+1}^0|^2 \, dx + (g_{k+1}^1, g_{k+1}^1)_h\right) / \left(\int_\Omega |g_0^0|^2 \, dx + (g_0^1, g_0^1)_h\right) \le \varepsilon^2$, *take* $\mathbf{f}_h^{\Delta t} = \mathbf{f}_{k+1}$; *else, compute*

$$\gamma_k = \frac{\int_\Omega |\nabla g_{k+1}^0|^2 \, dx + (g_{k+1}^1, g_{k+1}^1)_h}{\int_\Omega |g_k^0|^2 \, dx + (g_k^1, g_k^1)_h}, \tag{6.135}$$

and set

$$\mathbf{w}_{k+1} = \mathbf{g}_{k+1} + \gamma_k \mathbf{w}_k. \tag{6.136}$$

Do $k = k + 1$ *and go to* (6.125).

Remark 6.19 The above algorithm may seem a little bit complicated at first glance (21 statements); in fact, it is fairly easy to implement, since the only non-trivial part of it is the solution (on the coarse grid) of the discrete Dirichlet problems (6.122) and (6.130). An interesting feature of algorithm (6.116)–(6.136) is that the *forward integration* of the discrete wave equations (6.117) and (6.125) provides a very substantial computer memory saving. To illustrate this claim, let us consider the case where $\Omega = (0, 1) \times (0, 1)$, $\Gamma_0 = \Gamma$, $T = 2\sqrt{2}$, $h = 1/64$, $\Delta t = h/2\sqrt{2} = \sqrt{2}/256$; we have then – approximately – $(512)^2$ discretization points on Σ, therefore, in that specific case, using algorithm (6.116)–(6.136) avoids the storage of 6.52×10^5 real numbers. The saving would be even more substantial for larger T and would be an absolute necessity for three-dimensional problems. In fact, the above storage-saving strategy which is based on the *time reversibility* of the *wave equation* (6.1) cannot be applied to the control problems discussed in Chapters 1 and 2, these problems concerning systems modeled by *diffusion* equations which are *time irreversible*; however, the situation is far from being desperate, since the bi-section based memory saving method discussed in Chapter 1, Section 1.12, applies to those diffusion modeled controllability problems.

Remark 6.20 The above remark shows the interest of solving the *dual problem* from a computational point of view. In the original control problem, the unknown is the control u which is defined over Σ_0; for the dual problem the unknown is then the solution $\mathbf{f}$ of problem (6.50). If one considers again the particular case of Remark 6.19, that is, $\Omega = (0,1) \times (0,1)$, $\Gamma_0 = \Gamma$, $T = 2\sqrt{2}$, $h = 1/64$, $\Delta t = h/2\sqrt{2}$ the unknown u will be approximated by a finite-dimensional vector $u_h^{\Delta t}$ with 2.62×10^5 components, while $\mathbf{f}$ is approximated by $\mathbf{f}_h^{\Delta t}$ of dimension $2 \times (63)^2 = 7.938 \times 10^3$, a substantial saving indeed. Also, the dimension of $\mathbf{f}_h^{\Delta t}$ remains the same as T increases, while the dimension of $u_h^{\Delta t}$ is proportional to T.

Numerical results obtained using algorithm (6.116)–(6.136) will be discussed in Section 6.9.

6.8.8 *Solution of the approximate boundary controllability problem (6.24)*

Following the approach advocated for the exact boundary controllability problem, we shall address the *numerical solution* of the *approximate boundary controllability* problem (6.24) via the solution of its *dual problem*, namely problem (6.37), (6.38). This last problem can also be formulated as

$$\mathbf{\Lambda f} + \partial j(\mathbf{f}) \ni \begin{pmatrix} -z^1 \\ z^0 \end{pmatrix}, \tag{6.137}$$

where, in (6.137), the *convex functional* $j : H_0^1(\Omega) \times L^2(\Omega) \to \mathbb{R}$ is defined by

$$j(\mathbf{g}) = \beta_1 \|g^0\|_{H_0^1(\Omega)} + \beta_0 \|g^1\|_{L^2(\Omega)}, \ \forall \mathbf{g} = \{g^0, g^1\} \in H_0^1(\Omega) \times L^2(\Omega). \tag{6.138}$$

Following a strategy already used in preceding sections (see, for example, Chapter 1, Section 1.8.8) we associate with the "elliptic" problem (6.137) the following *initial value problem* (*flow* in the *dynamical system* terminology):

$$\begin{cases} \begin{pmatrix} -\Delta & 0 \\ 0 & I \end{pmatrix} \dfrac{\partial \mathbf{f}}{\partial \tau} + \mathbf{\Lambda f} + \partial j(\mathbf{f}) \ni \begin{pmatrix} -z^1 \\ z^0 \end{pmatrix}, \\ \mathbf{f}(0) = \mathbf{f}_0, \end{cases} \tag{6.139}$$

where τ is a pseudotime. The particular form of problem (6.139) strongly suggests time integration by *operator-splitting* (see, again, Chapter 1, Section 1.8.8). Concentrating on the *Peaceman–Rachford scheme*, we obtain – with $\Delta\tau (> 0)$ a pseudotime step – the following algorithm to compute the steady-state solution of problem (6.139), that is, the solution of problems (6.37), (6.38), and (6.137):

$$\mathbf{f}^0 = \mathbf{f}_0; \tag{6.140}$$

then, for $k \geq 0$, *assuming that* $\mathbf{f}^k$ *is known, we compute* $\mathbf{f}^{k+1/2}$ *and* $\mathbf{f}^{k+1}$ *via the solution of*

$$\begin{pmatrix} -\Delta & 0 \\ 0 & I \end{pmatrix} \frac{\mathbf{f}^{k+1/2} - \mathbf{f}^k}{\Delta\tau/2} + \mathbf{\Lambda f}^k + \partial j(\mathbf{f}^{k+1/2}) \ni \begin{pmatrix} -z^1 \\ z^0 \end{pmatrix}, \tag{6.141}$$

and

$$\begin{pmatrix} -\Delta & 0 \\ 0 & I \end{pmatrix} \frac{\mathbf{f}^{k+1} - \mathbf{f}^{k+1/2}}{\Delta\tau/2} + \mathbf{A}\mathbf{f}^{k+1} + \partial j(\mathbf{f}^{k+1/2}) \ni \begin{pmatrix} -z^1 \\ z^0 \end{pmatrix}. \tag{6.142}$$

Let us discuss the solution of the subproblems (6.141) and (6.142):

(i) Assuming that (6.141) has been solved, equation (6.142) can be formulated as

$$\begin{pmatrix} -\Delta & 0 \\ 0 & I \end{pmatrix} \frac{\mathbf{f}^{k+1} - 2\mathbf{f}^{k+1/2} + \mathbf{f}^{k}}{\Delta\tau/2} + \mathbf{A}\mathbf{f}^{k+1} = \mathbf{A}\mathbf{f}^{k},$$

that is,

$$\begin{pmatrix} -\Delta & 0 \\ 0 & I \end{pmatrix} \mathbf{f}^{k+1} + \frac{\Delta\tau}{2}\mathbf{A}\mathbf{f}^{k+1} = \begin{pmatrix} -\Delta & 0 \\ 0 & I \end{pmatrix} (2\mathbf{f}^{k+1/2} - \mathbf{f}^{k}) + \frac{\Delta\tau}{2}\mathbf{A}\mathbf{f}^{k}. \tag{6.143}$$

Problem (6.143) is a variant of problem (6.50) (a *regularized* one, in fact); it can be solved therefore by a *conjugate gradient algorithm* closely related to algorithm (6.58)–(6.73). We have to replace the bilinear functional

$$\{\mathbf{g}_1, \mathbf{g}_2\} \to \langle \mathbf{A}\mathbf{g}_1, \mathbf{g}_2 \rangle : (H_0^1(\Omega) \times L^2(\Omega))^2 \to \mathbb{R}$$

by

$$\{\mathbf{g}_1, \mathbf{g}_2\} \to \int_\Omega \nabla g_1^0 \cdot \nabla g_2^0 \, dx + \int_\Omega g_1^1 g_2^1 \, dx + \frac{1}{2}\Delta\tau \langle \mathbf{A}\mathbf{g}_1, \mathbf{g}_2 \rangle.$$

(ii) Concerning the solution of problem (6.141), we shall take advantage of the fact that the operator $\partial j(\cdot)$ is *diagonal* from $H_0^1(\Omega) \times L^2(\Omega)$ into $H^{-1}(\Omega) \times L^2(\Omega)$; solving problem (6.141) is then equivalent to solving the two following *uncoupled* minimization problems (where the notation is fairly obvious):

$$\min_{g^0 \in H_0^1(\Omega)} \left\{ \frac{1}{2}\int_\Omega |\nabla g^0|^2 \, dx + \beta_1 \frac{\Delta\tau}{2} \left(\int_\Omega |\nabla g^0|^2 \, dx \right)^{1/2} \right.$$
$$\left. + \frac{\Delta\tau}{2} \langle z^1 + (\mathbf{A}\mathbf{f}^k)^0, g^0 \rangle - \int_\Omega \nabla f^{0,k} \cdot \nabla g^0 \, dx \right\}, \tag{6.144}$$

and

$$\min_{g^1 \in L^2(\Omega)} \left\{ \frac{1}{2}\int_\Omega |g^1|^2 \, dx + \beta_0 \frac{\Delta\tau}{2} \|g^1\|_{L^2(\Omega)} \right.$$
$$\left. - \frac{\Delta\tau}{2} \int_\Omega (z^0 - (\mathbf{A}\mathbf{f}^k)^1) g^1 \, dx - \int_\Omega f^{1,k} g^1 \, dx \right\}. \tag{6.145}$$

Both problems (6.144), (6.145) have *closed form* solutions which can be obtained as in Chapter 1, Section 1.8.8 for the solution of problem (1.117). The solution of problems (6.144) and (6.145) clearly provides the two component of $\mathbf{f}^{k+1/2}$.

6.9 Experimental validation of the filtering procedure of Section 6.8.7 via the solution of the test problem of Section 6.8.5

We consider in this section the solution of the test problem of Section 6.8.5. The filtering technique discussed in Section 6.8.7 is applied with $\mathcal{T}_h$ a regular triangulation like the one on Figure 2.1 of Chapter 2, Section 2.6; we recall that $\mathcal{T}_h$ is used to approximate $\mathbf{f}_h^{\Delta t}$, while ψ and φ are approximated on $\mathcal{T}_{h/2}$ as shown in Section 6.8.7. Instead of taking h as the length of the largest edges of $\mathcal{T}_h$, it is convenient here to take h as the length of the edges adjacent to the right angles of $\mathcal{T}_h$. The approximate problems (6.106) have been solved by the conjugate gradient algorithm (6.116)–(6.136) of Section 6.8.7.3. This algorithm has been *initialized* with $f_0^0 = f_0^1 = 0$ and we have used

$$\frac{\int_\Omega |\nabla g_k^0|^2 \, dx + (g_k^1, g_k^1)_h}{\int_\Omega |\nabla g_0^0|^2 \, dx + (g_0^1, g_0^1)_h} \leq 10^{-14} \tag{6.146}$$

as the *stopping criterion* (for calculations on a CRAY X-MP).

Let us also mention that the functions z^0, z^1 and u of the test problem in Section 6.8.5 satisfy

$$\|z^0\|_{L^2(\Omega)} = 12.92\ldots, \quad \|z^1\|_{H^{-1}(\Omega)} = 11.77\ldots, \quad \|u\|_{L^2(\Sigma)} = 7.386\,68\ldots.$$

In the following, we shall denote by $\|\cdot\|_{0,\Omega}$, $|\cdot|_{1,\Omega}$, $\|\cdot\|_{-1,\Omega}$ and $\|\cdot\|_{0,\Sigma}$ the $L^2(\Omega)$, $H_0^1(\Omega)$, $H^{-1}(\Omega)$ and $L^2(\Sigma)$ norms, respectively (here $|v|_{1,\Omega} = (\int_\Omega |\nabla v|^2 \, dx)^{1/2}$ and $\|v\|_{-1,\Omega)} = |w|_{1,\Omega}$ where $w \in H_0^1(\Omega)$ is the solution of the Dirichlet problem $-\Delta w = v$ in Ω, $w = 0$ on Γ).

To approximate problem (6.50) by the discrete problem (6.106) we have been using $h = 1/4, 1/8, 1/16, 1/32, 1/64$ and $\Delta t = h/2\sqrt{2}$ (since the *wave equations* are solved on a space/time grid of step size $h/2$ for the space discretization and $h/2\sqrt{2}$ for the time discretization); we recall that $T = 15/4\sqrt{2}$.

Results of our numerical experiments have been summarized in Table 6.2. In this table f_*^0, f_*^1, and u_* are defined as in Section 6.8.5, and the new quantities z_*^0 and z_*^1 are the discrete analogues of $y(T)$ and $y_t(T)$, respectively, where y is the solution of (6.33b), associated via (6.33a), with the solution $\mathbf{f}$ of problem (6.50).

Remark 6.21 In Table 6.2 we have taken $h/2$ as discretization parameter to make easier comparisons with the results of Table 6.1 and Glowinski, Li, and J.L. Lions (1990, Section 10).

Comparing the above results to those in Table 6.1, the following facts appear quite clearly:

(i) The *filtering method* described in Section 6.8.7 has been a *very effective* cure to the *ill-posedness* of the approximate problem (6.88).
(ii) The number of *conjugate gradient* iterations necessary to achieve the convergence is (for h sufficiently small) essentially *independent* of h; if one realizes

Table 6.2. *Summary of numerical results (Nit is the number of conjugate gradient iterations).*

	$h/2$				
	$\frac{1}{8}$	$\frac{1}{16}$	$\frac{1}{32}$	$\frac{1}{64}$	$\frac{1}{128}$
Nit	7	10	12	12	12
CPU time(s) CRAY X-MP	0.1	0.6	2.8	14.8	83.9
$\frac{\Vert f^0-f^0_*\Vert_{0,\Omega}}{\Vert f^0\Vert_{0,\Omega}}$	9.6×10^{-2}	2.6×10^{-2}	2.2×10^{-2}	6.4×10^{-3}	1.5×10^{-3}
$\frac{\vert f^0-f^0_*\vert_{1,\Omega}}{\vert f^0\vert_{1,\Omega}}$	3.5×10^{-1}	1.8×10^{-1}	9×10^{-2}	4.4×10^{-2}	2.2×10^{-2}
$\frac{\Vert f^1-f^1_*\Vert_{0,\Omega}}{\Vert f^1\Vert_{0,\Omega}}$	1×10^{-1}	2.6×10^{-2}	1.5×10^{-2}	7×10^{-3}	3.2×10^{-3}
$\frac{\Vert z^0-z^0_*\Vert_{0,\Omega}}{\Vert z^0\Vert_{0,\Omega}}$	2.4×10^{-8}	3×10^{-8}	6×10^{-8}	8.3×10^{-8}	6.6×10^{-8}
$\frac{\Vert z^1-z^1_*\Vert_{-1,\Omega}}{\Vert z^1\Vert_{-1,\Omega}}$	6.9×10^{-7}	4.6×10^{-7}	9.4×10^{-6}	2×10^{-5}	8.5×10^{-5}
$\frac{\Vert u-u_*\Vert_{0,\Sigma}}{\Vert u\Vert_{0,\Sigma}}$	1.2×10^{-1}	4.3×10^{-2}	2×10^{-2}	7.6×10^{-3}	3.4×10^{-3}
$\Vert u_*\Vert_{0,\Sigma}$	7.271	7.386	7.453	7.405	7.381

that for $h = 1/64$ the number of unknowns is $2 \times (63)^2 = 7938$, converging in 12 iterations is a fairly good performance.

(iii) The target functions z^0 and z^1 have been reached within a fairly high accuracy.

The results of Table 6.2 compare favorably with those displayed in Tables 10.3 and 10.4 of Glowinski, Li, and J.L. Lions (1990, pp. 58 and 59) which were obtained using the *Tychonoff regularization* procedure briefly recalled in Section 6.8.6; in fact, fewer iterations are needed here, implying a smaller CPU time (actually, the CPU time seems to be – for the stopping criteria given by (6.146) – a *sublinear* function of h^{-3} which is – modulo a multiplicative constant – the number of points of the space/time-discretization grid). Table 6.2 also shows that the approximation errors satisfy (roughly)

$$\|f^0 - f^0_*\|_{L^2(\Omega)} = 0(h^2),\ \|f^0 - f^0_*\|_{H^1_0(\Omega)} = 0(h),\ \|f^1 - f^1_*\|_{L^2(\Omega)} = 0(h), \tag{6.147}$$

$$\|u - u_*\|_{L^2(\Sigma)} = 0(h). \tag{6.148}$$

Estimates (6.147) are of *optimal order* with respect to h in the sense that they have the order that we can expect when one approximates the solution of a boundary value problem, for a second-order linear elliptic operator, by piecewise linear finite element approximations; this result is not surprising after all since (from Section 6.8.2, relation (6.54)) the operator $\mathbf{\Lambda}$ associated with $\Omega = (0,1) \times (0,1)$ behaves for T *sufficiently large* like

$$2T\begin{pmatrix} -\Delta & 0 \\ 0 & I \end{pmatrix} \tag{6.149}$$

(we have here $x_0 = \{1/2, 1/2\}$ and $C = \frac{1}{2}$).

In order to visualize the influence of h we have plotted for $h = 1/4, 1/8, 1/16, 1/32, 1/64$ and $\Delta t = h/2\sqrt{2}$ the exact solutions f^0, f^1 and the corresponding computed solutions f_*^0, f_*^1. To be more precise, we have shown the graphs of the functions $x_1 \to f^0(x_1, 1/2)$, $x_1 \to -f^1(x_1, 1/2)$ (—) and of the corresponding computed functions $(\cdots)$. These results have been reported in Figures 6.11–6.15; the captions there are self-explanatory.

The above numerical experiments have been done with $T = 15/4\sqrt{2}$; in order to study the influence of T we have kept z^0 and z^1 as in the above experiments and taken $T = 20\sqrt{2}\,(= 28.2843)$. For $h = 1/64$ and $\Delta t = h/2\sqrt{2}$ we need just ten iterations of algorithm (6.116)–(6.136) to achieve convergence, the corresponding CRAY X-MP CPU time being then 800 s (the number of grid points for the space–time discretization

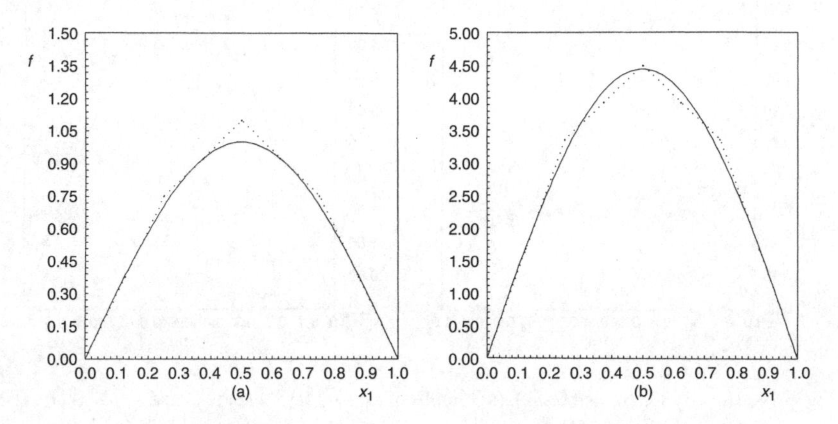

Fig. 6.11. $(h = 1/4, \Delta t = h/2\sqrt{2})$. (a) Graphs of $x_1 \to f^0(x_1, 1/2)$ (—) and $x_1 \to f_*^0(x_1, 1/2)$ $(\cdots)$. (b) Graphs of $x_1 \to -f^1(x_1, 1/2)$ (—) and $x_1 \to -f_*^1(x_1, 1/2)$ $(\cdots)$.

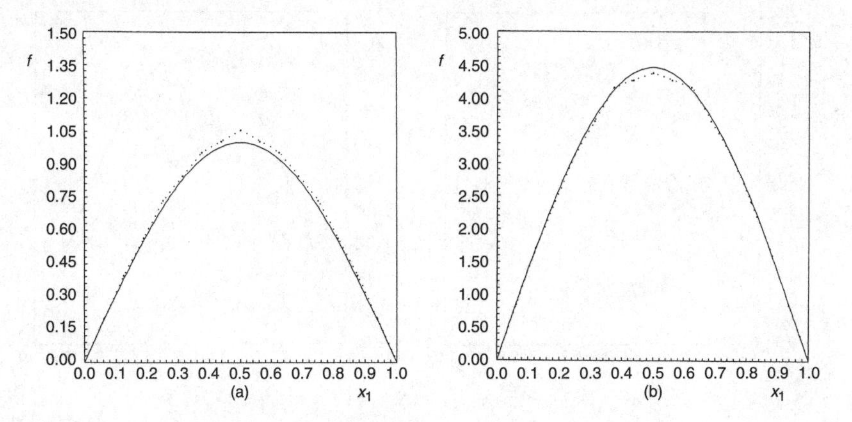

Fig. 6.12. $(h = 1/8, \Delta t = h/2\sqrt{2})$. (a) Graphs of $x_1 \to f^0(x_1, 1/2)$ (—) and $x_1 \to f_*^0(x_1, 1/2)$ $(\cdots)$. (b) Graphs of $x_1 \to -f^1(x_1, 1/2)$ (—) and $x_1 \to -f_*^1(x_1, 1/2)$ $(\cdots)$.

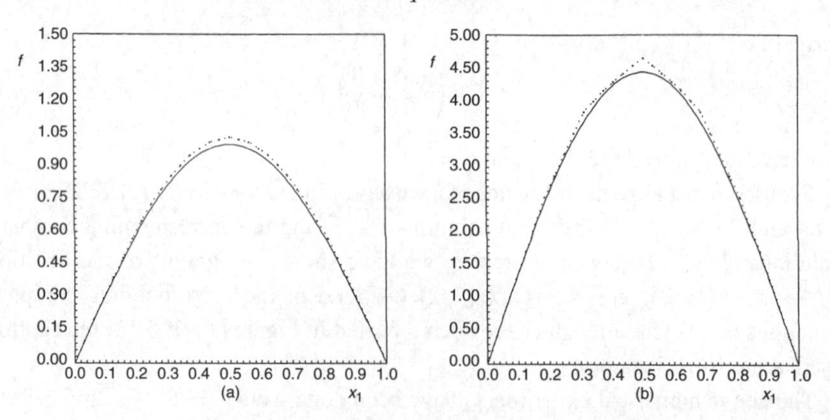

Fig. 6.13. $(h = 1/16, \Delta t = h/2\sqrt{2})$. (a) Graphs of $x_1 \to f^0(x_1, 1/2)$ (—) and $x_1 \to f_*^0(x_1, 1/2)$ (···). (b) Graphs of $x_1 \to -f^1(x_1, 1/2)$ (—) and $x_1 \to -f_*^1(x_1, 1/2)$ (···).

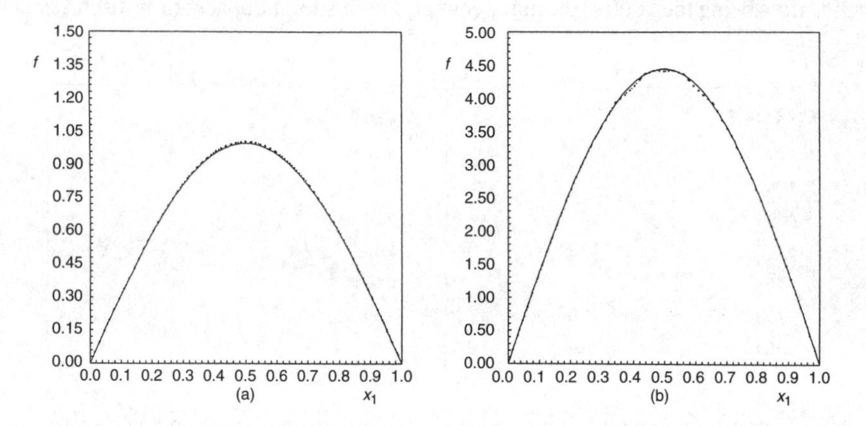

Fig. 6.14. $(h = 1/32, \Delta t = h/2\sqrt{2})$. (a) Graphs of $x_1 \to f^0(x_1, 1/2)$ (—) and $x_1 \to f_*^0(x_1, 1/2)$ (···). (b) Graphs of $x_1 \to -f^1(x_1, 1/2)$ (—) and $x_1 \to -f_*^1(x_1, 1/2)$ (···).

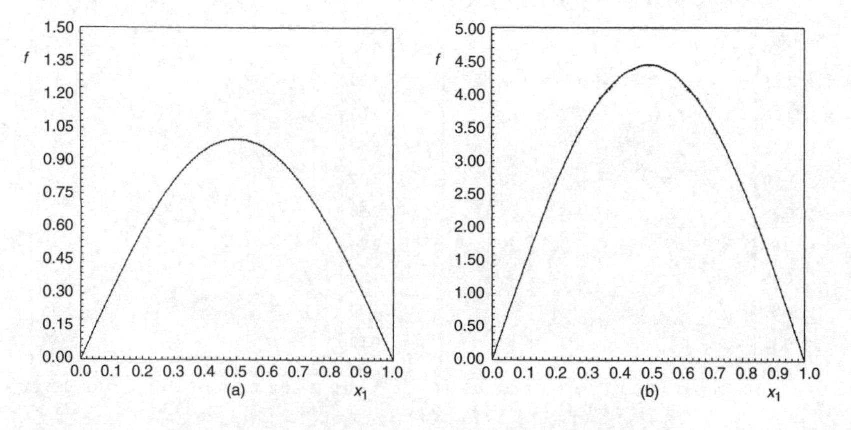

Fig. 6.15. $(h = 1/64, \Delta t = h/2\sqrt{2})$. (a) Graphs of $x_1 \to f^0(x_1, 1/2)$ (—) and $x_1 \to f_*^0(x_1, 1/2)$ (···). (b) Graphs of $x_1 \to -f^1(x_1, 1/2)$ (—) and $x_1 \to -f_*^1(x_1, 1/2)$ (···).

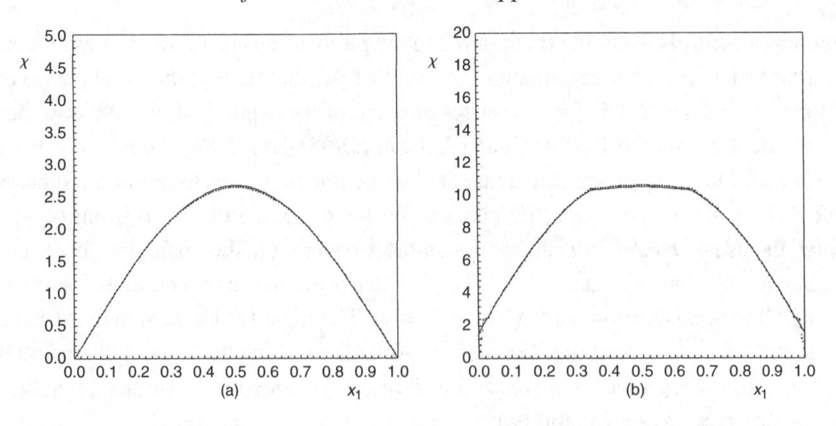

Fig. 6.16. ($h = 1/64$, $\Delta t = h/2\sqrt{2}$, $T = 20\sqrt{2}$). (a) Graphs of $x_1 \to \chi^0(x_1, 1/2)$ (—) and $x_1 \to Tf_*^0(x_1, 1/2)$ (···). (b) Graphs of $x_1 \to -\chi^1(x_1, 1/2)$ (—) and $x_1 \to -Tf_*^1(x_1, 1/2)$ (···).

is now $\approx 86 \times 10^6$). We have $\|u_*\|_{L^2(\Sigma)} = 2.32$, $\|z^0 - z_*^0\|_{L^2(\Omega)} = 5.8 \times 10^{-6}$, $\|z^1 - z_*^1\|_{-1,\Omega} = 1.6 \times 10^{-5}$. The most interesting results are the ones reported on Figure 6.16(a) and (b). There, we have compared Tf_*^0 and Tf_*^1 (for $T = 20\sqrt{2}$) with the corresponding theoretical limits χ^0 and χ^1 which, according to Section 6.8.2, relations (6.55)–(6.57), are given by

$$\Delta\chi^0 = z^1/2 \text{ in } \Omega, \quad \chi^0 = 0 \text{ on } \Gamma, \tag{6.150}$$

$$\chi^1 = z^0/2. \tag{6.151}$$

The *continuous* curves represent the variations of $x_1 \to \chi^0(x_1, 1/2)$ and of $x_1 \to -\chi^1(x_1, 1/2)$, while the *dotted* curves represent the variations of $x_1 \to Tf_*^0(x_1, 1/2)$ and $x_1 \to -Tf_*^1(x_1, 1/2)$.

In our opinion the above figures provide an excellent *numerical validation* of the convergence result (6.55) of Section 6.8.2 (we observe at $x_1 = 0$ and $x_1 = 1$ a (numerical) *Gibbs phenomenon* associated with the L^2 convergence of Tf_*^1 to χ^1). Conversely, these results provide a *validation* of the numerical methodology discussed here; they show that this methodology is particularly robust, accurate, non-dissipative and perfectly able to handle very long time intervals $[0, T]$. In fact, numerical experiments have shown that the above-mentioned qualities of the numerical methods discussed here persist for target functions z^0 and z^1 much rougher than those considered in this section.

Additional results can be found in Glowinski, Li, and J.L. Lions (1990, Section 4) and Glowinski, He, and J.L. Lions (2002); the second reference provides a numerical investigation of the controllability (and non-controllability) properties of wave models with *variable coefficients*.

6.10 Some references on alternative approximation methods

Not surprisingly, the numerical solution of boundary controllability problems for the wave equation has motivated many investigators, particularly the case of Dirichlet boundary controls. Keeping track of the many publications resulting of these investigations has become a time consuming task. We refer therefore to the excellent review article by Zuazua (2005) for references to some of these publications. Among them, let us mention the work of Asch and Lebeau (1998) on a *finite difference in space* variant of the finite element in space based methodology discussed in Sections 6.8 and 6.9. In the recent years, E. Zuazua and his collaborators have strongly advocated the use of *mixed finite element methods* to overcome the difficulties mentioned in Section 6.8 (see Zuazua, 2005, Section 7.3), a pioneering work in that direction being Glowinski, Kinton, and Wheeler (1989). The main idea here is to use different Galerkin spaces to approximate y and $\frac{\partial y}{\partial t}$. Actually, in Chapter 7, we will discuss the use of *mixed finite element methods* à la *Raviart–Thomas* for the numerical solution of another type of controllability problems for the wave equation.

Numerical methods for the boundary controllability of the wave equation are also discussed in Bardos, Bourquin, and Lebeau (1991).

6.11 Other boundary controls

6.11.1 Approximate Neumann boundary controllability

We consider now problems entirely similar to the previous ones but where we "exchange" the *Dirichlet conditions* for *Neumann conditions*. We define therefore the state function $y = y(v)$ by

$$y_{tt} + \mathcal{A}y = 0 \quad \text{in } Q(= \Omega \times (0, T)), \tag{6.152a}$$

$$y(0) = 0, \quad y_t(0) = 0, \tag{6.152b}$$

$$\frac{\partial y}{\partial n_{\mathcal{A}}} = v \text{ on } \Sigma_0(= \Gamma_0 \times (0, T)), \quad \frac{\partial y}{\partial n_{\mathcal{A}}} = 0 \text{ on } \Sigma \backslash \Sigma_0. \tag{6.152c}$$

To fix ideas, we still assume that

$$v \in L^2(\Sigma_0). \tag{6.153}$$

Using again *transposition*, we can show that problem (6.152) has a *unique* solution if (6.153) holds. In fact, the solution here is (slightly) smoother than the one in Section 6.1. We can, in any case, define an operator L from $L^2(\Sigma_0)$ into $H^{-1}(\Omega) \times L^2(\Omega)$ by

$$Lv = \{ -y_t(T; v), y(T; v) \}. \tag{6.154}$$

If the control function v is smooth, the solution $y(v)$ will also be smooth, assuming of course that the coefficients of the elliptic operator $\mathcal{A}$ are also smooth.

Let us study *approximate controllability* first.

We suppose thus that v is *smooth*; indeed, to fix ideas we assume that

$$v, \frac{\partial v}{\partial t} \in L^2(\Sigma_0), \quad v|_t = 0. \tag{6.155}$$

Then, $y(v)$ can be defined by a *variational formulation*, showing that

$$y \in C^0([0,T], H^1(\Omega)), \quad \frac{\partial y}{\partial t} \in C^0([0,T], L^2(\Omega)), \tag{6.156}$$

implying, in particular, that

$$Lv \in L^2(\Omega) \times L^2(\Omega). \tag{6.157}$$

Let us consider now $\mathbf{f}$ belonging to the orthogonal of the range of L, that is

$$\begin{cases} \mathbf{f} = \{f^0, f^1\} \in L^2(\Omega) \times L^2(\Omega), \\ -\int_\Omega f^0 y_t(T)\,\mathrm{d}x + \int_\Omega f^1 y(T)\,\mathrm{d}x = 0, \quad \forall\, v \text{ satisfying (6.155).} \end{cases} \tag{6.158}$$

Then, we introduce ψ defined by

$$\psi_{tt} + \mathcal{A}\psi = 0 \text{ in } Q, \quad \psi(T) = f^0, \ \psi_t(T) = f^1, \quad \frac{\partial \psi}{\partial n_{\mathcal{A}}} = 0 \text{ on } \Sigma. \tag{6.159}$$

Multiplying the wave equation in (6.159) by $y(v)$ and integrating by parts, we obtain

$$\langle Lv, \mathbf{f} \rangle = -\int_{\Sigma_0} \psi v \,\mathrm{d}\Sigma. \tag{6.160}$$

Therefore, (6.158) is equivalent to

$$\psi = 0 \text{ on } \Sigma_0. \tag{6.161}$$

If we assume (as in Section 6.3, relation (6.22)) that

$$\Sigma_0 \textit{ allows the application of the Holmgren's uniqueness theorem} \tag{6.162}$$

then (6.159), (6.161) imply that $\psi = 0$ in Q, so that $f = 0$; we have proved thus that

$$\text{assuming (6.162) the range of } L \text{ is dense in } L^2(\Omega) \times L^2(\Omega), \tag{6.163}$$

which implies in turn *approximate controllability*.

Remark 6.22 Suppose that Γ is a C^∞ manifold. If we can take

$$v \in \mathcal{D}(\Sigma_0) \tag{6.164}$$

($\mathcal{D}(\Sigma_0)$ being the space of the C^∞ functions with compact support in Σ_0), it can be proved that the range of L, for v describing $\mathcal{D}(\Sigma_0)$, is still dense in $L^2(\Omega) \times L^2(\Omega)$. □

We can now state the following control problem:

$$\inf_v \frac{1}{2} \int_{\Sigma_0} v^2 \,\mathrm{d}\Sigma; \quad y(T) \in z^0 + \beta_0 B, \ y_t(T) \in z^1 + \beta_1 B, \tag{6.165}$$

where, in (6.165), $\{y, v\}$ satisfies (6.152), (6.153), $\{z^0, z^1\} \in L^2(\Omega) \times L^2(\Omega)$, and where B denotes the closed unit ball of $L^2(\Omega)$ centered at 0.

Remark 6.23 We do *not* introduce $H^{-1}(\Omega)$ here for two reasons:

(i) In the present context the $H^{-1}(\Omega)$ space (not being the dual of $H^1(\Omega)$) is not natural.
(ii) The choice of the same norm, in (6.165), for both $y(T)$ and $y_t(T)$ shows the flexibility of the methodology.

Remark 6.24 Problem (6.165) has a unique solution. The *uniqueness* follows from the *strict convexity*. As far as *existence* is concerned let $\{u_n\}_{n\geq 0}$ be a *minimizing sequence*. Then $\{u_n\}_{n\geq 0}$ is *bounded* in $L^2(\Sigma_0)$. Let us set $y_n = y(u_n)$. By definition, $\{\{y_n(T), \partial y_n/\partial t(T)\}\}_{n\geq 0}$ remains in a bounded set of $L^2(\Omega)\times L^2(\Omega)$. We can therefore extract from $\{u_n\}_{n\geq 0}$ a subsequence, still denoted by $\{u_n\}_{n\geq 0}$, such that

$$u_n \to u \text{ weakly in } L^2(\Sigma_0), \tag{6.166}$$

$$\left\{y_n(T), \frac{\partial y_n}{\partial t}(T)\right\} \to \{\xi_0, \xi_1\} \text{ weakly in } L^2(\Omega) \times L^2(\Omega). \tag{6.167}$$

However,

$$\left\{y(T; u_n), \frac{\partial y}{\partial t}(T; u_n)\right\} \to \left\{y(T; u), \frac{\partial y}{\partial t}(T; u)\right\}$$

weakly in $L^2(\Omega) \times H^{-1}(\Omega)$, so that $\xi_0 = y(T; u)$, and $\xi_1 = y_t(T; u)$, which proves the existence of a solution, namely u, to problem (6.165). We can easily show that

$$\lim_{n\to+\infty} \int_{\Sigma_0} u_n^2 \, d\Sigma = \int_{\Sigma_0} u^2 \, d\Sigma$$

which combined with the uniqueness of the solution implies that the *whole* minimizing sequence converges strongly to u in $L^2(\Sigma_0)$.

Remark 6.25 The use of β_0 and β_1 allows avoiding the introduction of new (and complicated) function spaces. Unfortunately, these spaces cannot be avoided if we let β_0 and $\beta_1 \to 0$, as we shall see below.

6.11.2 Duality results: exact Neumann boundary controllability

Now, we will use *duality*, as already done in previous sections. We define then the functionals F_1 and F_2 by

$$F_1(v) = \frac{1}{2} \int_{\Sigma_0} v^2 \, d\Sigma, \tag{6.168}$$

and (with $\mathbf{g} = \{g^0, g^1\}$)

$$F_2(\mathbf{g}) = \begin{cases} 0 & \text{if } g^0 \in -z^1 + \beta_1 B \text{ and } g^1 \in z^0 + \beta_0 B, \\ +\infty & \text{otherwise on } L^2(\Omega) \times L^2(\Omega). \end{cases} \tag{6.169}$$

It follows then from (6.154) that the minimization problem (6.165) is equivalent to

$$\inf_{v\in L^2(\Sigma_0)} \left\{F_1(v) + F_2(Lv)\right\}. \tag{6.170}$$

Using *convex duality* arguments, we obtain

$$\inf_{v\in L^2(\Sigma_0)} \left\{F_1(v) + F_2(Lv)\right\} = -\inf_{\mathbf{g}\in L^2(\Omega)\times L^2(\Omega)} \left\{F_1^*(L^*\mathbf{g}) + F_2^*(-\mathbf{g})\right\}, \tag{6.171}$$

where we use L^* with L thought of as an unbounded operator. By virtue of (6.171), we have

$$L^*\mathbf{g} = -\psi_{\mathbf{g}}|_{\Sigma_0}, \tag{6.172}$$

where the function $\psi_{\mathbf{g}}$ is the solution of (6.159) when $\mathbf{f}$ is replaced by $\mathbf{g}$. We obtain then as *dual problem* of the control problem (6.165)

$$\inf_{\mathbf{g}\in L^2(\Omega)\times L^2(\Omega)} \left\{\frac{1}{2}\int_{\Sigma_0} \psi_{\mathbf{g}}^2 \,\mathrm{d}\Sigma + \int_\Omega (z^1 g_0 - z^0 g^1)\,\mathrm{d}x \right.$$
$$\left. + \beta_1\|g^0\|_{L^2(\Omega)} + \beta_0\|g^1\|_{L^2(\Omega)}\right\}. \tag{6.173}$$

Remark 6.26 We are going to give now an alternative formulation of the dual problem (6.173). This new formulation is particularly useful when $\{\beta_0, \beta_1\} \to \mathbf{0}$. Using the HUM approach, we introduce the operator $\mathbf{\Lambda}$ defined as follows:

The pair $\{g^0, g^1\}$ being given in, say, $L^2(\Omega) \times L^2(\Omega)$, we define $\psi_{\mathbf{g}}$ and $y_{\mathbf{g}}$ by

$$\frac{\partial^2\psi_{\mathbf{g}}}{\partial t^2} + \mathcal{A}\psi_{\mathbf{g}} = 0 \text{ in } Q,\ \psi_{\mathbf{g}}(T) = g^0,\ \frac{\partial\psi_{\mathbf{g}}}{\partial t}(T) = g^1,\ \frac{\partial\psi_{\mathbf{g}}}{\partial n_{\mathcal{A}}} = 0 \text{ on } \Sigma, \tag{6.174}$$

$$\begin{cases} \dfrac{\partial^2 y_{\mathbf{g}}}{\partial t^2} + \mathcal{A}y_{\mathbf{g}} = 0 \ \text{ in } Q, \quad y_{\mathbf{g}}(0) = \dfrac{\partial y_{\mathbf{g}}}{\partial t}(0) = 0, \\[2ex] \dfrac{\partial y_{\mathbf{g}}}{\partial n_{\mathcal{A}}} = -\psi_{\mathbf{g}} \ \text{ on } \Sigma_0, \quad \dfrac{\partial y_{\mathbf{g}}}{\partial n_{\mathcal{A}}} = 0 \ \text{ on } \Sigma\backslash\Sigma_0. \end{cases} \tag{6.175}$$

Let us denote by $\mathbf{g}$ the vector-valued function $\{g^0, g^1\}$; we set then:

$$\mathbf{\Lambda}\mathbf{g} = \{-\frac{\partial y_{\mathbf{g}}}{\partial t}(T), y_{\mathbf{g}}(T)\}. \tag{6.176}$$

Taking $\mathbf{g} = \mathbf{g}_1$ and $\mathbf{g}_2$, and denoting by ψ_1, ψ_2 the corresponding solutions of (6.174), we obtain from (6.174) and (6.175) that

$$\int_\Omega (\mathbf{\Lambda}\mathbf{g}_1)\cdot\mathbf{g}_2 \,\mathrm{d}x = \int_{\Sigma_0} \psi_1\psi_2 \,\mathrm{d}\Sigma. \tag{6.177}$$

It follows from (6.177) that

$$\text{operator } \mathbf{\Lambda} \text{ is symmetric and positive definite over } L^2(\Omega)\times L^2(\Omega). \tag{6.178}$$

It follows from (6.177) that problem (6.173) is *equivalent* to

$$\inf_{\mathbf{g}\in L^2(\Omega)\times L^2(\Omega)}\left\{\frac{1}{2}\int_\Omega(\mathbf{\Lambda g})\cdot\mathbf{g}\,\mathrm{d}x+\int_\Omega(z^1g^0-z^0g^1)\,\mathrm{d}x\right.$$
$$\left.+\beta_1\|g^0\|_{L^2(\Omega)}+\beta_0\|g^1\|_{L^2(\Omega)}\right\}. \tag{6.179}$$

In order to discuss the case $\beta_0=\beta_1=0$ in (6.179), we introduce over $L^2(\Omega)\times L^2(\Omega)$ the norm $[\cdot]$ defined by

$$[\mathbf{g}]=\left(\int_\Omega(\mathbf{\Lambda g})\cdot\mathbf{g}\,\mathrm{d}x\right)^{1/2},\quad\forall\,\mathbf{g}\in L^2(\Omega)\times L^2(\Omega). \tag{6.180}$$

We define next the space E by

$$E=\ \text{completion of}\ L^2(\Omega)\times L^2(\Omega)\ \text{ for the norm } [\cdot]. \tag{6.181}$$

Taking now the limit in (6.179) as $\{\beta_0,\beta_1\}\to\mathbf{0}$ we obtain – *formally* – the *dual problem* associated with *exact controllability*, namely

$$\inf_{\mathbf{g}\in E}\left\{\frac{1}{2}[\mathbf{g}]^2-\langle\boldsymbol{\sigma}\mathbf{z},\mathbf{g}\rangle\right\} \tag{6.182}$$

where, in (6.182), $\langle\cdot,\cdot\rangle$ denotes the duality pairing between E' and E, $\mathbf{z}=\{z^0,z^1\}$ and $\boldsymbol{\sigma}=\begin{pmatrix}0&-1\\1&0\end{pmatrix}$.

The problem (6.182) *has a solution (necessarily unique) if and only if*

$$\boldsymbol{\sigma}\mathbf{z}\in E'; \tag{6.183}$$

equivalently, exact controllability holds true if and only if condition (6.183) *is satisfied.*

Remark 6.27 Unlike the situation in Section 6.6, the space E, as defined by (6.181), has no simple interpretation. For further information concerning space E, we refer to J.L. Lions (1988b) and the references therein.

Remark 6.28 It is by now clear that the method employed here is general. It can, in particular, be applied to *other boundary conditions*.

6.11.3 A second approximate Neumann boundary controllability problem

Inspired by Chapters 1 and 2 we consider, for $k>0$, the following control problem:

$$\min_{v\in L^2(\Sigma_0)}\left\{\frac{1}{2}\int_{\Sigma_0}v^2\,\mathrm{d}\Sigma+\frac{k}{2}\left(\|y(T)-z^0\|^2_{L^2(\Omega)}+\|y_t(T)-z^1\|^2_{L^2(\Omega)}\right)\right\}, \tag{6.184}$$

where, in (6.184), y is – still – defined from v by the solution of the wave equation (6.152); problem (6.184) is clearly obtained by *penalization* of the conditions $y(T) = z^0, y_t(T) = z^1$.

Using the results of Section 6.10.1 it is quite easy to show that *problem* (6.184) *has a unique solution* (even if the condition (6.162) does not hold). If we denote by u the solution of problem (6.184), it is characterized by the existence of an *adjoint state function* p such that

$$\begin{cases} y_{tt} + \mathcal{A}y = 0 \text{ in } Q, \quad y(0) = y_t(0) = 0, \\ \dfrac{\partial y}{\partial n_{\mathcal{A}}} = u \text{ on } \Sigma_0, \quad \dfrac{\partial y}{\partial n_{\mathcal{A}}} = 0 \text{ on } \Sigma\backslash\Sigma_0, \end{cases} \tag{6.185}$$

$$\begin{cases} p_{tt} + \mathcal{A}p = 0 \text{ in } Q, \quad \dfrac{\partial p}{\partial n_{\mathcal{A}}} = 0 \text{ on } \Sigma, \\ p(T) = k(y_t(T) - z^1), \quad p_t(T) = -k(y(T) - z^0), \end{cases} \tag{6.186}$$

$$u = -p \text{ on } \Sigma_0. \tag{6.187}$$

Let us define $\mathbf{f} = \{f^0, f^1\} \in L^2(\Omega) \times L^2(\Omega)$ by

$$f^0 = p(T), \quad f^1 = p_t(T); \tag{6.188}$$

it follows then from (6.186), and from the definition of $\mathbf{\Lambda}$ (see Section 6.10.2) that

$$k^{-1}\mathbf{f} + \mathbf{\Lambda f} = \{-z^1, z^0\}. \tag{6.189}$$

Problem (6.189) is the *dual* problem of the control problem (6.184).

From the properties of operator $\mathbf{\Lambda}$ (*symmetry* and *positivity*) and from the $(L^2(\Omega))^2$-ellipticity of the bilinear functional associated with operator $k^{-1}\mathbf{I} + \mathbf{\Lambda}$, the dual problem (6.189) can be solved by a *conjugate gradient algorithm* operating in (the *Hilbert space*) $L^2(\Omega) \times L^2(\Omega)$; such an algorithm will be described in Section 6.11.4.

6.11.4 Conjugate gradient solution of the dual problem (6.189)

We can solve the dual problem (6.189) by the following variant of algorithm (6.58)–(6.73) (see Section 6.8.3):

$$\mathbf{f}_0 = \{f_0^0, f_0^1\} \text{ is given in } L^2(\Omega) \times L^2(\Omega); \tag{6.190}$$

solve then

$$\begin{cases} \dfrac{\partial^2 \psi_0}{\partial t^2} + \mathcal{A}\psi_0 = 0 \ \textit{in } Q, \quad \psi_0(T) = f_0^0, \\ \dfrac{\partial \psi_0}{\partial t}(T) = f_0^1, \quad \dfrac{\partial \psi_0}{\partial n_{\mathcal{A}}} = 0 \ \textit{on } \Sigma, \end{cases} \tag{6.191}$$

and

$$\begin{cases} \dfrac{\partial^2 \varphi_0}{\partial t^2} + \mathcal{A}\varphi_0 = 0 \ \ in\ Q, \quad \varphi_0(0) = \dfrac{\partial \varphi_0}{\partial t}(0) = 0, \\ \dfrac{\partial \varphi_0}{\partial n_{\mathcal{A}}} = -\psi_0 \ \ on\ \Sigma_0, \quad \dfrac{\partial \varphi_0}{\partial n_{\mathcal{A}}} = 0 \ \ on\ \Sigma\backslash\Sigma_0. \end{cases} \tag{6.192}$$

Define $\mathbf{g}_0 = \{g_0^0, g_0^1\} \in L^2(\Omega) \times L^2(\Omega)$ *by*

$$\begin{cases} \displaystyle\int_\Omega g_0^0 v\,dx = \frac{1}{k}\int_\Omega f_0^0 v\,dx + \int_\Omega \left(z^1 - \frac{\partial \varphi_0}{\partial t}(T)\right) v\,dx, \ \forall\, v \in L^2(\Omega), \\ \displaystyle\int_\Omega g_0^1 v\,dx = \frac{1}{k}\int_\Omega f_0^1 v\,dx + \int_\Omega (\varphi_0(T) - z^0) v\,dx, \quad \forall\, v \in L^2(\Omega), \end{cases} \tag{6.193}$$

and define $\mathbf{w}_0 = \{w_0^0, w_1^0\}$ *by*

$$\mathbf{w}_0 = \mathbf{g}_0. \tag{6.194}$$

Assuming that $\mathbf{f}_n$, $\mathbf{g}_n$, and $\mathbf{w}_n$ *are known (the last two different from* $\mathbf{0}$*), we obtain* $\mathbf{f}_{n+1}$, $\mathbf{g}_{n+1}$, and $\mathbf{w}_{n+1}$ *as follows:*
Solve

$$\begin{cases} \dfrac{\partial^2 \bar{\psi}_n}{\partial t^2} + \mathcal{A}\bar{\psi}_n = 0 \ \ in\ Q, \quad \bar{\psi}_n(T) = w_n^0, \\ \dfrac{\partial \bar{\psi}_n}{\partial t}(T) = w_n^1, \quad \dfrac{\partial \bar{\psi}_n}{\partial n_{\mathcal{A}}} = 0 \ \ on\ \Sigma, \end{cases} \tag{6.195}$$

and

$$\begin{cases} \dfrac{\partial^2 \bar{\varphi}_n}{\partial t^2} + \mathcal{A}\bar{\varphi}_n = 0 \ \ in\ Q, \quad \bar{\varphi}_n(0) = \dfrac{\partial \bar{\varphi}_n}{\partial t}(0) = 0, \\ \dfrac{\partial \bar{\varphi}_n}{\partial n_{\mathcal{A}}} = -\bar{\psi}_n \ \ on\ \Sigma_0, \quad \dfrac{\partial \bar{\varphi}_n}{\partial n_{\mathcal{A}}} = 0 \ \ on\ \Sigma\backslash\Sigma_0. \end{cases} \tag{6.196}$$

Define $\bar{\mathbf{g}}_n = \{\bar{g}_n^0, \bar{g}_n^1\} \in L^2(\Omega) \times L^2(\Omega)$ *by*

$$\begin{cases} \displaystyle\int_\Omega \bar{g}_n^0 v\,dx = k^{-1}\int_\Omega w_n^0 v\,dx - \int_\Omega \frac{\partial \bar{\varphi}_n}{\partial t}(T) v\,dx, \quad \forall\, v \in L^2(\Omega), \\ \displaystyle\int_\Omega \bar{g}_n^1 v\,dx = k^{-1}\int_\Omega w_n^1 v\,dx + \int_\Omega \bar{\varphi}_n(T) v\,dx, \quad \forall\, v \in L^2(\Omega). \end{cases} \tag{6.197}$$

Compute

$$\rho_n = \int_\Omega (|g_n^0|^2 + |g_n^1|^2)\,dx \Big/ \int_\Omega (\bar{g}_n^0 w_n^0 + \bar{g}_n^1 w_n^1)\,dx, \tag{6.198}$$

and then

$$\mathbf{f}_{n+1} = \mathbf{f}_n - \rho_n \mathbf{w}_n, \tag{6.199}$$

$$\mathbf{g}_{n+1} = \mathbf{g}_n - \rho_n \bar{\mathbf{g}}_n. \tag{6.200}$$

If $\|\mathbf{g}_{n+1}\|_{L^2(\Omega)\times L^2(\Omega)}/\|\mathbf{g}_0\|_{L^2(\Omega)\times L^2(\Omega)} \leq \epsilon$ *take* $\mathbf{f} = \mathbf{f}_{n+1}$; *else, compute*

$$\gamma_n = \frac{\|\mathbf{g}_{n+1}\|^2_{L^2(\Omega)\times L^2(\Omega)}}{\|\mathbf{g}_n\|^2_{L^2(\Omega)\times L^2(\Omega)}}, \tag{6.201}$$

and update $\mathbf{w}_n$ *by*

$$\mathbf{w}_{n+1} = \mathbf{g}_{n+1} + \gamma_n \mathbf{w}_n. \tag{6.202}$$

Do $n = n + 1$ *and go to* (6.195).

Remark 6.29 The *finite element* implementation of the above algorithm is just a variation of the one of algorithm (6.58)–(6.73) (it is in fact simpler). Actually, here too we can take advantage of the *reversibility* of the wave equations to reduce the storage requirements of the discrete analogues of algorithm (6.190)–(6.202).

Remark 6.30 In Glowinski and Li (1990), we can find a discussion of numerical methods for solving *exact Neumann boundary controllability* problems; the solution method is based on a combination of *finite element* approximations and of a *conjugate gradient* algorithm closely related to algorithm (6.190)–(6.202). One also discussed, in the above reference, the asymptotic behavior of the solution **f** of the dual problem when $T \to +\infty$; there too the analytical results confirmed the numerical ones, validating thus the computational methodology.

6.11.5 Application to the solution of the dual problem (6.179)

Assuming that β_0 and β_1 are both *positive* the *dual problem* (6.179) can also be written as

$$\mathbf{\Lambda f} + \partial j(\mathbf{f}) \ni \begin{pmatrix} -z^1 \\ z^0 \end{pmatrix}, \tag{6.203}$$

where the functional $j : L^2(\Omega) \times L^2(\Omega) \to \mathbb{R}$ is defined by

$$j(\mathbf{g}) = \beta_1 \|g^0\|_{L^2(\Omega)} + \beta_0 \|g^1\|_{L^2(\Omega)}, \ \forall \mathbf{g} = \{f^0, f^1\} \in L^2(\Omega) \times L^2(\Omega). \tag{6.204}$$

As in Section 6.8.8, we associate with (6.203) the following *initial value problem* (*flow* in the *Dynamical System* terminology):

$$\begin{cases} \dfrac{\partial \mathbf{f}}{\partial \tau} + \mathbf{\Lambda f} + \partial j(\mathbf{f}) \ni \begin{pmatrix} -z^1 \\ z^0 \end{pmatrix}, \\ \mathbf{f}(0) = \mathbf{f}_0. \end{cases} \tag{6.205}$$

Applying, for example, the *Peaceman–Rachford scheme* to the (pseudo)time discretization of the initial value problem (6.205), we obtain:

$$\mathbf{f}^0 = \mathbf{f}_0; \tag{6.206}$$

then for $k \geq 0$, assuming that $\mathbf{f}^k$ is known, we compute $\mathbf{f}^{k+1/2}$ and $\mathbf{f}^{k+1}$ via

$$\frac{\mathbf{f}^{k+1/2} - \mathbf{f}^k}{\Delta\tau/2} + \mathbf{\Lambda} f^k + \partial j(\mathbf{f}^{k+1/2}) \ni \begin{pmatrix} -z^1 \\ z^0 \end{pmatrix}, \tag{6.207}$$

$$\frac{\mathbf{f}^{k+1} - \mathbf{f}^{k+1/2}}{\Delta\tau/2} + \mathbf{\Lambda} f^{k+1} + \partial j(\mathbf{f}^{k+1/2}) \ni \begin{pmatrix} -z^1 \\ z^0 \end{pmatrix}. \tag{6.208}$$

Problem (6.207) is fairly easy to solve (see Section 6.8.8) since the operator $\partial j(\cdot)$ is *diagonal*. On the other hand, once $\mathbf{f}^{k+1/2}$ is known, problem (6.208) can be rewritten as

$$\frac{\mathbf{f}^{k+1} - 2\mathbf{f}^{k+1/2} + \mathbf{f}^k}{\Delta\tau/2} + \mathbf{\Lambda} f^{k+1} = \mathbf{\Lambda} f^k,$$

which shows that problem (6.208) is a particular case of problem (6.189) (with $k = \Delta\tau/2$); it can be solved therefore by the conjugate gradient algorithm (6.190)–(6.202).

6.12 Distributed controls for wave equations

Let us consider $\mathcal{O} \subset \Omega$ and let the *state equation* be

$$y_{tt} + \mathcal{A}y = v\chi_{\mathcal{O}} \text{ in } Q, \quad y(0) = y_t(0) = 0, \quad y = 0 \text{ on } \Sigma. \tag{6.209}$$

We choose

$$v \in L^2(\mathcal{O} \times (0, T)). \tag{6.210}$$

The solution of the wave problem (6.209) is *unique* and it satisfies

$$\{y, y_t\} \text{ is continuous from } [0, T] \text{ into } H_0^1(\Omega) \times L^2(\Omega). \tag{6.211}$$

Let us investigate when

$$\{y(T), y_t(T)\} \text{ spans a dense subset of } H_0^1(\Omega) \times L^2(\Omega). \tag{6.212}$$

We consider $\mathbf{f} = \{f^0, f^1\} \in L^2(\Omega) \times H^{-1}(\Omega)$ such that

$$-\int_\Omega y_t(T) f^0 \, dx + \langle f^1, y(T)\rangle = 0, \quad \forall v \in L^2(\mathcal{O} \times (0, T)), \tag{6.213}$$

where, in (6.213), $\langle \cdot, \cdot \rangle$ denotes the duality pairing between $H^{-1}(\Omega)$ and $H_0^1(\Omega)$. Next, we introduce ψ, the solution of

$$\psi_{tt} + \mathcal{A}\psi = 0 \text{ in } Q, \quad \psi(T) = f^0, \ \psi_t(T) = f^1, \quad \psi = 0 \text{ on } \Sigma. \tag{6.214}$$

Then

$$-\int_\Omega y_t(T) f^0 \, dx + \langle f^1, y(T)\rangle = \int_{\mathcal{O}\times(0,T)} \psi v \, dx \, dt. \tag{6.215}$$

It follows from (6.215) that (6.213) is equivalent to

$$\psi = 0 \text{ on } \mathcal{O} \times (0,T). \tag{6.216}$$

Let assume that we can apply the Holmgren's uniqueness theorem to $\mathcal{O} \times (0,T)$; then $\psi \equiv 0$ and $f = 0$, so that (6.212) holds true.

We can then consider the following *approximate controllability* problem:

$$\inf_v \frac{1}{2} \iint_{\mathcal{O}\times(0,T)} v^2 \, dx \, dt; \;\; y(T;v) \in z^0 + \beta_0 B_1, \;\; y_t(T;v) \in z^1 + \beta_1 B, \tag{6.217}$$

where, in (6.217), $y(v)$ is obtained from v via (6.209), $\{z^0, z^1\}$ is given in $H_0^1(\Omega) \times L^2(\Omega)$, and B_1 (respectively B) is the closed unit ball of $H_0^1(\Omega)$ (respectively $L^2(\Omega)$).

Considerations similar to everything which has been said in the previous sections can be adapted to the present situation, from either a *purely mathematical point of view* (see J.L. Lions, 1988b) or a *numerical point of view*.

Remark 6.31 One can also consider *pointwise control*, as in

$$\begin{cases} y_{tt} + \mathcal{A}y = v(t)\delta(x-b) \;\; \text{in } Q, \\ y(0) = y_t(0) = 0, \quad y = 0 \;\; \text{on } \Sigma \text{ (to fix ideas).} \end{cases} \tag{6.218}$$

Control problems for systems modeled by (6.218) have been discussed in J.L. Lions (1988b, Volume I, Chapter 7). Interesting phenomena appear concerning the role of $b \in \Omega$. Methods of *harmonic analysis* have been used in this respect by Meyer (Private Communication, 1989) and further developed by Haraux and Jaffard (1991) and I. Joó (1991). The *numerical implementation of HUM* for the *exact controllability* of the wave system (6.128) is discussed in Foss (2006), in the particular case where $\Omega = (0,1)$ and $\mathcal{A} = -\frac{\partial^2}{\partial x^2}$; exact controllability holds in the above case if $b \notin \Omega \cap \mathbb{Q}$, where $\mathbb{Q}$ is the set of the *rational* real numbers.

6.13 Dynamic programming

We are going to apply *Dynamic Programming* to the situations described in Section 6.11. The approach is *formal* and somewhat similar to the one in Chapter 5.

Remark 6.32 We could have applied dynamic programming to the situation described in Section 6.1 or 6.11, but the situation is simpler for the control problems described in Section 6.12.

For s given in $(0,T)$ we consider

$$\begin{cases} y_{tt} + \mathcal{A}y = v\chi_{\mathcal{O}} \quad \text{in } \Omega \times (s,T), \\ y(s) = h^0, \quad y_t(s) = h^1, \quad y = 0 \;\; \text{on } \Sigma_s = \Gamma \times (s,T), \end{cases} \tag{6.219}$$

with $\{h^0, h^1\} \in H_0^1(\Omega) \times L^2(\Omega)$. We introduce

$$\phi(h^0, h^1, s) = \inf_v \iint_{\mathcal{O}\times(s,T)} v^2 \, dx \, dt \tag{6.220}$$

where, in (6.220), v is such that $\{y(v), v\}$ satisfies (6.219) and

$$y(T; v) \in z^0 + \beta_0 B_1, \quad y_t(T; v) \in z^1 + \beta_1 B. \tag{6.221}$$

The quantity $\phi(h^0, h^1, s)$ is finite for every

$$z^0 \in H_0^1(\Omega), \, z^1 \in L^2(\Omega), \quad \beta_0 > 0, \, \beta_1 > 0$$

if and only if the Holmgren's uniqueness theorem applies for $\mathcal{O} \times (s, T)$ in $\Omega \times (s, T)$. *This is true for* $s < s_0$, s_0 being a suitable number in $(0, T)$. In that case, the infimum in (6.221) is *finite* for $s < s_0$, implying that the function ϕ is *defined* over $H_0^1(\Omega) \times L^2(\Omega) \times (0, s_0)$.

Let us write now the *Hamilton–Jacobi–Bellmann (HJB) equation* (namely, the partial differential equation satisfied by ϕ); we take

$$v(x, t) = w(x) \text{ in } \mathcal{O} \times (s, s + \varepsilon). \tag{6.222}$$

With this choice of v, $\{y(t), y_t(t)\}$ varies on the time interval $(s, s + \varepsilon)$ from $\{h^0, h^1\}$ to

$$\{h^0 + \varepsilon h^1, h^1 + \varepsilon w \chi_{\mathcal{O}} - \varepsilon \mathcal{A} h^0\} + 0(\varepsilon^2)$$

(assuming that $h^0 \in H_0^1(\Omega) \cap H^2(\Omega)$ and that the coefficients of the elliptic operator $\mathcal{A}$ are smooth enough). Then, according to the *optimality principle*, we have

$$\phi(h^0, h^1, s) = \inf_w \left\{ \frac{\varepsilon}{2} \int_{\mathcal{O}} w^2 \, dx \right. \\ \left. + \phi(h^0 + \varepsilon h^1, h^1 + \varepsilon w \chi_{\mathcal{O}} - \varepsilon \mathcal{A} h^0, s + \varepsilon) \right\} + 0(\varepsilon^2). \tag{6.223}$$

Expanding ϕ we obtain

$$\frac{\partial \phi}{\partial s} + \left(\frac{\partial \phi}{\partial h^0}, h^1 \right) - \left(\frac{\partial \phi}{\partial h^1}, \mathcal{A} h^0 \right) \\ + \inf_{w \in L^2(\mathcal{O})} \left\{ \frac{1}{2} \int_{\mathcal{O}} w^2 \, dx + \left(\frac{\partial \phi}{\partial h^1}, w \chi_{\mathcal{O}} \right) \right\} = 0. \tag{6.224}$$

The minimization problem in (6.224) has a unique solution given by

$$u(s) = -\frac{\partial \phi}{\partial h^1}(h^0, h^1, s) \chi_{\mathcal{O}}; \tag{6.225}$$

combining (6.224) with (6.225) yields the following *HJB equation*:

$$-\frac{\partial \phi}{\partial s} - \left(\frac{\partial \phi}{\partial h^0}, h^1\right) + \left(\frac{\partial \phi}{\partial h^1}, \mathcal{A}h^0\right) + \frac{1}{2}\int_{\mathcal{O}} \left|\frac{\partial \phi}{\partial h^1}\right|^2 dx = 0. \tag{6.226}$$

The above functional equation has to be completed by the following "final" condition:

$$\phi(h^0, h^1, s_0) = \begin{cases} 0 & \text{if } \{h^0, h^1\} \in E, \\ +\infty & \text{otherwise,} \end{cases} \tag{6.227}$$

where E is the set described by $\{y(s_0; v), y_t(s_0; v)\}$ when y satisfies

$$y_{tt} + \mathcal{A}y = v\chi_{\mathcal{O}} \text{ in } \Omega \times (s_0, T), \quad v \in L^2(\mathcal{O} \times (s_0, T)), \quad y = 0 \text{ on } \Gamma \times (s_0, T)$$

and when the two conditions in (6.221) hold true.

Remark 6.33 We emphasize once more that the above approach is fairly *formal*.

Remark 6.34 The *real time optimal policy* is given at time $t \in (0, s_0)$ by

$$u(t) = -\frac{\partial \phi}{\partial h^1}(h^0, h^1, t)\chi_{\mathcal{O}}. \tag{6.228}$$

How to proceed for $t \in (s_0, T)$ seems to be an open question, even from a conceptual point of view.

7

On the application of controllability methods to the solution of the Helmholtz equation at large wave numbers

7.1 Introduction

Stealth technologies have enjoyed a considerable growth of interest during the last two decades both for aircraft and space applications. Due to the *very high frequencies* used by modern radars the computation of the *Radar Cross Section (RCS)* of a full aircraft using the *Maxwell equations* is still a *great challenge* (see Talflove, 1992). From the fact that *boundary integral methods* (see Nedelec, 2001 and the references therein for a discussion of boundary integral methods) are not well suited to general heterogeneous media and coated materials, *field approaches* seem to provide an alternative which is worth exploring.

In this chapter (which follows closely Section 6.13 of the original *Acta Numerica* article and Bristeau, Glowinski, and Périaux, 1998), we consider a particular application of *controllability methods* to the solution of the *Helmholtz equations* obtained when looking for the *monochromatic solutions of linear wave problems*. The idea here is to go back to the original wave equation and to apply techniques, inspired by controllability studies, in order to find its *time-periodic* solutions. Indeed, this method (introduced in Bristeau, Glowinski, and Périaux, 1993a,b) is a competitor – and is related – to the one in which the wave equation is integrated from 0 to $+\infty$ in order to obtain asymptotically a time-periodic solution; it is well-known (from, for example, Lax and Phillips, 1989) that if the scattering body is *convex*, then the solution will converge *exponentially* to the time-periodic solution as $t \to +\infty$. On the other hand, for *nonconvex* reflectors (which is quite a common situation) the convergence can be very slow; the method described in this chapter improves substantially the speed of convergence of its asymptotic "relative", particularly for stiff problems where internal rays can be trapped by successive reflections.

7.2 The Helmholtz equation and its equivalent wave problem

Let us consider a scattering body B, of boundary $\partial B = \gamma$, "illuminated" by an *incident monochromatic wave* of frequency $f = k/2\pi$ (see Figure 7.1).

In the case of the wave equation $u_{tt} - \Delta u = 0$ with a time-periodic solution $u = \mathrm{Re}\,(U\mathrm{e}^{-\mathrm{i}kt})$, the associated *Helmholtz equation*, satisfied by the coefficient

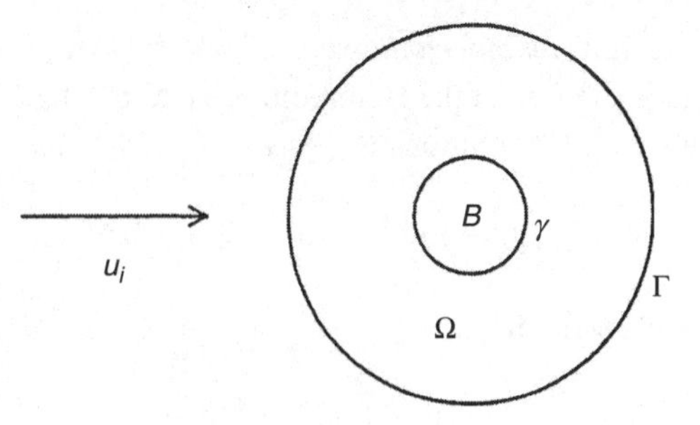

Fig. 7.1. Geometrical description (u_i is the incident field).

$U(x)$ of e^{-ikt} is given by

$$\Delta U + k^2 U = 0 \quad \text{in } \mathbb{R}^d \backslash \bar{B} \ (d = 2, 3), \tag{7.1}$$

$$U = G \quad \text{on } \gamma. \tag{7.2}$$

In practice, we bound $\mathbb{R}^d \backslash \bar{B}$ by an artificial boundary Γ on which we prescribe, for example, an *approximate first-order Sommerfeld radiation condition* such as

$$\frac{\partial U}{\partial n} - i k U = 0 \quad \text{on } \Gamma; \tag{7.3}$$

now, equation (7.1) is prescribed on Ω only, where Ω is this portion of $\mathbb{R}^d \backslash \bar{B}$ between γ and Γ, as shown in Figure 7.1. In the above equations, U is the *scattered field*, $-G$ is the incident field, U and G are *complex valued functions*.

Remark 7.1 More complicated (and efficient) absorbing boundary conditions than (7.3) have been coupled to the controllability method described hereafter; they allow the use of smaller computational domains. The resulting methodology is described in Bristeau, Glowinski, and Périaux (1993c, 1994).

Systems (7.1)–(7.3) is related to the T-periodic solutions ($T = 2\pi/k$) of the following wave equation and associated boundary conditions:

$$u_{tt} - \Delta u = 0 \quad \text{in } Q \ (= \Omega \times (0, T)), \tag{7.4}$$

$$u = g \quad \text{on } \sigma \ (= \gamma \times (0, T)), \tag{7.5}$$

$$\frac{\partial u}{\partial n} + \frac{\partial u}{\partial t} = 0 \quad \text{on } \Sigma \ (= \Gamma \times (0, T)), \tag{7.6}$$

where, in (7.5), $g(x, t)$ is a time-periodic function related to G by $g(x, t) = \operatorname{Re}(e^{-ikt} G(x))$. If we denote

$$G(x) = G_r(x) + i\, G_{im}(x),$$

g satisfies

$$g(x, t) = G_r(x) \cos kt + G_{im}(x) \sin kt.$$

Our goal, here, is to find *time-periodic solutions* to system (7.4)–(7.6) in order to solve (indirectly, in some sense) the Helmholtz system (7.1)–(7.3). More precisely, in the following, we look for solutions of (7.4)–(7.6) satisfying

$$u(0) = u(T), \quad u_t(0) = u_t(T). \tag{7.7}$$

These solutions can be written

$$u(x,t) = \mathrm{Re}\,(\mathrm{e}^{-\mathrm{i}kt}U(x)),$$

or

$$u(x,t) = U_{\mathrm{r}} \cos kt + U_{\mathrm{im}} \sin kt,$$

where $U = U_{\mathrm{r}} + \mathrm{i}\,U_{\mathrm{im}}$ is the solution of (7.1)–(7.3); we have thus

$$u(0) = U_{\mathrm{r}} \text{ and } u_t(0) = kU_{\mathrm{im}}.$$

7.3 Exact controllability methods for the calculation of time-periodic solutions to the wave equation

In order to solve problem (7.4)–(7.7) we advocate the following approach whose main merit is to reduce the above problem to an *exact controllability* one, close to those problems whose solution is discussed in Chapter 6, Sections 6.1 to 6.12. Indeed, problem (7.4)–(7.7) is clearly equivalent to the following one:

Find $\mathbf{e} = \{e^0, e^1\}$ *such that*

$$u_{tt} - \Delta u = 0 \quad \text{in } Q, \tag{7.8}$$

$$u = g \quad \text{on } \sigma, \tag{7.9}$$

$$\frac{\partial u}{\partial n} + \frac{\partial u}{\partial t} = 0 \quad \text{on } \Sigma, \tag{7.10}$$

$$u(0) = e^0, \quad u_t(0) = e^1, \quad u(T) = e^0, \quad u_t(T) = e^1. \tag{7.11}$$

Problem (7.8)–(7.11) is an *exact controllability problem* which can be solved by methods directly inspired by those in Chapter 6, Sections 6.1–6.6. We shall not address here the *existence* and *uniqueness* of solutions to problem (7.8)–(7.11) (these two issues are addressed in Bardos and Rauch, 1994); instead we shall focus on the practical calculation of such solutions, assuming they do exist.

7.4 Least-squares formulation of the problem (7.8)–(7.11)

In order to be able to apply controllability methods to the solution of problem (7.8)–(7.11) the appropriate choice for the space E containing $\mathbf{e} = \{e^0, e^1\}$ is of fundamental importance. We advocate

$$E = V_{\mathbf{g}} \times L^2(\Omega), \tag{7.12}$$

where

$$V_{\mathbf{g}} = \{\varphi \mid \varphi \in H^1(\Omega),\ \varphi = g(0) \text{ on } \gamma\}. \tag{7.13}$$

In order to solve (7.8)–(7.11), we use the following *least-squares* formulation (where y plays the role of u in (7.8)–(7.11)):

$$\min_{\mathbf{v}\in E} J(\mathbf{v}), \tag{7.14}$$

with

$$J(\mathbf{v}) = \frac{1}{2}\int_{\Omega}\left(|\nabla(y(T) - v^0)|^2 + |y_t(T) - v^1|^2\right) dx, \quad \forall\, \mathbf{v} = \{v^0, v^1\}, \tag{7.15}$$

where, in (7.15), the function y is the solution of

$$y_{tt} - \Delta y = 0 \quad \text{in } Q, \tag{7.16}$$

$$y = g \quad \text{on } \sigma, \tag{7.17}$$

$$\frac{\partial y}{\partial n} + \frac{\partial y}{\partial t} = 0 \quad \text{on } \Sigma, \tag{7.18}$$

$$y(0) = v^0, \quad y_t(0) = v^1. \tag{7.19}$$

The choice of the cost function J is directly related to the fact that the *natural energy* of the system under consideration is given by

$$\mathcal{E}(t) = \frac{1}{2}\int_{\Omega}\left(|\nabla y(t)|^2 + |y_t(t)|^2\right) dx. \tag{7.20}$$

Assuming that $\mathbf{e}$ is the solution of problem (7.14), it will satisfy the following (Euler–Lagrange) equation:

$$\langle J'(\mathbf{e}), \mathbf{v}\rangle = 0, \quad \forall\, \mathbf{v} \in E_0, \tag{7.21}$$

where, in (7.21), $E_0 = V_0 \times L^2(\Omega)$ (with $V_0 = \{\varphi \mid \varphi \in H^1(\Omega),\ \varphi = 0 \text{ on } \gamma\}$) and where $\langle\cdot,\cdot\rangle$ denotes the duality pairing between E_0' and E_0 (E_0': dual space of E_0). In (7.21), J' denotes the *differential* of J.

The (clearly) *linear* problem (7.21) can be solved by a *conjugate gradient algorithm* (described in Section 7.6) operating in E and E_0; in order to implement this algorithm, we need to be able to compute $J'(\mathbf{v})$, $\forall\, \mathbf{v} \in E$; this most important issue will be addressed in the following section.

Remark 7.2 The well-posedness of problem (7.14) is discussed in Bardos and Rauch (1994).

7.5 Calculation of J'

To compute J' we use a *perturbation analysis*: starting from (7.15), we obtain

$$\begin{aligned}\delta J(\mathbf{v}) &= \langle J'(\mathbf{v}), \delta\mathbf{v}\rangle \\ &= \int_\Omega \nabla(v^0 - y(T)) \cdot \nabla\delta v^0 \,\mathrm{d}x + \int_\Omega (v^1 - y_t(T))\delta v^1 \,\mathrm{d}x \\ &\quad + \int_\Omega \nabla(y(T) - v^0) \cdot \nabla\delta y(T)\,\mathrm{d}x + \int_\Omega (y_t(T) - v^1)\delta y_t(T)\,\mathrm{d}x. \end{aligned} \tag{7.22}$$

We also have from (7.16)–(7.19):

$$\delta y_{tt} - \Delta\delta y = 0 \quad \text{in } Q, \tag{7.23}$$

$$\delta y = 0 \quad \text{on } \sigma, \tag{7.24}$$

$$\left(\frac{\partial}{\partial n} + \frac{\partial}{\partial t}\right)\delta y = 0 \quad \text{on } \Sigma, \tag{7.25}$$

$$\delta y(0) = \delta v^0, \quad \delta y_t(0) = \delta v^1. \tag{7.26}$$

Consider now a function p of x and t such that the function $p(t) : x \to p(x,t)$ vanishes on γ; next, multiply both sides of (7.23) by p, integrate over Q and then by parts over $(0, T)$. It follows then from the boundary conditions (7.24) and (7.25) that

$$\begin{aligned}&\int_\Omega \delta y_t\, p\,\mathrm{d}x\Big|_0^T - \int_\Omega \delta y\, p_t\,\mathrm{d}x\Big|_0^T + \int_Q p_{tt}\,\delta y\,\mathrm{d}x\,\mathrm{d}t + \int_Q \nabla p\cdot\nabla\delta y\,\mathrm{d}x\,\mathrm{d}t \\ &\quad + \int_\Gamma \delta y\, p\,\mathrm{d}\Gamma\Big|_0^T - \int_\Sigma p_t\,\delta y\,\mathrm{d}\Gamma\,\mathrm{d}t = 0.\end{aligned} \tag{7.27}$$

Suppose that p satisfies

$$\begin{cases} \displaystyle\int_\Omega (p_{tt}z + \nabla p\cdot\nabla z)\,\mathrm{d}x - \int_\Gamma p_t z\,\mathrm{d}\Gamma = 0, \ \forall z \in V_0, \ \text{a.e. on } (0,T), \\ p = 0 \quad \text{on } \sigma; \end{cases} \tag{7.28}$$

there is *equivalence* between (7.28) and

$$p_{tt} - \Delta p = 0 \quad \text{in } Q, \tag{7.29}$$

$$p = 0 \quad \text{on } \sigma, \tag{7.30}$$

$$\frac{\partial p}{\partial n} - \frac{\partial p}{\partial t} = 0 \quad \text{on } \Sigma. \tag{7.31}$$

Using (7.26) and (7.28), equation (7.27) reduces then to

$$\begin{aligned}\int_\Omega \delta y_t(T)p(T)\,dx - \int_\Omega \delta y(T)p_t(T)\,dx + \int_\Gamma \delta y(T)p(T)\,d\Gamma \\ = \int_\Omega \delta y_t(0)p(0)\,dx - \int_\Omega \delta y(0)p_t(0)\,dx + \int_\Gamma \delta y(0)p(0)\,d\Gamma \\ = \int_\Omega p(0)\delta v^1\,dx - \int_\Omega p_t(0)\delta v^0\,dx + \int_\Gamma p(0)\delta v^0\,d\Gamma.\end{aligned} \tag{7.32}$$

Let us define $p(T)$ and $p_t(T)$ by

$$p(T) = y_t(T) - v^1 \tag{7.33}$$

and

$$\int_\Omega p_t(T)z\,dx = \int_\Gamma (y_t(T) - v^1)z\,d\Gamma - \int_\Omega \nabla(y(T) - v^0)\cdot\nabla z\,dx, \quad \forall z \in V_0, \tag{7.34}$$

respectively. Finally, using (7.22) and (7.32)–(7.34), with $z = \delta y(T)$, shows that

$$\begin{aligned}\langle J'(\mathbf{v}), \mathbf{w}\rangle = \int_\Omega \nabla(v^0 - y(T))\cdot\nabla w^0\,dx - \int_\Omega p_t(0)w^0\,dx + \int_\Gamma p(0)w^0\,d\Gamma \\ + \int_\Omega p(0)w^1\,dx + \int_\Omega (v^1 - y_t(T))w^1\,dx, \quad \forall \mathbf{w} = \{w^0, w^1\} \in E_0,\end{aligned} \tag{7.35}$$

where, in (7.35), p is the solution of the *adjoint equation* (7.29)–(7.31), completed by the “final conditions” (7.33) and (7.34).

Remark 7.3 Relations (7.34) and (7.35) are largely formal since some of the integrals in them should be replaced by duality pairings. However, it is worth mentioning that the discrete variants of the above two relations make sense and lead to algorithms with fast convergence properties.

7.6 Conjugate gradient solution of the least-squares problem (7.14)

Following Chapter 1, Section 1.8.2 (see also Glowinski, 2003, Chapter 3) we advocate a *conjugate gradient algorithm* for the solution of the *linear variational problem* (7.21) (equivalent to problem (7.14)). This conjugate gradient algorithm reads as follows:

Step 0: Initialization.

$$\mathbf{e}_0 = \{e_0^0, e_0^1\} \text{ is given in } E. \tag{7.36}$$

Solve the following forward wave problem:

$$\frac{\partial^2 y_0}{\partial t^2} - \Delta y_0 = 0 \quad \text{in } Q, \tag{7.37a}$$

$$y_0 = g \quad \text{on } \sigma, \tag{7.37b}$$

$$\frac{\partial y_0}{\partial n} + \frac{\partial y_0}{\partial t} = 0 \quad \text{on } \Sigma, \tag{7.37c}$$

$$y_0(0) = e_0^0, \quad \frac{\partial y_0}{\partial t}(0) = e_0^1. \tag{7.37d}$$

Solve the following backward wave problem:

$$\frac{\partial^2 p_0}{\partial t^2} - \Delta p_0 = 0 \quad \text{in } Q, \tag{7.38a}$$

$$p_0 = 0 \quad \text{on } \sigma, \tag{7.38b}$$

$$\frac{\partial p_0}{\partial n} - \frac{\partial p_0}{\partial t} = 0 \quad \text{on } \Sigma, \tag{7.38c}$$

with $p_0(T)$ *and* $\frac{\partial p_0}{\partial t}(T)$ *given by*

$$p_0(T) = \frac{\partial y_0}{\partial t}(T) - e_0^1, \tag{7.38d}$$

$$\int_\Omega \frac{\partial p_0}{\partial t}(T) z \, dx = \int_\Gamma p_0(T) z \, d\Gamma - \int_\Omega \nabla(y_0(T) - e_0^0) \cdot \nabla z \, dx, \quad \forall z \in V_0, \tag{7.38e}$$

respectively.

Define next $\mathbf{g}_0 = \{g_0^0, g_0^1\} \in E_0 \ (= V_0 \times L^2(\Omega))$ *by*

$$\int_\Omega \nabla g_0^0 \cdot \nabla z \, dx = \int_\Omega \nabla(e_0^0 - y_0(T)) \cdot \nabla z \, dx - \int_\Omega \frac{\partial p_0}{\partial t}(0) z \, dx + \int_\Gamma p_0(0) z \, d\Gamma, \quad \forall z \in V_0, \tag{7.39a}$$

$$g_0^1 = p_0(0) + e_0^1 - \frac{\partial y_0}{\partial t}(T), \tag{7.39b}$$

and then

$$\mathbf{w}^0 = \mathbf{g}^0. \tag{7.40}$$

For $k \geq 0$, suppose that $\mathbf{e}_k$, $\mathbf{g}_k$, $\mathbf{w}_k$ are known (the last two different from $\mathbf{0}$); we compute then $\mathbf{e}_{k+1}$, $\mathbf{g}_{k+1}$, $\mathbf{w}_{k+1}$ as follows:

Step 1: Descent.
Solve the following forward wave problem:

$$\frac{\partial^2 \bar{y}_k}{\partial t^2} - \Delta \bar{y}_k = 0 \quad \text{in } Q, \tag{7.41a}$$

$$\bar{y}_k = 0 \quad \text{on } \sigma, \tag{7.41b}$$

$$\frac{\partial \bar{y}_k}{\partial t} + \frac{\partial \bar{y}_k}{\partial n} = 0 \quad \text{on } \Sigma, \tag{7.41c}$$

$$\bar{y}_k(0) = w_k^0, \quad \frac{\partial \bar{y}_k}{\partial t}(0) = w_k^1. \tag{7.41d}$$

Solve then the following backward wave problem:

$$\frac{\partial^2 \bar{p}_k}{\partial t^2} - \Delta \bar{p}_k = 0 \quad \text{in } Q, \tag{7.42a}$$

$$\bar{p}_k = 0 \quad \text{on } \sigma, \tag{7.42b}$$

$$\frac{\partial \bar{p}_k}{\partial t} - \frac{\partial \bar{p}_k}{\partial n} = 0 \quad \text{on } \Sigma, \tag{7.42c}$$

with $\bar{p}_k(T)$ and $\frac{\partial \bar{p}_k}{\partial t}(T)$ given by

$$\bar{p}_k(T) = \frac{\partial \bar{y}_k}{\partial t}(T) - w_k^1, \tag{7.42d}$$

$$\int_\Omega \frac{\partial \bar{p}_k}{\partial t}(T) z \, dx = \int_\Gamma \bar{p}_k(T) z \, d\Gamma - \int_\Omega \nabla(\bar{y}_k(T) - w_k^0) \cdot \nabla z \, dx, \quad \forall z \in V_0, \tag{7.42e}$$

respectively. Define next $\bar{\mathbf{g}}_k = \{\bar{g}_k^0, \bar{g}_k^1\} \in V_0 \times L^2(\Omega)$ by

$$\int_\Omega \nabla \bar{g}_k^0 \cdot \nabla z \, dx = \int_\Omega \nabla (w_k^0 - \bar{y}_k(T)) \cdot \nabla z \, dx - \int_\Omega \frac{\partial \bar{p}_k}{\partial t}(0) z \, dx$$
$$+ \int_\Gamma \bar{p}_k(0) z \, d\Gamma, \quad \forall z \in V_0, \tag{7.43a}$$

$$\bar{g}_k^1 = \bar{p}_k(0) + w_k^1 - \frac{\partial \bar{y}_k}{\partial t}(T), \tag{7.43b}$$

and then ρ_k by

$$\rho_k = \int_\Omega (|\nabla g_k^0|^2 + |g_k^1|^2) \, dx \Big/ \int_\Omega (\nabla \bar{g}_k^0 \cdot \nabla w_k^0 + \bar{g}_k^1 w_k^1) \, dx. \tag{7.44}$$

We update then $\mathbf{e}_k$ *and* $\mathbf{g}_k$ *by*

$$\mathbf{e}_{k+1} = \mathbf{e}_k - \rho_k \mathbf{w}_k, \tag{7.45}$$

$$\mathbf{g}_{k+1} = \mathbf{g}_k - \rho_k \bar{\mathbf{g}}_k. \tag{7.46}$$

Step 2: Test of the convergence and construction of the new descent direction.
If $\int_\Omega (|\nabla g_{k+1}^0|^2 + |g_{k+1}^1|^2)\,dx / \int_\Omega (|\nabla g_0^0|^2 + |g_0^1|^2)\,dx \le \epsilon^2$ *take* $\mathbf{e} = \mathbf{e}_{k+1}$; *otherwise, compute*

$$\gamma_k = \int_\Omega (|\nabla g_{k+1}^0|^2 + |g_{k+1}^1|^2)\,dx \Big/ \int_\Omega (|\nabla g_k^0|^2 + |g_k^1|^2)\,dx \tag{7.47}$$

and update $\mathbf{w}_k$ *by*

$$\mathbf{w}_{k+1} = \mathbf{g}_{k+1} + \gamma_k \mathbf{w}_k. \tag{7.48}$$

Do $k = k + 1$ *and go to* (7.41).

Remark 7.4 Algorithm (7.36)–(7.48) looks complicated at first glance. In fact, it is not that complicated to implement since each iteration requires basically the solution of *two wave equations* such as (7.41) and (7.42), and of an *elliptic problem* such as (7.43a). The above problems are classical ones for which efficient solution methods already exist.

Remark 7.5 Algorithm (7.36)–(7.48) can be seen as a variation of the *asymptotic method* mentioned in Section 7.1; there, we integrate the periodically excited wave equation until we reach a periodic solution (namely, a *limit cycle*). What algorithm (7.36)–(7.48) does indeed is to *periodically* measure the *lack* (or *defect*) of *periodicity* and use the result of this measure as a *residual* to speed up the convergence to a periodic solution giving to the overhaul operation an *impulse control* flavor (incidentally, it has also the flavor of a *shooting method* for the solution of *two point – in time – boundary value problems*). Actually, a similar idea was used in Auchmuty, Dean, Glowinski, and Zhang (1987) to compute the periodic solutions of systems of *stiff* nonlinear ordinary differential equations (including cases where the period itself is an unknown parameter of the problem). Other applications can be found in Auchmuty and Zhai (1996).

7.7 A finite element–finite difference implementation

The practical implementation of the previously discussed control-based method is straightforward. It relies on a *time discretization* by the *centered second-order accurate explicit finite difference scheme*, already employed in Chapter 6, Sections 6.8 and 6.9. This scheme is combined to *piecewise linear finite element approximations* (as in Chapter 6, Sections 6.8 and 6.9) for the space variables; we use *mass lumping* – through numerical integration by the *trapezoïdal rule* – to obtain a *diagonal* mass matrix for the acceleration term. The fully discrete scheme has to satisfy a *stability condition* such as $\Delta t \leq Ch$, where C is a (positive) constant. To obtain accurate solutions, we need to have h at least *ten times smaller* than the wavelength; in fact, h has to be even smaller ($h \approx \lambda/20$) in those regions where internal rays are trapped by successive reflections. For the *mesh generation*, the advancing front method proposed by J.A. George (1971) has been used; this method (implemented at INRIA by P.L. George and Seveno) gives *homogeneous* triangulations (see the following figures).

7.8 Numerical experiments

In order to validate the methods discussed in the above sections, we have considered the solution of three test problems of increasing difficulty, namely, the scattering of planar incident waves by a *disk*, then by a *nonconvex* reflector which can be seen as a semiopen cavity (a kind of – very idealized – air intake) and finally the scattering of a planar wave by a *two-dimensional aircraft like body*. For these different cases the artificial boundary Γ is located at a distance of 3λ from B and we assume that the boundary of the reflector is *perfectly conducting*.

The following results have been obtained by M.O. Bristeau at INRIA (see Bristeau, Glowinski, and Périaux, 1993a,b,c, 1998 for further numerical experiments and details):

Before discussing our numerical experiments, let us observe that the model (7.8)–(7.10) assumes, implicitly, that the waves it describes propagate with speed 1, implying that, here the wave length is equal to the period. If c (> 0) is different from 1, we shall rescale x and t, so that $c = 1$. In the following examples, the data are given in the MKSA system before rescaling.

Example 1 (Scattering by a disk) For this problem, B is a disk of radius 0.25 m, $k = 2\pi f$ with $f = 2.4$ GHz, implying that the wave length is 0.125 m; the disk is illuminated by an incident *planar wave* coming from the left and propagating horizontally. The artificial boundary Γ is a circle located at a distance of 3λ from B. The *finite element triangulation* has 18 816 vertices and 36 970 triangles; the mean length of the edges is $\lambda/15$, the minimal value being $\lambda/28$, while the maximal one is $\lambda/10$. The value of Δt is $T/35$. To obtain convergence of the iterative method, 74 iterations of (the discrete variant of) algorithm (7.36)–(7.48) were needed (with $\epsilon = 5 \times 10^{-5}$ for the stopping criterion) corresponding to a 3 min computation on a CRAY2.

We have shown on Figure 7.2 the contours of the scattered field e^0 (*real component* of the Helmholtz problem solution). For this test problem, where the exact solution is known, we have compared on Figures 7.3 and 7.4 the *computed solution* (—) with the exact one ($\cdots$) on two cross sections (incident direction and opposite to incident direction, respectively). Of course, for this "simple" problem, the *asymptotic method* (just integrating the wave equation from 0 to nT, with n "large") is less CPU time consuming (twice faster, approximately), but it has to be understood that this example was chosen for validation purposes, essentially (to test the accuracy of this novel approach, in particular).

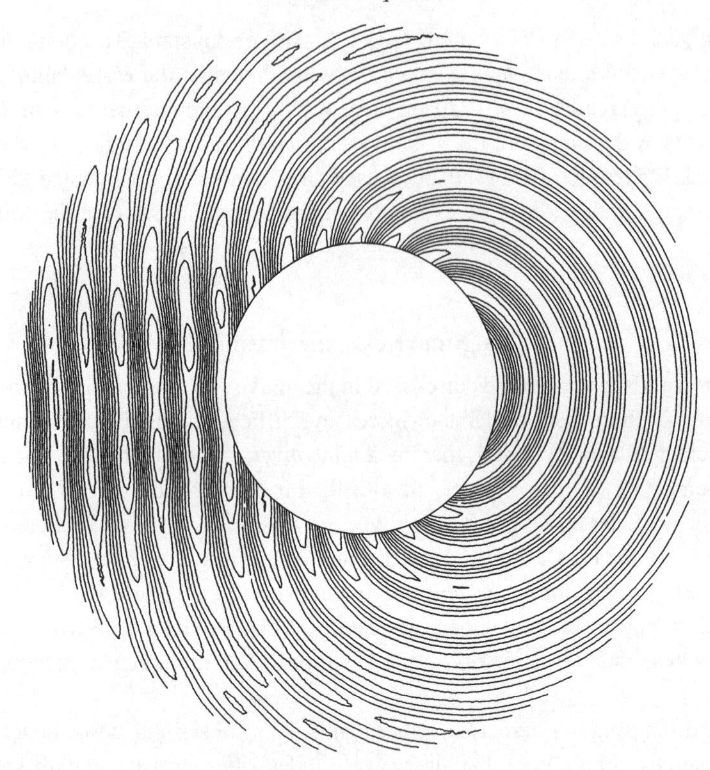

Fig. 7.2. *First test problem*: Contours of the scattered field (real component).

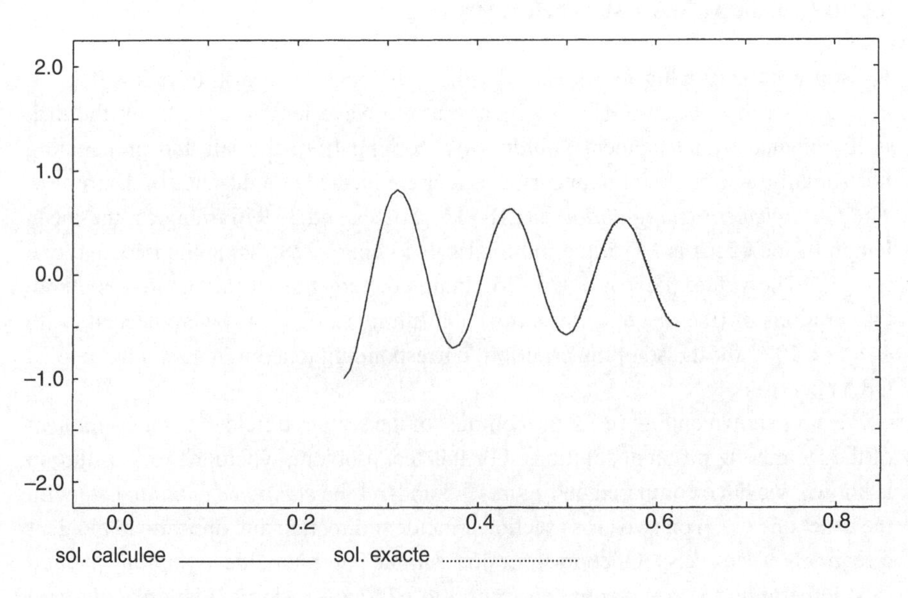

Fig. 7.3. *First test problem*: Comparison between exact (···) and computed (—) scattered fields $e^0(x_1, 0)$ (incident side).

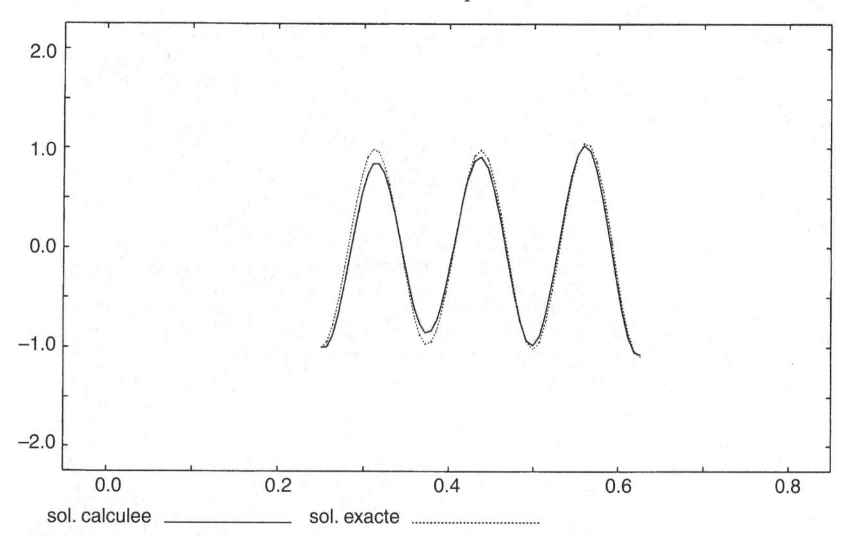

Fig. 7.4. *First test problem*: Comparison between exact (· · ·) and computed (—) scattered fields $e^0(x_1, 0)$ (shadow side).

Remark 7.6 One can substantially increase the accuracy by using on Γ instead of (7.6), *second-order absorbing boundary conditions* like those discussed in Bristeau, Glowinski, and Périaux (1993c) and (1994).

Example 2 (Scattering by semiopen cavities) We have considered two semiopen cavities. We choose $f = 3$ GHz implying that the wave length is 0.10 m. For the first cavity, the inside dimensions are $4\lambda \times \lambda$ and the thickness of the wall is $\lambda/5$. The finite element triangulation has 22 951 vertices and 44 992 triangles and we took $\Delta t = T/40$. We consider an illuminating monochromatic wave of incidence $\alpha = 30°$, coming from the right. The contours of the scattered fields e^0 (real part) and e^1/k (imaginary part) are shown on Figures 7.5 and 7.6, respectively. The convergence is reached with 136 conjugate gradient iterations (with, again, $\epsilon = 5 \times 10^{-5}$ for the stopping criterion), corresponding to 8 min of CPU time on a CRAY 2. Figure 7.7 shows the convergence of the *residuals* $R(e_k^0)$ and $R(e_k^1)$ associated to the controllability method; these residuals are defined by

$$R(e_k^0) = \frac{\|e_{k+1}^0 - e_k^0\|_{L^2(\Omega)}}{\|e_1^0 - e_0^0\|_{L^2(\Omega)}} \quad \text{and} \quad R(e_k^1) = \frac{\|e_{k+1}^1 - e_k^1\|_{L^2(\Omega)}}{\|e_1^1 - e_0^1\|_{L^2(\Omega)}}.$$

The *asymptotic method* produces the same solution, essentially, but for this *nonconvex* obstacle, the convergence is significantly *slower* (800 iterations, 18 min of CPU time on a CRAY 2) than the convergence of the controllability method, as shown on Figure 7.8.

We have considered a second semiopen cavity for the same frequency and wave length; the inside dimensions of this larger cavity are $20\lambda \times 5\lambda$, the wall thickness being λ. For this test problem where many reflections take place inside the cavity, one needs a fine triangulation. The one used here has 208 015 vertices and 412 028

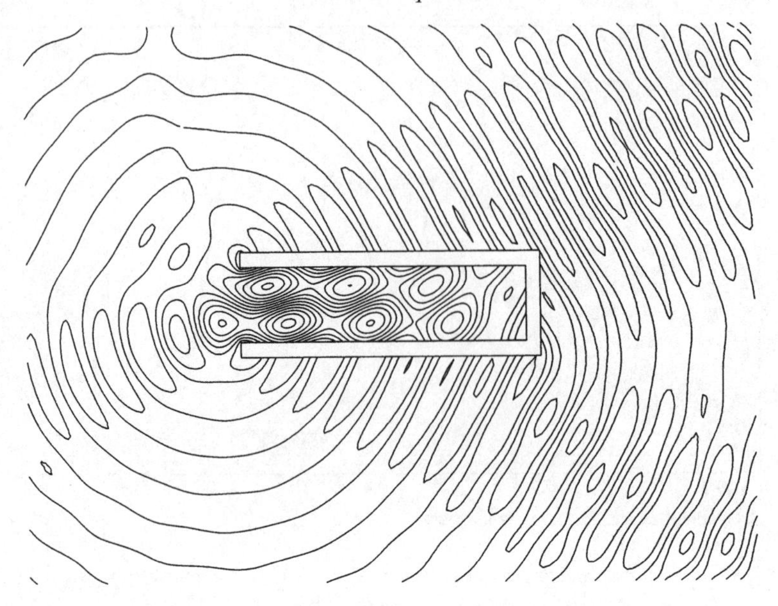

Fig. 7.5. *Second test problem* (semiopen cavity # 1): Contours of the scattered field (real part; $\alpha = 30°$).

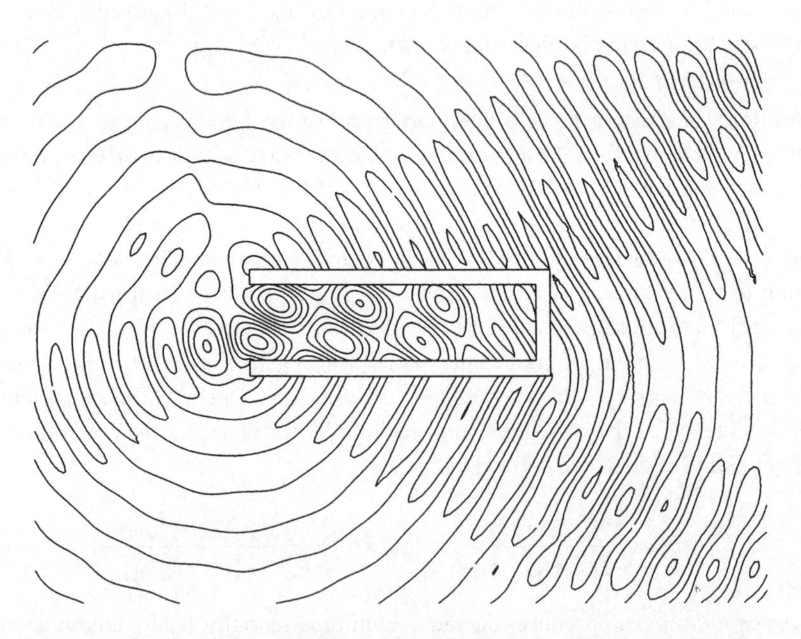

Fig. 7.6. *Second test problem* (semiopen cavity # 1): Contours of the scattered field (imaginary part; $\alpha = 30°$).

triangles, with $\lambda/30$ as the mean length of the edges inside the cavity ($\lambda/20$ outside). We have taken $\Delta t = T/70$. The test problem corresponds to an illuminating wave of incidence $\alpha = 30°$, coming from the right. The contours of the total field related to e^0 are shown on Figure 7.9. Figure 7.10 shows the convergence of the discrete

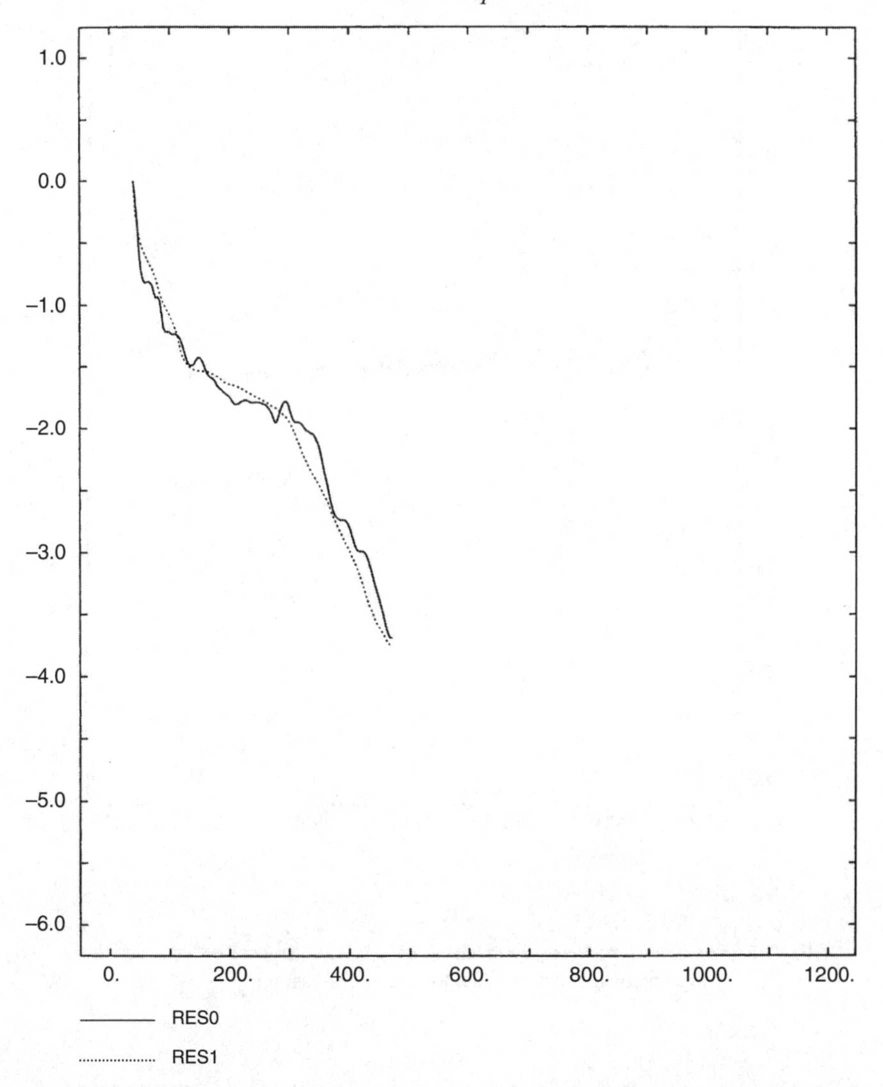

Fig. 7.7. *Second test problem* (semiopen cavity # 1): Convergence of the residuals for the controllability method (residual for y (—), residual for $\frac{\partial y}{\partial t}$ $(\cdots)$).

analogue of $J(\mathbf{e}_k)$, with $J(\cdot)$ defined by (7.15); we have also shown on Figure 7.10 the convergence to zero of the two components of this cost function (the one related to $y(T)$, and the one related to $y_t(T)$). For this "difficult" case the convergence is slower than for the above cavity problem (300 iterations instead of 136). We have shown on Figure 7.11 some details of the finite element triangulation close to the wall at the entrance of the cavity.

Example 3 (scattering by a two-dimensional aircraft like body) The reflector that we consider is defined as the cross section of a Dassault Aviation Falcon 50 by its symmetry plane; we have artificially closed the air-intake in order to enhance reflections inside it. The aircraft length is about 18 m, while its height is 6 m. We take

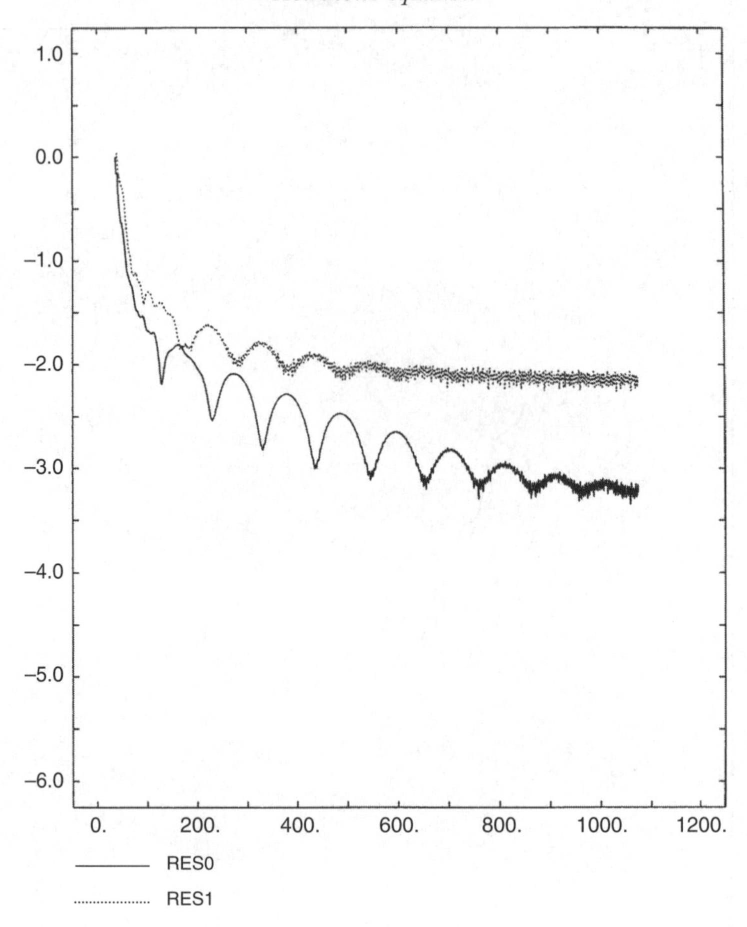

Fig. 7.8. *Second test problem* (semiopen cavity # 1): Convergence of the residuals for asymptotic method (residual for y (—), residual for $\frac{\partial y}{\partial t}$ $(\cdots)$).

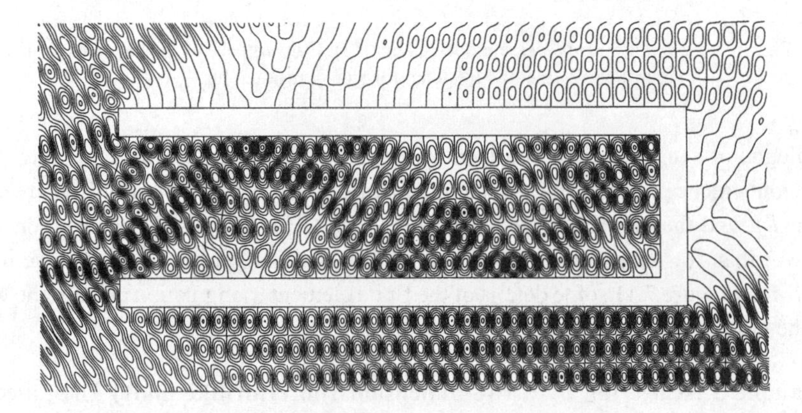

Fig. 7.9. *Second test problem* (semiopen cavity # 2): Contours of the real part of the total field ($\alpha = 30°$).

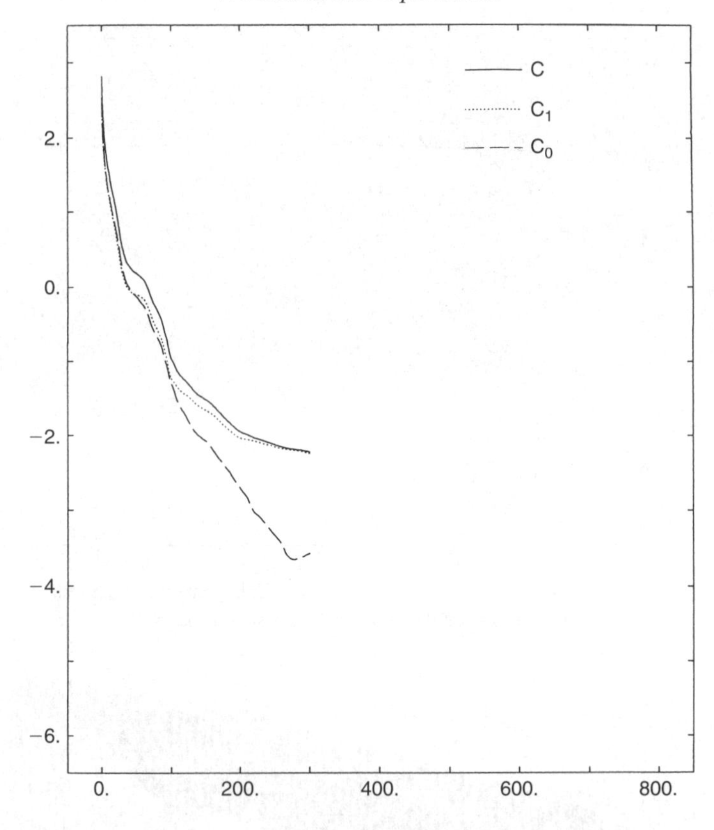

Fig. 7.10. *Second test problem* (semiopen cavity # 2): Convergence of $J(\mathbf{e}_k)$ (—), of the $y(T)$ component of $J(\mathbf{e}_k)$ (- -), and of the $y_t(T)$ component of $J(\mathbf{e}_k)$ $(\cdots)$.

Fig. 7.11. *Second test problem* (semiopen cavity # 2): Enlargement of the finite element mesh close to the cavity intake.

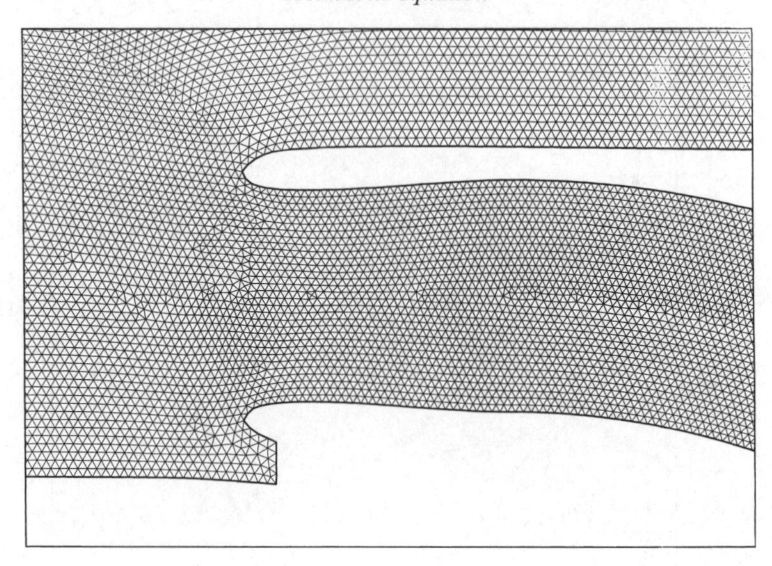

Fig. 7.12. *Third test problem* (aircraft like body): Enlargement of the mesh close to the air intake (by courtesy of Dassault Aviation).

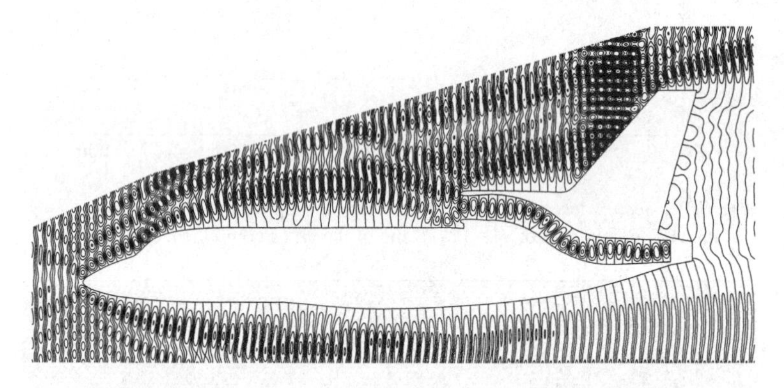

Fig. 7.13. *Third test problem* (aircraft like body): Contours of the real part of the total field ($\alpha = 0°$) (by courtesy of Dassault Aviation).

$f = 0.6$ GHz, so that $\lambda = 0.5$ m. The finite element mesh has 143 850 vertices and 283 873 triangles. Figure 7.12 shows an enlargement of the mesh near the air intake. We have used $\Delta t = T/40$. We consider an illuminating wave coming from the left with $\alpha = 0°$ as angle of incidence. The contours of the real part of the total field are shown on Figure 7.13; we observe the *shadow region* behind the aircraft. The convergence (for $\epsilon = 5 \times 10^{-5}$) of the discrete analogue of algorithm (7.36)–(7.48) is obtained after 260 iterations, corresponding to 90 min of CPU time on a CRAY 2; Figure 7.14 shows the convergence of $J(\mathbf{e}_k)$ to 0 as $k \to +\infty$.

Remark 7.7 For all the test problems discussed above, we have used a direct method based on a *sparse Cholesky solver* to solve the (discrete) *elliptic problem* encountered at each iteration of the discrete analogue of algorithm (7.36)–(7.48). Despite the "respectable" size of these systems (up to 2×10^5 unknowns) this part of the algorithm

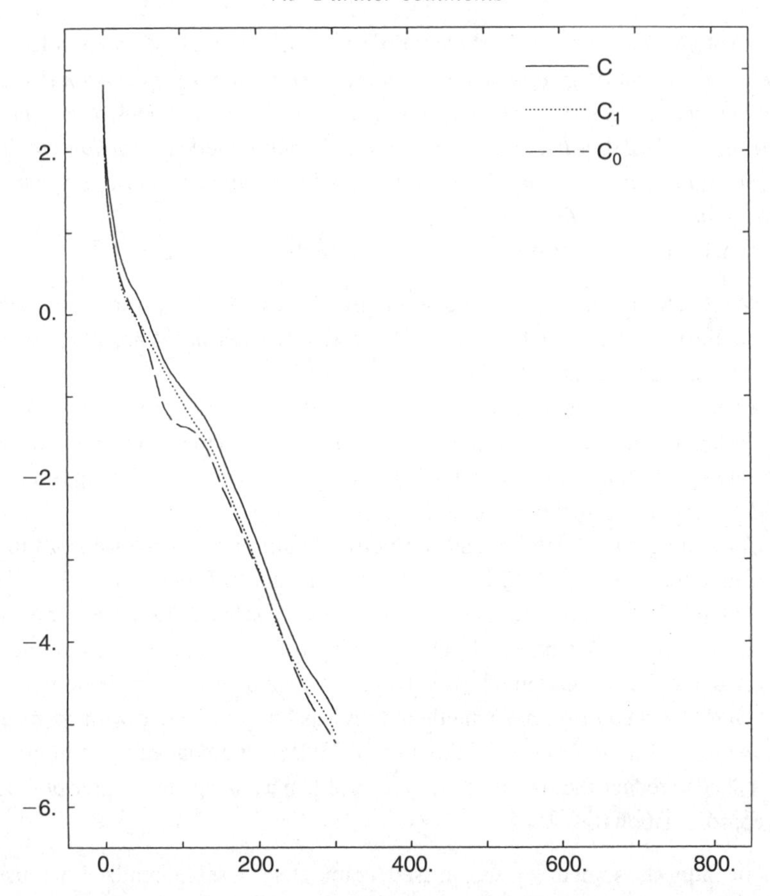

Fig. 7.14. *Third test problem* (aircraft like body): Convergence of $J(\mathbf{e}_k)$ (—), of the $y(T)$ component of $J(\mathbf{e}_k)$ (- -), and of the $y_t(T)$ component of $J(\mathbf{e}_k)$ $(\cdots)$.

never took more than a *few percent* of the total computational effort. Indeed, most of the computational time is spent integrating the *forward* and *backward wave equations*; fortunately, this is the easiest part to *parallelize* (as shown and done in Bristeau, Ehrel, Glowinski, and Périaux, 1993; Bristeau, Ehrel, Feat, Glowinski, and Périaux, 1995) since it is based on an *explicit time-discretization scheme*.

7.9 Further comments. Description of a mixed formulation based variant of the controllability method

7.9.1 Further comments

We have discussed in Sections 7.2–7.8 of this chapter a *controllability* based approach for solving the Helmholtz (and two-dimensional harmonic Maxwell) equations for *large wave numbers* and complicated geometries. The resulting method appears to be more efficient than traditional computational methods which are based on time asymptotic behavior and, or linear algebra algorithms for large indefinite linear systems.

In our *Acta Numerica* 1995 article (Glowinski and Lions, 1995) we wrote:

"The new methodology appears to be promising for the three-dimensional Maxwell equations and for heterogeneous media, including dissipative ones. For very large problems, we shall very likely have to combine the above method with domain decomposition and/or fictitious domain methods, and also to higher order approximations, to reduce the number of grid points."

It seems that our forecasting has been reasonably accurate since:

(i) The controllability based methodology discussed in the previous sections has been applied to the solution of the *three-dimensional harmonic Maxwell equations* in Bristeau (1997).

(ii) An application involving *heterogeneous media* (*scattering* by *coated obstacles*) has been discussed in Bristeau, Dean et al. (1997). In the above publication, the solution method relies on combining the above *controllability* approach with *domain decomposition methods*.

(iii) Combining *controllability* and *fictitious domain methods* is discussed in Bristeau, Girault et al. (1997), Glowinski, Périaux, and J. Toivanen (2003) and Chen et al. (2005) (for an introduction to *fictitious domain methods* see, for example, Glowinski (2003, Chapter 8) and the many references therein, including some related to the simulation of *wave propagation* and *scattering* phenomena).

(iv) "Our" *exact controllability* methodology has been combined with *spectral element approximations* in Heikkola et al. (2005) (the goal of the authors being indeed to reduce the size of the discrete problem by using high degree polynomial approximations).

To be honest, something we missed completely was to realize that using a well-chosen *mixed formulation* we can derive a controllability problem whose least-squares/conjugate gradient solution *does not require* the solution of an elliptic problem at each iteration, as it is the case with algorithm (7.36)–(7.48). This alternative approach will be discussed in the following sections (following Glowinski and Rossi, 2006).

7.9.2 A mixed formulation of the wave problem (7.4)–(7.7)

Let us consider again the wave problem (7.4)–(7.7), namely,

$$u_{tt} - \Delta u = 0 \quad \text{in } Q,$$

$$u = g \text{ on } \sigma, \quad \frac{\partial u}{\partial t} + \frac{\partial u}{\partial n} = 0 \quad \text{on } \Sigma,$$

$$u(0) = u(T), \quad u_t(0) = u_t(T).$$

To obtain a mixed formulation of problem (7.4)–(7.7), we introduce the functions v and $\mathbf{p}$ defined by

$$v = \frac{\partial u}{\partial t}, \qquad \mathbf{p} = \nabla u. \tag{7.49}$$

The pair $\{v, \mathbf{p}\}$ verifies:

$$\frac{\partial v}{\partial t} - \nabla \cdot \mathbf{p} = 0 \text{ in } Q, \quad \frac{\partial \mathbf{p}}{\partial t} - \nabla v = \mathbf{0} \text{ in } Q, \tag{7.50}$$

$$v = \frac{\partial g}{\partial t} \text{ on } \sigma, \quad v + \mathbf{p} \cdot \mathbf{n} = 0 \text{ on } \Sigma_{\text{ext}}, \tag{7.51}$$

$$v(0) = v(T), \quad \mathbf{p}(0) = \mathbf{p}(T); \tag{7.52}$$

in (7.51), $\mathbf{n}$ is the outward unit normal vector at the boundary $\Gamma = \partial\Omega$. The first-order system (7.50) has a Maxwell equation "flavor," albeit simpler. A (mixed) *variational formulation* of this system is given by

$$\int_\Omega \left(\frac{\partial v}{\partial t} - \nabla \cdot \mathbf{p} \right) w \, dx = 0, \quad \forall w \in L^2(\Omega), \text{ a.e. on } (0, T), \tag{7.53}$$

$$\int_\Omega \left(\frac{\partial \mathbf{p}}{\partial t} \cdot \mathbf{q} + v \nabla \cdot \mathbf{q} \right) dx + \int_{\Gamma_{\text{ext}}} (\mathbf{p} \cdot \mathbf{n})(\mathbf{q} \cdot \mathbf{n}) \, d\Gamma = \int_\gamma \frac{\partial g}{\partial t} \mathbf{q} \cdot \mathbf{n} \, d\Gamma,$$
$$\forall \mathbf{q} \in \mathbf{Q}, \text{ a.e. on } (0, T), \tag{7.54}$$

completed by (7.52). In (7.54), the space $\mathbf{Q}$ is defined by

$$\mathbf{Q} = \{ \mathbf{q} \mid \mathbf{q} \in H(\Omega; \text{div}), \ \mathbf{q} \cdot \mathbf{n}|_{\Gamma_{\text{ext}}} \in L^2(\Gamma_{\text{ext}}) \}.$$

We look for solutions $\{v, \mathbf{p}\}$ which belong at least to $L^2(0, T; L^2(\Omega) \times \mathbf{Q}) \cap C^0([0, T]; L^2(\Omega) \times (L^2(\Omega))^d)$, a reasonable assumption if the function $\frac{\partial g}{\partial t}$ is smooth enough.

7.9.3 An exact controllability formulation of problem (7.50)–(7.52)

Let us define the (*control*) space $\mathbf{E}$ by

$$\mathbf{E} = L^2(\Omega) \times (L^2(\Omega))^d, \tag{7.55}$$

equipped with the following scalar product and corresponding norm:

$$(\mathbf{f}, \mathbf{g})_{\mathbf{E}} = \int_\Omega (f_0 g_0 + \mathbf{f}_1 \cdot \mathbf{g}_1) \, dx, \quad \forall \mathbf{f} = \{f_0, \mathbf{f}_1\}, \ \mathbf{g} = \{g_0, \mathbf{g}_1\} \in \mathbf{E}. \tag{7.56}$$

An exact controllability problem (in $\mathbf{E}$) equivalent to (7.50)-(7.52) reads as follows:

Find $\mathbf{e} = \{e_0, \mathbf{e}_1\} \in \mathbf{E}$ *such that*

$$\frac{\partial v}{\partial t} - \nabla \cdot \mathbf{p} = 0 \text{ in } Q, \quad \frac{\partial \mathbf{p}}{\partial t} - \nabla v = \mathbf{0} \text{ in } Q, \tag{7.57}$$

$$v = \frac{\partial g}{\partial t} \text{ on } \sigma, \quad v + \mathbf{p} \cdot \mathbf{n} = 0 \text{ on } \Sigma_{\text{ext}}, \tag{7.58}$$

$$v(0) = e_0, \qquad \mathbf{p}(0) = \mathbf{e}_1, \tag{7.59}$$

implies

$$v(0) = v(T), \quad \mathbf{p}(0) = \mathbf{p}(T). \tag{7.60}$$

A least-squares/conjugate gradient method for the solution of problem (7.57)–(7.60) will be discussed in the following sections.

7.9.4 A least-squares formulation of the controllability problem (7.57)–(7.60)

In order to solve problem (7.57)–(7.60) we introduce the following least-squares formulation

$$\begin{cases} \textit{Find } \mathbf{e} \in \mathbf{E} \textit{ such that} \\ J(\mathbf{e}) \leq J(\mathbf{f}), \quad \forall \mathbf{f} \in \mathbf{E}, \end{cases} \tag{7.61}$$

where

$$J(\mathbf{f}) = \frac{1}{2} \int_\Omega \left(|v(T) - f_0|^2 + |\mathbf{p}(T) - \mathbf{f}_1|^2 \right) \mathrm{d}x, \tag{7.62}$$

the functions v and $\mathbf{p}$ being obtained from $\mathbf{f} = \{f_0, \mathbf{f}_1\}$ via the solution of the following initial value problem:

$$\frac{\partial v}{\partial t} - \boldsymbol{\nabla} \cdot \mathbf{p} = 0 \text{ in } Q, \quad \frac{\partial \mathbf{p}}{\partial t} - \nabla v = \mathbf{0} \text{ in } Q, \tag{7.63}$$

$$v = \frac{\partial g}{\partial t} \text{ on } \sigma, \quad v + \mathbf{p} \cdot \mathbf{n} = 0 \text{ on } \Sigma_{\text{ext}}, \tag{7.64}$$

$$v(0) = f_0, \qquad \mathbf{p}(0) = \mathbf{f}_1. \tag{7.65}$$

In order to solve the least-squares problem (7.61) we will use a *conjugate gradient algorithm* operating in the Hilbert space $\mathbf{E}$ equipped with the scalar product defined by (7.56) and the corresponding norm. The implementation of such an algorithm is greatly facilitated by the knowledge of the *differential* $J'(\mathbf{f})$ of J at $\mathbf{f}$, $\forall \mathbf{f} \in \mathbf{E}$. This issue will be addressed in the following section.

7.9.5 On the computation of $J'(\mathbf{f})$

Using (for example) the *perturbation method* already used in various parts of this book, we can show that the differential $J'(\mathbf{f})$ of J at $\mathbf{f}$ is given by

$$(J'(\mathbf{f}), \mathbf{g})_{\mathbf{E}} = \int_\Omega \left(\left(f_0 - v(T) + v^*(0)\right) g_0 + \left(\mathbf{f}_1 - \mathbf{p}(T) + \mathbf{p}^*(0)\right) \cdot \mathbf{g}_1 \right) \mathrm{d}x,$$

$$\forall \mathbf{f} = \{f_0, \mathbf{f}_1\}, \; \mathbf{g} = \{g_0, \mathbf{g}_1\} \in \mathbf{E}, \tag{7.66}$$

or, equivalently, by

$$(J'(\mathbf{f}),\mathbf{g})_{\mathbf{E}} = \int_\Omega \left(\left(v^*(0) - v^*(T)\right) g_0 + \left(\mathbf{p}^*(0) - \mathbf{p}^*(T)\right) \cdot \mathbf{g}_1 \right) dx,$$

$$\forall \mathbf{f} = \{f_0, \mathbf{f}_1\},\ \mathbf{g} = \{g_0, \mathbf{g}_1\} \in \mathbf{E}, \quad (7.67)$$

where the pair $\{v^*, \mathbf{p}^*\}$ is the unique solution of the following adjoint system:

$$\frac{\partial v^*}{\partial t} - \nabla \cdot \mathbf{p}^* = 0 \text{ in } Q, \quad \frac{\partial \mathbf{p}^*}{\partial t} - \nabla v^* = \mathbf{0} \text{ in } Q, \quad (7.68)$$

$$v^* = 0 \text{ on } \sigma, \quad v^* - \mathbf{p}^* \cdot \mathbf{n} = 0 \text{ on } \Sigma_{\text{ext}}, \quad (7.69)$$

$$v^*(T) = v(T) - f_0, \qquad \mathbf{p}^*(T) = \mathbf{p}(T) - \mathbf{f}_1. \quad (7.70)$$

7.9.6 Conjugate gradient solution of the least-squares problem (7.61)

Suppose that $\mathbf{e} = \{e_0, \mathbf{e}_1\}$ is solution of the least-squares problem (7.61); we have then,

$$\begin{cases} \mathbf{e} \in \mathbf{E}, \\ (J'(\mathbf{e}), \mathbf{f})_{\mathbf{E}} = 0, \quad \forall \mathbf{f} \in \mathbf{E}; \end{cases} \quad (7.71)$$

conversely, any solution of (7.71) solves problem (7.61). Since the operator J' is clearly *affine continuous* over $\mathbf{E}$, with its linear part *positive semidefinite* (at least), problem (7.71) is a *linear variational problem* in the Hilbert space $\mathbf{E}$. From its analogies with the linear variational problems whose conjugate gradient solution is discussed in, for example, Glowinski (2003, Chapter 3), we are going to describe such an algorithm, operating in $\mathbf{E}$ for the solution of problem (7.61), (7.71). This algorithm reads as follows:

Step 0: Initialization.

$$\mathbf{e}^0 \ (= \{e_0^0, \mathbf{e}_1^0\}) \text{ is given in } \mathbf{E}; \quad (7.72)$$

solve

$$\frac{\partial v^0}{\partial t} - \nabla \cdot \mathbf{p}^0 = 0 \text{ in } Q, \quad \frac{\partial \mathbf{p}^0}{\partial t} - \nabla v^0 = \mathbf{0} \text{ in } Q, \quad (7.73)$$

$$v^0 = \frac{\partial g}{\partial t} \text{ on } \sigma, \quad v^0 + \mathbf{p}^0 \cdot \mathbf{n} = 0 \text{ on } \Sigma_{\text{ext}}, \quad (7.74)$$

$$v^0(0) = e_0^0, \qquad \mathbf{p}^0(0) = \mathbf{e}_1^0, \quad (7.75)$$

and then the adjoint system

$$\frac{\partial v^{*0}}{\partial t} - \nabla \cdot \mathbf{p}^{*0} = 0 \text{ in } Q, \quad \frac{\partial \mathbf{p}^{*0}}{\partial t} - \nabla v^{*0} = \mathbf{0} \text{ in } Q, \tag{7.76}$$

$$v^{*0} = 0 \text{ on } \sigma, \quad v^{*0} - \mathbf{p}^{*0} \cdot \mathbf{n} = 0 \text{ on } \Sigma_{\text{ext}}, \tag{7.77}$$

$$v^{*0}(T) = v^0(T) - e_0^0, \qquad \mathbf{p}^{*0}(T) = \mathbf{p}^0(T) - \mathbf{e}_1^0. \tag{7.78}$$

Next, define $\mathbf{g}^0 = \{g_0^0, \mathbf{g}_1^0\}$ *and* $\mathbf{w}^0 = \{w_0^0, \mathbf{w}_1^0\}$ *by*

$$g_0^0 = v^{*0}(0) - v^{*0}(T), \quad \mathbf{g}_1^0 = \mathbf{p}^{*0}(0) - \mathbf{p}^{*0}(T), \tag{7.79}$$

and

$$\mathbf{w}^0 = \mathbf{g}^0, \tag{7.80}$$

respectively.

For $n \geq 0$, *assuming that* $\mathbf{e}^n$, $\mathbf{g}^n$, *and* $\mathbf{w}^n$ *are known, the last two different from* $\mathbf{0}$, *we compute* $\mathbf{e}^{n+1}$, $\mathbf{g}^{n+1}$, *and if necessary,* $\mathbf{w}^{n+1}$ *as follows:*

Step 1: Descent.
Solve

$$\frac{\partial \bar{v}^n}{\partial t} - \nabla \cdot \bar{\mathbf{p}}^n = 0 \text{ in } Q, \quad \frac{\partial \bar{\mathbf{p}}^n}{\partial t} - \nabla \bar{v}^n = \mathbf{0} \text{ in } Q, \tag{7.81}$$

$$\bar{v}^n = 0 \text{ on } \sigma, \quad \bar{v}^n + \bar{\mathbf{p}}^n \cdot \mathbf{n} = 0 \text{ on } \Sigma_{\text{ext}}, \tag{7.82}$$

$$\bar{v}^n(0) = w_0^n, \qquad \bar{\mathbf{p}}^n(0) = \mathbf{w}_1^n, \tag{7.83}$$

and then the adjoint system

$$\frac{\partial \bar{v}^{*n}}{\partial t} - \nabla \cdot \bar{\mathbf{p}}^{*n} = 0 \text{ in } Q, \quad \frac{\partial \bar{\mathbf{p}}^{*n}}{\partial t} - \nabla \bar{v}^{*n} = \mathbf{0} \text{ in } Q, \tag{7.84}$$

$$\bar{v}^{*n} = 0 \text{ on } \sigma, \quad \bar{v}^{*n} - \bar{\mathbf{p}}^{*n} \cdot \mathbf{n} = 0 \text{ on } \Sigma_{\text{ext}}, \tag{7.85}$$

$$\bar{v}^{*n}(T) = \bar{v}^n(T) - w_0^n, \qquad \bar{\mathbf{p}}^{*n}(T) = \bar{\mathbf{p}}^n(T) - \mathbf{w}_1^0. \tag{7.86}$$

Define then $\bar{\mathbf{g}}^n = \{\bar{g}_0^n, \bar{\mathbf{g}}_1^n\}$ *by*

$$\bar{g}_0^n = \bar{v}^{*n}(0) - \bar{v}^{*n}(T), \quad \bar{\mathbf{g}}_1^n = \bar{\mathbf{p}}^{*n}(0) - \bar{\mathbf{p}}^{*n}(T), \tag{7.87}$$

and compute

$$\rho_n = \frac{(\mathbf{g}^n, \mathbf{g}^n)_{\mathbf{E}}}{(\bar{\mathbf{g}}^n, \mathbf{w}^n)_{\mathbf{E}}}, \tag{7.88}$$

$$\mathbf{e}^{n+1} = \mathbf{e}^n - \rho_n \mathbf{w}^n, \tag{7.89}$$

$$\mathbf{g}^{n+1} = \mathbf{g}^n - \rho_n \bar{\mathbf{g}}^n. \tag{7.90}$$

Step 2: Testing of the convergence. Construction of the new descent direction.
If $(\mathbf{g}^{n+1}, \mathbf{g}^{n+1})_{\mathbf{E}} / (\mathbf{g}^0, \mathbf{g}^0)_{\mathbf{E}} \leq \epsilon^2$, *take* $\mathbf{e} = \mathbf{e}^{n+1}$; *else compute*

$$\gamma_n = \frac{(\mathbf{g}^{n+1}, \mathbf{g}^{n+1})_{\mathbf{E}}}{(\mathbf{g}^n, \mathbf{g}^n)_{\mathbf{E}}}, \tag{7.91}$$

$$\mathbf{w}^{n+1} = \mathbf{g}^{n+1} + \gamma_n \mathbf{w}^n. \tag{7.92}$$

Do $n = n + 1$ *and return to (7.81)–(7.83).*

7.9.7 Some remarks concerning the implementation of the controllability approach

Remark 7.8 The conjugate gradient algorithm (7.72)–(7.92) is particularly easy to implement, using for example the *lowest-order Raviart–Thomas mixed finite element approximation* (and may be also the one next to the lowest order). No (complicated) preconditioning is needed since we operate in $((L^2(\Omega))^{d+1}$. Actually, we can always use over $\mathbf{E}_h$ approximating $\mathbf{E}$ a scalar product, obtained by numerical integration, associated with a diagonal matrix. Two basic references concerning *mixed finite element approximations* are Brezzi and Fortin (1991) and Roberts and Thomas (1991).

Remark 7.9 The *time integration* is particularly easy if one uses a *staggered mesh*. Suppose that $\Delta t = T/N$ and denote $n\Delta t$ by t^n and $(n + \frac{1}{2})\Delta t$ by $t^{n+\frac{1}{2}}$. We can for example discretize v over the set $\{t^n\}_{n=0}^N$ and $\mathbf{p}$ over the set $\{t^{n+\frac{1}{2}}\}_{n=0}^N$ (or the other way around) and use t^0 $(= 0)$ and t^N $(= T)$ to impose, respectively, the initial and final conditions on v, while (and without loss of accuracy) the initial and final conditions on $\mathbf{p}$ will be imposed at $t^{\frac{1}{2}}$ $(= \frac{1}{2}\Delta t)$ and $t^{N+\frac{1}{2}}$ $(= T + \frac{1}{2}\Delta t)$, respectively. The coupled system can be easily discretized using a centered second-order accurate scheme which is "almost" explicit (almost only since we will use a Crank–Nicolson type scheme to treat the radiation condition in the $\mathbf{p}$-equation). We can expect a *stability condition* such as $\Delta t \leq Ch$ (which is classical for this type of problems).

7.10 A final comment

In Périaux, Mantel, and Chen (1997) and Bristeau et al. (1999) one has combined the *controllability method* discussed in this chapter with *genetic algorithms*, in order to solve *direct* and *inverse scattering problems*. Some of these problems concern the *reduction* of the *Radar Cross Section* of airfoil shaped reflectors.

8

Other wave and vibration problems. Coupled systems

8.1 Generalities and further references

In Chapters 6 and 7, we have discussed *controllability* issues concerning *wave equations* such as

$$u_{tt} - c^2 \Delta u = 0; \tag{8.1}$$

a basic tool for studying exact or approximate controllability for equations such as (8.1) has been the *Hilbert Uniqueness Method* (HUM). Actually, HUM has been applied in Lagnese (1989) to prove the exact boundary controllability of the *Maxwell equations*

$$\varepsilon \frac{\partial \mathbf{E}}{\partial t} - \nabla \times \mathbf{H} = \mathbf{0}, \ \ \mu \frac{\partial \mathbf{H}}{\partial t} + \nabla \times \mathbf{E} = \mathbf{0} \quad \text{in } Q, \tag{8.2}$$

$$\nabla \cdot \mathbf{E} = 0, \ \ \nabla \cdot \mathbf{H} = 0 \quad \text{in } Q \tag{8.3}$$

(see also Bensoussan, 1990); other related references are Kapitanov (1994), Nicaise (2000), Kapitanov and Perla-Menzela (2003). To the best of our knowledge, most computational aspects still have to be explored.

The HUM has been applied in J.L. Lions (1988b) and Lagnese and Lions (1988) to the exact or approximate controllability of systems (mostly from Elasticity) modeled by *Petrowskys type equations*. Concerning the numerical application of HUM to the exact controllability of those Petrowsky-type equations modeling *elastic shells* vibrations we refer to Marini, Testa, and Valente (1994). *Linear elastic beams* are mechanical systems whose vibrations are modeled by Petrowsky type equations as shown in, for example, Lagnese and Lions (1988). Considering the important role played by beams in various areas of the engineering sciences, it is not surprising that the *control of their vibrations* has motivated the work of many investigators starting with the pioneering work of Lagnese and Lions (1988) and Littman and Markus (1988); concerning related numerical investigations let us mention Carlsson (1991), where HUM is numerically implemented in order to solve the controllability problems discussed in Lagnese and Lions (1988) and Littman and Markus (1988), using either the *Euler–Bernouilli* or the *Mindlin–Timoshenko* models describing the low amplitude motions of elastic beams; other references worth mentioning (among many others) concerning the control of beams and related structures are Tucsnak (1996), León and Zuazua (2002) and the many publications by F. Bourquin and his collaborators, among them Bourquin (1994), (1995), (2001), Bourquin, Collet, and Joly (1999), Briffaut (1999), and Collet, Bourquin, Joly, and Ratier (2004). Actually, a most interesting discussion concerning a *pre-HUM* approach to the exact controllability of *linear elastic beams* can be found in Ball, Marsden, and Slemrod (1982) (see also Section 6.7 of Marsden and Hughes, 1983 and 1994); the exact controllability problem under consideration in the above three references reads as follows:

Find a control function p, depending of t only, such that

$$w_{tt} + w_{xxxx} + p\, w_{xx} = 0 \quad \text{in } (0,1) \times (0,T), \tag{8.4}$$

$$w(0,t) = w_{xx}(0,t) = 0, \quad w(1,t) = w_{xx}(1,t) = 0 \quad \text{in } (0,T), \tag{8.5}$$

$$w(0) = w_0, \quad w_t(0) = w_1 \tag{8.6}$$

implies

$$w(T) = z_0, \quad w_t(T) = z_1. \tag{8.7}$$

Relation (8.4) corresponds to the *Euler–Bernouilli* linear beam model (written in nondimensional form). A main difference between the exact controllability problem (8.4)–(8.7) and those discussed in the previous chapters of the present book is the *nonlinear dependence* between w and p despite the fact that the state system is linear with respect to w. The controllability properties of the (*bilinear*) system (8.4)–(8.6) are discussed in detail in Ball, Marsden, and Slemrod (1982).

Remark 8.1 System (8.4)–(8.6) is prototypical of those systems known as *bilinear distributed systems*, the control of such systems being called *bilinear control* by most control specialists. As shown below, bilinear control is a most natural approach concerning the control of *Schrödinger equation*. □

The controllability of *linear elastic plates* is another popular issue related to Petrowskys type equations; it has been systematically addressed in Lagnese and Lions (1988) and then for various types of controls in Jaffard (1988), Zuazua (1988), Haraux (1989), Haraux and Jaffard (1991), and Burq (1993); numerical aspects are considered in Bourquin, Namar, and Urquiza (1998). Actually, mathematical results and techniques concerning the *controllability of linear plates* can be (and have been) used to address the controllability of systems modeled by *Schrödinger equations*. Showing the relationships between *linear plate models* and *Schrödinger equations* is quite easy. Indeed, let us consider a single particle moving into space; if no external force acts on the particle its *wave function* ψ verifies the following *Schrödinger equation* (written in nondimensional form):

$$\mathrm{i}\,\frac{\partial \psi}{\partial t} + \Delta \psi = 0. \tag{8.8}$$

Time-differentiating both sides of (8.8) and combining with (8.8) we obtain that ψ is also a solution of

$$\frac{\partial^2 \psi}{\partial t^2} + \Delta^2 \psi = 0. \tag{8.9}$$

If ψ is a real valued function of t and of two space variables, equation (8.9) is nothing but the celebrated *Kirchoff equation* (written in nondimensional form) modeling the vibrations of thin elastic plates (see, for example, Lagnese and Lions, 1988) for its derivation). Suppose now that one wishes to control the evolution of the above wave function ψ; a sensible way to reach that goal is to introduce a well-chosen *potential* v. The Schrödinger equation takes then the following form:

$$\mathrm{i}\,\frac{\partial \psi}{\partial t} + \Delta \psi = v\,\psi; \tag{8.10}$$

this shows that *bilinear control* is the right way to control physical and chemical systems modeled by Schrödinger equations (including these many situations where the Schrödinger equation includes a given potential V; we have then to replace (8.8) by

$$\mathrm{i}\,\frac{\partial \psi}{\partial t} + \Delta \psi = V\,\psi$$

and (8.10) by

$$\mathrm{i}\,\frac{\partial \psi}{\partial t} + \Delta \psi = (V + v)\,\psi).$$

The control of Schrödinger equations is part of a larger research area known as *quantum control*. Quantum control has known a flurry of activity these last two decades; this is not surprising considering the importance of quantum modeling in material science, drug design, molecular biology, and so on. The number of related publications is so large that the interested reader will have to go to Internet if he or she wants to have an idea of the existing literature on this highly popular topic. Focusing on those publications we are aware of, let us mention. Peirce, Dahleh, and Rabitz (1988), Machtyngier (1990), (1994), Lasiecka and Triggiani (1992), Lebeau (1992), Machtyngier and Zuazua (1994), Brumer and Shapiro (1995), Shapiro and Brumer (1997), Kime (1995), Allibert (1998), Zhu and Rabitz (1998), Phung (2001), Turinici and Rabitz (2001), Li, Turinici et al. (2002), Maday and Turinici (2003), Rabitz, Turinici, and Brown (2003), and Zuazua (2003), (2005, pp. 231, 232) (see, of course, the many references therein, particularly in the last two publications, which are, among other things, review articles).

In Glowinski and Lions (1995, p. 312) we wrote,

> Finally, very little is known about the exact or approximate controllability of those (nonlinear) wave (or Petrowskys type) equations modeling the vibrations of nonlinear systems; we intend, however, to explore the solution of these problems in the near future.

The above quotation calls for several comments:

(i) When Glowinski and Lions (1995) appeared, it should have mentioned (among very few others) Cirina (1969), Fattorini (1975), Chewning (1976), Lasiecka and Triggiani (1991), Lasiecka and Tataru (1993), and Zuazua (1990a,b), (1993a).

(ii) The situation depicted in the quotation has not changed much this past ten years; let us mention, however, Dehman, Lebeau, and Zuazua (2003), and Li and Zhang (1998), Li and Rao (2000), (2002), (2003), Li, Rao, and Yi (2001), Li and Xu (2003).

(iii) When we mentioned, in Glowinski and Lions (1995), our intention to explore the solution of controllability problems for nonlinear vibrating systems, we had in mind (as far as we remember) their numerical solution. The boundary controllability of the *Von Kármán equations* modeling the vibrations of nonlinear elastic thin plates was part of these objectives. Actually, the realization of this ambitious project never took place, since the authors of this book started putting at the time most of their research effort on *flow control* (see Part III, Chapter 9 for the *control of incompressible viscous flow* modeled by the *Navier–Stokes equations*).

8.2 Coupled Systems (I): a problem from thermo-elasticity

In Chapters 1–7 we have discussed controllability issues for *diffusion* and *wave* equations. The control of systems obtained by the coupling of *different types* of equations brings new difficulties which are worth discussing, justifying therefore the present section. The *numerical aspects* will not be addressed here, but in our opinion this section can be a starting point for investigations in this direction.

In this section, we will focus on the controllability of a simplified system from *Thermo-Elasticity* but it is likely that the techniques described here can be applied to systems modeled by more complicated equations.

8.2.1 Formulation of the problem

Let Ω be a *bounded* domain of $\mathbb{R}^d$, $d \leq 3$, with a smooth boundary Γ. Motivated by applications from *Thermo-Elasticity* we consider the following system:

$$\frac{\partial^2 \mathbf{y}}{\partial t^2} - \Delta \mathbf{y} + \alpha \nabla \theta = \mathbf{0} \quad \text{in } Q\,(= \Omega \times (0, T)), \tag{8.11}$$

$$\frac{\partial \theta}{\partial t} - \Delta \theta + \alpha \nabla \cdot \frac{\partial \mathbf{y}}{\partial t} = 0 \quad \text{in } Q, \tag{8.12}$$

where $\mathbf{y} = \{y_i\}_{i=1}^d$ and $\alpha \geq 0$. In (8.11), (8.12), $\mathbf{y}$ (respectively θ) denotes an *elastic displacement* (respectively a *temperature*) function of x and t. Scaling has been made so that the constants in front of $-\Delta$ are equal to 1 in both equations. The *initial conditions* are

$$\mathbf{y}(0) = \mathbf{0}, \quad \frac{\partial \mathbf{y}}{\partial t}(0) = 0, \tag{8.13}$$

$$\theta(0) = 0. \tag{8.14}$$

We suppose that the control is on the boundary of Ω, actually on $\Gamma_0 \subset \Gamma$. Moreover, it is applied only on the component $\mathbf{y}$ of the state vector $\{\mathbf{y}, \theta\}$.

Concerning the *boundary conditions*, we shall consider the two following cases:

Case I

$$\mathbf{y} = \begin{cases} \mathbf{v} & \text{on } \Sigma_0 = \Gamma_0 \times (0,T), \text{ with } \mathbf{v} = \{v_i\}_{i=1}^d, \\ \mathbf{0} & \text{on } \Sigma \backslash \Sigma_0, \text{ with } \Sigma = \Gamma \times (0,T), \end{cases} \tag{8.15}$$

and

$$\theta = \theta_0 \quad \text{on } \Sigma. \tag{8.16}$$

Case II

$$\frac{\partial \mathbf{y}}{\partial n} = \begin{cases} \mathbf{v} & \text{on } \Sigma_0, \\ \mathbf{0} & \text{on } \Sigma \backslash \Sigma_0 \end{cases} \tag{8.17}$$

with (8.16) unchanged.

Remark 8.2 One can consider a variety of other types of boundary conditions and controls. The corresponding problems can be treated by methods very close to those given below.

Remark 8.3 In order to simplify the proofs and formulae below, we shall take

$$\theta_0 = 0, \tag{8.18}$$

but this is just a technical detail. □

In the following sections, we will study the spaces described by $\mathbf{y}(T)$, $\frac{\partial \mathbf{y}}{\partial t}(T)$ and $\theta(T)$; we will show that under "reasonable" conditions, one can control $\mathbf{y}(T)$ and $\frac{\partial \mathbf{y}}{\partial t}(T)$ but not $\theta(T)$.

Remark 8.4 Controllability for equations (8.11)–(8.15) has been studied in J.L. Lions (1988b, Volume 2) (see also Narukawa, 1983). We follow here a slightly different approach, our goal being here to obtain constructive approximation methods.

8.2.2 The limit cases $\alpha \to 0$ and $\alpha \to +\infty$

In order to obtain a better understanding of the problem under consideration, it is worthwhile to look at the limit cases $\alpha \to 0$ and $\alpha \to +\infty$. Moreover, they have intrinsic mathematical interest, particularly when $\alpha \to +\infty$.

8.2.2.1 The case $\alpha \to 0$

This case is simple. The coupled system (8.11)–(8.16) (or its variant (8.11)–(8.14), (8.16), (8.17)) reduces to *uncoupled* wave and heat equations. The control acts only on the $\mathbf{y}$ component of the state vector $\{\mathbf{y}, \theta\}$; we have then

$$\begin{cases} \dfrac{\partial^2 \mathbf{y}}{\partial t^2} - \Delta \mathbf{y} = \mathbf{0} \quad \text{in } Q, \\ \mathbf{y}(0) = \mathbf{0}, \quad \dfrac{\partial \mathbf{y}}{\partial t}(0) = \mathbf{0}, \\ \mathbf{y} = \mathbf{v} \quad \text{on } \Sigma_0, \quad \mathbf{y} = \mathbf{0} \quad \text{on } \Sigma \backslash \Sigma_0. \end{cases} \tag{8.19}$$

Since the Laplace operator Δ is *diagonal* we have recovered cases discussed in Chapter 6.

Remark 8.5 The *general linear elasticity* system (with the operator Δ replaced by

$$\mathbf{z} \to \lambda \Delta \mathbf{z} + \mu \nabla (\nabla \cdot \mathbf{z}),$$

λ and μ being Lamé coefficients) would lead to similar considerations, with more complicated technical details.

Remark 8.6 Similar considerations apply when (8.15) is replaced by (8.17).

8.2.2.2 The case $\alpha \to +\infty$ (with boundary conditions (8.15))

We shall assume (this is *necessary* for what follows) that

$$\int_{\Gamma_0} \mathbf{v} \cdot \mathbf{n} \, d\Gamma = 0. \tag{8.20}$$

Then, assuming that $\mathbf{v}$ is smooth enough (a condition which does not restrict the generality, since we are going to consider *approximate controllability*) and

$$\mathbf{v} = \mathbf{0} \text{ on } \partial \Gamma_0 \times (0, T), \quad \mathbf{v}|_{t=0} = \mathbf{0}, \quad \left. \frac{\partial \mathbf{v}}{\partial t} \right|_{t=0} = \mathbf{0}, \tag{8.21}$$

one can construct a vector-valued function ϕ such that

$$\begin{cases} \phi \text{ is smooth in } \bar{\Omega} \times (0, T), \quad \nabla \cdot \phi = 0 \text{ in } \Omega \times (0, T), \\ \phi = \mathbf{v} \text{ on } \Sigma_0, \quad \phi = \mathbf{0} \text{ on } \Sigma \backslash \Sigma_0. \end{cases} \tag{8.22}$$

Then, if we introduce $\mathbf{z} = \mathbf{y} - \phi$, we obtain

$$\frac{\partial^2 \mathbf{z}}{\partial t^2} - \Delta \mathbf{z} + \alpha \nabla \theta = -\left(\frac{\partial^2 \phi}{\partial t^2} - \Delta \phi \right) \; (\equiv \mathbf{f}) \text{ in } Q, \tag{8.23}$$

$$\frac{\partial \theta}{\partial t} - \Delta \theta + \alpha \nabla \cdot \frac{\partial \mathbf{z}}{\partial t} = 0 \text{ in } Q, \tag{8.24}$$

$$\mathbf{z} = \mathbf{0} \text{ and } \theta = 0 \text{ on } \Sigma, \quad \mathbf{z}(0) = \mathbf{0} \; \frac{\partial \mathbf{z}}{\partial t}(0) = \mathbf{0}, \quad \theta(0) = 0. \tag{8.25}$$

Multiply now both sides of (8.23) (respectively (8.24)) by $\frac{\partial \mathbf{z}}{\partial t}$ (respectively, θ). We obtain with obvious notation ($\|\cdot\| = \|\cdot\|_{L^2(\Omega)}$) that

$$\frac{1}{2}\frac{\mathrm{d}}{\mathrm{d}t}\left(\left\|\frac{\partial \mathbf{z}}{\partial t}\right\|^2 + \|\nabla \mathbf{z}\|^2 + \|\theta\|^2\right) + \|\nabla\theta\|^2$$

$$+\,\alpha\left(\nabla\theta, \frac{\partial \mathbf{z}}{\partial t}\right) + \alpha\left(\nabla\cdot\frac{\partial \mathbf{z}}{\partial t}, \theta\right) = \left(\mathbf{f}, \frac{\partial \mathbf{z}}{\partial t}\right). \tag{8.26}$$

But

$$\left(\nabla\theta, \frac{\partial \mathbf{z}}{\partial t}\right) + \left(\nabla\cdot\frac{\partial \mathbf{z}}{\partial t}, \theta\right) = 0,$$

so that relations (8.25) and (8.26) lead to *a priori estimates which are independent of* α. It follows then from these estimates that if we denote by $\{\mathbf{z}_\alpha, \theta_\alpha\}$ the solution of (8.23)–(8.25) one has when $\alpha \to +\infty$

$$\begin{cases} \left\{\mathbf{z}_\alpha, \dfrac{\partial \mathbf{z}_\alpha}{\partial t}\right\} \to \left\{\mathbf{z}, \dfrac{\partial \mathbf{z}}{\partial t}\right\} \text{ weakly* in } L^\infty(0,T;H_0^1(\Omega)\times L^2(\Omega)), \\ \theta_\alpha \to 0 \text{ weakly in } L^2(0,T;H_0^1(\Omega)) \text{ and weakly* in } L^\infty(0,T;L^2(\Omega)) \end{cases} \tag{8.27}$$

Returning to the notation $\{\mathbf{y}, \theta\}$, we have that $\mathbf{y}_\alpha \to \mathbf{y}$, where $\mathbf{y}$ is the solution of

$$\begin{cases} \dfrac{\partial^2 \mathbf{y}}{\partial t^2} - \Delta\mathbf{y} + \nabla p = \mathbf{0} \text{ in } Q, \quad \nabla\cdot\mathbf{y} = 0 \text{ in } Q, \\ \mathbf{y}(0) = \mathbf{0}, \quad \dfrac{\partial \mathbf{y}}{\partial t}(0) = \mathbf{0}, \quad \mathbf{y} = \mathbf{v} \text{ on } \Sigma_0, \quad \mathbf{y} = \mathbf{0} \text{ on } \Sigma\backslash\Sigma_0. \end{cases} \tag{8.28}$$

We clearly see why the flux relation (8.20) is necessary (from the *divergence theorem*). We observe also that at the limit when $\alpha \to +\infty$, since $\theta \equiv 0$ the best we can hope is the controllability of $\{\mathbf{y}(T), \frac{\partial \mathbf{y}}{\partial t}(T)\}$, but not, of course, the controllability of $\theta(T)$ (unless, one is looking for $\theta(T) = 0$ (null-controllability)).

Remark 8.7 One can find in J.L. Lions (1992b) a discussion of the controllability of system (8.28) under strict geometrical conditions.

Remark 8.8 Similar results hold when (8.15) is replaced by (8.17). However, in that case no additional condition such as (8.20) is needed.

8.2.3 Approximate partial controllability

We now return to the case $0 < \alpha < +\infty$. We assume that the following property holds:

$$\begin{cases} \text{when } \mathbf{v} \text{ spans } (L^2(\Sigma_0))^d, \text{ then } \left\{\mathbf{y}(T;\mathbf{v}), \dfrac{\partial \mathbf{y}}{\partial t}(T;\mathbf{v})\right\} \\ \text{spans a } \textit{dense} \text{ subset of } (L^2(\Omega)\times H^{-1}(\Omega))^d, \end{cases} \tag{8.29}$$

where $\{\mathbf{y}(\mathbf{v}), \theta(\mathbf{v})\}$ is the solution of (8.11)–(8.15).

Remark 8.9 *Sufficient* conditions for (8.29) to be true are given in Chapter 1 of J.L. Lions (1988b, Volume 2); they are of the following type:

(i) Σ_0 is "sufficiently large".
(ii) $0 < \alpha < \alpha_0$.

Necessary and sufficient conditions for (8.29) to be true were not known at the time of Glowinski and Lions (1995), the most complete results in that direction, known by the authors at that time, being those in Zuazua (1993b) and (1995). □

We can then consider the following optimal control problem:

$$\begin{cases} \inf_{\mathbf{v}} \dfrac{1}{2}\displaystyle\int_{\Sigma_0} |\mathbf{v}|^2 \, d\Sigma, \quad \mathbf{v} \in (L^2(\Sigma_0))^d \text{ such that} \\ \mathbf{y}(T;\mathbf{v}) \in \mathbf{z}^0 + \beta_0 B, \quad \dfrac{\partial \mathbf{y}}{\partial t}(T;\mathbf{v}) \in \mathbf{z}^1 + \beta_1 B_{-1}, \end{cases} \tag{8.30}$$

where B (respectively B_{-1}) denotes the unit ball of $(L^2(\Omega))^d$ (respectively of $(H^{-1}(\Omega))^d$).

The control problem (8.30) has a unique solution; it can be characterized by a *variational inequality* which can be obtained either directly or by duality methods. Here we use duality, because (among other things) it will be convenient for the next section (where we introduce penalty arguments).

Formulation of a dual problem We follow the same approach as in Chapter 6, Section 6.4. We define thus an operator $\mathbf{L}$ from $(L^2(\Sigma_0))^d$ into $(H^{-1}(\Omega))^d \times (L^2(\Omega))^d$ by

$$\mathbf{L}\mathbf{v} = \left\{ -\frac{\partial \mathbf{y}}{\partial t}(T;\mathbf{v}), \mathbf{y}(T;\mathbf{v}) \right\}. \tag{8.31}$$

We define next F_1 and F_2 by

$$F_1(\mathbf{v}) = \frac{1}{2}\int_{\Sigma_0} |\mathbf{v}|^2 \, d\Sigma, \tag{8.32}$$

and, with $\mathbf{g} = \{\mathbf{g}^0, \mathbf{g}^1\}$,

$$F_2(\mathbf{g}) = \begin{cases} 0 & \text{if } \mathbf{g}^0 \in -\mathbf{z}^1 + \beta_1 B_{-1} \text{ and } \mathbf{g}^1 \in \mathbf{z}^0 + \beta_0 B, \\ +\infty & \text{otherwise.} \end{cases} \tag{8.33}$$

The control problem (8.30) is then equivalent to

$$\inf_{\mathbf{v} \in (L^2(\Sigma_0))^d} [F_1(\mathbf{v}) + F_2(\mathbf{L}\mathbf{v})]. \tag{8.34}$$

By *duality*, we obtain

$$\inf_{\mathbf{v}\in(L^2(\Sigma_0))^d} \{F_1(\mathbf{v}) + F_2(\mathbf{L}\mathbf{v})\}$$
$$= - \inf_{\mathbf{g}\in(H_0^1(\Omega))^d\times(L^2(\Omega))^d} \{F_1^*(\mathbf{L}^*\mathbf{g}) + F_2^*(-\mathbf{g})\}, \tag{8.35}$$

with

$$F_1^*(\mathbf{v}) = \frac{1}{2}\int_{\Sigma_0} |\mathbf{v}|^2 \, d\Sigma, \quad \forall \mathbf{v} \in (L^2(\Omega))^d \tag{8.36}$$

and (with the notation essentially like in Chapter 6, Section 6.4)

$$F_2^*(\mathbf{g}) = -\langle \mathbf{z}^1, \mathbf{g}^0\rangle + \int_\Omega \mathbf{z}^0 \cdot \mathbf{g}^1 \, dx + \beta_1 \left\|\mathbf{g}^0\right\|_{(H_0^1(\Omega))^d} + \beta_0 \left\|\mathbf{g}^1\right\|_{(L^2(\Omega))^d},$$
$$\forall \mathbf{g} = \{\mathbf{g}^0, \mathbf{g}^1\} \in (H_0^1(\Omega))^d \times (L^2(\Omega))^d. \tag{8.37}$$

The operator L^* is defined as follows: We introduce $\{\boldsymbol{\varphi}_\mathbf{g}, \psi_\mathbf{g}\}$ solution of

$$\begin{cases} \dfrac{\partial^2 \boldsymbol{\varphi}_\mathbf{g}}{\partial t^2} - \Delta \boldsymbol{\varphi}_\mathbf{g} + \alpha \nabla \dfrac{\partial \psi_\mathbf{g}}{\partial t} = 0 & \text{in } Q, \\ -\dfrac{\partial \psi_\mathbf{g}}{\partial t} - \Delta \psi_\mathbf{g} - \alpha \nabla \cdot \boldsymbol{\varphi}_\mathbf{g} = 0 & \text{in } Q, \end{cases} \tag{8.38}$$

$$\boldsymbol{\varphi}_\mathbf{g}(T) = \mathbf{g}^0 \in (H_0^1(\Omega))^d, \quad \frac{\partial \boldsymbol{\varphi}_\mathbf{g}}{\partial t}(T) = \mathbf{g}^1 \in (L^2(\Omega))^d, \quad \psi_\mathbf{g}(T) = 0, \tag{8.39}$$

$$\boldsymbol{\varphi}_\mathbf{g} = \mathbf{0} \quad \text{and} \quad \psi_\mathbf{g} = 0 \text{ on } \Sigma. \tag{8.40}$$

Then if $\mathbf{g} = \{\mathbf{g}^0, \mathbf{g}^1\}$, we have

$$L^*\mathbf{g} = \frac{\partial \boldsymbol{\varphi}_\mathbf{g}}{\partial n} \quad \text{on } \Sigma_0. \tag{8.41}$$

We obtain thus as dual problem (that is, for the minimization problem in the right-hand side of (8.35))

$$\inf_\mathbf{g} \left\{ \frac{1}{2}\int_{\Sigma_0} \left|\frac{\partial \boldsymbol{\varphi}_\mathbf{g}}{\partial n}\right|^2 d\Sigma + \langle \mathbf{z}^1, \mathbf{g}^0\rangle - \int_\Omega \mathbf{z}^0 \cdot \mathbf{g}^1 \, dx \right.$$
$$\left. + \beta_1 \left\|\mathbf{g}^0\right\|_{(H_0^1(\Omega))^d} + \beta_0 \left\|\mathbf{g}^1\right\|_{(L^2(\Omega))^d} \right\}. \tag{8.42}$$

Remark 8.10 The same considerations apply to the Neumann controls (that is, of type (8.17)).

8.2.4 Approximate controllability via penalty

We consider again (8.11)–(8.15) (with $\theta_0 = 0$) and we introduce (with obvious notation):

$$J_k(\mathbf{v}) = \frac{1}{2}\int_{\Sigma_0} |\mathbf{v}|^2 \,d\Sigma + \frac{k_0}{2}\left\|\mathbf{y}(T;\mathbf{v}) - \mathbf{z}^0\right\|_{L^2}^2 + \frac{k_1}{2}\left\|\frac{\partial \mathbf{y}}{\partial t}(T;\mathbf{v}) - \mathbf{z}^1\right\|_{H^{-1}}^2. \tag{8.43}$$

In (8.43) we have

$$\mathbf{k} = \{k_0, k_1\}, \quad k_i > 0 \text{ and “large” for } i = 0 \text{ and } 1.$$

The optimal control problem

$$\inf_{\mathbf{v}\in(L^2(\Sigma_0))^d} J_k(\mathbf{v}), \tag{8.44}$$

has a *unique* solution, $\mathbf{u}_k$.

Considerations similar to those of Chapter 7 apply; thus we shall have

$$\mathbf{y}(T;\mathbf{u}_k) \in \mathbf{z}^0 + \beta_0 B, \quad \frac{\partial \mathbf{y}}{\partial t}(T;\mathbf{u}_k) \in \mathbf{z}^1 + \beta_1 B_{-1}, \tag{8.45}$$

for $\mathbf{k}$ “*large enough*”, with the “large enough” not being defined in a constructive way, so far. In order to obtain some estimates on the choice of $\mathbf{k}$, we will look first for a well-chosen *dual problem* of the control problem (8.44). Therefore, we consider again the operator $\mathbf{L}$ defined by

$$\mathbf{L}\mathbf{v} = \left\{-\frac{\partial \mathbf{y}}{\partial t}(T;\mathbf{v}), \mathbf{y}(T;\mathbf{v})\right\},$$

and we introduce

$$F_3(\mathbf{g}) = \frac{k_0}{2}\left\|\mathbf{g}^1 - \mathbf{z}^0\right\|_{L^2}^2 + \frac{k_1}{2}\left\|\mathbf{g}^0 + \mathbf{z}^1\right\|_{H^{-1}}^2. \tag{8.46}$$

With $F_1(\cdot)$ still defined by (8.32), we clearly have

$$\inf_{\mathbf{v}\in(L^2(\Sigma_0))^d} J_k(\mathbf{v}) = \inf_{\mathbf{v}\in(L^2(\Sigma_0))^d} \{F_1(\mathbf{v}) + F_3(\mathbf{L}\mathbf{v})\}. \tag{8.47}$$

By duality, it follows from (8.47) that

$$\inf_{\mathbf{v}\in(L^2(\Sigma_0))^d} J_k(\mathbf{v}) = -\inf_{\mathbf{g}} \{F_1^*(\mathbf{L}^*\mathbf{g}) + F_3^*(-\mathbf{g})\}, \tag{8.48}$$

with $\mathbf{g} = \{\mathbf{g}^0, \mathbf{g}^1\} \in (H_0^1(\Omega))^d \times (L^2(\Omega))^d$ in (8.48). After some calculations, we obtain

$$\inf_{\mathbf{v}\in(L^2(\Sigma_0))^d} J_k(\mathbf{v}) = - \inf_{\mathbf{g}\in(H_0^1(\Omega))^d\times(L^2(\Omega))^d} \left\{ \frac{1}{2}\int_{\Sigma_0} \left|\frac{\partial \boldsymbol{\varphi}_{\mathbf{g}}}{\partial n}\right|^2 \mathrm{d}\Sigma + \langle \mathbf{z}^1, \mathbf{g}^0\rangle \right.$$

$$\left. - \int_\Omega \mathbf{z}^0 \cdot \mathbf{g}^1 \, \mathrm{d}x + \frac{1}{2k_1}\left\|\mathbf{g}^0\right\|^2_{(H_0^1(\Omega))^d} + \frac{1}{2k_0}\left\|\mathbf{g}^1\right\|^2_{(L^2(\Omega))^d} \right\}. \tag{8.49}$$

The dual problem to the control problem (8.44) is therefore

$$\inf_{\mathbf{g}\in(H_0^1(\Omega))^d\times(L^2(\Omega))^d} \left\{ \frac{1}{2}\int_{\Sigma_0} \left|\frac{\partial \boldsymbol{\varphi}_{\mathbf{g}}}{\partial n}\right|^2 \mathrm{d}\Sigma + \langle \mathbf{z}^1, \mathbf{g}^0\rangle - \int_\Omega \mathbf{z}^0 \cdot \mathbf{g}^1 \, \mathrm{d}x \right.$$

$$\left. + \frac{1}{2k_1}\left\|\mathbf{g}^0\right\|^2_{(H_0^1(\Omega))^d} + \frac{1}{2k_0}\left\|\mathbf{g}^1\right\|^2_{(L^2(\Omega))^d} \right\}. \tag{8.50}$$

Let us denote by $\mathbf{f}_k$ the solution of (8.50); it is characterized (with obvious notation) by

$$\begin{cases} \mathbf{f}_k = \{\mathbf{f}_k^0, \mathbf{f}_k^1\} \in (H_0^1(\Omega))^d \times (L^2(\Omega))^d, \\ \displaystyle\int_{\Sigma_0} \frac{\partial \boldsymbol{\varphi}_k}{\partial n}\cdot\frac{\partial \boldsymbol{\varphi}_{\mathbf{g}}}{\partial n} \, \mathrm{d}\Sigma + \frac{1}{k_0}\int_\Omega \mathbf{f}_k^1 \cdot \mathbf{g}^1 \, \mathrm{d}x + \frac{1}{k_1}\int_\Omega \nabla \mathbf{f}_k^0 : \nabla \mathbf{g}^0 \, \mathrm{d}x = \\ \displaystyle\int_\Omega \mathbf{z}^0 \cdot \mathbf{g}^1 \, \mathrm{d}x - \langle \mathbf{z}^1, \mathbf{g}^0\rangle, \quad \forall\, \mathbf{g} = \{\mathbf{g}^0, \mathbf{g}^1\} \in (H_0^1(\Omega))^d \times (L^2(\Omega))^d. \end{cases} \tag{8.51}$$

Similarly, the solution $\mathbf{f}_\beta$ of problem (8.42) is characterized by the following *variational inequality*:

$$\begin{cases} \mathbf{f}_\beta = \{\mathbf{f}_\beta^0, \mathbf{f}_\beta^1\} \in (H_0^1(\Omega))^d \times (L^2(\Omega))^d, \ \forall\, \mathbf{g} \in (H_0^1(\Omega))^d \times (L^2(\Omega))^d, \\ \displaystyle\int_{\Sigma_0} \frac{\partial \boldsymbol{\varphi}_\beta}{\partial n}\cdot\frac{\partial}{\partial n}(\boldsymbol{\varphi}_{\mathbf{g}} - \boldsymbol{\varphi}_\beta) \, \mathrm{d}\Sigma + \beta_1\left(\left\|\mathbf{g}^0\right\|_{H_0^1} - \left\|\mathbf{f}_\beta^0\right\|_{H_0^1}\right) \\ \displaystyle + \beta_0\left(\left\|\mathbf{g}^1\right\|_{L^2} - \left\|\mathbf{f}_\beta^1\right\|_{L^2}\right) \geq \int_\Omega \mathbf{z}^0 \cdot (\mathbf{g}^1 - \mathbf{f}_\beta^1) \, \mathrm{d}x - \langle \mathbf{z}^1, \mathbf{g}^0 - \mathbf{f}_\beta^0\rangle. \end{cases} \tag{8.52}$$

Taking $\mathbf{g} = \mathbf{f}_k$ in (8.51) (respectively $\mathbf{g} = \mathbf{0}$ and $\mathbf{g} = 2\mathbf{f}_\beta$ in (8.52)) we obtain

$$\int_{\Sigma_0} \left|\frac{\partial \boldsymbol{\varphi}_k}{\partial n}\right|^2 \mathrm{d}\Sigma + \frac{1}{k_0}\int_\Omega |\mathbf{f}_k^1|^2 \, \mathrm{d}x + \frac{1}{k_1}\int_\Omega |\nabla \mathbf{f}_k^0|^2 \, \mathrm{d}x$$

$$= \int_\Omega \mathbf{z}^0 \cdot \mathbf{f}_k^1 \, \mathrm{d}x - \langle \mathbf{z}^1, \mathbf{f}_k^0\rangle, \tag{8.53}$$

and

$$\int_{\Sigma_0} \left| \frac{\partial \varphi_\beta}{\partial n} \right|^2 d\Sigma + \beta_0 \left\| \mathbf{f}_\beta^1 \right\|_{L^2} + \beta_1 \left\| \mathbf{f}_\beta^0 \right\|_{H_0^1} = \int_\Omega \mathbf{z}^0 \cdot \mathbf{f}_\beta^1 \, dx - \langle \mathbf{z}^1, \mathbf{f}_\beta^0 \rangle. \tag{8.54}$$

Assuming that problems (8.51) and (8.52) have the same solution, it follows from (8.53) and (8.54) that

$$\frac{1}{k_0} \left\| \mathbf{f}_k^1 \right\|_{L^2}^2 + \frac{1}{k_1} \left\| \mathbf{f}_k^0 \right\|_{H_0^1}^2 = \beta_0 \left\| \mathbf{f}_k^1 \right\|_{L^2} + \beta_1 \left\| \mathbf{f}_k^0 \right\|_{H_0^1}. \tag{8.55}$$

Relation (8.55) suggests the following simple rule (may be too simple, indeed): adjust k_0 and k_1 so that

$$\frac{1}{k_0} \left\| \mathbf{f}_k^1 \right\|_{L^2} = \beta_0 \quad \text{and} \quad \frac{1}{k_1} \left\| \mathbf{f}_k^0 \right\|_{H_0^1} = \beta_1. \tag{8.56}$$

Numerical experiments still have to be performed to validate (8.56).

8.2.5 Further references on the controllability of thermo-elastic systems

Thermo-elastic systems occurring in many practical situations, it is not surprising that their stabilization and control has motivated an abundant literature, mostly from mathematical nature, computationally oriented publications lagging way behind (to the best of our knowledge). In addition to those publications already mentioned (namely, Narukawa, 1983, J.L. Lions, 1988b, Vol. 2, Zuazua, 1993b and 1995) let us mention, among several others, de Teresa and Zuazua (1996), Zuazua (1996), Hansen and Zhang (1997), Lebeau and Zuazua (1998), Avalos (2000), Avalos and Lasiecka (2000), Benabdallah and Naso (2000), Eller, Lasiecka, and Triggiani (2000), Muñoz-Rivera and Naso (2003) (see also the many references therein).

In some of these publications (see in particular Lebeau and Zuazua, 1998), it is shown that if the wave equation part of the thermo-elastic model is exactly controllable in the sense of Chapter 6, the full thermo-elastic system is *exactly null-controllable*, that is, starting from $\left\{\mathbf{y}(0), \frac{\partial \mathbf{y}}{\partial t}(0), \theta(0)\right\}$ in "suitable" energy spaces we can control the system so that $\left\{\mathbf{y}(T), \frac{\partial \mathbf{y}}{\partial t}(T), \theta(T)\right\} = \{\mathbf{0}, \mathbf{0}, 0\}$ if T is large enough (see the above reference for details).

8.3 Coupled systems (II): Other systems

The *stabilization* and *controllability* of *thermo-elastic* systems being most popular research topics it was making sense to dedicate them a full section, namely Section 8.2. On the other hand there are many other situations involving the coupling of chemico-physical phenomena of different nature, such as elasticity and electromagnetism, elasticity and fluid mechanics, heat transfer and fluid mechanics, heat transfer and electromagnetism, electromagnetism and fluid mechanics, and so on. Considering the practical importance of many of these coupled phenomena, it is not surprising that the controllability of the related mathematical models has motivated the research of a fair number of investigators. Among the related publications, let us mention:

- J.L. Lions and Zuazua (1996), Zhang and Zuazua (2003) concerning the control of coupled *fluid-structure* phenomena.
- Fursikov and Imanuvilov (1998), Crépeau (2003) concerning the control of the Boussinesq equations modeling *thermal convection* (these equations form a system coupling the incompressible Navier–Stokes equations with the heat equation; see, for example, Glowinski, 2003, Chapter 9, Section 46 for details).
- Kapitanov and Perla-Menzala (2003), Kapitanov and Raupp (2003) concerning the control of coupled *elasticity-electromagnetic* phenomena (from *piezoelectricity* to be more precise).
- Masserey (2003) concerning the control of systems modeled by coupled Maxwell and heat equations (occurring in aluminum industry).

Part III

Flow Control

9

Optimal control of systems modelled by the Navier–Stokes equations: Application to drag reduction

9.1 Introduction. Synopsis

We will conclude this volume by discussing some aspects (mostly computational) of the optimal control of systems governed by the *Navier–Stokes equations* modeling *unsteady incompressible Newtonian viscous fluids*. This chapter can be viewed as a sequel of Chapter 3, where we addressed the controllability of Stokes flow. The methods and results presented hereafter were not available at the time of Glowinski and J.L. Lions (1995), explaining thus the need for a new chapter.

To begin with, let us say that engineers have not waited for mathematicians to successfully address flow control problems (see, for example, Gad Hel Hak, 1989; Buschnell and Hefner, 1990 for a review of flow control from the Engineering point of view); indeed Prandtl as early as 1915 was concerned with flow control and was designing ingenious systems to suppress or delay *boundary layer separation* (see Prandtl, 1925). The last two decades have seen an explosive growth of investigations and publications of mathematical nature concerning various aspects of the control of viscous flow, good examples of these publications being Gunzburger (1995) and Sritharan (1998). Actually, the above two references also contain articles related to the computational aspects of the optimal control of viscous flow, but, usually, the geometry of the flow region is fairly simple and *Reynolds numbers* rather low. Some publications of computational nature are Hou and Ravindran (1996), Ghattas and Bark (1997), and Ito and Ravindran (1998); however, in those articles, once again the geometry is simple and/or the Reynolds number is low (more references will be given at the end of this section).

The main goal of this chapter, which follows closely He, Chevalier et al. (2000), He, Glowinski et al. (2000), He et al. (2002), and Glowinski (2003, Section 53), is to investigate computational methods for the active control and drag optimization of incompressible viscous flow past cylinders, using the two-dimensional Navier–Stokes equations as the flow model. The computational methodology relies on the following ingredients: space discrimination of the Navier–Stokes equations by finite element approximations, time discretization by a second-order accurate two-step implicit/explicit finite difference scheme, calculation of the cost function gradient by the adjoint equation approach, and minimization of the discrete cost function by a quasi-Newton method à la BFGS. Motivated in part by the experimental work of Tokumaru and Dimotakis (1991), the above methods have been applied to the *boundary control by rotation* of the flow around a circular cylinder, leading to 30–60% drag reduction, compared to the fixed cylinder configuration, for Reynolds numbers in the range of 200–1000. Next, we apply the same methodology to drag reduction for flow past a circular cylinder, using this time *control by blowing and suction*. Using only three blowing-suction slots, we have been able to suppress completely the formation of the (celebrated) *Von Kármán vortex street* up to Re = 200 with a further net drag reduction compared to control by rotation.

From a methodological point of view, some of the methods used here are clearly related to those employed by Glowinski's former PhD student, M. Berggren, in Berggren (1998), for the boundary control by blowing and suction of incompressible viscous flow in bounded cavities.

The organization of the remainder of this chapter is as follows: In Section 9.2 we formulate the flow control problem, and address its time discrimination in Section 9.3. The important problem of the space discretization by a finite element method is discussed in Section 9.4; a special attention is given there to velocity spaces which are *discretely divergence free* in order to reduce the number of algebraic constraints in the control problem. The full discretization of the control problem is addressed in Section 9.5. Since we intend to use solution methods based on a *quasi-Newton algorithm à la BFGS* (briefly discussed in Section 9.6) attention is focused in Section 9.5 on the derivation of the gradient of the fully discrete cost function via the classical *adjoint equation* method. The *flow simulator* (actually a Navier–Stokes equations solver) is further discussed in Section 9.7 where it is validated, for various values of the Reynolds number, on well documented flow around circular cylinder test problems. Finally, the results of various numerical experiments for flow control past a circular cylinder are discussed in Sections 9.8–9.10; they definitely show that a *substantial drag reduction can be obtained* using either an oscillatory rotation, or blowing and suction.

As can be expected, flow control and in particular those controllability issues associated with the Navier–Stokes equations have generated a quite large literature addressing both the mathematical and computational aspects; in addition to the few references given above, let us mention among many others: Abergel and Temam (1990), Fursikov and Imanuvilov (1994), Moin and Bewley (1994), Fursikov (1995), Coron (1996a,b), J.L. Lions and Zuazua (1998), Fernandez-Cara (1999), Hinze and Kunisch (2001), Collis et al. (2002), Homescu, Navon and Li (2002), Moubachir (2002), Collis, Ghayour, and Heinkenschloss (2003), Fernandez-Cara et al. (2004), Hintermüller et al. (2004), and Raymond (2006) (see also the references therein); further references will be given in the following sections of this chapter, but before, we would like to mention that it is the opinion of the authors of this book that very few people have contributed more to *Flow Control* than M. D. Gunzburger and his collaborators, an evidence of this statement being the string of remarkable publications they have produced these last two decades; let us mention among several others Gunzburger, Hou, and Svobodny (1989, 1992), Gunzburger and Lee (1996), Gunzburger, Fursikov, and Hou (1998, 2005), Gunzberger, Hou et al. (1998), Gunzberger and Manservisi (1999, 2000a,b), and Gunzberger (2000).

9.2 Formulation of the control problem

9.2.1 Fluid flow formulation

Let Ω be a region of $\mathbb{R}^d$ ($d = 2, 3$ in practice); we denote by Γ the boundary $\partial\Omega$ of Ω. We suppose that Ω is filled with a *Newtonian incompressible viscous fluid* of *density* ρ and *viscosity* μ; we suppose that the temperature is constant. Under these circumstances the flow of such a fluid is modeled by the following system of *Navier–Stokes equations*:

$$\rho\,[\partial_t \mathbf{y} + (\mathbf{y}\cdot\nabla)\mathbf{y}] = \nabla\cdot\boldsymbol{\sigma} + \rho\,\mathbf{f} \quad \text{in } \Omega\times(0,T), \tag{9.1}$$

$$\nabla\cdot\mathbf{y} = 0 \quad \text{in } \Omega\times(0,T) \quad \text{(incompressibility condition)}. \tag{9.2}$$

In (9.1), (9.2), $\mathbf{y} = \{y_i\}_{i=1}^d$ denotes the *velocity* field, π the *pressure*, $\mathbf{f}$ a *density of external forces* per mass unit, and $\boldsymbol{\sigma}$ $(= \boldsymbol{\sigma}(\mathbf{y},\pi))$ the *stress tensor* defined by

$$\boldsymbol{\sigma} = 2\mu\,\mathbf{D}(\mathbf{y}) - \pi\,\mathbf{I}$$

with the *rate of deformation tensor* $\mathbf{D}(\mathbf{y})$ being defined by

$$\mathbf{D}(\mathbf{y}) = \frac{1}{2}(\nabla\mathbf{y} + \nabla\mathbf{y}^t).$$

We also have

$$\partial_t = \frac{\partial}{\partial t}, \quad \nabla^2 = \sum_{i=1}^d \frac{\partial^2}{\partial x_i^2},$$

$$\nabla\cdot\mathbf{y} = \sum_{i=1}^d \frac{\partial y_i}{\partial x_i}, \quad (\mathbf{y}\cdot\nabla)\mathbf{z} = \left\{\sum_{j=1}^d y_j \frac{\partial z_i}{\partial x_j}\right\}_{i=1}^d.$$

In (9.1) and (9.2), $(0,T)$ is the *time interval* during which the flow is considered (observed).

Equations (9.1), (9.2) have to be completed by further conditions, such as the following *initial condition*:

$$\mathbf{y}(0) = \mathbf{y}_0 \quad (\text{with } \nabla\cdot\mathbf{y}_0 = 0) \tag{9.3}$$

and *boundary conditions*. Let us consider the typical situation of interest to us described in Figure 9.1, corresponding to an external flow around a cylinder of cross section B; we assume that the classical two-dimensional reduction holds.

In fact $\Gamma_u \cup \Gamma_d \cup \Gamma_N \cup \Gamma_S$ is an *artificial* boundary which has to be taken sufficiently far from B, so that the corresponding flow is a good approximation of the unbounded external flow around B. Typical boundary conditions are (if $d = 2$)

$$\mathbf{y} = \mathbf{y}_\infty \quad \text{on } (\Gamma_u \cup \Gamma_N \cup \Gamma_S)\times(0,T), \tag{9.4}$$

$$\boldsymbol{\sigma}\,\mathbf{n} = \mathbf{0} \quad \text{on } \Gamma_d\times(0,T) \quad \text{(downstream boundary condition)}, \tag{9.5}$$

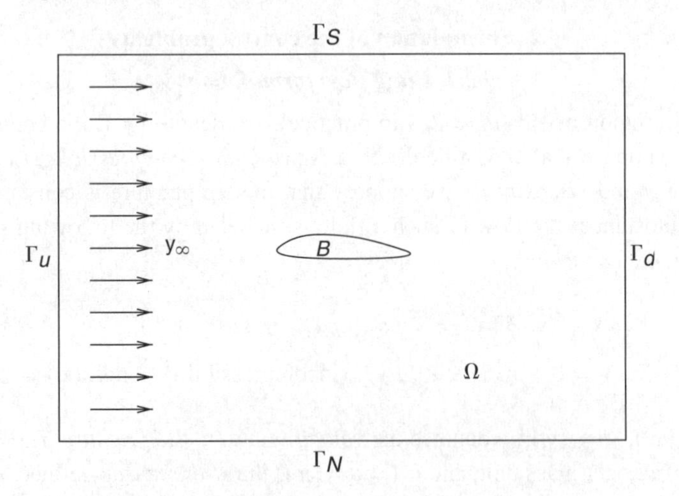

Fig. 9.1. External flow around a cylinder of cross-section B.

with, as usual, $\mathbf{n}$ the *unit vector* of the outward normal on Γ. We are voluntarily vague, concerning the boundary conditions on ∂B, since they will be part of the control process. Let us conclude this section by recalling that the *Reynolds number* Re is classically defined by

$$\mathrm{Re} = \rho\, UL/\mu, \tag{9.6}$$

with U a *characteristic velocity* ($|\mathbf{y}_\infty|$, here) and L a *characteristic length* (the thickness of B, for example).

Our goal in this chapter is to prescribe on ∂B boundary conditions of the *Dirichlet type* (that is, velocity prescribed on ∂B) so that some flow related performance criterion (the cost function) will be minimized under reasonable constraints on the control variables.

Remark 9.1 The *existence* and *uniqueness* of solutions to the *Navier–Stokes equations* (9.1), (9.2), completed by the *initial conditions* (9.3) and various type of *boundary conditions*, are topics which have generated a very large number of publications; focusing on few books let us mention Ladyzenskaya (1969), J.L. Lions (1969, 2002), Temam (1977, 2001), Tartar (1978, 2006), P.L. Lions (1996) (see also the many references contained in these publications).

9.2.2 *Formulation of the control problem*

The flow control problem to be discussed in this chapter consists of minimizing a *drag* related *cost function* via controls acting on ∂B; this problem can be formulated as follows (using classical control formalism):

$$\begin{cases} \mathbf{u} \in \mathcal{U}, \\ J(\mathbf{u}) \le J(\mathbf{v}), \quad \forall \mathbf{v} \in \mathcal{U}, \end{cases} \tag{9.7}$$

where, in (9.7), the *control space* $\mathcal{U}$ is a vector space of vector-valued functions $\mathbf{v}$ defined on $\partial B \times (0, T)$ and satisfying

$$\int_{\partial B} \mathbf{v}(t) \cdot \mathbf{n} \, \mathrm{d}s = 0 \quad \text{for } t \in (0, T), \tag{9.8}$$

and where the *cost function* J is defined by

$$J(\mathbf{v}) = \frac{\epsilon}{2}\left(\int_0^T \|\mathbf{v}(t)\|_\alpha^2 \, \mathrm{d}t + \int_0^T \int_{\partial B} |\partial_t \mathbf{v}(x,t)|^2 \, \mathrm{d}s \, \mathrm{d}t \right) + \int_0^T P_d(t) \, \mathrm{d}t, \tag{9.9}$$

with ϵ (≥ 0) a *regularization* parameter.

In (9.8), (9.9), we have used the following notation:

- $\varphi(t)$ for the function $x \to \varphi(x,t)$.
- $\mathrm{d}s$ for the superficial measure on ∂B.
- $\|\cdot\|_\alpha$ for a norm defined on ∂B, involving space derivatives of order α, with α possibly noninteger (the readers afraid of these mathematical complications should not worry, since in this chapter we will only consider boundary controls functions of t or smooth functions of s and t).
- $P_d(t)$ is the power needed to overcome, at time t, the *drag* exerted on B in the opposite direction of $\mathbf{y}_\infty$; $P_d(t)$ is defined by

$$P_d(t) = \int_{\partial B} \boldsymbol{\sigma} \, \mathbf{n} \cdot (\mathbf{v} - \mathbf{y}_\infty) \, \mathrm{d}s. \tag{9.10}$$

- Finally, $\mathbf{y}$ is the solution to the following Navier–Stokes system:

$$\rho \, [\partial_t \mathbf{y} + (\mathbf{y} \cdot \nabla)\mathbf{y}] = \nabla \cdot \boldsymbol{\sigma} + \rho \, \mathbf{f} \quad \text{in } \Omega \times (0, T), \tag{9.11}$$

$$\nabla \cdot \mathbf{y} = 0 \quad \text{in } \Omega \times (0, T), \tag{9.12}$$

$$\mathbf{y}(0) = \mathbf{y}_0 \quad (\text{with } \nabla \cdot \mathbf{y}_0 = 0), \tag{9.13}$$

$$\boldsymbol{\sigma} \, \mathbf{n} = \mathbf{0} \quad \text{on } \Gamma_d \times (0, T), \tag{9.14}$$

$$\mathbf{y} = \mathbf{y}_\infty \quad \text{on } (\Gamma_u \cup \Gamma_N \cup \Gamma_S) \times (0, T) \quad (\text{if } d = 2), \tag{9.15}$$

$$\mathbf{y} = \mathbf{v} \quad \text{on } \partial B \times (0, T). \tag{9.16}$$

It make sense to assume that $\mathbf{v}(0) = \mathbf{0}$.

Remark 9.2 The flux condition (9.8) is not essential and can be easily relaxed if, for example, the downstream boundary conditions are of the *Neumann type* (like those in (9.14)).

Remark 9.3 The momentum equation (9.11) can also be written as:

$$\rho \, [\partial_t \mathbf{y} + (\mathbf{y} \cdot \nabla)\mathbf{y}] - \mu \, \nabla^2 \mathbf{y} + \nabla \pi = \rho \, \mathbf{f} \quad \text{in } \Omega \times (0, T),$$

however, formulation (9.11) is better suited to the drag reduction problem (9.7), since the cost function $J(\cdot)$ involves also the stress tensor $\boldsymbol{\sigma}$ (via relation (9.10)).

Remark 9.4 In the particular case of an *incompressible viscous flow* we have

$$\int_0^T P_d(t)\,\mathrm{d}t = \int_0^T E_d(t)\,\mathrm{d}t + (K(T) - K(0)) + \int_0^T P_c(t)\,\mathrm{d}t + \int_0^T P_\infty(t)\,\mathrm{d}t - \int_0^T P_f(t)\,\mathrm{d}t \qquad (9.17)$$

where, in (9.17),

- $E_d(t) = 2\mu \int_\Omega |\mathbf{D}(\mathbf{y} - \mathbf{y}_\infty)|^2\,\mathrm{d}x$ is the *viscous dissipation energy*.
- $K(t) = \frac{\rho}{2} \int_\Omega |\mathbf{y}(t) - \mathbf{y}_\infty|^2\,\mathrm{d}x$ is a *kinetic energy*.
- $P_c(t) = \frac{\rho}{2} \int_{\partial B} |\mathbf{v} - \mathbf{y}_\infty|^2 \mathbf{v} \cdot \mathbf{n}\,\mathrm{d}s$ is a *control related power*.
- $P_\infty(t) = \frac{\rho}{2} \int_{\Gamma_\infty^d} |\mathbf{y} - \mathbf{y}_\infty|^2 \mathbf{y} \cdot \mathbf{n}\,\mathrm{d}s$ is a *downstream boundary related power*.
- $P_f(t) = \rho \int_\Omega \mathbf{f}(t) \cdot (\mathbf{y}(t) - \mathbf{y}_\infty)\,\mathrm{d}x$ is the *external forcing power*.

Some observations are in order, such as:

(i) To obtain (9.17) from (9.10)–(9.16), one applies the *divergence theorem* on the right-hand side of equation (9.10).
(ii) If Γ_d "goes to infinity" in the Ox_1 direction, then $\mathbf{y} \to \mathbf{y}_\infty$ which implies in turn that $P_\infty \to 0$.
(iii) Whenever the control is absent, that is, $\mathbf{v} = \mathbf{0}$, we have $P_c(t) = 0$.

We can summarize relation (9.17) by noting that:

> *the drag work is equal to the energy dissipated by viscosity + the kinetic energy variation between* 0 *and* T *+ the control associated work + the downstream boundary associated work – the external forcing work.*

The above observation has the following consequence: instead of minimizing $J(\cdot)$ defined by (9.9) we can minimize the cost function defined by

$$J(\mathbf{v}) = \frac{\epsilon}{2}\left(\int_0^T \|\mathbf{v}(t)\|_\alpha^2\,\mathrm{d}t + \int_0^T \int_{\partial B} |\partial_t \mathbf{v}(x,t)|^2\,\mathrm{d}s\,\mathrm{d}t\right) + \int_0^T E_d(t)\,\mathrm{d}t + (K(T) - K(0)) + \int_0^T P_c(t)\,\mathrm{d}t + \int_0^T P_\infty(t)\,\mathrm{d}t - \int_0^T P_f(t)\,\mathrm{d}t;$$

written this way, the cost function is simpler than when defined by (9.9) since it does not involve boundary integrals of stress-tensor related quantities. However, we shall keep working with the cost function defined by (9.9) since it seems to lead to more accurate results.

In order to solve the control problem (9.7) we will apply a *quasi-Newton method à la BFGS* to the solution of the (necessary) *optimality condition*

$$\nabla J_h^{\Delta t}(\mathbf{u}_h^{\Delta t}) = \mathbf{0},$$

where the cost function $J_h^{\Delta t}(\cdot)$ is obtained from the full space–time approximation of the continuous cost function $J(\cdot)$ defined by relations (9.9)–(9.16), and where $\mathbf{u}_h^{\Delta t}$ is a solution of the fully discrete control problem. It is then of crucial importance to know how to compute the *gradient* $\nabla J_h^{\Delta t}(\cdot)$ of the cost function $J_h^{\Delta t}(\cdot)$. The calculation of $\nabla J_h^{\Delta t}$ will be discussed in Section 9.5.

9.3 Time discretization of the control problem

9.3.1 Generalities

We are going to discuss first the *time-discretization* issue. The space and consequently the full space/time-discretization issues will be addressed in Section 9.4. This approach of fractioning (splitting) the computational difficulties has the definite advantage that some practitioners will be able to use the material in this chapter for other types of space approximations than the finite element ones discussed in Section 9.4 (one may prefer spectral methods or finite volume methods, for example).

9.3.2 Formulation of the time-discrete control problem

Assuming that $0 < T < +\infty$, we define first a time-discretization step Δt by $\Delta t = T/N$, with N a (large) positive integer. We approximate then the control problem (9.7) by

$$\begin{cases} \mathbf{u}^{\Delta t} \in \mathcal{U}^{\Delta t}, \\ J^{\Delta t}(\mathbf{u}^{\Delta t}) \leq J^{\Delta t}(\mathbf{v}), \quad \forall \mathbf{v} \in \mathcal{U}^{\Delta t}, \end{cases} \tag{9.18}$$

with

$$\mathcal{U}^{\Delta t} = \mathbf{\Lambda}^N, \tag{9.19}$$

$$\mathbf{\Lambda} = \{ \boldsymbol{\lambda} \mid \boldsymbol{\lambda} \in (H^{\alpha}(\partial\Omega))^d, \int_{\partial B} \boldsymbol{\lambda} \cdot \mathbf{n} \, ds = 0 \}, \tag{9.20}$$

and

$$J^{\Delta t}(\mathbf{v}) = \frac{\epsilon}{2} \Delta t \sum_{n=1}^{N} \left(\|\mathbf{v}^n\|_{\alpha}^2 + \left\| \frac{\mathbf{v}^n - \mathbf{v}^{n-1}}{\Delta t} \right\|_0^2 \right) + \Delta t \sum_{n=1}^{N} P_d^n, \tag{9.21}$$

where, in (9.21), P_d^n is the discrete drag power defined (with obvious notation) by

$$P_d^n = \int_{\partial B} \boldsymbol{\sigma}(\mathbf{y}^n, \pi^n) \mathbf{n} \cdot (\mathbf{v}^n - \mathbf{y}_{\infty}) \, ds,$$

with $\{(\mathbf{y}^n, \pi^n)\}_{n=1}^N$ obtained from $\mathbf{v}$ via the solution of the following *semidiscrete* Navier–Stokes equations:

$$\mathbf{y}^0 = \mathbf{y}_0, \tag{9.22}$$

then

$$\rho\left(\frac{\mathbf{y}^1-\mathbf{y}^0}{\Delta t}+(\mathbf{y}^0\cdot\nabla)\mathbf{y}^0\right)=\nabla\cdot\boldsymbol{\sigma}(\frac{2}{3}\mathbf{y}^1+\frac{1}{3}\mathbf{y}^0,\pi^1)+\rho\,\mathbf{f}^1 \quad \text{in } \Omega, \tag{9.23}$$

$$\nabla\cdot\mathbf{y}^1=0 \quad \text{in } \Omega, \tag{9.24}$$

$$\boldsymbol{\sigma}\left(\frac{2}{3}\mathbf{y}^1+\frac{1}{3}\mathbf{y}^0,\pi^1\right)\mathbf{n}=\mathbf{0} \quad \text{on } \Gamma_d, \tag{9.25}$$

$$\mathbf{y}^1=\mathbf{y}_\infty \quad \text{on } \Gamma_u\cup\Gamma_N\cup\Gamma_S, \tag{9.26}$$

$$\mathbf{y}^1=\mathbf{v}^1 \quad \text{on } \partial B, \tag{9.27}$$

and for $n=2,\dots,N$

$$\rho\left(\frac{(3/2)\mathbf{y}^n-2\mathbf{y}^{n-1}+(1/2)\mathbf{y}^{n-2}}{\Delta t}+((2\mathbf{y}^{n-1}-\mathbf{y}^{n-2})\cdot\nabla)(2\mathbf{y}^{n-1}-\mathbf{y}^{n-2})\right)$$
$$=\nabla\cdot\boldsymbol{\sigma}(\mathbf{y}^n,\pi^n)+\rho\,\mathbf{f}^n \quad \text{in } \Omega, \tag{9.28}$$

$$\nabla\cdot\mathbf{y}^n=0 \quad \text{in } \Omega, \tag{9.29}$$

$$\boldsymbol{\sigma}(\mathbf{y}^n,\pi^n)\,\mathbf{n}=\mathbf{0} \quad \text{on } \Gamma_d, \tag{9.30}$$

$$\mathbf{y}^n=\mathbf{y}_\infty \quad \text{on } \Gamma_u\cup\Gamma_N\cup\Gamma_S, \tag{9.31}$$

$$\mathbf{y}^n=\mathbf{v}^n \quad \text{on } \partial B. \tag{9.32}$$

The above scheme is a *semi-implicit, second-order accurate two-step scheme*.

Anticipating the finite element approximation to be discussed in Section 9.4, we can rewrite system (9.23)–(9.32) in *variational form*. We obtain thus

$$\rho\int_\Omega\frac{\mathbf{y}^1-\mathbf{y}^0}{\Delta t}\cdot\mathbf{z}\,dx+2\mu\int_\Omega\mathbf{D}\left(\frac{2}{3}\mathbf{y}^1+\frac{1}{3}\mathbf{y}^0\right):\mathbf{D}(\mathbf{z})\,dx$$
$$+\rho\int_\Omega(\mathbf{y}^0\cdot\nabla)\mathbf{y}^0\cdot\mathbf{z}\,dx$$
$$-\int_\Omega\pi^1\nabla\cdot\mathbf{z}\,dx=\rho\int_\Omega\mathbf{f}^1\cdot\mathbf{z}\,dx, \quad \forall\mathbf{z}\in\mathbf{V}_0, \tag{9.33}$$

$$\int_\Omega\nabla\cdot\mathbf{y}^1 q\,dx=0, \quad \forall q\in L^2(\Omega), \tag{9.34}$$

$$\mathbf{y}^1=\mathbf{y}_\infty \quad \text{on } \Gamma_u\cup\Gamma_N\cup\Gamma_S, \tag{9.35}$$

$$\mathbf{y}^1=\mathbf{v}^1 \quad \text{on } \partial B; \tag{9.36}$$

and for $n = 2, \dots, N$

$$\rho \int_\Omega \frac{(3/2)\mathbf{y}^n - 2\mathbf{y}^{n-1} + (1/2)\mathbf{y}^{n-2}}{\Delta t} \cdot \mathbf{z}\, dx + 2\mu \int_\Omega \mathbf{D}(\mathbf{y}^n) : \mathbf{D}(\mathbf{z})\, dx$$
$$+ \rho \int_\Omega ((2\mathbf{y}^{n-1} - \mathbf{y}^{n-2}) \cdot \nabla)(2\mathbf{y}^{n-1} - \mathbf{y}^{n-2}) \cdot \mathbf{z}\, dx$$
$$- \int_\Omega \pi^n \nabla \cdot \mathbf{z}\, dx = \rho \int_\Omega \mathbf{f}^n \cdot \mathbf{z}\, dx, \quad \forall \mathbf{z} \in \mathbf{V}_0, \tag{9.37}$$

$$\int_\Omega \nabla \cdot \mathbf{y}^n q\, dx = 0, \quad \forall q \in L^2(\Omega), \tag{9.38}$$

$$\mathbf{y}^n = \mathbf{y}_\infty \quad \text{on } \Gamma_u \cup \Gamma_N \cup \Gamma_S, \tag{9.39}$$
$$\mathbf{y}^n = \mathbf{v}^n \quad \text{on } \partial B. \tag{9.40}$$

In (9.33) and (9.37), we have used the notation

$$\mathbf{T} : \mathbf{S} = \sum_{i=1}^{d} \sum_{j=1}^{d} t_{ij} s_{ij},$$

to denote the scalar product in $\mathbb{R}^{d^2}$ of the two tensors $\mathbf{T} = \{t_{ij}\}_{i,j=1}^d$ and $\mathbf{S} = \{s_{ij}\}_{i,j=1}^d$; the space $\mathbf{V}_0$ is defined by

$$\mathbf{V}_0 = \{\mathbf{z} \,|\, \mathbf{z} \in (H^1(\Omega))^d,\ \mathbf{z} = \mathbf{0} \text{ on } \Gamma_u \cup \Gamma_N \cup \Gamma_S \cup \partial B\}. \tag{9.41}$$

9.3.3 Comments on the time discretization of the control problem (9.7)

Since the time-discretization scheme used in Section 9.3.2 is a two-step one, a *starting procedure* is required; the one we advocate, namely, (9.23)–(9.27), leads to a generalized Stokes problem to obtain $\{\mathbf{y}^1, \pi^1\}$; this last problem has the same coefficients as the generalized Stokes problem providing $\{\mathbf{y}^n, \pi^n\}$ from $\mathbf{v}^n$ and $\mathbf{y}^{n-1}$, $\mathbf{y}^{n-2}$. As we shall see later on, scheme (9.22)–(9.32) albeit partly explicit has shown very good robustness properties when applied to the solution of drag reduction problems.

9.4 Full discretization of the control problem

9.4.1 Synopsis

From now on, we assume that $\Omega \subset \mathbf{R}^2$. In order to *spatially discretize* the control problem (9.7), we are going to use a *finite element approximation* closely related to those discussed in Glowinski (2003, Chapter 5); indeed, finite element approximations are well suited to the handling of complicated boundary and boundary conditions. The discretization to be used combines a *continuous* piecewise Q_2 approximation for the *velocity* and a *discontinuous* piecewise P_1 approximation for the *pressure* (we recall that P_1 (respectively, Q_2) denotes the space of those polynomials of degree

≤ 1 (respectively, ≤ 2) with respect to all the variables (respectively, with respect to each variable)). This approximation satisfies a *discrete inf-sup condition* (see, for example, Brezzi and Fortin, 1991 for this notion), implying that the discrete problems are well-posed and the approximation is convergent (see, for example, Girault and Raviart, 1986; Pironneau, 1989; Gunzburger, 1989, 1999; Gresho and Sani, 1998; Turek, 1999; Rannacher 2000; Glowinski, 2003 for the finite element approximation of the Navier–Stokes equations modeling incompressible viscous flow).

9.4.2 Discrete flow model

In order to fully discretize the semidiscrete model (9.22)–(9.32) we are going to mimic its equivalent variational formulation. Doing so we obtain

$$\mathbf{y}_h^0 = \mathbf{y}_{0h} \quad \text{(a convenient approximation of } \mathbf{y}_0\text{)}; \tag{9.42}$$

and

$$\begin{aligned}
&\rho \int_{\Omega_h} \frac{\mathbf{y}_h^1 - \mathbf{y}_h^0}{\Delta t} \cdot \mathbf{z}\, dx + 2\mu \int_{\Omega_h} \mathbf{D}\left(\frac{2}{3}\mathbf{y}_h^1 + \frac{1}{3}\mathbf{y}_h^0\right) : \mathbf{D}(\mathbf{z})\, dx \\
&+ \rho \int_{\Omega_h} (\mathbf{y}_h^0 \cdot \nabla)\mathbf{y}_h^0 \cdot \mathbf{z}\, dx \\
&- \int_{\Omega_h} \pi_h^1 \nabla \cdot \mathbf{z}\, dx = \rho \int_{\Omega_h} \mathbf{f}^1 \cdot \mathbf{z}\, dx, \quad \forall \mathbf{z} \in \mathbf{V}_{0h},
\end{aligned} \tag{9.43}$$

$$\int_{\Omega_h} \nabla \cdot \mathbf{y}_h^1 q\, dx = 0, \quad \forall q \in \mathbf{P}_h, \tag{9.44}$$

$$\mathbf{y}_h^1 = \mathbf{y}_\infty \quad \text{on } \Gamma_{u,h} \cup \Gamma_{N,h} \cup \Gamma_{S,h}, \tag{9.45}$$

$$\mathbf{y}_h^1 = \mathbf{v}^1 \quad \text{on } \partial B_h; \tag{9.46}$$

then for $n = 2, \ldots, N$

$$\begin{aligned}
&\rho \int_{\Omega_h} \frac{(3/2)\mathbf{y}_h^n - 2\mathbf{y}_h^{n-1} + (1/2)\mathbf{y}_h^{n-2}}{\Delta t} \cdot \mathbf{z}\, dx + 2\mu \int_{\Omega_h} \mathbf{D}(\mathbf{y}_h^n) : \mathbf{D}(\mathbf{z})\, dx \\
&+ \rho \int_{\Omega_h} ((2\mathbf{y}_h^{n-1} - \mathbf{y}_h^{n-2}) \cdot \nabla)(2\mathbf{y}_h^{n-1} - \mathbf{y}_h^{n-2}) \cdot \mathbf{z}\, dx \\
&- \int_{\Omega_h} \pi_h^n \nabla \cdot \mathbf{z}\, dx = \rho \int_{\Omega_h} \mathbf{f}^n \cdot \mathbf{z}\, dx, \quad \forall \mathbf{z} \in \mathbf{V}_{0h},
\end{aligned} \tag{9.47}$$

$$\int_{\Omega_h} \nabla \cdot \mathbf{y}_h^n q\, dx = 0, \quad \forall q \in \mathbf{P}_h, \tag{9.48}$$

$$\mathbf{y}_h^n = \mathbf{y}_\infty \quad \text{on } \Gamma_{u,h} \cup \Gamma_{N,h} \cup \Gamma_{S,h}, \tag{9.49}$$

$$\mathbf{y}_h^n = \mathbf{v}^n \quad \text{on } \partial B_h. \tag{9.50}$$

In formulation (9.42)–(9.50) we require

$$\mathbf{y}_h^n \in \mathbf{V}_h, \quad \forall n = 0, 1, \ldots, N, \tag{9.51}$$

$$\pi_h^n \in \mathbf{P}_h, \quad \forall n = 1, \ldots, N. \tag{9.52}$$

The spaces $\mathbf{V}_h$ and $\mathbf{P}_h$ are defined as follows:

$$\mathbf{V}_h = \{\mathbf{z} \mid \mathbf{z} \in (C^0(\bar{\Omega}))^2,\ \mathbf{z}|_K \in Q_{2K}^2,\ \forall K \in \mathcal{Q}_h\}, \tag{9.53}$$

$$\mathbf{P}_h = \{q \mid q \in L^2(\Omega_h),\ q|_K \in P_1,\ \forall K \in \mathcal{Q}_h\}, \tag{9.54}$$

where

- $\mathcal{Q}_h$ is a *"quadrangulation"* of Ω_h (that is, $K \in \mathcal{Q}_h$ implies that K is a convex quadrilateral contained in Ω_h).
- Ω_h is a finite element approximation of Ω.
- The space Q_{2K} is defined by

$$Q_{2K} = \{\varphi \mid \varphi \circ F_K \in Q_2\}, \tag{9.55}$$

with Q_2 the space of the polynomials in x_1, x_2 of degree ≤ 2 with respect to *each* variable, that is,

$$Q_2 = \{q \mid q(x_1,x_2) = \sum_{k,l=0,1,2} a_{kl} x_1^k x_2^l,\ a_{kl} \in \mathbf{R},\ \forall k,l = 0,1,2\};$$

we clearly have $P_2 \subset Q_2 \subset P_4$, each inclusion being a strict one. Concerning F_K, let us say that it is a well-chosen one-to-one mapping from $[0,1]^2$ onto K, such that $F_K \in Q_2^2$ (making thus the above finite element approximation *isoparametric* in the sense of, for example, Ciarlet, 1978).

The space $\mathbf{V}_{0h}$ ($\subset \mathbf{V}_h$) is defined as follows:

$$\mathbf{V}_{0h} = \{\mathbf{z} \mid \mathbf{z} \in \mathbf{V}_h,\ \mathbf{z} = \mathbf{0} \text{ on } \Gamma_{u,h} \cup \Gamma_{N,h} \cup \Gamma_{S,h} \cup \partial B_h\}. \tag{9.56}$$

Let us introduce now the following subspaces of $\mathbf{V}_h$ and $\mathbf{V}_{0h}$

$$\mathbf{W}_h = \{\mathbf{z} \mid \mathbf{z} \in \mathbf{V}_h,\ \int_K q \nabla \cdot \mathbf{z}\, dx = 0,\ \forall q \in \mathbf{P}_h,\ \forall K \in \mathcal{Q}_h\}, \tag{9.57}$$

$$\mathbf{W}_{0h} = \mathbf{W}_h \cap \mathbf{V}_{0h}. \tag{9.58}$$

An equivalent formulation to (9.42)–(9.50) is given by

$$\mathbf{y}_h^0 = \mathbf{y}_{0h} \quad \text{in } \mathbf{W}_h; \tag{9.59}$$

and

$$\rho \int_{\Omega_h} \frac{\mathbf{y}_h^1 - \mathbf{y}_h^0}{\Delta t} \cdot \mathbf{z}\, dx + 2\mu \int_{\Omega_h} \mathbf{D}\left(\frac{2}{3}\mathbf{y}_h^1 + \frac{1}{3}\mathbf{y}_h^0\right) : \mathbf{D}(\mathbf{z})\, dx$$
$$+ \rho \int_{\Omega_h} (\mathbf{y}_h^0 \cdot \nabla)\mathbf{y}_h^0 \cdot \mathbf{z}\, dx$$
$$= \rho \int_{\Omega_h} \mathbf{f}^1 \cdot \mathbf{z}\, dx, \quad \forall \mathbf{z} \in \mathbf{W}_{0h}, \quad \mathbf{y}_h^1 \in \mathbf{W}_h, \tag{9.60}$$

$$\mathbf{y}_h^1 = \mathbf{y}_\infty \quad \text{on } \Gamma_{u,h} \cup \Gamma_{N,h} \cup \Gamma_{S,h}, \tag{9.61}$$
$$\mathbf{y}_h^1 = \mathbf{v}^1 \quad \text{on } \partial B_h; \tag{9.62}$$

then for $n = 2, \ldots, N$

$$\rho \int_{\Omega_h} \frac{(3/2)\mathbf{y}_h^n - 2\mathbf{y}_h^{n-1} + (1/2)\mathbf{y}_h^{n-2}}{\Delta t} \cdot \mathbf{z}\, dx + 2\mu \int_{\Omega_h} \mathbf{D}(\mathbf{y}_h^n) : \mathbf{D}(\mathbf{z})\, dx$$
$$+ \rho \int_{\Omega_h} ((2\mathbf{y}_h^{n-1} - \mathbf{y}_h^{n-2}) \cdot \nabla)(2\mathbf{y}_h^{n-1} - \mathbf{y}_h^{n-2}) \cdot \mathbf{z}\, dx$$
$$= \rho \int_{\Omega_h} \mathbf{f}^n \cdot \mathbf{z}\, dx, \quad \forall \mathbf{z} \in \mathbf{W}_{0h}, \quad \mathbf{y}_h^n \in \mathbf{W}_h, \tag{9.63}$$

$$\mathbf{y}_h^n = \mathbf{y}_\infty \quad \text{on } \Gamma_{u,h} \cup \Gamma_{N,h} \cup \Gamma_{S,h}, \tag{9.64}$$
$$\mathbf{y}_h^n = \mathbf{v}^n \quad \text{on } \partial B_h. \tag{9.65}$$

The unknown discrete pressure function has been eliminated, at the price of having to use finite element spaces defined by nontrivial linear equality constraints and also the necessity to construct vector bases of $\mathbf{W}_h$ and $\mathbf{W}_{0h}$ satisfying these constraints. The construction of *discretely divergence-free vector bases* is a fairly complicated issue; despite the fact that most practitioners do not like this approach for the numerical simulation of incompressible viscous flow we decided to try it, considering that it has the advantage of eliminating the pressure (we also benefited from the most helpful advices of F. Hecht, one of the pioneers and leading experts of this kind of method). We will say no more on approximately divergence-free bases in this book, sending the interested readers to, for example, Pironneau (1989, Chapter 4), Brezzi and Fortin (1991, Chapter 6, Section 8), and Gresho and Sani (1998, Chapter 3, Section 3.13.7).

9.4.3 Formulation of the fully-discrete control problem

Using the above approximation of the Navier–Stokes equations yields the following discrete control problem:

$$\begin{cases} \mathbf{u}_h^{\Delta t} \in \mathcal{U}_h^{\Delta t}, \\ J_h^{\Delta t}(\mathbf{u}_h^{\Delta t}) \leq J_h^{\Delta t}(\mathbf{v}), \quad \forall \mathbf{v} \in \mathcal{U}_h^{\Delta t}, \end{cases} \tag{9.66}$$

with

$$\mathcal{U}_h^{\Delta t} = \mathbf{\Lambda}_{0h}^N, \tag{9.67}$$

$$\mathbf{\Lambda}_{0h} = \{\boldsymbol{\lambda} \mid \int_{\partial B} \boldsymbol{\lambda} \cdot \mathbf{n} \, ds = 0, \, \boldsymbol{\lambda} = \tilde{\boldsymbol{\lambda}}|_{\partial B}, \, \tilde{\boldsymbol{\lambda}} \in \mathbf{W}_h \}, \tag{9.68}$$

and

$$J_h^{\Delta t}(\mathbf{v}) = \frac{\epsilon}{2} \Delta t \sum_{n=1}^{N} \left(\|\mathbf{v}^n\|_\alpha^2 + \left\| \frac{\mathbf{v}^n - \mathbf{v}^{n-1}}{\Delta t} \right\|_0^2 \right) + \Delta t \sum_{n=1}^{N} P_{d,h}^n, \tag{9.69}$$

with $\mathbf{v}^0 = \mathbf{0}$, and where the *discrete drag power* $P_{d,h}^n$ is defined by

$$\begin{aligned} P_{d,h}^1 = \rho \int_{\Omega_h} \frac{\mathbf{y}_h^1 - \mathbf{y}_h^0}{\Delta t} \cdot \mathbf{y}_b^1 \, dx + 2\mu \int_{\Omega_h} \mathbf{D}\left(\frac{2}{3}\mathbf{y}_h^1 + \frac{1}{3}\mathbf{y}_h^0\right) : \mathbf{D}(\mathbf{y}_b^1) \, dx \\ + \rho \int_{\Omega_h} (\mathbf{y}_h^0 \cdot \nabla)\mathbf{y}_h^0 \cdot \mathbf{y}_b^1 \, dx - \rho \int_{\Omega_h} \mathbf{f}^1 \cdot \mathbf{y}_b^1 \, dx, \end{aligned} \tag{9.70}$$

and for $n = 2, \ldots, N$

$$\begin{aligned} P_{d,h}^n = \rho \int_{\Omega_h} \frac{(3/2)\mathbf{y}_h^n - 2\mathbf{y}_h^{n-1} + (1/2)\mathbf{y}_h^{n-2}}{\Delta t} \cdot \mathbf{y}_b^n \, dx + 2\mu \int_{\Omega_h} \mathbf{D}(\mathbf{y}_h^n) : \mathbf{D}(\mathbf{y}_b^n) \, dx \\ + \rho \int_{\Omega_h} ((2\mathbf{y}_h^{n-1} - \mathbf{y}_h^{n-2}) \cdot \nabla)(2\mathbf{y}_h^{n-1} - \mathbf{y}_h^{n-2}) \cdot \mathbf{y}_b^n \, dx \\ - \rho \int_{\Omega_h} \mathbf{f}^n \cdot \mathbf{y}_b^n \, dx; \end{aligned} \tag{9.71}$$

in (9.71), (9.71), $\mathbf{y}_b^n$ can be *any function* belonging to $\mathbf{W}_h$ and such that $\mathbf{y}_b^n|_{\partial B} = \mathbf{v}^n - \mathbf{y}_\infty$, $\mathbf{y}_b^n = \mathbf{0}$ on $\Gamma_{u,h} \cup \Gamma_{N,h} \cup \Gamma_{S,h}$. If this is the case, $\forall n = 1, \ldots, N$, P_{dh}^n approximates the drag power P_d^n defined in Section 9.3.2. In (9.70) and (9.71), $\{\mathbf{y}_h^n\}_{n=1}^N$ is obtained from $\{\mathbf{v}^n\}_{n=1}^N$ via the solution of the system (9.59)–(9.65).

Remark 9.5 Using the facts that $\mathbf{y}_h^n$ and $\mathbf{y}_b^n$ are approximately divergence free, that $\mathbf{y}_b^n$ satisfies $\mathbf{y}_b^n|_{\partial B} = \mathbf{v}^n - \mathbf{y}_\infty$ and $\mathbf{y}_b^n = \mathbf{0}$ on $\Gamma_{u,h} \cup \Gamma_{N,h} \cup \Gamma_{S,h}$, and that $\mathbf{y}_h^n$ satisfies (9.59)–(9.65), it can be shown that the relations (9.70) and (9.71) are discrete analogues of the *drag power* (9.10). For more details on the use of variational approximation of boundary fluxes in the context of the Navier–Stokes equations, see, for example, Tabata and Itakura (1998) and Gunzburger (1999).

9.5 Gradient calculation

Computing the gradient $\nabla J_h^{\Delta t}$ of functional $J_h^{\Delta t}$ is at the same time straightforward and complicated; let us comment on this apparently paradoxical statement: computing the gradient is straightforward in the sense that it relies on a well-established and systematical methodology which has been discussed in many publications (see, for example, Glowinski (1991, Section 7), Glowinski and J.L. Lions (1994, 1995), Berggren (1998), He and Glowinski (1998) and He, Glowinski, Metcalfe, and Périaux (1998); see also the preceding chapters of this book); on the other hand the relative complication of the discrete state equations (9.59)–(9.65) makes the calculation of $\nabla J_h^{\Delta t}$ a bit tedious. After some hesitation, we decided to skip these calculations in this book, sending the interested readers to Glowinski (2003, Chapter 10, Section 53, pp. 1001–1010) (to the best of our knowledge, no detail has been omitted there; on the other hand we found some typos left uncorrected). It follows from the above reference that the gradient $\nabla J_h^{\Delta t}(\mathbf{v})$ of the functional $J_h^{\Delta t}$ at $\mathbf{v}$ is given by

$$\left\{ \begin{aligned} & < \nabla J_h^{\Delta t}(\mathbf{v}), \mathbf{w} > = \Delta t \sum_{n=1}^{N} \int_{\partial B_h} \left(\mathbf{g}_h^n - \frac{\int_{\partial B_h} \mathbf{g}_h^n \cdot \mathbf{n} \, ds}{\int_{\partial B_h} ds} \mathbf{n} \right) \cdot \mathbf{w}^n \, ds, \\ & \forall \mathbf{v} = \{\mathbf{v}^n\}_{n=1}^N, \ \mathbf{w} = \{\mathbf{w}^n\}_{n=1}^N \in \mathcal{U}_h^{\Delta t}, \end{aligned} \right. \tag{9.72}$$

where, in (9.72), $\mathbf{g}_h^n$ is, $\forall n = 1, \ldots, N$, the unique element of

$$\mathbf{\Lambda}_h = \{ \boldsymbol{\lambda} \,|\, \boldsymbol{\lambda} = \tilde{\boldsymbol{\lambda}}|_{\partial B_h}, \ \tilde{\boldsymbol{\lambda}} \in \mathbf{W}_h \}, \tag{9.73}$$

defined by

$$\begin{aligned} \int_{\partial B_h} \mathbf{g}_h^1 \cdot \mathbf{w} \, ds = {} & \epsilon \, (\mathbf{v}^1, \mathbf{w})_\alpha + \epsilon \int_{\partial B_h} \frac{2\mathbf{v}^1 - \mathbf{v}^0 - \mathbf{v}^2}{(\Delta t)^2} \cdot \mathbf{w} \, ds \\ & + \rho \int_{\Omega_h} \frac{\mathbf{p}_h^1 - 2\mathbf{p}_h^2 + (1/2)\mathbf{p}_h^3}{\Delta t} \cdot \tilde{\mathbf{w}} \, dx + \frac{4}{3}\mu \int_{\Omega_h} \mathbf{D}(\mathbf{p}_h^1) : \mathbf{D}(\tilde{\mathbf{w}}) \, dx \\ & + 2\rho \int_{\Omega_h} ((2\mathbf{y}_h^1 - \mathbf{y}_h^0) \cdot \nabla)\tilde{\mathbf{w}} \cdot \mathbf{p}_h^2 \, dx \\ & - \rho \int_{\Omega_h} ((2\mathbf{y}_h^2 - \mathbf{y}_h^1) \cdot \nabla)\tilde{\mathbf{w}} \cdot \mathbf{p}_h^3 \, dx \\ & + 2\rho \int_{\Omega_h} (\tilde{\mathbf{w}} \cdot \nabla)(2\mathbf{y}_h^1 - \mathbf{y}_h^0) \cdot \mathbf{p}_h^2 \, dx \end{aligned}$$

$$
\begin{aligned}
&- \rho \int_{\Omega_h} (\tilde{\mathbf{w}} \cdot \nabla)(2\mathbf{y}_h^2 - \mathbf{y}_h^1) \cdot \mathbf{p}_h^3 \, dx \\
&+ \rho \int_{\Omega_h} \frac{\mathbf{y}_h^1 - \mathbf{y}_h^0}{\Delta t} \cdot \tilde{\mathbf{w}} \, dx + \rho \int_{\Omega_h} (\mathbf{y}_h^0 \cdot \nabla)(\mathbf{y}_h^0) \cdot \tilde{\mathbf{w}} \, dx \\
&+ \frac{4}{3}\mu \int_{\Omega_h} \mathbf{D}\left(\mathbf{y}_h^1 + \frac{1}{2}\mathbf{y}_h^0\right) : \mathbf{D}(\tilde{\mathbf{w}}) \, dx - \rho \int_{\Omega_h} \mathbf{f}_h^1 \cdot \tilde{\mathbf{w}} \, dx, \\
&\forall \mathbf{w} \in \mathbf{\Lambda}_h,
\end{aligned} \tag{9.74}
$$

then, for $n = 2, \ldots, N-2$,

$$
\begin{aligned}
\int_{\partial B_h} \mathbf{g}_h^n \cdot \mathbf{w} \, ds = {} & \epsilon \, (\mathbf{v}^n, \mathbf{w})_\alpha + \epsilon \int_{\partial B_h} \frac{2\mathbf{v}^n - \mathbf{v}^{n-1} - \mathbf{v}^{n+1}}{(\Delta t)^2} \cdot \mathbf{w} \, ds \\
&+ \rho \int_{\Omega_h} \frac{(3/2)\mathbf{p}_h^n - 2\mathbf{p}_h^{n+1} + (1/2)\mathbf{p}_h^{n+2}}{\Delta t} \cdot \tilde{\mathbf{w}} \, dx \\
&+ 2\mu \int_{\Omega_h} \mathbf{D}(\mathbf{p}_h^n) : \mathbf{D}(\tilde{\mathbf{w}}) \, dx \\
&+ 2\rho \int_{\Omega_h} \big((2\mathbf{y}_h^n - \mathbf{y}_h^{n-1}) \cdot \nabla\big)\tilde{\mathbf{w}} \cdot \mathbf{p}_h^{n+1} \, dx \\
&- \rho \int_{\Omega_h} \big((2\mathbf{y}_h^{n+1} - \mathbf{y}_h^n) \cdot \nabla\big)\tilde{\mathbf{w}} \cdot \mathbf{p}_h^{n+2} \, dx \\
&+ 2\rho \int_{\Omega_h} (\tilde{\mathbf{w}} \cdot \nabla)(2\mathbf{y}_h^n - \mathbf{y}_h^{n-1}) \cdot \mathbf{p}_h^{n+1} \, dx \\
&- \rho \int_{\Omega_h} (\tilde{\mathbf{w}} \cdot \nabla)(2\mathbf{y}_h^{n+1} - \mathbf{y}_h^n) \cdot \mathbf{p}_h^{n+2} \, dx \\
&+ \rho \int_{\Omega_h} \frac{(3/2)\mathbf{y}_h^n - 2\mathbf{y}_h^{n-1} + (1/2)\mathbf{y}_h^{n-2}}{\Delta t} \cdot \tilde{\mathbf{w}} \, dx \\
&+ \rho \int_{\Omega_h} \big((2\mathbf{y}_h^{n-1} - \mathbf{y}_h^{n-2}) \cdot \nabla\big)(2\mathbf{y}_h^{n-1} - \mathbf{y}_h^{n-2}) \cdot \tilde{\mathbf{w}} \, dx \\
&+ 2\mu \int_{\Omega_h} \mathbf{D}(\mathbf{y}_h^n) : \mathbf{D}(\tilde{\mathbf{w}}) \, dx - \rho \int_{\Omega_h} \mathbf{f}_h^n \cdot \tilde{\mathbf{w}} \, dx, \\
&\forall \mathbf{w} \in \mathbf{\Lambda}_h,
\end{aligned} \tag{9.75}
$$

and, finally,

$$
\begin{aligned}
\int_{\partial B_h} \mathbf{g}_h^{N-1} \cdot \mathbf{w} \, ds = {} & \epsilon \, (\mathbf{v}^{N-1}, \mathbf{w})_\alpha + \epsilon \int_{\partial B_h} \frac{2\mathbf{v}^{N-1} - \mathbf{v}^{N-2} - \mathbf{v}^N}{(\Delta t)^2} \cdot \mathbf{w} \, ds \\
&+ \rho \int_{\Omega_h} \frac{(3/2)\mathbf{p}_h^{N-1} - 2\mathbf{p}_h^N}{\Delta t} \cdot \tilde{\mathbf{w}} \, dx + 2\mu \int_{\Omega_h} \mathbf{D}(\mathbf{p}_h^{N-1}) : \mathbf{D}(\tilde{\mathbf{w}}) \, dx
\end{aligned}
$$

$$
\begin{aligned}
&+ 2\rho \int_{\Omega_h} ((2\mathbf{y}_h^{N-1} - \mathbf{y}_h^{N-2}) \cdot \nabla)\tilde{\mathbf{w}} \cdot \mathbf{p}_h^N \, dx \\
&+ 2\rho \int_{\Omega_h} (\tilde{\mathbf{w}} \cdot \nabla)(2\mathbf{y}_h^{N-1} - \mathbf{y}_h^{N-2}) \cdot \mathbf{p}_h^N \, dx \\
&+ \rho \int_{\Omega_h} \frac{(3/2)\mathbf{y}_h^{N-1} - 2\mathbf{y}_h^{N-2} + (1/2)\mathbf{y}_h^{N-3}}{\Delta t} \cdot \tilde{\mathbf{w}} \, dx \\
&+ \rho \int_{\Omega_h} ((2\mathbf{y}_h^{N-2} - \mathbf{y}_h^{N-3}) \cdot \nabla)(2\mathbf{y}_h^{N-2} - \mathbf{y}_h^{N-3}) \cdot \tilde{\mathbf{w}} \, dx \\
&+ 2\mu \int_{\Omega_h} \mathbf{D}(\mathbf{y}_h^{N-1}) : \mathbf{D}(\tilde{\mathbf{w}}) \, dx - \rho \int_{\Omega_h} \mathbf{f}_h^{N-1} \cdot \tilde{\mathbf{w}} \, dx, \\
&\forall \mathbf{w} \in \mathbf{\Lambda}_h,
\end{aligned}
\tag{9.76}
$$

and

$$
\begin{aligned}
\int_{\partial B_h} \mathbf{g}_h^N \cdot \mathbf{w} \, ds = {} & \epsilon \, (\mathbf{v}^N, \mathbf{w})_\alpha + \epsilon \int_{\partial B_h} \frac{\mathbf{v}^N - \mathbf{v}^{N-1}}{(\Delta t)^2} \cdot \mathbf{w} \, ds \\
&+ \frac{3}{2}\rho \int_{\Omega_h} \frac{\mathbf{p}_h^N}{\Delta t} \cdot \tilde{\mathbf{w}} \, dx + 2\mu \int_{\Omega_h} \mathbf{D}(\mathbf{p}_h^N) : \mathbf{D}(\tilde{\mathbf{w}}) \, dx \\
&+ \rho \int_{\Omega_h} \frac{(3/2)\mathbf{y}_h^N - 2\mathbf{y}_h^{N-1} + (1/2)\mathbf{y}_h^{N-2}}{\Delta t} \cdot \tilde{\mathbf{w}} \, dx \\
&+ \rho \int_{\Omega_h} ((2\mathbf{y}_h^{N-1} - \mathbf{y}_h^{N-2}) \cdot \nabla)(2\mathbf{y}_h^{N-1} - \mathbf{y}_h^{N-2}) \cdot \tilde{\mathbf{w}} \, dx \\
&+ 2\mu \int_{\Omega_h} \mathbf{D}(\mathbf{y}_h^N) : \mathbf{D}(\tilde{\mathbf{w}}) \, dx - \rho \int_{\Omega_h} \mathbf{f}_h^N \cdot \tilde{\mathbf{w}} \, dx, \\
&\forall \mathbf{w} \in \mathbf{\Lambda}_h.
\end{aligned}
\tag{9.77}
$$

In (9.75)–(9.77) $\tilde{\mathbf{w}}$ is any function belonging to $\mathbf{W}_h$, such that $\tilde{\mathbf{w}}|_{\partial B_h} = \mathbf{w}$ and $\tilde{\mathbf{w}} = \mathbf{0}$ on $\Gamma_{uh} \cup \Gamma_{Nh} \cup \Gamma_{Sh}$; moreover, $\{\mathbf{p}_h^n\}_{n=1}^N$ $(\in (\mathbf{W}_h)^N)$ is the solution of the following *discrete adjoint system*:

$$
\left\{
\begin{aligned}
&\mathbf{p}_h^N \in \mathbf{W}_h, \\
&\frac{3}{2}\rho \int_{\Omega_h} \frac{\mathbf{p}_h^N}{\Delta t} \cdot \mathbf{z} \, dx + 2\mu \int_{\Omega_h} \mathbf{D}(\mathbf{p}_h^N) : \mathbf{D}(\mathbf{z}) \, dx = 0, \quad \forall \mathbf{z} \in \mathbf{W}_{0h}, \\
&\mathbf{p}_h^N = \mathbf{v}_h^N - \mathbf{y}_\infty \quad \text{on } \partial B_h, \quad \mathbf{p}_h^N = \mathbf{0} \quad \text{on } \Gamma_{uh} \cup \Gamma_{Nh} \cup \Gamma_{Sh},
\end{aligned}
\right.
\tag{9.78}
$$

and

$$\begin{cases} \mathbf{p}_h^{N-1} \in \mathbf{W}_h, \\ \rho \int_{\Omega_h} \dfrac{(3/2)\mathbf{p}_h^{N-1} - 2\mathbf{p}_h^N}{\Delta t} \cdot \mathbf{z}\, dx + 2\mu \int_{\Omega_h} \mathbf{D}(\mathbf{p}_h^{N-1}) : \mathbf{D}(\mathbf{z})\, dx \\ \quad +2\rho \int_{\Omega_h} \big((2\mathbf{y}_h^{N-1} - \mathbf{y}_h^{N-2}) \cdot \nabla\big)\mathbf{z} \cdot \mathbf{p}_h^N\, dx \\ \quad +2\rho \int_{\Omega_h} (\mathbf{z} \cdot \nabla)(2\mathbf{y}_h^{N-1} - \mathbf{y}_h^{N-2}) \cdot \mathbf{p}_h^N\, dx = 0, \quad \forall \mathbf{z} \in \mathbf{W}_{0h}, \\ \mathbf{p}_h^{N-1} = \mathbf{v}_h^{N-1} - \mathbf{y}_\infty \quad \text{on } \partial B_h, \quad \mathbf{p}_h^{N-1} = \mathbf{0} \quad \text{on } \Gamma_{uh} \cup \Gamma_{Nh} \cup \Gamma_{Sh}, \end{cases} \tag{9.79}$$

then, for $n = N-2, \ldots, 2$,

$$\begin{cases} \mathbf{p}_h^{n} \in \mathbf{W}_h, \\ \rho \int_{\Omega_h} \dfrac{(3/2)\mathbf{p}_h^{n} - 2\mathbf{p}_h^{n+1} + (1/2)\mathbf{p}_h^{n+2}}{\Delta t} \cdot \mathbf{z}\, dx + 2\mu \int_{\Omega_h} \mathbf{D}(\mathbf{p}_h^{n}) : \mathbf{D}(\mathbf{z})\, dx \\ \quad +2\rho \int_{\Omega_h} \big((2\mathbf{y}_h^{n} - \mathbf{y}_h^{n-1}) \cdot \nabla\big)\mathbf{z} \cdot \mathbf{p}_h^{n+1}\, dx \\ \quad -\rho \int_{\Omega_h} \big((2\mathbf{y}_h^{n+1} - \mathbf{y}_h^{n}) \cdot \nabla\big)\mathbf{z} \cdot \mathbf{p}_h^{n+2}\, dx \\ \quad +2\rho \int_{\Omega_h} (\mathbf{z} \cdot \nabla)(2\mathbf{y}_h^{n} - \mathbf{y}_h^{n-1}) \cdot \mathbf{p}_h^{n+1}\, dx \\ \quad -\rho \int_{\Omega_h} (\mathbf{z} \cdot \nabla)(2\mathbf{y}_h^{n+1} - \mathbf{y}_h^{n}) \cdot \mathbf{p}_h^{n+2}\, dx = 0, \quad \forall \mathbf{z} \in \mathbf{W}_{0h}, \\ \mathbf{p}_h^{n} = \mathbf{v}_h^{n} - \mathbf{y}_\infty \quad \text{on } \partial B_h, \quad \mathbf{p}_h^{n} = \mathbf{0} \quad \text{on } \Gamma_{uh} \cup \Gamma_{Nh} \cup \Gamma_{Sh}, \end{cases} \tag{9.80}$$

and, finally, for $n = 1$,

$$\begin{cases} \mathbf{p}_h^{1} \in \mathbf{W}_h, \\ \rho \int_{\Omega_h} \dfrac{\mathbf{p}_h^{1} - 2\mathbf{p}_h^{2} + \frac{1}{2}\mathbf{p}_h^{3}}{\Delta t} \cdot \mathbf{z}\, dx + \dfrac{4}{3}\mu \int_{\Omega_h} \mathbf{D}(\mathbf{p}_h^{1}) : \mathbf{D}(\mathbf{z})\, dx \\ \quad +2\rho \int_{\Omega_h} \big((2\mathbf{y}_h^{1} - \mathbf{y}_h^{0}) \cdot \nabla\big)\mathbf{z} \cdot \mathbf{p}_h^{2}\, dx \\ \quad -\rho \int_{\Omega_h} \big((2\mathbf{y}_h^{2} - \mathbf{y}_h^{1}) \cdot \nabla\big)\mathbf{z} \cdot \mathbf{p}_h^{3}\, dx \\ \quad +2\rho \int_{\Omega_h} (\mathbf{z} \cdot \nabla)(2\mathbf{y}_h^{1} - \mathbf{y}_h^{0}) \cdot \mathbf{p}_h^{2}\, dx \\ \quad -\rho \int_{\Omega_h} (\mathbf{z} \cdot \nabla)(2\mathbf{y}_h^{2} - \mathbf{y}_h^{1}) \cdot \mathbf{p}_h^{3}\, dx = 0, \quad \forall \mathbf{z} \in \mathbf{W}_{0h}, \\ \mathbf{p}_h^{1} = \mathbf{v}_h^{1} - \mathbf{y}_\infty \quad \text{on } \partial B_h, \quad \mathbf{p}_h^{1} = \mathbf{0} \quad \text{on } \Gamma_{uh} \cup \Gamma_{Nh} \cup \Gamma_{Sh}, \end{cases} \tag{9.81}$$

Once $\nabla J_h^{\Delta t}$ is known, via the solution of the above adjoint system, we can derive optimality conditions which enable us to solve the discrete control problem (9.66) by various kinds of descent methods such as conjugate gradient, BFGS, and so on. Here, the optimality conditions take the following form:

$$< \nabla J_h^{\Delta t}(\mathbf{u}_h^{\Delta t}) , \mathbf{w} >= 0, \quad \forall \mathbf{w} \in \mathcal{U}_h^{\Delta t}, \tag{9.82}$$

that is, from (9.72) and with obvious notation,

$$\int_{\partial B_h} \left(\mathbf{g}_h^n - \frac{\mathbf{n} \int_{\partial B_h} \mathbf{g}_h^n \cdot \mathbf{n} \, \mathrm{d}s}{\int_{\partial B_h} \mathrm{d}s} \right) \cdot \mathbf{w} \, \mathrm{d}s = 0, \quad \forall \mathbf{w} \in \mathbf{\Lambda}_{0h}, \quad \forall n = 1, \ldots, N. \tag{9.83}$$

The BFGS solution of problem (9.66), via (9.82), (9.83), will be discussed in Section 9.6 hereafter.

9.6 A BFGS algorithm for solving the discrete control problem

In order to solve the discrete control problem (9.66), via the optimality conditions (9.82) and (9.83), we shall employ a *quasi-Newton method* à la BFGS (here, B, F, G, and S, stand for Broyden, Fletcher, Goldfarb, and Schanno, the coinventors of the method in the early seventies); for a discussion of the convergence properties of BFGS algorithms, and practical details of on their implementation, see, for example, Dennis and Schnabel (1983, Chapter 9 and Appendix A), (1996, Chapter 9 and Appendix A), Liu and Nocedal (1989), and Nocedal (1992). Let us consider now the following generic optimization problem in finite dimension:

$$\begin{cases} x \in \mathbb{R}^l, \\ f(x) \leq f(y), \ \forall y \in \mathbb{R}^l. \end{cases} \tag{9.84}$$

If f is smooth enough the solution of problem (9.84) verifies also

$$\nabla f(x) = 0. \tag{9.85}$$

The BFGS methodology applied to the solution of (9.84), (9.85) leads to the following algorithm:

$$x^0 \in \mathbb{R}^l, H^0 \in \mathcal{L}(\mathbb{R}^l, \mathbb{R}^l) \quad \text{are given,} \tag{9.86}$$

$$g^0 = \nabla f(x^0). \tag{9.87}$$

For $k \geq 0$, assuming that x^k, H^k, and g^k are known, we proceed as follows:

$$d^k = -H^k g^k, \tag{9.88}$$

$$\begin{cases} \rho_k \in \mathbb{R}, \\ f(x^k + \rho_k d^k) \leq f(x^k + \rho d^k), \quad \forall \rho \in \mathbb{R}, \end{cases} \tag{9.89}$$

$$x^{k+1} = x^k + \rho_k d^k, \tag{9.90}$$

$$g^{k+1} = \nabla f(x^{k+1}), \tag{9.91}$$

$$s^k = x^{k+1} - x^k, \tag{9.92}$$

$$y^k = g^{k+1} - g^k, \tag{9.93}$$

$$H^{k+1} = H^k + \frac{(s^k - H^k y^k) \otimes s^k + s^k \otimes (s^k - H^k y^k)}{(y^k, s^k)} - \frac{(s^k - H^k y^k, y^k)}{(y^k, s^k)^2} s^k \otimes s^k. \tag{9.94}$$

Set $k = k + 1$ and return to (9.88).

The tensor product $u \otimes v$ of the two vectors u and v, both belonging to $\mathbb{R}^l$, is defined as usual as the linear operator from $\mathbb{R}^l \times \mathbb{R}^l$ into $\mathbb{R}^l$ such that

$$(u \otimes v)w = (v, w)u, \quad \forall w \in \mathbb{R}^l, \tag{9.95}$$

which implies in turn that

$$(u \otimes v)_{ij} = u_i v_j, \quad 1 \leq i, j \leq l \tag{9.96}$$

if the scalar product in (9.95) is the dot product of $\mathbb{R}^l$. Actually, the Euclidian scalar product used in (9.94) and (9.95) is not necessarily the above dot product (we may have $(v, w) = Sv \cdot w$ with S being a $l \times l$ matrix, symmetric, positive definite, and possibly different from the identity matrix).

Applying algorithm (9.86)–(9.94) to the solution of problem (9.94) is (almost) straightforward.

9.7 Validation of the flow simulator

9.7.1 Motivation

An important issue for the flow control problems discussed in this chapter is the quality of the flow simulator, that is, of the methodology which will be used to solve the Navier–Stokes equations modeling the flow (and also the adjoint equations) in order to compute $\nabla J_h^{\Delta t}$. For the validation of the flow simulator we have chosen as test problem the flow past a circular cylinder at various Reynolds numbers. This test problem has the advantage of combining a simple geometry with a complex flow dynamics and it has always been a traditional benchmarking problem for incompressible viscous flow simulators (see, for example, Fornberg, 1980; Braza, Chassaing, and Minh, 1986; Ingham and Tang, 1990; Badr et al., 1990; see also the many references in the above publications). Also, this particular geometry has motivated the work of several flow investigators from the experimental point of view (see, for example, Roshko, 1955; Williamson, 1989; Tokumaru and Dimotakis, 1991).

(a)

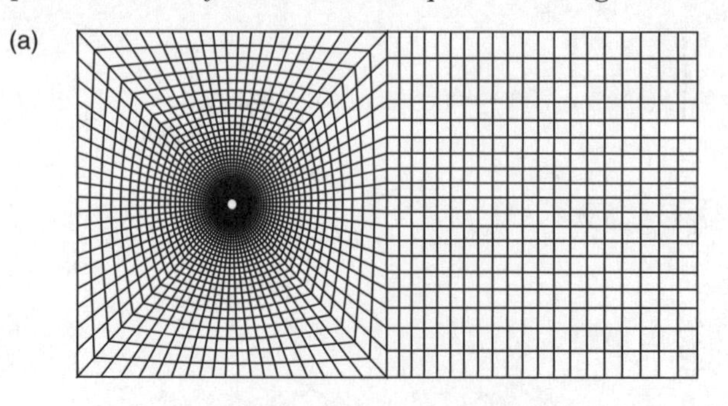

(b)

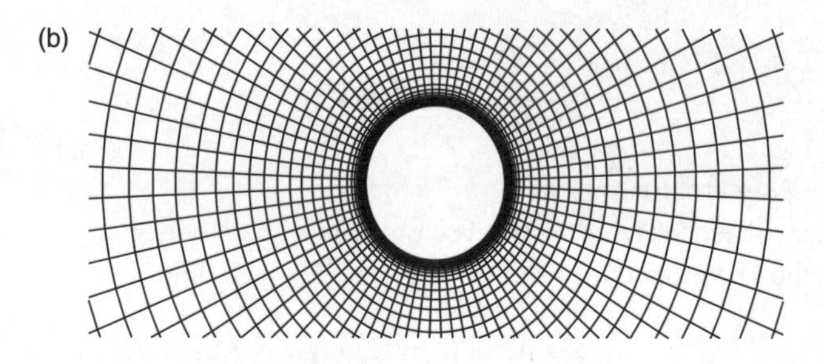

Fig. 9.2. Mesh used at Re = 1000.

9.7.2 Description of the mesh and other parameters

In order to validate our incompressible viscous flow simulator, we have chosen as computational domain the two-dimensional region Ω so that

$$\Omega = \Pi \backslash \bar{B}$$

where Π is the rectangle $(-15, 45) \times (-15, 15)$ and B is the (open) disk of center $(0, 0)$ and of radius $a = 0.5$. The diameter of B will be taken as characteristic length, implying that the Reynolds number is defined by

$$\text{Re} = \frac{2a\rho|\mathbf{y}_\infty|}{\mu}.$$

The simulations will be done with $\mu = 1/200$ and $1/1000$ implying that Re = 200 and 1000 for $\rho = 1$ and $|\mathbf{y}_\infty| = 1$. The finite element mesh used for the calculations at Re = 1000 has been shown in Figure 9.2, where we have also visualized the mesh in the neighborhood of B; we observe the refinement used close to ∂B in order to better capture the boundary layer.

Actually, further information concerning both Re = 200 and Re = 1000 calculations is provided by Table 9.1.

9.7.3 Numerical results and comparisons

The goal of the computational experiments reported here is to simulate the development of the *vortex street* in an unforced laminar wake behind the circular cylinder

Table 9.1. *Discretization parameters used at Re = 200 and 1000.*

Re	Δt	Number of finite elements	Number of vertices	Number of nodes	Number of unknowns
200	5×10^{-3}	2 226	2 297	9 046	11 415
1000	10^{-3}	4 104	4 200	16 608	20 905

of Figure 9.2, for Re = 200 and 1000. Although the simulations at Re = 1000 are two-dimensional and do not include the effects of three-dimensional instabilities and turbulence, these high Reynolds number two-dimensional simulations are still of interest; indeed they will allow us comparisons with the results of other two-dimensional simulations and will provide us with some key insight on the dynamics of those large two-dimensional vortices, that clearly dominate the high Reynolds number flow experiments reported in, for example, Tokumaru and Dimotakis (1991). Actually, for values of Re below 40–50, a stable steady flow is observed with formation of a symmetric recirculation region in the wake. The length of this recirculation region increases with the Reynolds number and beyond a certain critical value of Re the flow becomes unstable. Alternating eddies are formed in an asymmetrical pattern which generates an alternating separation of vortices. These vortices are advected and diffused downstream forming the well-known *Kármán vortex street*. In "actual life" the *symmetry breaking* is triggered by various causes such as disturbances in the initial and/or boundary conditions. In our simulations, the computational mesh and the boundary conditions are perfectly symmetric. As initial condition we have taken the symmetric steady-state solution obtained (asymptotically) from a time dependent Navier–Stokes computation where symmetry is systematically enforced at each time step by averaging. This symmetric solution (unstable for Re sufficiently large) is itself used as initial condition for a simulation where the symmetry constraint has been removed. The symmetry breaking occurring for Re sufficiently large can be explained by the various truncation and rounding errors taking place in the computation. At the initial stage of the symmetry breaking, the growth of the perturbation is linear and the drag coefficient grows first very fast up to a point where the growth become oscillating and a saturation is observed. In Figures 9.3 and 9.4 we have represented the time variations of the drag and the lift for Re = 200 and 1000, respectively.

The periodic regime which is reached, asymptotically, is characterized by the frequency f_n at which the vortices are shed. For comparison purposes it has been found convenient to introduce the *Strouhal number*

$$S_n = \frac{2a}{|\mathbf{y}_\infty|} f_n; \tag{9.97}$$

S_n is a nondimensional representation of the shedding frequency. For various values of Re, we compare, in Table 9.2, the Strouhal numbers obtained from our simulations

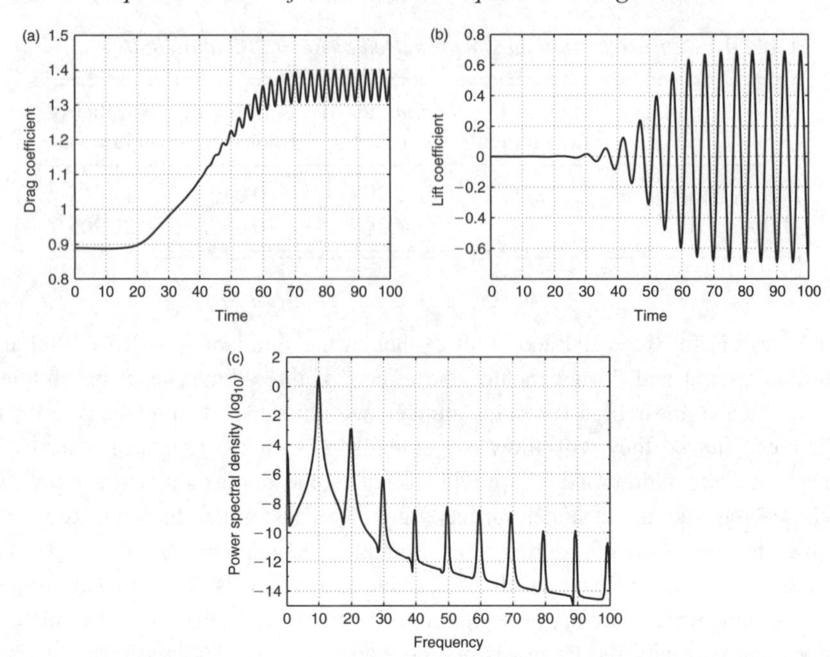

Fig. 9.3. Case of a fixed circular cylinder in a uniform free-stream flow at Re = 200. (a) Drag coefficient, (b) lift coefficient, (c) history of the power spectrum of the lift coefficient. The Strouhal number is 0.1978.

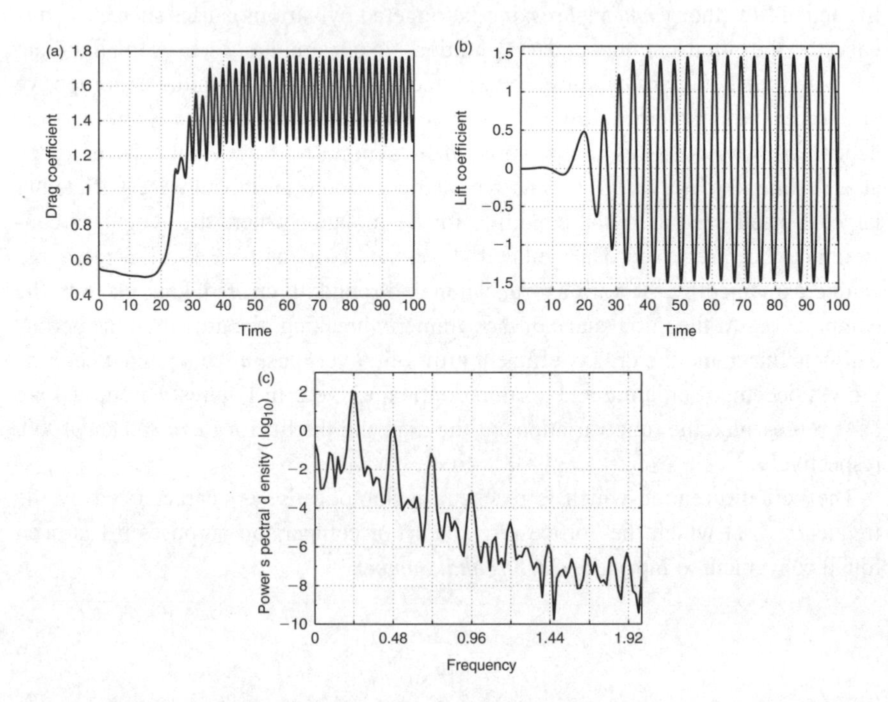

Fig. 9.4. Case of a fixed circular cylinder in a uniform free-stream flow at Re = 1000. (a) Drag coefficient, (b) lift coefficient, (c) history of the power spectrum of the lift coefficient. The Strouhal number is 0.2392.

Table 9.2. *Strouhal number for different Reynolds numbers. Comparison with Roshko (1954), Braza, Chassaing, and Minh (1986), Williamson (1989), and Henderson (1997).*

			S_n		
Re	Present	Henderson	Williamson	Roshko	Braza
60	0.1353	0.1379	0.1356	0.1370	—
80	0.1526	0.1547	0.1521	0.1557	—
100	0.1670	0.1664	0.1640	0.1670	0.16
200	0.1978	0.1971	—	—	0.20
400	0.2207	0.2198	—	—	—
600	0.2306	0.2294	—	—	—
800	0.2353	0.2343	—	—	—
1000	0.2392	0.2372	—	—	0.21

Table 9.3. *Drag coefficient for different Reynolds numbers. Comparison with Fornberg (1980), Braza, Chassaing, and Minh (1986), and Henderson (1997).*

			C_D	
Re	Present work	Henderson	Braza et al.	Fornberg
20	2.0064	2.0587	2.19	2.0001
40	1.5047	1.5445	1.58	1.4980
60	1.3859	1.4151	1.35	—
80	1.3489	1.3727	—	—
100	1.3528	1.3500	1.36	—
200	1.3560	1.3412	1.39	—
400	1.4232	1.4142	—	—
600	1.4641	1.4682	—	—
800	1.4979	1.4966	—	—
1000	1.5191	1.5091	1.198	—

with those obtained experimentally and computationally by various authors, namely, Roshko (1954), Fornberg (1980), Braza, Chassaing, and Minh (1986), Williamson (1989), and Henderson (1997). The agreement with the Henderson's computational results and Williamson experimental ones is very good for Re between 60 and 1000 (for more details on these comparisons see Nordlander, 1998). Similarly, in Table 9.3, the time-averaged drag coefficient is clearly in very good agreement with Henderson's results for the steady and periodic solutions.

A well-known effect of having just two dimensions in numerical simulations as opposed to three is that the drag tends to be overpredicted for high Reynolds number flow, where three-dimensional instabilities would occur. For more details on these drag comparisons, see, again, Nordlander (1998).

9.8 Active control by rotation

9.8.1 Synopsis

In this section we will use simulations to investigate various strategies for the *active control by rotation* of the flow around a cylinder. In Section 9.8.2 we consider the dynamical behavior of the flow under the effect of forced sinusoidal rotation of the cylinder. Then in Section 9.8.3 we present the results obtained when applying the optimal control strategy discussed in Sections 9.2–9.6.

9.8.2 Active control by forced sinusoidal rotation

The active control discussed in this section is based on *oscillatory rotation* as in the experiments of Tokumaru and Dimotakis (1991). If the forcing is *sinusoidal* there are *two degrees of freedom*, namely, the *frequency* f_e and the *amplitude* ω_1 of the *angular velocity*. The forcing Strouhal number is defined as

$$S_e = 2af_e/|\mathbf{y}_\infty|$$

which yields the following forcing angular velocity:

$$\omega(t) = \omega_1 \sin(2\pi S_e t).$$

A series of simulations with different forcing frequencies S_e varying from 0.35 to 1.65 was performed at Re $= 200$. The amplitude ω_1 of the forcing angular velocity was held fixed to the value 6 for all simulations. Once the transients have died out, a spectral analysis of the (time-dependent) drag minus its time-averaged value was performed, leading to the results shown in Figure 9.5(a), (b), and (c), which correspond to $S_e = 0.75$, 1.25, and 1.65, respectively. Several comments are in order:

(i) At $S_e = 0.75$ a perfect lock-in to the forcing frequency can be observed, in which the forcing frequency dominates the dynamics of the flow (in simple terms: *the flow oscillates at the forcing frequency*).
(ii) At $S_e = 1.25$, there is a competition between the forcing frequency and the natural fundamental shedding frequency. The dynamics corresponds to a *quasi periodic* state.
(iii) At $S_e = 1.65$ the flow dynamics is dominated by the natural shedding frequency (≈ 0.2 according to Table 9.2); the forcing frequency has little influence on the flow dynamics.

These results agree with those in Karniadakis and Triantafyllou (1989) where one discusses the active control of flow around cylinders by sinusoidal transversal motions (a kind of *chattering control*; reminiscent of Chapter 3, Section 3.11.7).

Similar experiments were performed at Re $= 1000$, with $\omega_1 = 5.5$ and $S_e = 0.625$, 1.325, and 1.425. The corresponding results are reported in Figure 9.6(a), (b), and (c).

The computed results suggest the existence of a critical amplitude for the forcing; we need to operate beyond this threshold for the flow to "feel" the forcing. It was

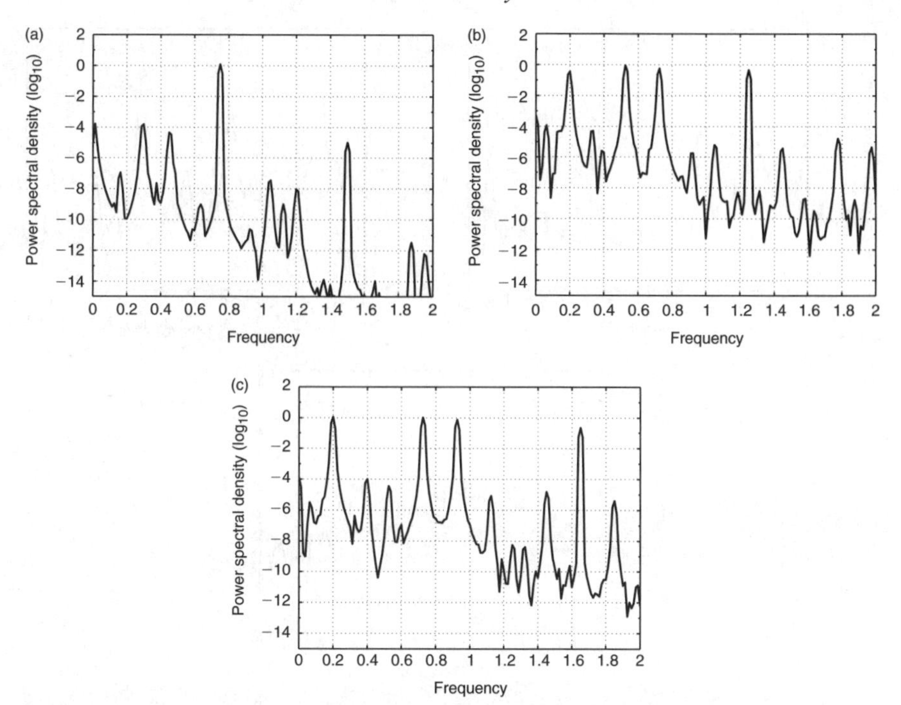

Fig. 9.5. Case of a sinusoidal rotating cylinder in a uniform free-stream flow (Re $= 200$). The power spectral density of the lift coefficient is shown for: (a) lock-in ($S_e = 0.75$), (b) quasiperiodic ($S_e = 1.25$), and (c) nonreceptive state ($S_e = 1.65$). The natural Strouhal number (S_n) is 0.1978.

further observed that this threshold is a function of the forcing frequency: higher frequencies require higher amplitude for the control to stay effective.

The above results suggest looking for *optimal pairs* $\{\omega_1, S_e\}$ for the *drag minimization*. To be more precise, we consider the drag as a function of $\{\omega_1, S_e\}$ and try to minimize it for $\{\omega_1, S_e\}$ varying in a "reasonable" subset of $\mathbb{R}^2$. For Re $= 200$ a hybrid method coupling *direct search* and a BFGS algorithm gives $\omega_1 = 6$ and $S_e = 0.74$, which corresponds to the lock-in case previously described. In Figure 9.7 we have visualized the contours of the drag, considered as a function of ω_1, and S_e, in the neighborhood of the optimal solution. In Figure 9.8(a) we have represented the variation versus time of the optimal sinusoidal control whose action started at time $t = 0$. The transition to low drag is visualized in Figure 9.8(b), which also shows the shedding frequency transition. The drag reduction was found to be of the order of 30%. Finally the lift coefficient is visualized in Figure 9.8(c); we observe that the amplitude of the lift oscillations is substantially reduced. Finally, in Figure 9.9(a) and (b) we have shown snapshots of the uncontrolled flow and of the optimally forced flow. The significant vortex-shedding phenomenon in Figure 9.9(a) has been substantially reduced and the flow has been "almost" symmetrized. This is qualitatively similar to the effects observed by Tokumaru and Dimotakis (1991).

Details of the vortex shedding for various values of t are reported in Figure 9.10; the above figure clearly shows the important reduction of vortex shedding in the

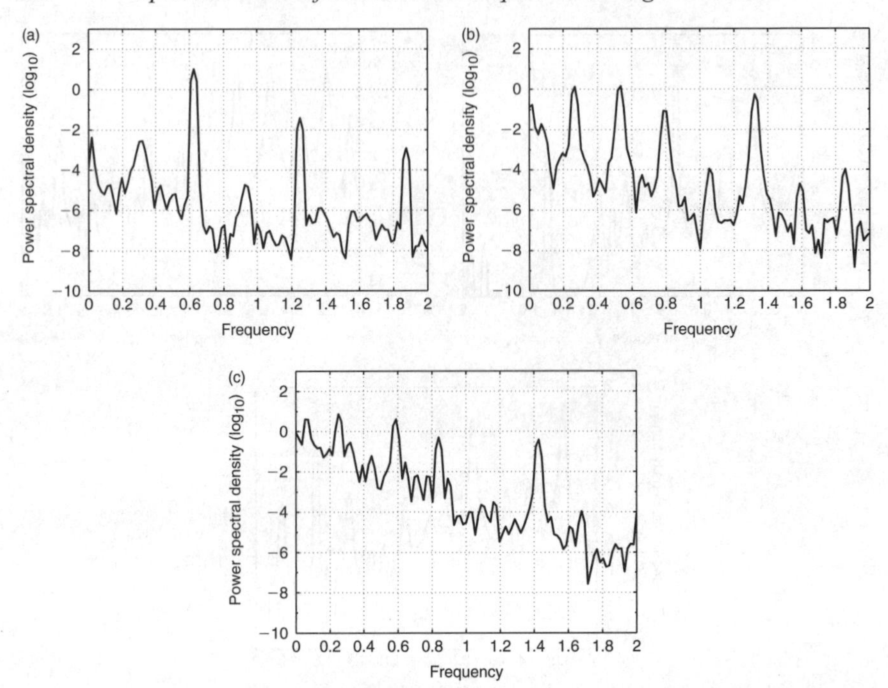

Fig. 9.6. Case of a sinusoidal rotating cylinder in a uniform free-stream flow, Re = 1000. The power spectral density of the lift coefficient is shown for: (a) lock-in ($S_e = 0.625$), (b) quasiperiodic ($S_e = 1.325$), and (c) nonreceptive state ($S_e = 1.425$). The natural Strouhal number (S_n) is 0.2392.

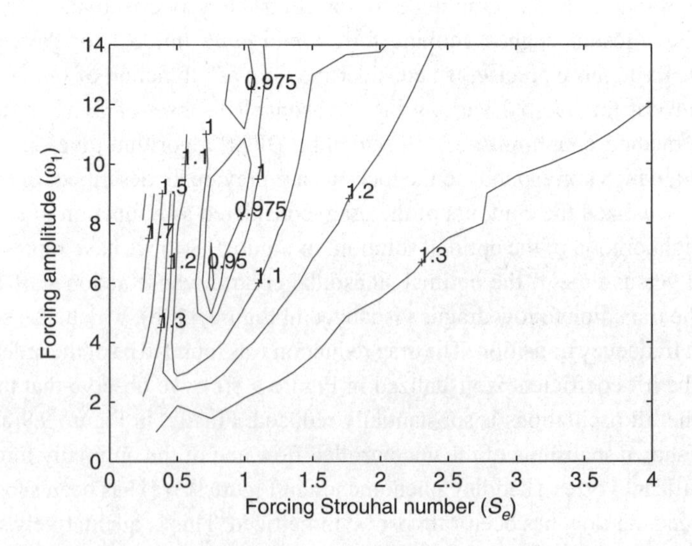

Fig. 9.7. Variation of the drag C_D with S_e and ω_1 at Re = 200.

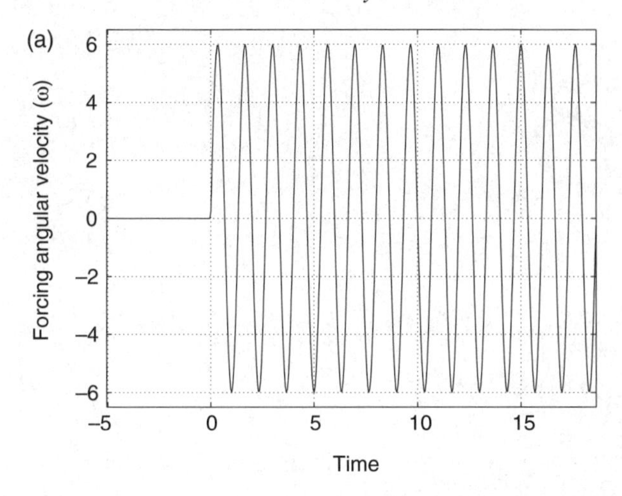

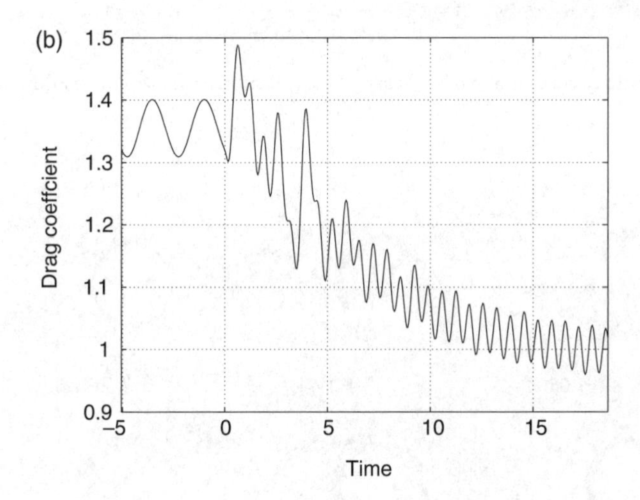

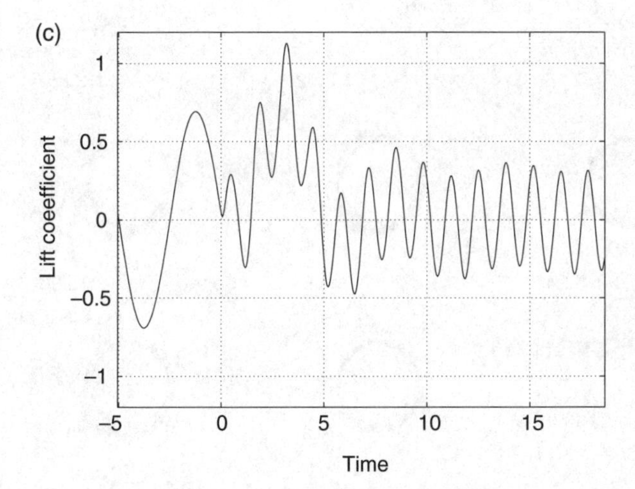

Fig. 9.8. The time evolution of the (a) sinusoidal-optimal forcing, with $S_e = 0.75$ and $\omega_1 = 6.00$, (b) drag C_D, and (c) lift C_L, at Re $= 200$. Forcing was started at time $t = 0$.

Fig. 9.9. Vorticity contour plot of the wake of the (a) unforced and (b) forced flow at Re = 200.

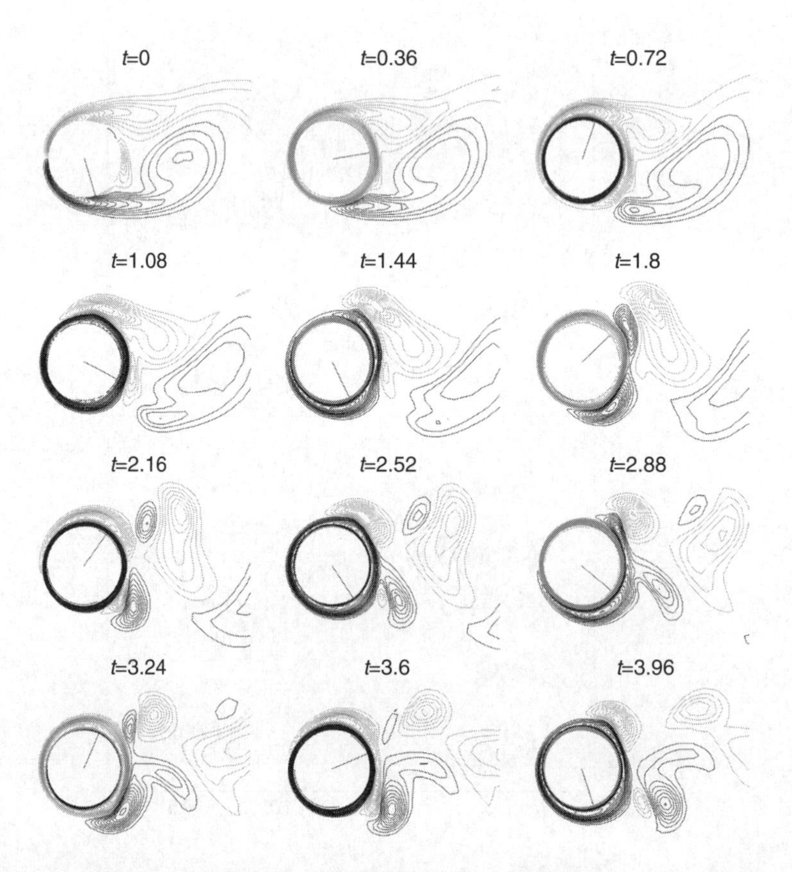

Fig. 9.10. Near-wake region: forced vortex shedding at Re = 200 with $S_e = 0.75$ and $\omega_1 = 6$. The sequence represents the first three forcing period.

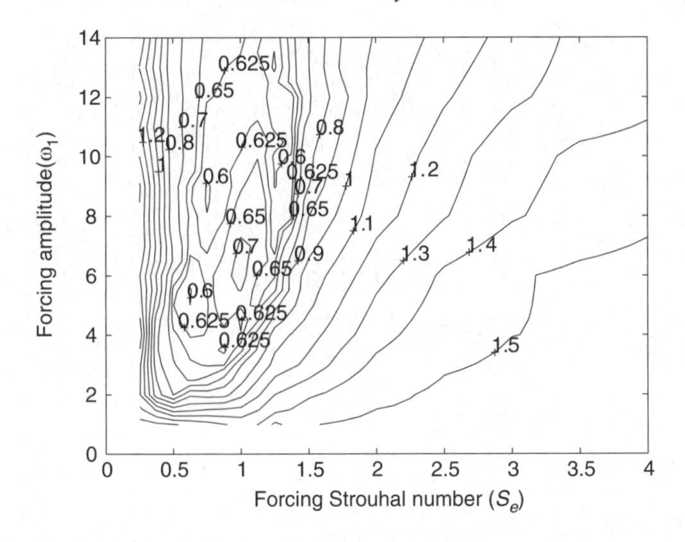

Fig. 9.11. Variation of the drag C_D with S_e and ω_1 at Re = 1000.

cylinder wake. We recall that the *vorticity vector* is defined by $\nabla \times \mathbf{y}$, namely, for a two-dimensional flow, by $\frac{\partial y_2}{\partial x_1} - \frac{\partial y_1}{\partial x_2}$.

Similar experiments have been carried out for Re $= 1000$. From a qualitative point of view, the simulated phenomena are identical to those observed for Re $= 200$; however, the drag reduction is, this time, of the order of 60%, the optimal amplitude and frequency being this time $\omega_1 = 5.5$ and $S_e = 0.625$. The results shown in Figures 9.11–9.14 are self-explanatory.

9.8.3 Drag reduction by optimal control

In this section we will present the results obtained by applying the methods discussed in Sections 9.2–9.6 to active flow control by rotation, the cost function being essentially the drag, since the following results have been obtained with $\epsilon = 0$ in (9.9). The values of Re are as in Section 9.8.2, namely, 200 and 1000. The discretization parameters are also chosen as in Section 9.8.2. As initial guess for the optimal control computation we have used the quasi-optimal forcing obtained in Section 9.8.2. Typically, convergence was obtained in 20 iterations of the BFGS algorithm using

$$\frac{(g^{k+1}, g^{k+1})}{(g^0, g^0)} \leq 10^{-6}$$

as stopping criterion. Let us comment first on the results obtained for Re $= 200$. In Figure 9.15(a) we have represented the computed optimal control (–) as a function of t and compared it to the optimal sinusoidal control (– –) obtained in Section 9.8.2. We observe that the fundamental frequency of the optimal control is very close to the optimal frequency for the sinusoidal control. The power spectral density of the optimal control is shown in Figure 9.15(b).

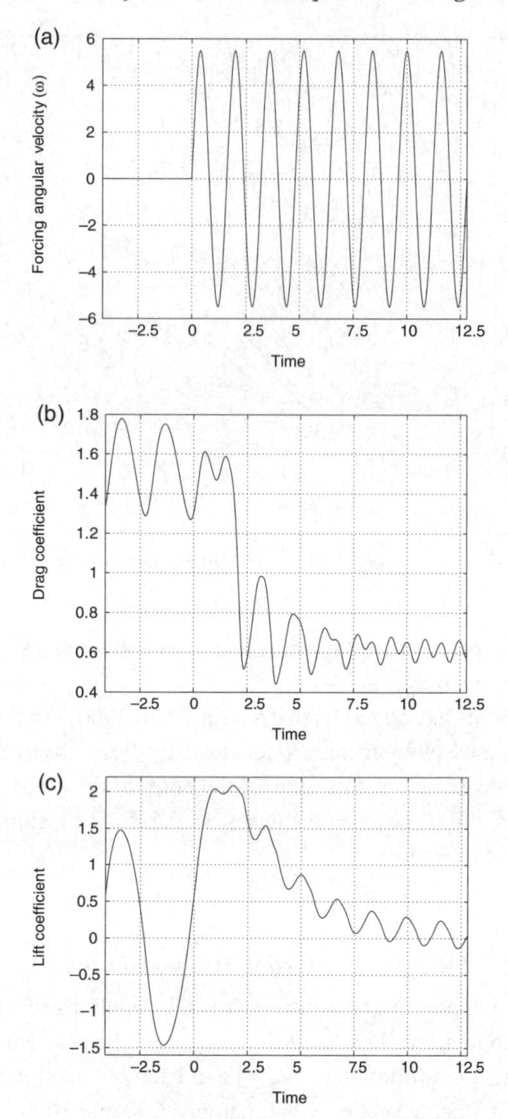

Fig. 9.12. The time evolution of the (a) sinusoidal-optimal forcing, with $S_e = 0.625$ and $\omega_1 = 5.5$, (b) drag C_D, and (c) lift C_L, at $\mathrm{Re} = 1000$. Forcing was started at time $t = 0$.

Similarly, we have shown in Figure 9.16(a) and (b) the results corresponding to $\mathrm{Re} = 1000$. From these figures we observe that the fundamental frequency of the optimal control and the optimal frequency for the sinusoidal control are even closer than for $\mathrm{Re} = 200$.

From these simulations it follows that

(i) The fundamental frequency of the optimal control is very close to the optimal frequency obtained by the methods of Section 9.8.2.
(ii) The optimal control has one fundamental frequency and several harmonics whose frequencies are *odd* multiples of the fundamental frequency.

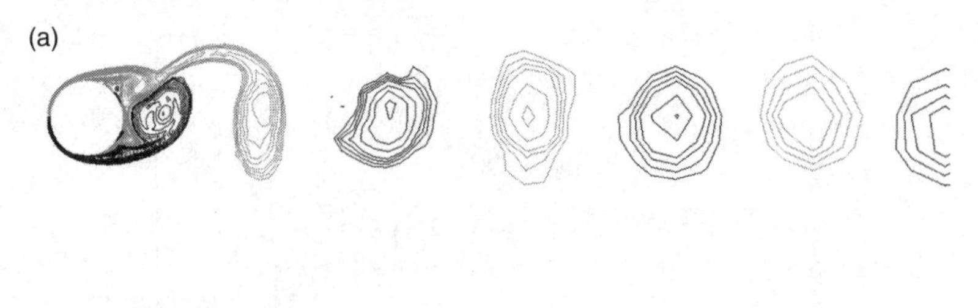

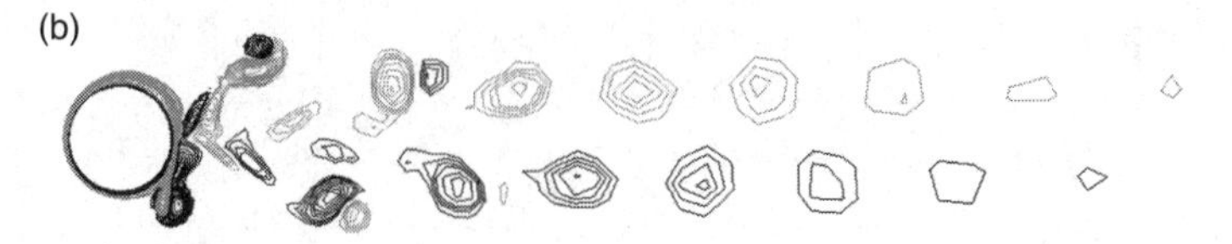

Fig. 9.13. Vorticity contour plot of the wake of the (a) unforced and (b) forced flow at Re = 1000.

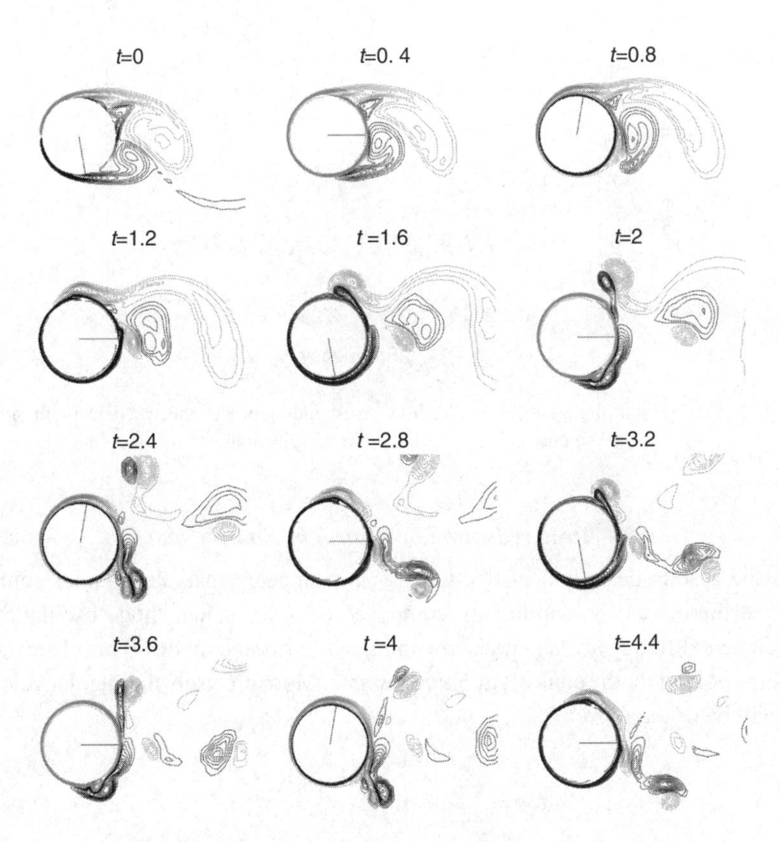

Fig. 9.14. Near-wake region: vortex shedding frequency shifted from its natural shedding frequency ($S_n = 0.2398$) to the forcing frequency ($S_e = 0.625$), Re = 1000 and $\omega_1 = 5.5$. The sequence represents the first three forcing period.

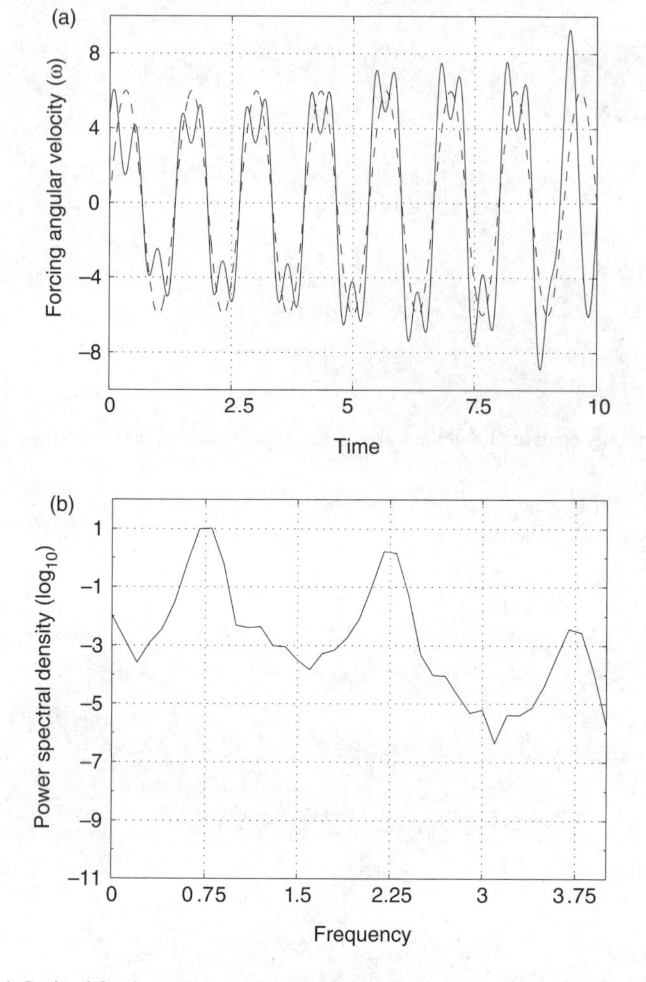

Fig. 9.15. (a) Optimal forcing at Re = 200. (b) Corresponding power spectral density. In (a) the dashed curve represents the optimal sinusoidal control.

9.8.4 Drag reduction by control in Fourier space

From the results described in Section 9.8.3, it appears that the optimal controls obtained there were predominantly composed of a sinusoidal mode oscillating at a fundamental frequency superposed with higher harmonic modes. This observation suggests looking for the controls in *Fourier space*. More precisely the angular velocity $\omega(t)$ will be of the form:

$$\omega(t) = \sum_{k=1}^{K} \omega_k \sin(2k\pi S_e t - \delta_k). \tag{9.98}$$

At Re $=$ 200, in order to see what effect additional harmonics may have on the drag reduction, the optimal forcing was sought in the space described by (9.98) with three different values of K, namely, 1, 3, and 10. The time interval $(0, T)$ for the

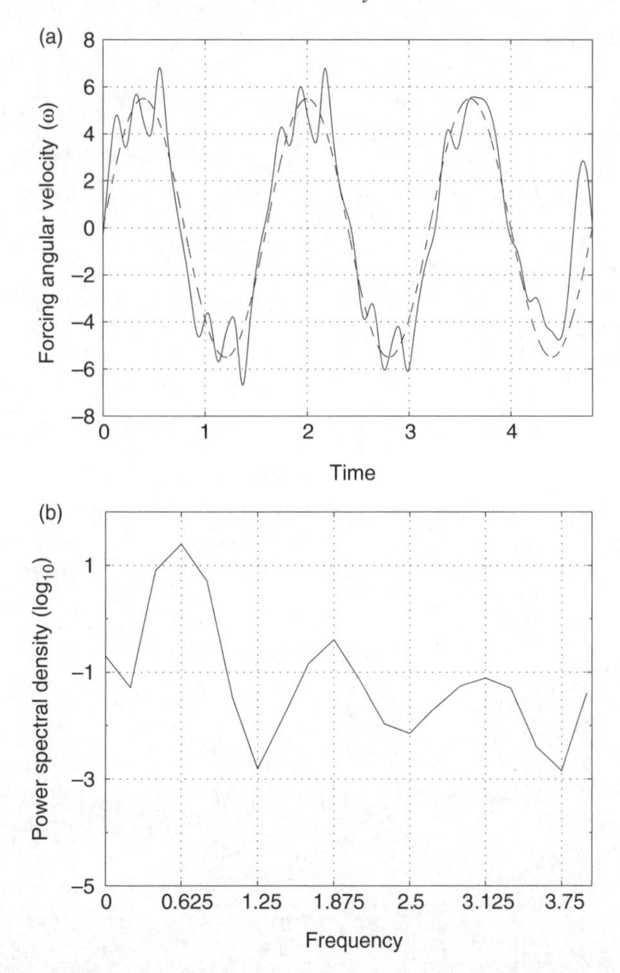

Fig. 9.16. (a) Optimal periodic forcing at Re = 1000. (b) Corresponding power spectral density. In (a) the dashed curve represent the optimal sinusoidal control.

control was chosen so that $T = 3T_f$, with $T_f = 1/S_e$ the forcing period. For large time intervals (including $(0, +\infty)$) we used a *piecewise optimal control strategy*, that is, assuming that these large intervals are of the form $\cup_{m=0}^{M-1}\big(mT, (m+1)T\big]$ (with, possibly, $M = +\infty$) we solved on interval $\big(mT, (m+1)T\big)$ an optimal control problem taking as initial flow velocity the velocity provided by the initial condition $\mathbf{y}(0) = \mathbf{y}_0$ if $m = 0$, and by the solution of the previous control problem if $m \geq 1$. The computational results show that the effect of the phase shifts δ_k is small, suggesting to take $\delta_k = 0$ in (9.98).

The computational experiments reported in Figures 9.17–9.19 correspond to the following scenario:

- From $t = -T$ to $t = 0$, the cylinder is fixed, there is no control and the flow oscillates at its natural frequency f_n.
- At $t = 0$, we start controlling using periodic optimal control in the class given by relation (9.98).

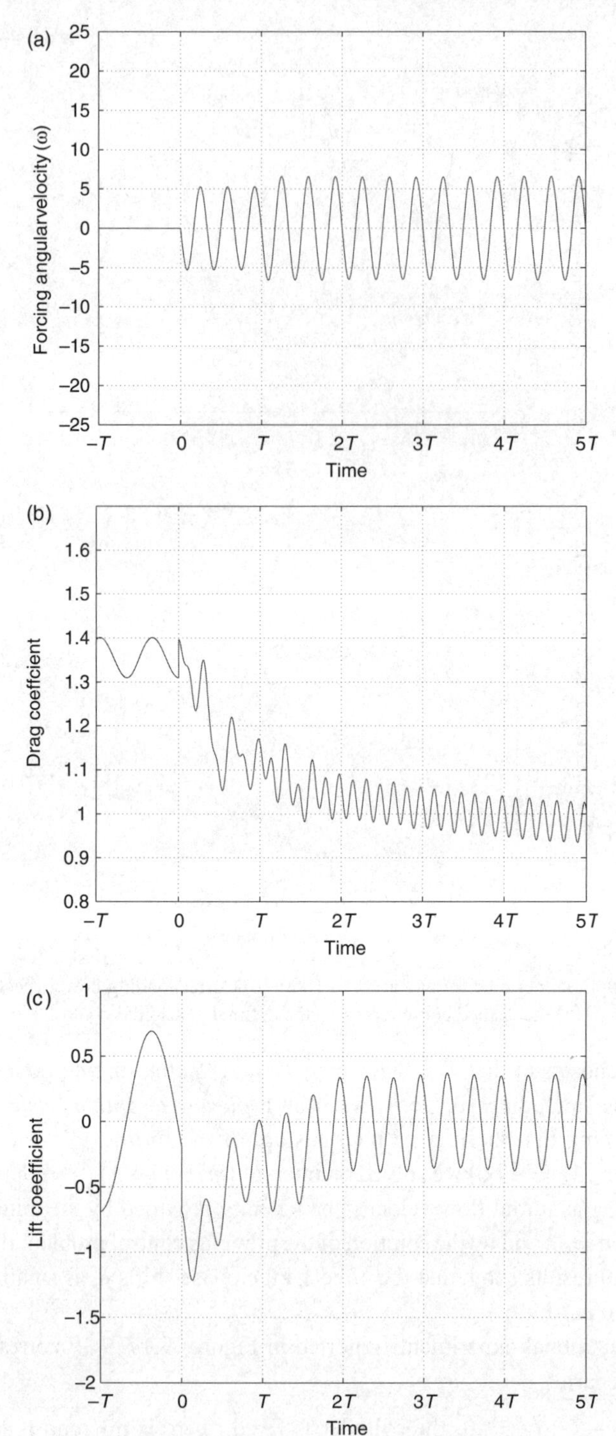

Fig. 9.17. Optimal periodic control (Re $= 200$, $K = 1$). (a) Time evolution of the control, (b) time evolution of the drag coefficient, and (c) time evolution of the lift coefficient.

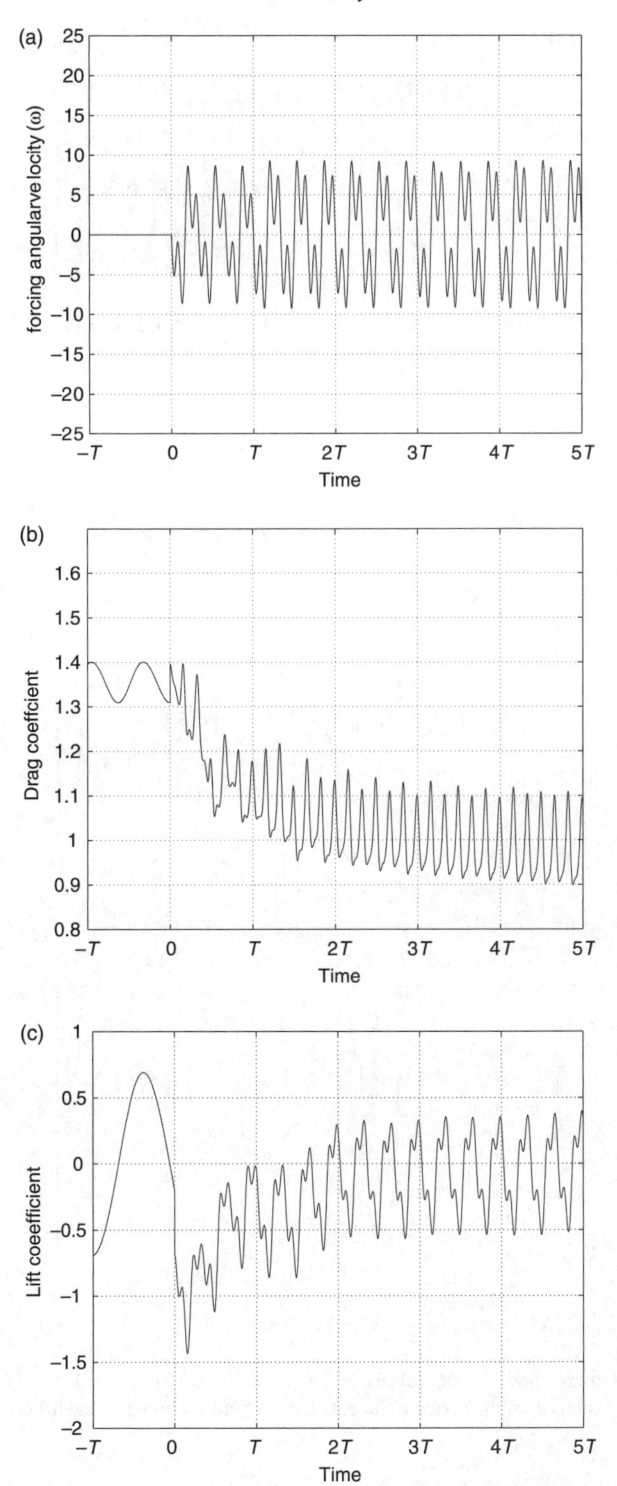

Fig. 9.18. Optimal periodic control (Re = 200, $K = 3$). (a) Time evolution of the control, (b) time evolution of the drag coefficient, and (c) time evolution of the lift coefficient.

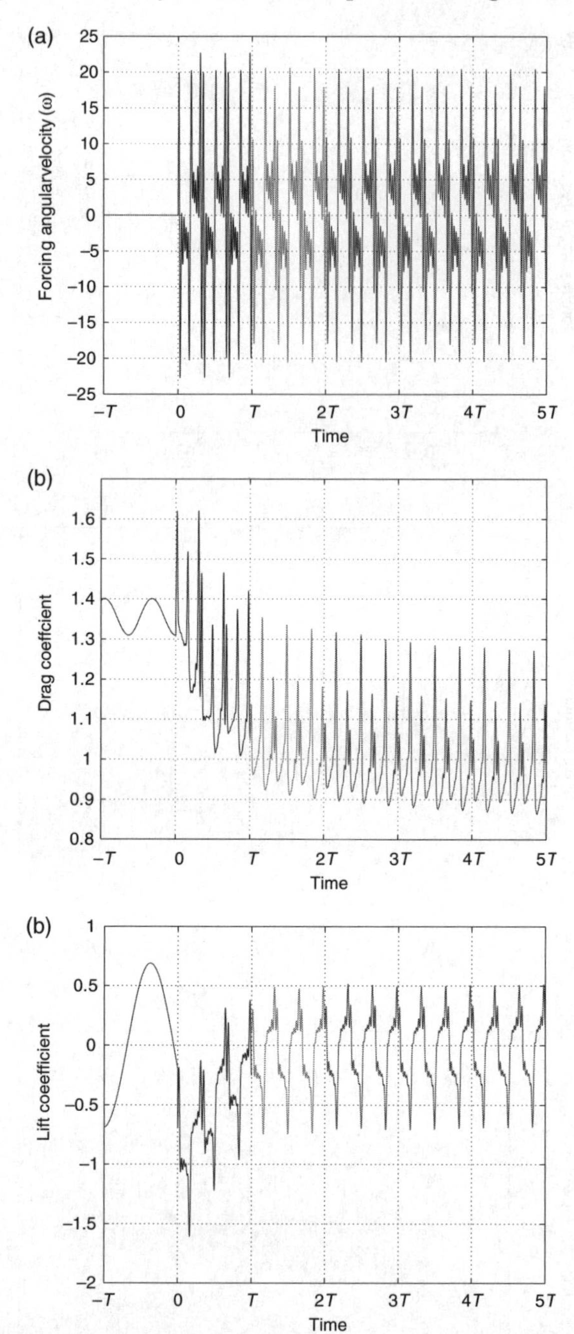

Fig. 9.19. Optimal periodic control (Re $= 200$, $K = 10$). (a) Time evolution of the control, (b) time evolution of the drag coefficient, and (c) time evolution of the lift coefficient.

The optimal control with $K = 1$ (respectively, $K = 3$ and 10) is shown in Figure 9.17(a) (respectively, Figures 9.18(a) and 9.19(a)) and the corresponding drag and lift are shown in Figure 9.17(b) and (c) (respectively, Figures 9.18(b) and (c), 9.19(b) and (c)). The highly oscillatory controls, drags and lifts observed in Figures 9.17–9.19 as K increases can be explained by the fact that they were computed with $\epsilon = 0$ in the cost function (9.9); introducing a positive regularization parameter ϵ into the cost function would have the immediate effect of smoothing the above results.

The optimal periodic control obtained during the tenth piecewise control loop has been used to stabilize the system beyond that loop: the effectiveness of this approach relies on the fact that most transitional effects have been damped out before one reached that loop. A deeper analysis of the optimal periodic state which has been reached is in order: we observe, from Figure 9.20, that when the peak angular velocity is reached, it corresponds to a minimum of the drag (taking place at $t = 0.6$, 1.25, and 1.9, here).

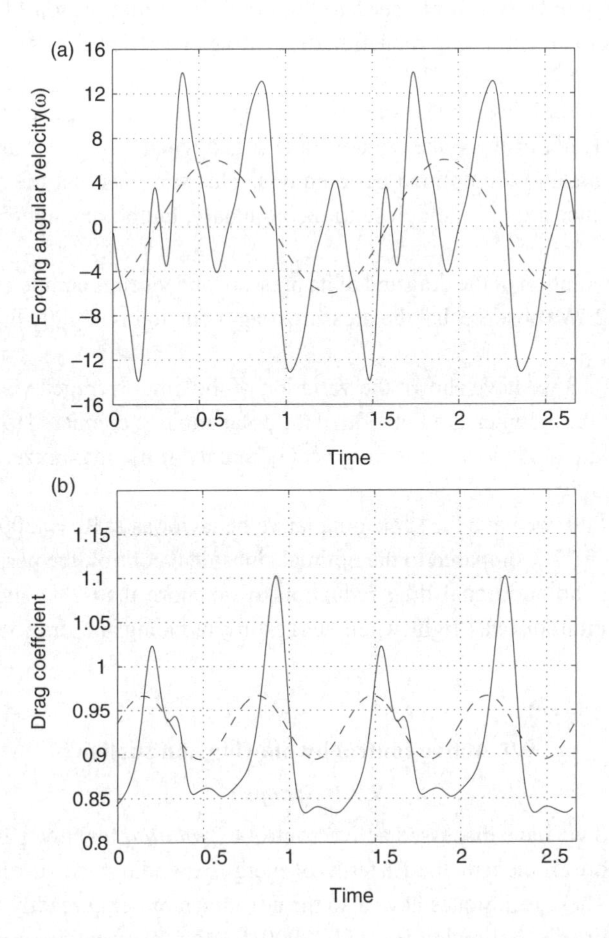

Fig. 9.20. Optimal periodic control (Re = 200, $K = 3$). (a) Comparison between the optimal sinusoidal control (−−) and the optimal periodic one (–), (b) comparison of the corresponding drags.

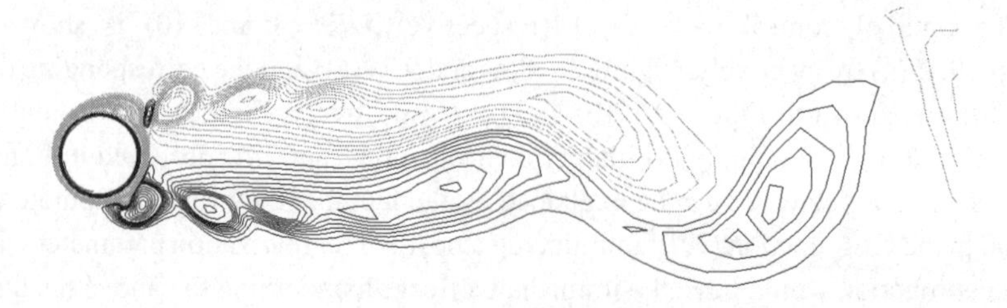

Fig. 9.21. Optimal periodic control (Re = 200, $K = 3$). Vorticity contours.

We observe that the effect of the optimal control is to flatten the drag in the neighborhood of its minima and to sharpen it in the neighborhood of its maxima. This can be seen in Figure 9.20(b) at $t = 0.5, 1.15$, and 1.85. Indeed, the sharp peaks in the drag correspond to times when the forcing change direction, that is, crosses zero. A very interesting feature can be seen at $t = 0.2$ and 1.5, where a "zig–zag" forcing motion corresponds to a lower peak in the drag. The optimization of the periodic forcing leads to an extra drag reduction of the drag coefficient, from 0.932 to 0.905 (if $K = 3$), or 2.87%.

From Figure 9.21 (which shows vorticity snapshots), it can be seen that quantitatively the structure of the wake remains unchanged when "we go" from the optimal sinusoidal control to the optimal periodic one. This suggests that the effects of the higher harmonics are only felt close to the boundary, but do not significantly affect the wake.

The time evolution of the drag and of its pressure and viscous components is shown in Figure 9.22. We observe that the pressure drag reduction is slightly higher than the viscous one.

In Figure 9.23 we have shown the variation of the time-averaged viscous drag on the surface of the cylinder as a function of the polar angle θ; compared to the unforced case the reduction of the viscous drag occurs mainly at the maximizers, namely, at $\theta = 60°$ and 300°.

At Re = 1000, we have the same qualitative behavior as at Re = 200 as shown in Figures 9.24–9.27. Compared to the optimal sinusoidal control, the periodic optimal control brings an additional drag reduction of no more than 2%, suggesting that engineering intuition was right when suggesting reducing the drag via sinusoidal control.

9.9 Active control by blowing and suction

9.9.1 Synopsis

In Section 9.8 we have discussed active control by *rotation* for flow past a cylinder. Actually, a more efficient method in terms of energy expenditure is to use local *blowing* and *suction*. These techniques have also the advantage of being readily applicable to *noncircular* airfoils. Following He et al. (2000, Section 8) we will present the results of simulations performed with two and three blowing and suction slots, respectively, and compare these results to existing experimental ones.

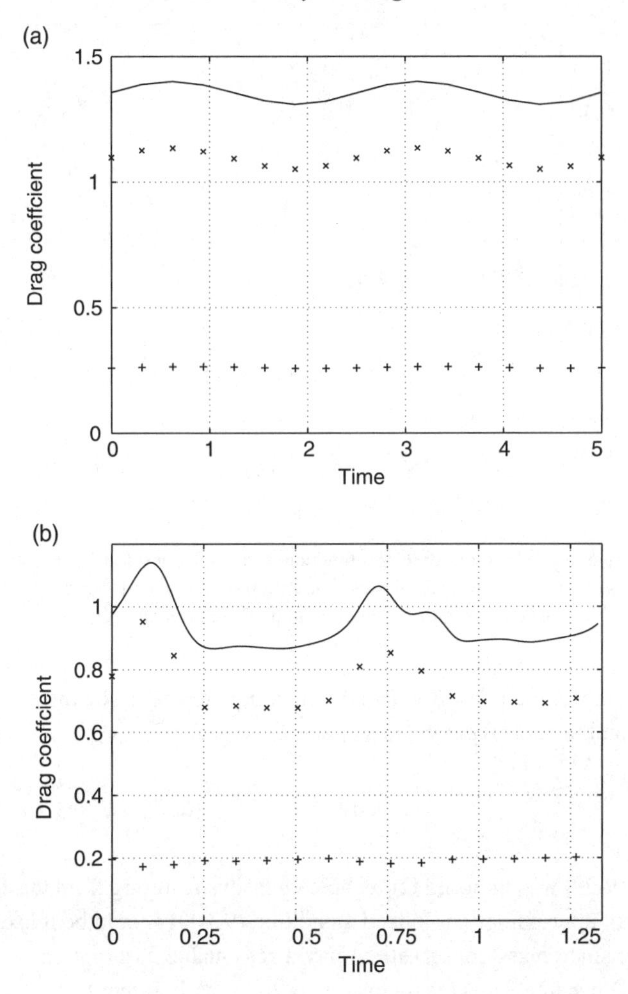

Fig. 9.22. Time evolution of the total (−), pressure (×), and viscous (+) drags at Re = 200. (a) Unforced flow, (b) optimal periodic control ($K = 3$).

9.9.2 Simulation configuration

The flow region being as described in Section 9.7, we consider the configuration of depicted in Figure 9.28 for the blowing and suction simulations. At time $t = 0$, we assume that the flow is fully established (essentially periodic for Re large enough) and we start controlling by injection and suction of fluid on the boundary. The fluid is injected or sucked in a direction making an angle θ_i with the normal at the boundary (i is the slot number) and several slots may be used simultaneously. The above angles θ_i's can be either fixed or be flow control parameters by themselves. The angle α_i denotes the angular distance between the leading edge and the ith slot. The slots have the same parabolic outflow velocity profile $h_i(\mathbf{x})$, scaled by the flow rate parameter $c_i(t)$; the corresponding boundary conditions read as follows:

$$\mathbf{y} = c_i(t)\big(\cos\theta_i(t)\mathbf{n} + \sin\theta_i(t)\mathbf{t}\big)h_i(\mathbf{x}), \quad \text{on } \partial B_i \times (0, T),$$

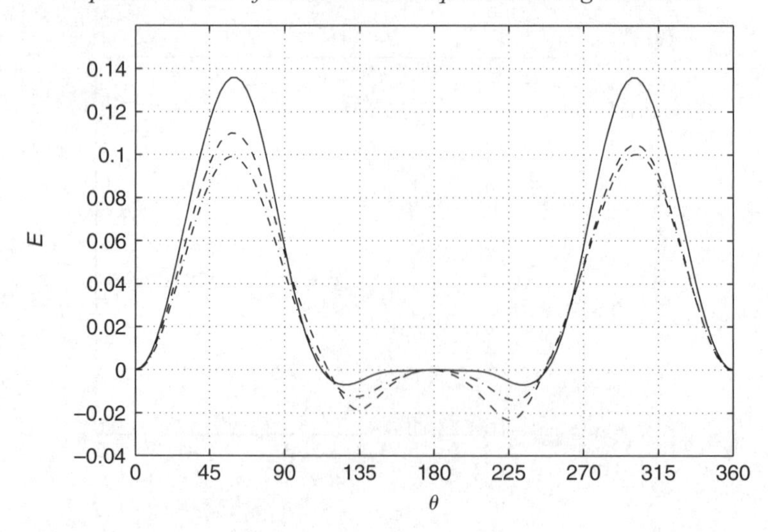

Fig. 9.23. Time-averaged viscous drag as a function of the polar angle at Re = 200. The solid line represents the unforced case, the dashed line the optimal sinusoidal forced case, and the dashed-dotted the optimal periodic forced case for $K = 3$.

where ∂B_i denotes the part ∂B where the ith slot is located. The only constraint that we impose on the control parameters c_i and h_i is that

$$\int_{\partial B} \mathbf{y}(t) \cdot \mathbf{n} \, \mathrm{d}s = 0, \quad \forall t \in (0, T). \tag{9.99}$$

Relation (9.99) was imposed for technical reasons resulting from the choice of our finite element approximation. Indeed, condition (9.100) should be relaxed in further simulations, since mass conservation may a reasonable assumption over long time intervals, but instantaneous mass conservation is clearly rather restrictive.

The slot aperture β_i is chosen to be 10° for all slots throughout the simulations and will be denoted just as β in the sequel. A smaller angle would give too few grid points on the slots and larger slots would not have sufficient local forcing, besides being not very realistic in practice.

9.9.3 Blowing and suction with two slots

9.9.3.1 Antisymmetrical forcing at Re = 470

In order to validate our software and to insure that it operates properly for the blowing and suction set up described in the above paragraph we have simulated some of the experiments reported in Williams, Mansy, and Amato (1992). The experiments reported in the above reference were performed in a water tank with a cylinder with small holes drilled along side in two rows at ±45° from the upstream stagnation point. The "bleeding" from the two hole rows could be either in phase (symmetrical forcing) or 180° out of phase (antisymmetrical forcing). The Reynolds number in the Williams et al. experiments was set to 470 and we have used the same value in our

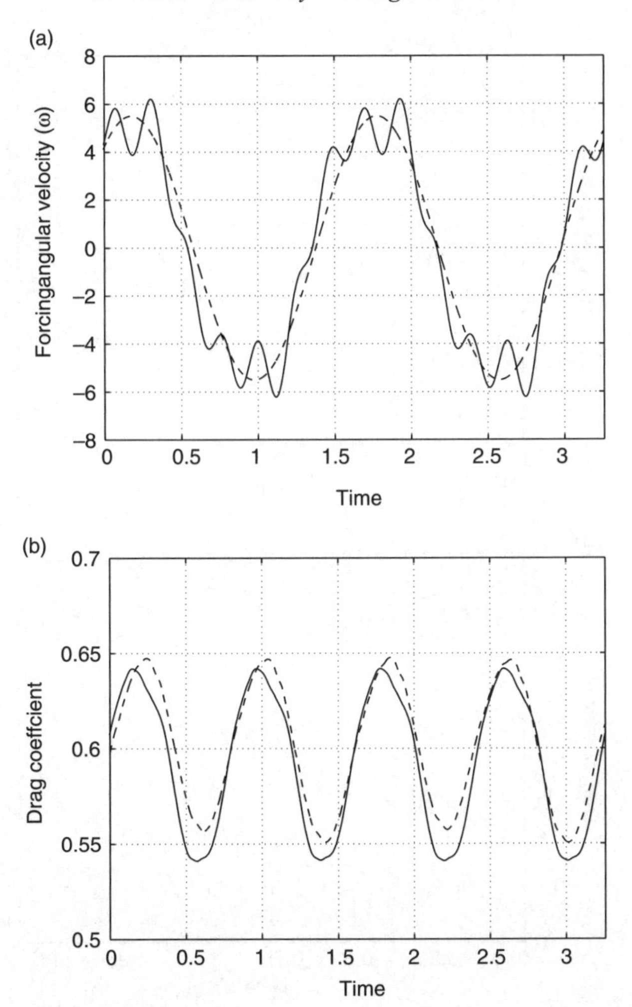

Fig. 9.24. Optimal periodic control (Re = 1000, $K = 3$). (a) Comparison between the optimal sinusoidal control (−−) and the optimal periodic one (−), (b) comparison of the corresponding drags.

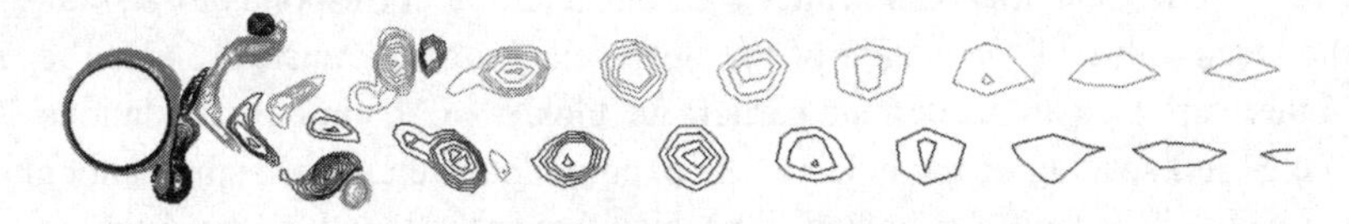

Fig. 9.25. Optimal periodic control (Re = 1000, $K = 3$). Vorticity contours.

simulations. To measure the forcing, Williams et al. introduced the *bleed coefficient* C_b defined by

$$C_b = \frac{\bar{y}^2 d_j}{y_\infty^2 d}, \tag{9.100}$$

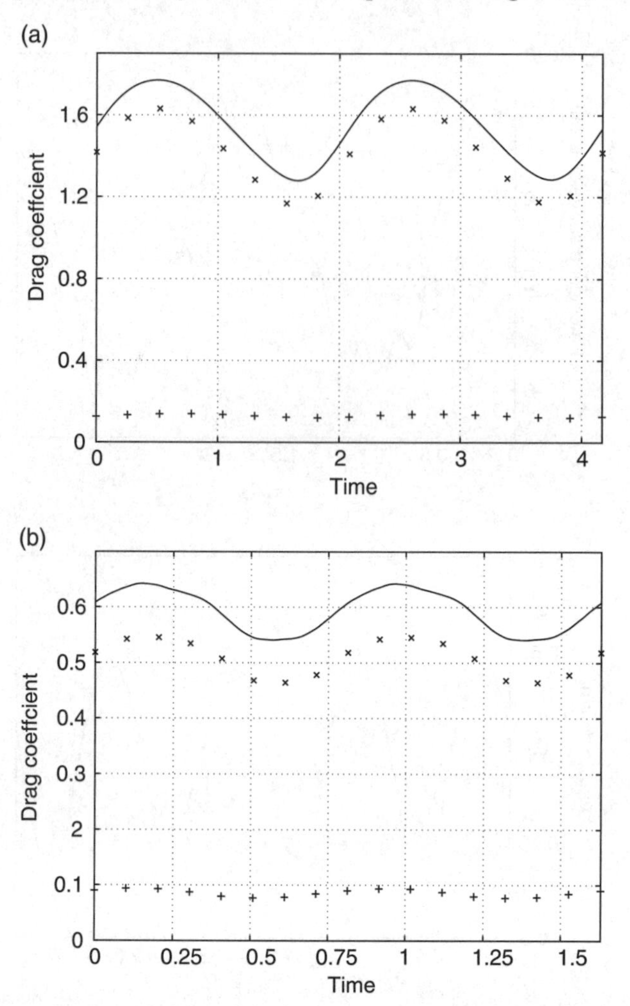

Fig. 9.26. Time evolution of the total (–), pressure (×), and viscous (+) drags at Re = 1000. (a) Unforced flow, (b) optimal periodic control ($K = 3$).

where d_j is the hole diameter which with our notation corresponds to $\beta \approx 8.9°$, $\bar{y}$ is the mean value of the square of the flow velocity of the unsteady bleed jet and the other variables are as defined earlier; we took $\beta = 10°$ in our simulations. The diameter and spacing of the holes in the experimental setup were small enough for the effective disturbance to be two-dimensional over the length of the cylinder.

Williams et al. tested both symmetrical and antisymmetrical forcing and concluded that the symmetrical one was the more efficient way to tame the wake, but we have focused here on the antisymmetrical forcing, for the reasons mentioned above. We follow here the same scenario as in Williams et al., that is, we look at the flow for four different situations

(i) No forcing ($C_b = 0$).
(ii) Low amplitude forcing ($C_b = 4 \times 10^{-3}$).

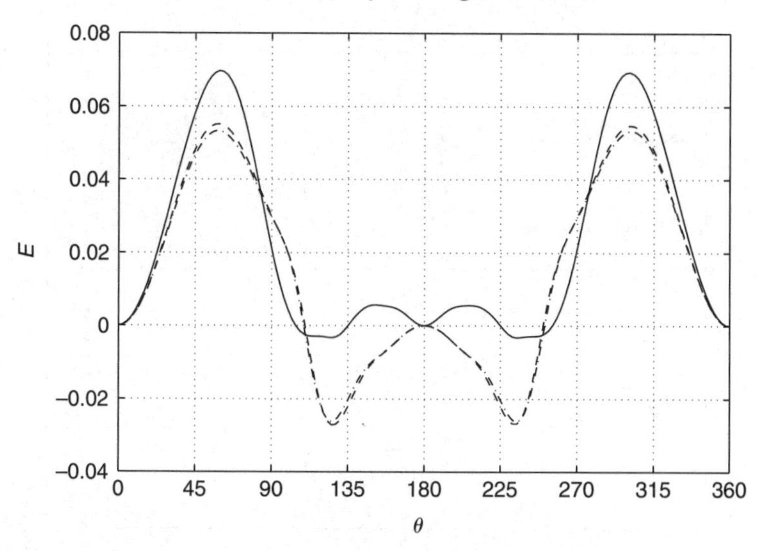

Fig. 9.27. Time-averaged viscous drag as a function of the polar angle at Re = 1000. The solid line represents the unforced case, the dashed line the optimal sinusoidal forced case, and the dashed-dotted the optimal periodic forced case for $K = 3$.

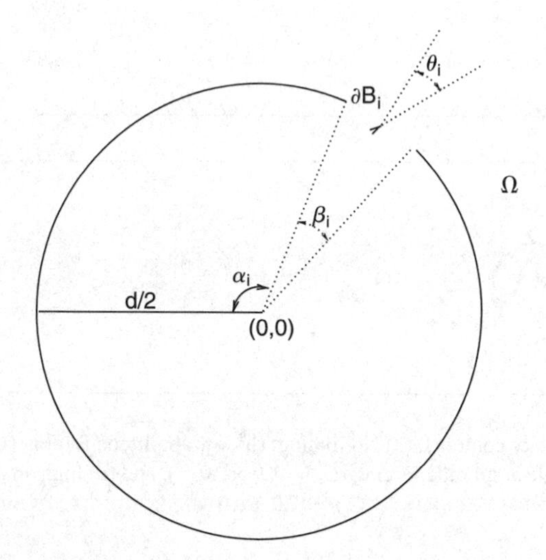

Fig. 9.28. Blowing and suction configuration. The index i denotes the ith slot, each slot having its own configuration: θ_i denotes the angle of incidence of the velocity profile, α_i the angular distance from the leading edge, β_i the aperture of the slot, ∂B_i the part of ∂B occupied by the ith slot.

(iii) Intermediate amplitude forcing ($C_b = 1.9 \times 10^{-2}$).
(iv) High amplitude forcing ($C_b = 7.3 \times 10^{-2}$).

The excitation frequency was fixed at $S_{f_e} = 8.85 \times S$ during the simulation; here S is the unforced Strouhal number. For the present Reynolds number (Re = 470) we have $S = 0.226$.

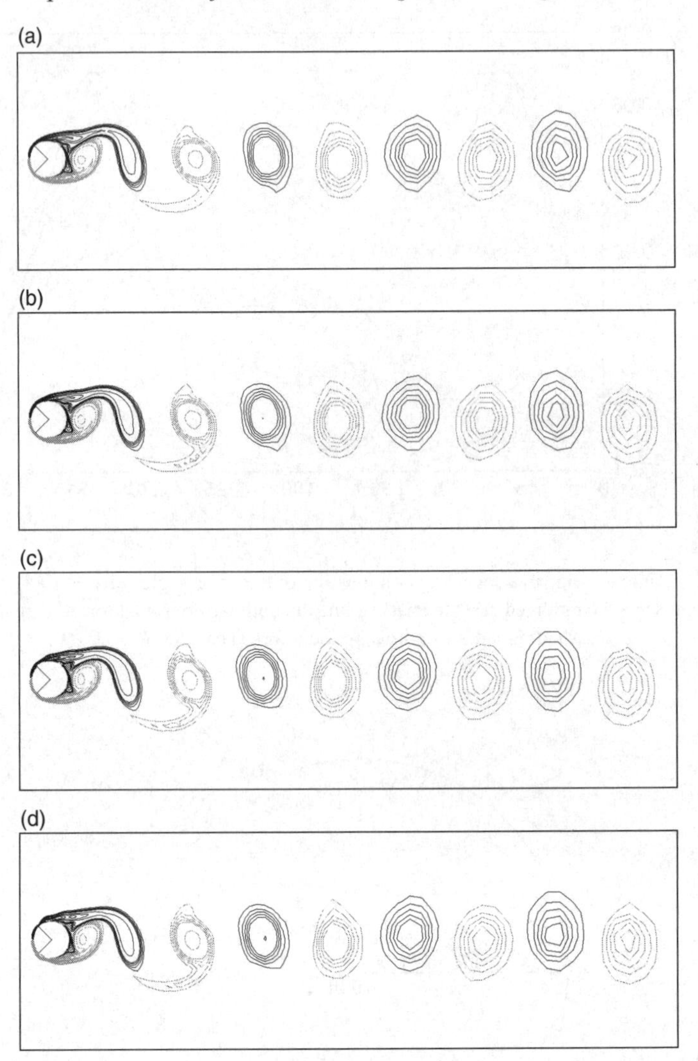

Fig. 9.29. Vorticity contours. (a) No forcing, (b) low amplitude forcing ($C_b = 4 \times 10^{-3}$), (c) intermediate amplitude forcing ($C_b = 1.9 \times 10^{-2}$), and (d) high amplitude forcing ($C_b = 7.3 \times 10^{-2}$).

For these four different cases, the main structures in the flow are essentially the same as can be seen in Figure 9.29. We can conclude these simulations by stating that *antisymmetrical forcing with two slots is too weak to tame the wake*. Williams et al. came to the same conclusion.

9.9.3.2 Flow stabilization at Re $= 60$

It has been shown in, for example, Roussopoulos (1993) and Park, Ladd, and Hendricks (1994) that feedback signals from the wake can stabilize the flow and prevent vortex shedding up to some Reynolds number depending on the control method. Our main goal here is to use an optimal control approach and extend if possible the range of Reynolds numbers for which stable, low drag flow can be achieved.

For these low Reynolds numbers separation takes place around $\pm 120°$ from the leading stagnation point. Park et al. showed, via numerical simulation, that at $\mathrm{Re} = 60$ they could stabilize the flow via a single feedback sensor located in the wake. They used two slots located at $\pm 110°$ where the vertical component of the velocity (that is, y_2) at a point $\mathbf{x}_s$ downstream was providing the feedback signal, via the relation

$$f(t) = \gamma \frac{y_2(\mathbf{x} = \mathbf{x}_s, t)}{y_{2_{\max}}(t)}, \tag{9.101}$$

where

$$y_{2_{\max}}(t) = \max_{\tau \leq t} |y_2(\mathbf{x} = \mathbf{x}_s, \tau)|,$$

γ is a scaling factor.

The feedback signal $f(t)$ was used as a scaling factor of the velocity profiles at the slots, and, like us, Park et al. used an antisymmetrical forcing. For Reynolds numbers up to 80, they were able to suppress the primary vortex shedding, but excited at the same time a secondary instability mode, triggering thus a new vortex shedding.

Our approach is somewhat different; instead of using a simple feedback as above, we are going to use once more the optimal control approach (with $\epsilon = 0$ in (9.9) and (9.21)). For the simulations reported below we have used *B-splines* to approximate the various functions of t occurring in the control process (namely, the C_i's and θ_i's), reducing thus the dimension of the control space, compared to the method used for the optimal control by rotation (see, for example, Stoer and Bulirsh, 1993, Chapter 2; Quarteroni, Sacco, and Saleri, 2000, Chapter 8 for a discussion of *B*-spline approximations; see also the references therein). Taking $T = 32$, we were able to stabilize and symmetrize the flow at $\mathrm{Re} = 60$. The power P_c necessary to control the flow decreases quickly with time, as seen in Figure 9.30(a). In Figure 9.30(b), we have plotted the function $t \to c(t)$; this function behaves essentially like the feedback signal $t \to f(t)$ in Park et al. as soon as the control starts acting, $c(t)$ decreases since the wake become more symmetrical. After less than 15 time units the amplitude of the control is less than 1% of its initial value (its seems also that "our" control stabilizes the flow faster than the one in Park et al; this is not surprising, since, after all, our control is *optimal* (in some sense)). From the above reference, we know that stabilization is possible up to $\mathrm{Re} = 60$, which is confirmed by the simulations reported here. In Figure 9.31(a) and (b) we have visualized the drag and lift corresponding to the optimal control. As long as the control acts the drag goes down and levels out close to the drag corresponding to the (unstable) symmetric steady flow at $\mathrm{Re} = 60$. On the other hand, the lift goes down and becomes quite small (nearly zero). In Figure 9.32(a) and (b) we have visualized, respectively, the vorticity contours of the uncontrolled flow and of the controlled flow at $\mathrm{Re} = 60$. Both Park et al. and Roussopoulos reported that the amount of feedback necessary to maintain non-vortex shedding at $\mathrm{Re} = 60$ is very low. It follows from Figure 9.30 that our results lead to a similar conclusion.

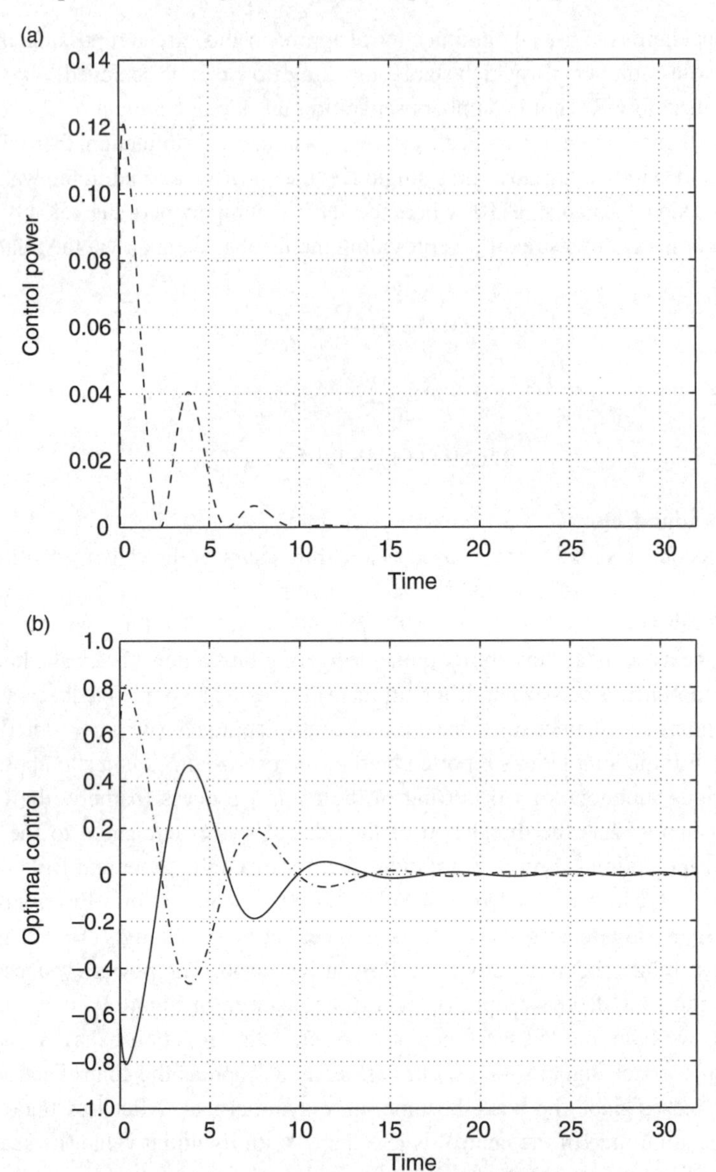

Fig. 9.30. Optimal control by blowing and suction at Re = 60. (a) Power control P_c, (b) optimal controls c_1 (−) and c_2 ($= -c_1$). The slots were located at ±110° from the leading edge.

9.9.4 Control by blowing and suction at Re = 200

Using two slots we have been able to stabilize the flow at Re = 60 and as a side effect the wake was symmetrized along the x_1-axis. To improve the ability of stabilizing at higher Reynolds numbers a third slot was added on the boundary of the cylinder. Actually, this additional slot was located at the trailing edge, the other two being located symmetrically with respect to the x_1-axis.

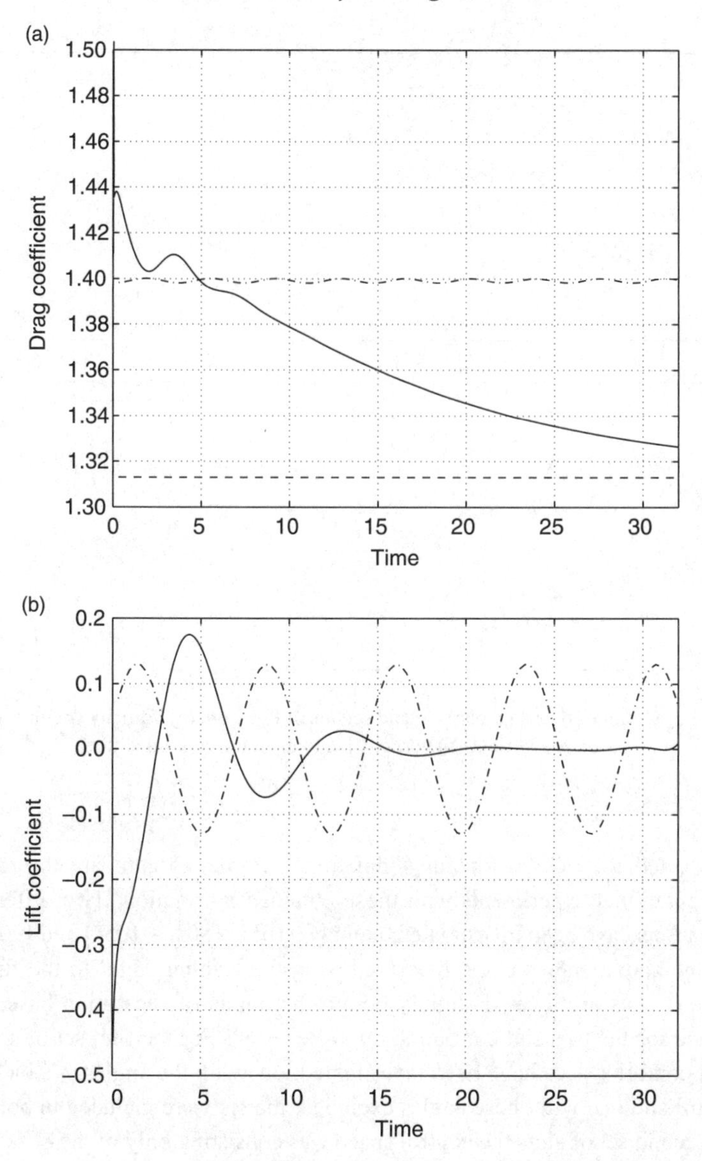

Fig. 9.31. Optimal control by blowing and suction at Re = 60. (a) Drag coefficients for the uncontrolled flow (− · −), the controlled flow (−, and the (unstable) symmetric steady flow (−−), (b) lift coefficients for the uncontrolled flow (− · −), and the controlled flow (−).

Introducing a third slot brings a completely different structure to the wake compared to the two slot case. This can be seen in Figure 9.33 which shows snapshots of the vorticity at Re = 200, for t varying from 0.05 to 10.05. It turns out that the additional slot operates always in the blowing mode, while the other two operate in the suction mode. We observe also that the additional slot has essentially the effect of a splitter plate located behind the cylinder and preventing the interaction between the upper and lower parts of the flow (see, for example, Kwon and Choi, 1996).

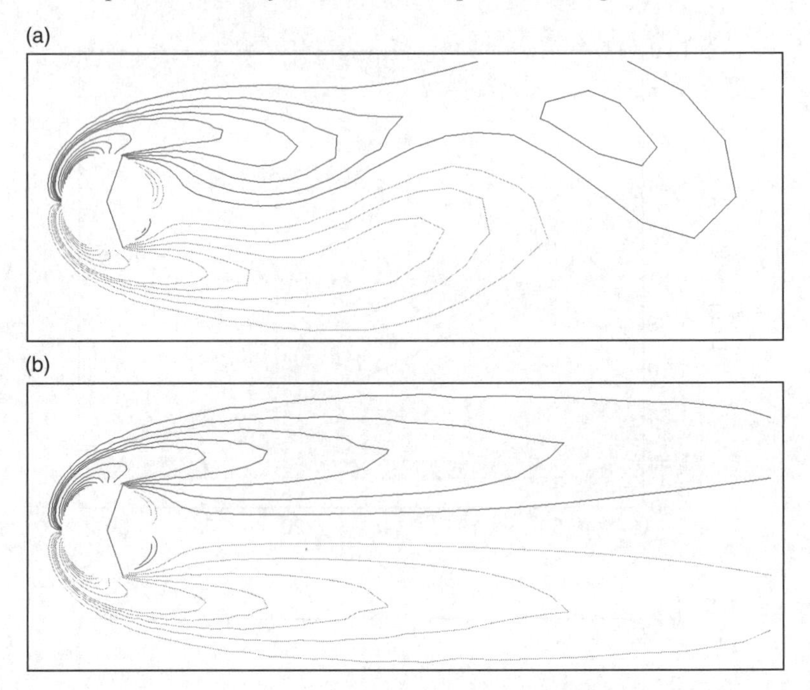

Fig. 9.32. Optimal control by blowing and suction at Re = 60. (a) Vorticity contours of the uncontrolled flow, (b) vorticity contours of the controlled flow.

We selected Re = 200 for our simulations in order to compare the results of our computational experiments with those obtained by rotation. Two different slot configurations have been investigated, namely, $\{105°, 180°, -105°\}$ and $\{90°, 180°, -90°\}$, the above angles being measured from the leading edge. In the first case, the off-axis slots are located slightly before the points of separation based on the experience for the two slot configuration at Re = 60. For the first set of slots, two different control spaces have been investigated: one with the angles θ_i's included in the control and one with these angles excluded; the c_is were included in both cases. For the second set of slots the control space was consisting only of the c_i's.

A *piecewise optimal control strategy* has been used for $t \in (0, 90)$, each time interval being of length 10. In Figures 9.34, 9.36, and 9.38 we have visualized the time evolution of the control parameters and of the corresponding total control power P_c for different control configurations. The corresponding drag and lift have been reported in Figures 9.35, 9.37, and 9.39. We observe that the inclusion of the angles θ_is brings just a little improvement from a stabilization point of view; similarly, moving two slots from $\{105°, -105°\}$ to $\{90°, -90°\}$ improves only slightly the performances of the control system. On Figure 9.36(b) we observe that the optimal value for the angles θ_is associated to the off-axis slots is close to 35° inwards towards the x_1-axis. Drag reduction of up to 37% was achieved, and by taking into account the power used to drive the control, we could get a net drag reduction of up to 32%.

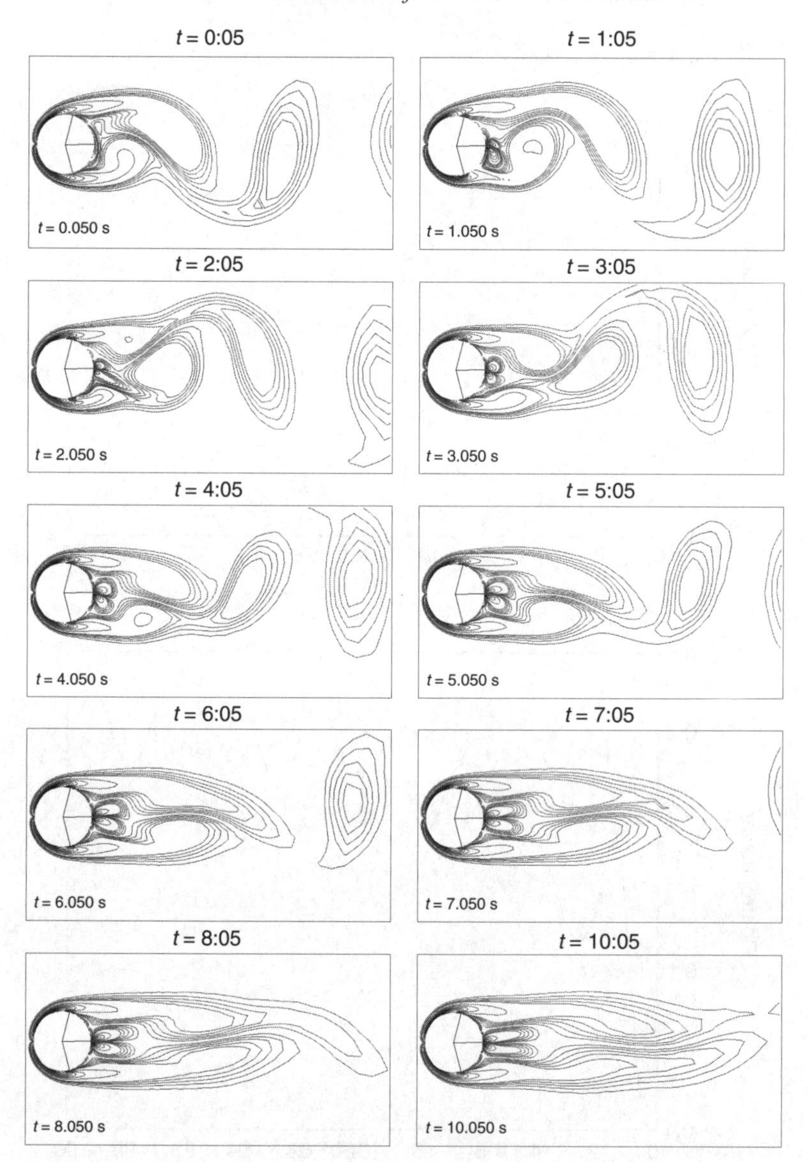

Fig. 9.33. Vorticity contour plot with the optimal control applied from $t = 0$ (Re = 200). The angles θ_i are included in the control.

9.10 Further comments on flow control and conclusion

Through a parametric search in the frequency-amplitude space, minima were found for the reduction of the drag coefficient for the flow around a spinning cylinder at Re = 200 and 1000. These minima correspond to a drag reduction of 31% at Re = 200 and 61% at Re = 1000. These results are quantitatively consistent with the experimental drag reduction of 80% at Re = 15 000 reported in Tokumaru and Dimotakis (1991). This suggests possibilities for further significant drag reduction, at least up to the critical Reynolds number of 300 000.

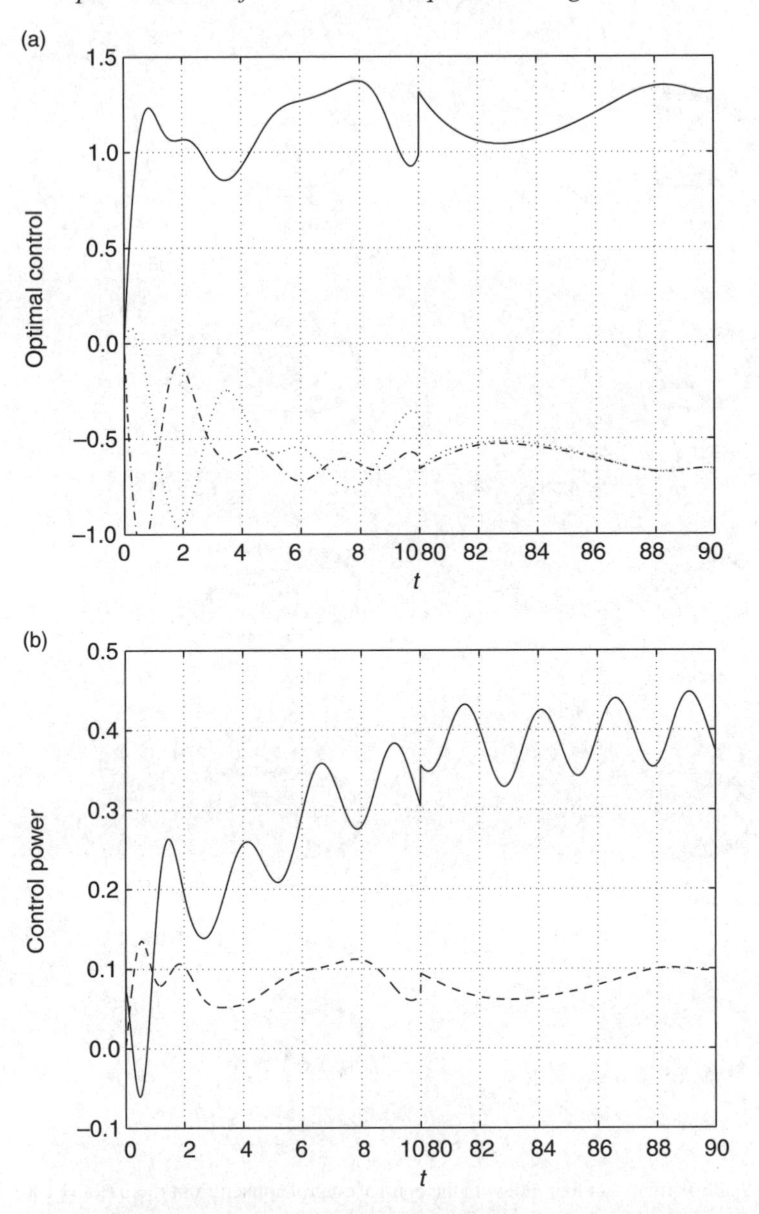

Fig. 9.34. Optimal control by blowing and suction at Re = 200 with slots located at $\{105°, 180°, -105°\}$; the angles θ_is are fixed at 0°. (a) Optimal controls c_i for the slots located at 105° (− · −), −105° (· · ·), and 180° (−), (b) power P_c (−) and power saving due to the control (−−).

Under condition of optimal forcing, it was observed that the wakes are smaller, less energetic, and have smaller spreading angles compared with the uncontrolled case. Also, to generate the flow field necessary for maximum drag reduction, increased amplitude of forcing was required as the oscillation frequencies increased. The

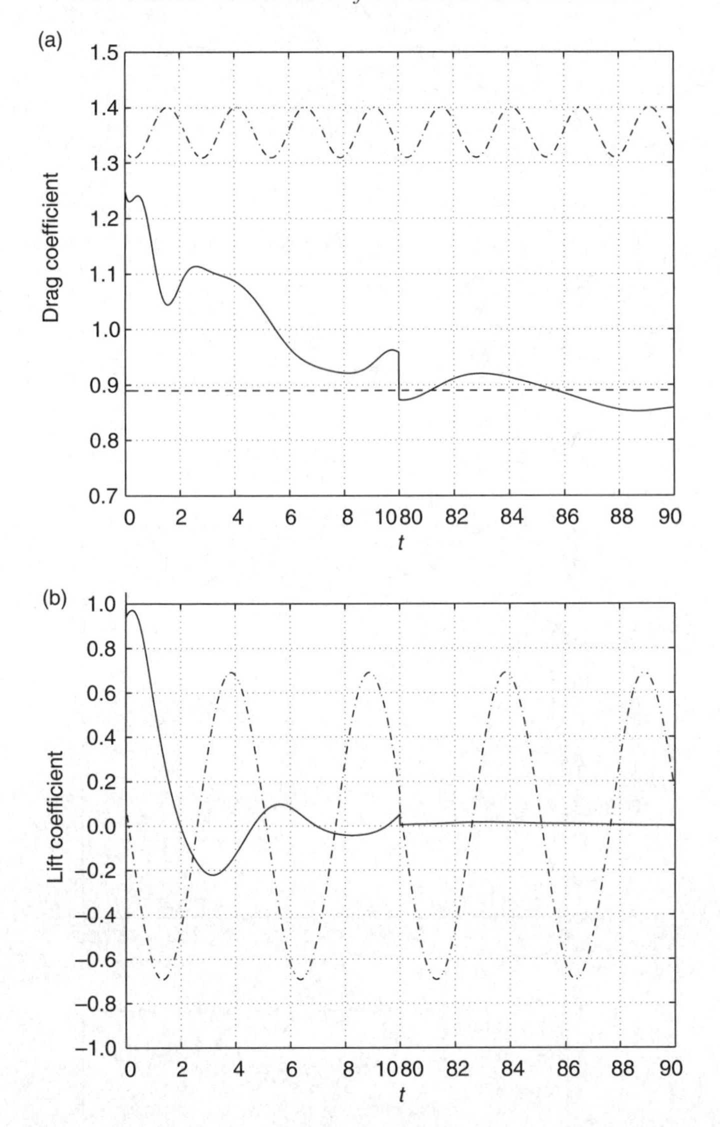

Fig. 9.35. Optimal control by blowing and suction at Re $= 200$ with slots located at $\{105°, 180°, -105°\}$; the angles θ_is are fixed at 0°. (a) Drag coefficients for uncontrolled flow ($-\cdot-$), steady symmetric flow ($--$), and controlled flow ($-$). (b) Lift coefficients for uncontrolled flow ($-\cdot-$) and controlled flow ($-$).

quasi-optimal forcing condition determined by parametric search agreed closely with those found by application of optimal control theory. The theory predicted, and this was confirmed by simulation, that further drag reduction could be obtained by adding higher harmonics to the forcing oscillations. This was achieved by extending the time interval of minimum drag at the expense of slightly higher, narrower peaks of maximum drag (see Figure 9.20(b)); however, the improvement is fairly small.

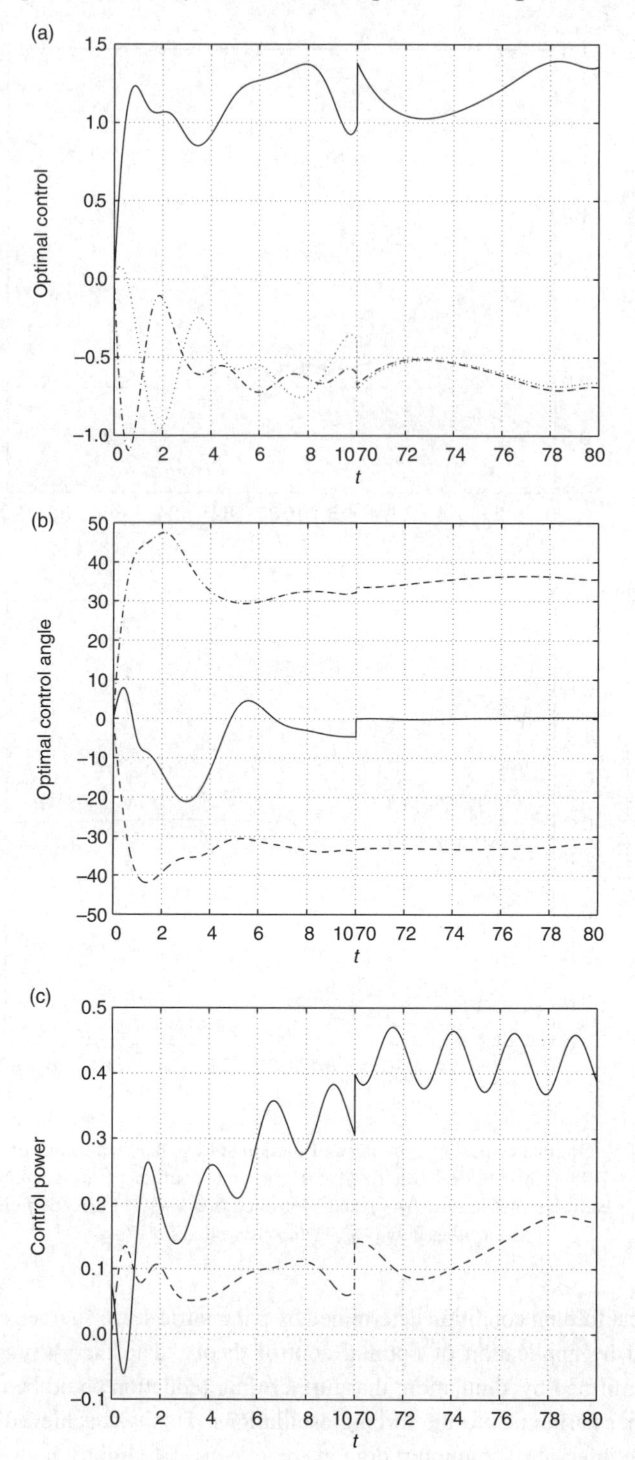

Fig. 9.36. Optimal control by blowing and suction at Re = 200 with slots located at {105°, 180°, −105°}; the angles θ_is are parts of the control parameters. (a) Optimal controls c_i for the slots located at 105° (− · −), −105° (· · ·), and 180° (–). (b) Optimal angles θ_is for the slots located at 105° (− · −), −105° (· · ·), and 180° (–). (c) Power P_c (–) and power saving due to the control (−−).

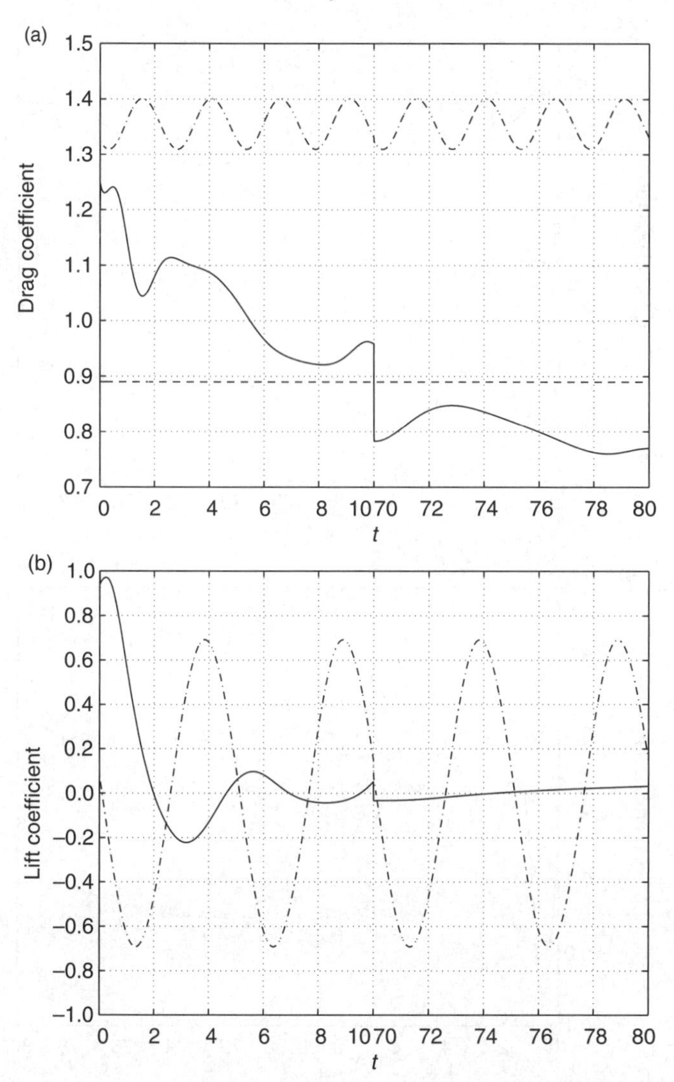

Fig. 9.37. Optimal control by blowing and suction at Re = 200 with slots located at $\{105°, 180°, -105°\}$; the angles θ_is are are parts of the control parameters. (a) Drag coefficients for uncontrolled flow (− · −), steady symmetric flow (−−), and controlled flow (−). (b) Lift coefficients for uncontrolled flow (− · −) and controlled flow (−).

A more efficient forcing technique (measured in terms of energy expenditure) than controlling the drag by rotation is to use local blowing and suction. Using two slots located at $\pm 110°$ from the leading edge we have been able to stabilize the flow and to prevent vortex shedding at Re = 60. The control power necessary to control the flow decreases quickly with time as shown in Figure 9.30. The optimal control forcing features the same behavior that Park et al. reported for the feedback signal approach, namely, as soon as the control starts having effect, the amplitude of the control goes down as the wake become more symmetric; actually, "our" control

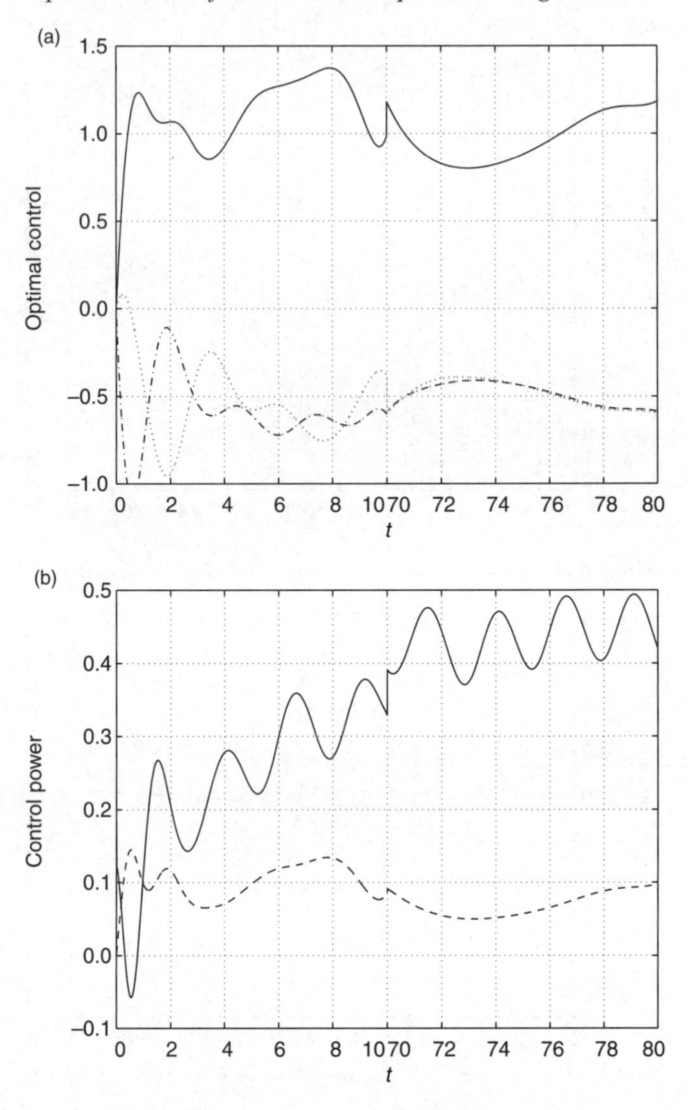

Fig. 9.38. Optimal control by blowing and suction at Re = 200 with slots located at $\{90°, 180°, -90°\}$; the angles θ_is are fixed at 0°. (a) Optimal controls c_i for the slots located at 105° (− · −), −105° (· · ·), and 180° (–). (b) Power P_c (–) and power saving due to the control (——).

seems to stabilize the flow faster than in Park et al. (after less than 15 time units, the control amplitude is less than 1% of its initial value and the control power necessary to maintain non-vortex shedding is very small). Using three blowing-suction slots we have been able to suppress completely the formation of Von Kàrmàn vortex streets up to Re = 200, with a further net drag reduction, compared to control by rotation.

To conclude, let us say that while drag reduction for flow around cylinders – using either an oscillatory rotation or blowing and suction – provides an excellent

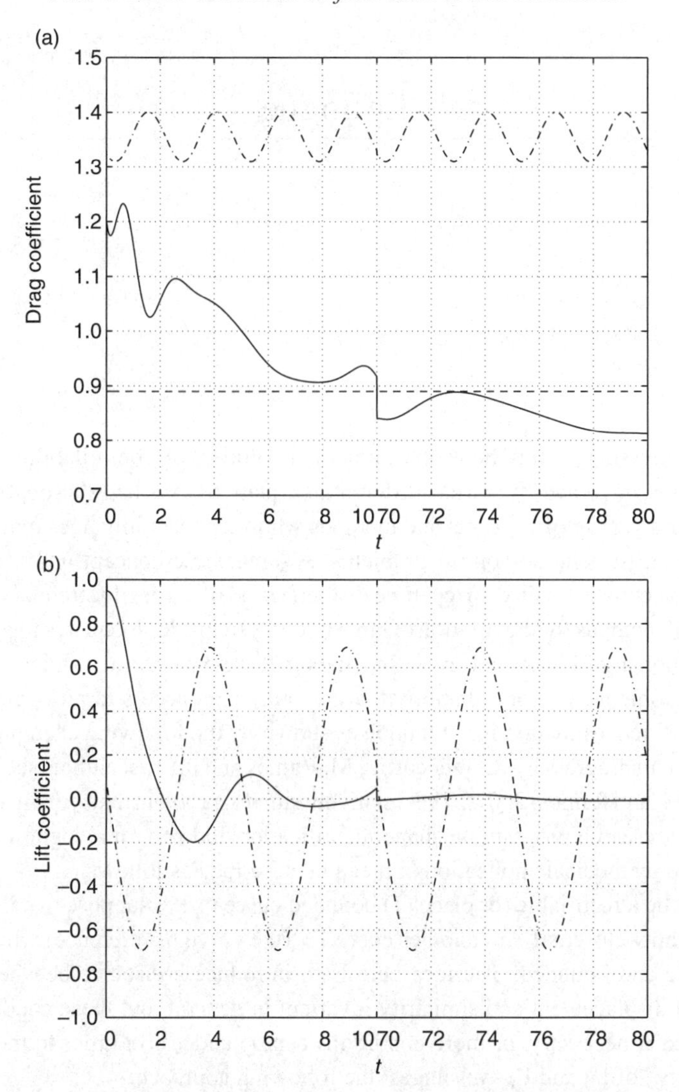

Fig. 9.39. Optimal control by blowing and suction at Re = 200 with slots located at $\{90°, 180°, -90°\}$; the angles θ_is are fixed at 0°. (a) Drag coefficients for uncontrolled flow ($- \cdot -$), steady symmetric flow ($--$), and controlled flow ($-$). (b) Lift coefficients for uncontrolled flow ($- \cdot -$) and controlled flow ($-$).

demonstration of the possibilities of the optimal control approach, it is clearly of little practical importance. However, the application of this theory to more complex shapes like airfoils is not particularly difficult and should lead to some interesting issues and possibly to new forcing strategies.

Epilogue

While addressing in this book the numerical solution of controllability problems for systems governed by partial differential equations, we had the opportunity to encounter a variety of concepts and methods whose applicability goes much beyond the solution of genuine control problems. Among these concept and methods let us mention convex duality, space–time discretization of partial differential equations, numerical methods for the solution of large linear systems, least-squares formulations, optimization algorithms, and so on. In Chapter 7, we have shown while formulating a given problem as a controllability that one may gain access to powerful solution methods. Such a situation is not unique as shown by the following example inspired from work in progress by R. Azencott, A.M. Ramos and the first author (see Azencott, Glowinski, and Ramos, 2007). The (relatively simple) problem that we consider is part of a large research program on shape identification and pattern recognition (largely motivated by medical applications); it can be described as follows:

Let Γ_0 be a rectifiable (or piece of) bounded curve in $\mathbb{R}^2$; suppose that one wishes to know how close is Γ_0 to another curve Γ_R (the curve of reference) which is also rectifiable and bounded. The idea here is to introduce a distance between Γ_0 and Γ_R (rigid displacement and similarity invariant in general, but these conditions can be relaxed if necessary, or more conditions can be added). In order to measure the "proximity" of Γ_0 and Γ_R we suggest the following approach:

(i) Solve the following minimization problem:

$$\begin{cases} \mathbf{u} \in \mathcal{U}_f, \\ J(\mathbf{u}) \leq J(\mathbf{v}), \quad \forall \mathbf{v} \in \mathcal{U}_f, \end{cases} \tag{E.1}$$

with

$$J(\mathbf{v}) = \frac{1}{2} \int_0^1 |\mathbf{v}(t)|_\alpha^2 \, dt, \tag{E.2}$$

and

$$\mathcal{U}_f = \{ \mathbf{v} \mid \mathbf{v} \in L^2(0, 1; \mathbf{V}_\alpha), \ \Gamma_1 = \Gamma_R \}. \tag{E.3}$$

In (E.2), (E.3)

- $|\cdot|_\alpha$ is a norm (or a seminorm) over the Hilbert–Sobolev space $(H^\alpha(\mathbb{R}^2))^2$ ($\alpha \geq 2$ is a reasonable choice).
- As earlier in this book, $f(t)$ denotes the function $\mathbf{x} \to f(\mathbf{x}, t)$.
- $\mathbf{V}_\alpha = (H^\alpha(\mathbb{R}^2))^2$ or is a closed subspace of the above Sobolev space.
- $\Gamma_1 = \{\, \mathbf{y}(\mathbf{x}, 1) \mid \mathbf{x} \in \Gamma_0 \,\}$, where the vector-valued function $\mathbf{y}$ is solution of the following initial value problem:

$$\begin{cases} \dfrac{\partial \mathbf{y}}{\partial t} = \mathbf{v}(\mathbf{y}, t) \quad \text{on } \mathbb{R}^2 \times (0, 1), \\ \mathbf{y}(\mathbf{x}, 0) = \mathbf{x}, \quad \mathbf{x} \in \mathbb{R}^2; \end{cases} \tag{E.4}$$

in (E.4), $\mathbf{v}(\mathbf{y}, t)$ denotes the function $\{\mathbf{x}, t\} \to \mathbf{v}(\mathbf{y}(\mathbf{x}, t), t)$.

(ii) Define the distance $\operatorname{dist}(\Gamma_0, \Gamma_R)$ between Γ_0 and Γ_R by

$$\operatorname{dist}(\Gamma_0, \Gamma_R) = \left(2J(\mathbf{u})\right)^{\frac{1}{2}}. \tag{E.5}$$

The minimization problem (E.1) has clearly the structure of an *exact controllability* problem, with $\mathbf{v}$ playing here the role of the control variable, $\mathcal{U}_f$ the role of the control set, $\mathbf{y}_b$ the role of the state variable, and (E.4) the role of the state equation.

Following an approach already used in several of the preceding parts of this book, we are going to relax the condition $\Gamma_1 = \Gamma_R$ in (E.3) by approximating the minimization problem (E.1) by *penalty*, obtaining thus

$$\begin{cases} \mathbf{u}_\varepsilon \in \mathcal{U}, \\ J_\varepsilon(\mathbf{u}_\varepsilon) \leq J_\varepsilon(\mathbf{v}), \quad \forall \mathbf{v} \in \mathcal{U}, \end{cases} \tag{E.6}$$

with

$$J_\varepsilon(\mathbf{v}) = J(\mathbf{v}) + \frac{1}{2\varepsilon}\langle S(\psi_1 - \psi_R), \psi_1 - \psi_R \rangle, \tag{E.7}$$

and

$$\mathcal{U} = L^2(0, 1; \mathbf{V}_\alpha).$$

In (E.7)

- ε is a "small" positive parameter (the penalty parameter).
- Operator S is a duality isomorphism between $H^2(\mathbb{R}^2)$ and $H^{-2}(\mathbb{R}^2)$ (and is therefore self-adjoint), and $\langle \cdot, \cdot \rangle$ denotes the duality pairing between the above two spaces.

- ψ_1 (respectively, ψ_R) is the unique solution of the following *linear variational problem*:

$$\begin{cases} \psi_1 \in H^2(\mathbb{R}^2), \\ \langle S\psi_1, \varphi \rangle = \int_{\Gamma_1} \varphi \, d\Gamma_1, \quad \forall \varphi \in H^2(\mathbb{R}^2), \end{cases} \tag{E.8}$$

(respectively,

$$\begin{cases} \psi_R \in H^2(\mathbb{R}^2), \\ \langle S\psi_R, \varphi \rangle = \int_{\Gamma_R} \varphi \, d\Gamma_R, \quad \forall \varphi \in H^2(\mathbb{R}^2)). \end{cases} \tag{E.9}$$

Assuming the existence of such $\mathbf{u}_\varepsilon$ and the differentiability of $J_\varepsilon(\cdot)$, $\mathbf{u}_\varepsilon$ verifies the following optimality condition:

$$DJ_\varepsilon(\mathbf{u}_\varepsilon) = \mathbf{0}, \tag{E.10}$$

where $DJ_\varepsilon(\mathbf{u}_\varepsilon)$ denotes the differential of $J_\varepsilon(\cdot)$ at $\mathbf{u}_\varepsilon$.

If one wants to solve (by iterative methods, in general) the approximate problem (E.6), we know from the above chapters of this book that it may be quite useful to know how to compute $DJ_\varepsilon(\mathbf{v})$, $\forall \mathbf{v} \in \mathcal{U}$ (of course, other approaches are possible, like those based on automatic differentiation as advocated in Griewank, 1992); the computation of $DJ_\varepsilon(\mathbf{v})$ is addressed in Azencott, Glowinski, and Ramos (2007).

The above problem has many variants, also worth to be considered. We definitely think that the methods discussed in this book will contribute to the solution of these problems. Future will tell if this prediction will be fulfilled.

We will conclude by mentioning that many issues addressed in this book are further discussed in Lasiecka and Triggiani (2000a, 2000b); the above references contain also control related topics not considered in our book.

Further Acknowledgments

The authors would like to acknowledge the help of the following individuals for friendly discussion and/or collaboration: A. Bamberger, C. Bardos, A. Bensoussan, M. Berggren, M.O. Bristeau, C. Carthel, E.J. Dean, T. Dupont, C. Fabre, V. Fraysse, P. Grisvard, P. Joly, W. Kinton, Y.M. Kuo, T. Lachand-Robert, J. Lagnese, I. Lasiecka, G. Lebeau, C.H. Li, G.M. Nasser, J. Périaux, J.-P. Puel, J. Rauch, D.L. Russel, H. Stève, R. Triggiani, M.F. Wheeler, and E. Zuazua.

The support of the following corporations or institutions is also acknowledged: AWARE, CERFACS, Collège de France, CRAY Research, Dassault Aviation, INRIA, LCC-Rio de Janeiro, Rice University, University of Colorado, University of Houston, Université P. & M. Curie. We also benefited from the support of DARPA (Contracts AFOSR F49620-89-C-0125 and AFOSR 90-0334), DRET, NASA, NSF (Grant INT 8612680), Texas Higher Education Coordinating Board (Grant ARP 003652156), and the Faculty Development Program, University of Houston (special thanks are due to Glenn Aumann, John Bear, and Garret J. Etgen).

Finally, special thanks are due to the editors of *Acta Numerica* for suggesting to us to write the two-part article Glowinski and J.L. Lions (1994, 1995), and to D. Bidois, C. Demars, L. Ruprecht, and J. Wilson for processing a preliminary version of it.

References

Abergel, F. and R. Temam (1990), On some control problems in fluid mechanics, *Theoret. Comput. Fluid Dyn.*, **1**, 303–326.

Allibert, B. (1998), Contrôle analytique de l'équation des ondes et de l'équation de Schrödinger sur des surfaces de révolution, *Comm. Partial Diff. Equations*, **23**(9–10), 1493–1556.

Antoulas, A.C. (2005), *Approximation of Large-Scale Dynamical Systems*, SIAM, Philadelphia, PA.

Armbruster, D., J. Guckenheimer, and P. Holmes (1989), Kuramoto-Sivashinski dynamics on the center-unstable manifold, *SIAM J. Appl. Math.*, **49**, 676–691.

Ash, M. and G. Lebeau (1998), Geometrical aspects of exact boundary controllability of the wave equation. A numerical study, *ESAIM Contrôle Opt. Calc. Var.*, **3**, 163–212.

Auchmuty, G., E.J. Dean, R. Glowinski, and S.C. Zhang (1987), Control methods for the numerical computation of periodic solutions of autonomous differential equations. In *Control Problems for Systems Described by Partial Differential Equations and Applications*, I. Lasiecka and R. Triggiani, eds., Lecture Notes in Control and Information, Vol. 97, Springer-Verlag, Berlin, 64–89.

Auchmuty, G. and C. Zhai (1996), Computation of rotating wave solutions of reaction–diffusion equations on a circle, *East–West J. Num. Anal.*, **4**, 151–163.

Avalos, G. (2000), Exact controllability of a thermo-elastic system with control in the thermal component only, *Diff. Int. Equ.*, **13**(4–6), 613–630.

Avalos, G. and I. Lasiecka (2000), Boundary controllability of thermo-elastic plates via the free boundary condition, *SIAM J. Control Opt.*, **38**(2), 337–383.

Azencott, R., R. Glowinski, and A.M. Ramos (2007), A controllability approach to shape identification, *Appl. Math. Lett.* (*to appear*).

Badr, H.M., S.C. Coutanceau, R. Dennis, and C. Ménard (1990), Unsteady flow past a circular cylinder at Reynolds numbers 10 and 104, *J. Fluid Mech.*, **220**, 459–484.

Ball, J.M., J.E. Marsden, and M. Slemrod (1982), Controllability for distributed bilinear systems, *SIAM J. Control Opt.*, **20**(4), 575–597.

Bamberger, A. (1977), Private communication.

Barck-Holst, S. (1996), *On the Numerical Simulation and Control of Systems Modeled by Partial Differential Equations*, Master Thesis dissertation, Department of Mechanical Engineering, University of Houston, Houston, TX.

Bardos, C., F. Bourquin, and G. Lebeau (1991), Calcul de dérivées normales et méthodes de Galerkin appliquées au problème de contrôlabilité exacte, *C. R. Acad. Sci. Paris, Série I, Math.*, **313**, 757–760.

Bardos, C., G. Lebeau, and J. Rauch (1988), Contrôle et stabilisation dans les problèmes hyperboliques. Appendix 2 of *Contrôlabilité Exacte, Perturbation et Stabilisation des Systèmes Distribués*, Vol. 1, J.L. Lions, ed., Masson, Paris, 492–537.

Bardos, C. and J. Rauch (1994), Variational algorithms for the Helmholtz equation using time evolution and artificial boundaries, *Asymptotic Anal.*, **9**, 101–117.

Bebernes, J. and D. Eberly (1989), *Mathematical Problems from Combustion Theory*, Springer-Verlag, New York, NY.

Benabdallah, B. and M.G. Naso (2000), Null-controllability of a thermo-elastic plate, *Abstr. Appl. Anal.*, **7**, 585–599.

Bensoussan, A. (1990), On the general theory of exact controllability for skew-symmetric operators, *Acta Mathematicae Applicandae*, **20**, 197–229.

Berggren, M. (1992), *Control and Simulation of Advection–Diffusion Systems*, Master Thesis dissertation, Department of Mechanical Engineering, University of Houston, Houston, TX.

Berggren, M. (1995), *Optimal Control of Time Evolution Systems: Controllability Investigations and Numerical Algorithms*, PhD Thesis dissertation, Department of Computational and Applied Mathematics, Rice University, Houston, TX.

Berggren, M. (1998), Numerical solution of a flow control problem: vorticity reduction by dynamic boundary action, *SIAM J. Sci. Comput.*, **19**, 829–860.

Berggren, M., R. Glowinski, and J.L. Lions (1996a), A computational approach to controllability issues for flow related models. (I) Pointwise control of the viscous Burgers equation, *J. Comput. Fluid Dynamics*, **7**, 237–252.

Berggren, M., R. Glowinski, and J.L. Lions (1996b), A computational approach to controllability issues for flow related models. (II) Control of two-dimensional linear advection–diffusion and Stokes models, *J. Comput. Fluid Dynamics*, **6**, 253–274.

Bittanti, S., A.J. Laub, and J.C. Willems (1991), *The Riccati Equation*, Springer-Verlag, New York, NY.

Bourquin, F. (1993), Approximation theory for the problem of exact controllability of the wave equation with boundary control. In *Mathematical and Numerical Aspects of Wave Propagation*, R. Kleinman, Th. Angell, D. Colton, F. Santosa, and I. Stackgold, eds., SIAM, Philadelphia, PA, 103–112.

Bourquin, F. (1994), A numerical approach to the exact controllability of Euler–Navier–Bernouilli beams. In *Proceedings of the 1st World Conference on Structural Control, August 3–5, 1994*, Pasadena, California, 120–129.

Bourquin, F. (1995), A numerical controllability test for distributed systems, *J. Struct. Control*, **2**, 5–23.

Bourquin, F. (2001), Numerical methods for the control of flexible structures, *J. Struct. Control*, **8**(1), 83–103.

Bourquin, F., M. Collet, and M. Joly (1999), Komornik's feedback stabilization: an experimental realization, *J. Acoust. Soc. of America*, **105**(2), 1240.

Bourquin, F., R. Namar, and J. Urquiza (1998), Discretization of the controllability Gramian in view of the exact boundary control: the case of thin plates. In *Optimal Control of Partial Differential Equations*, K.H. Hoffmann, ed., Birkhäuser, Basel, 53–62.

Braza, M., P. Chassaing, and H.H. Minh (1986), Numerical study and physical analysis of the pressure and velocity fields in the near wake of a circular cylinder, *J. Fluid Mech.*, **165**, 79–130.

Brenner, S.C. and L.R. Scott (2002), *The Mathematical Theory of Finite Element Methods*, Springer-Verlag, New York, NY.

Brezzi, F. and M. Fortin (1991), *Mixed and Hybrid Finite Element Methods*, Springer-Verlag, New York, NY.

Briffaut, J.S. (1999), *Méthodes Numériques pour le Contrôle et la Stabilisation Rapides des Grandes Structures Flexibles*, PhD dissertation, Ecole Nationale des Ponts et Chaussées, Paris, France.

Bristeau, M.O. (1997), Exact controllability methods for the calculation of 3-D time-periodic Maxwell solutions. In *Computational Science for the 21st Century*, M.O. Bristeau, G. Etgen, W. Fitzgibbon, J.L. Lions, J. Périaux, and M.F. Wheeler, eds., Wiley, Chichester, 492–503.

Bristeau, M.O., E.J. Dean, R. Glowinski, V. Kwok, and J. Periaux (1997), Exact controllability and domain decomposition methods with non-matching grids for the computation of waves. In *Domain Decomposition Methods in Science and Engineering*, R. Glowinski, J. Périaux, Z.C. Shi, and O. Widlund, eds., J. Wiley, Chichester, 291–308.

Bristeau, M.O., J. Ehrel, P. Feat, R. Glowinski and J. Périaux (1995), Solving the Helmholtz equation at high wave numbers on a parallel computer with a shared virtual memory, *Int. J. Supercomput. Ap.*, **9**(1), 18–29.

Bristeau, M.O., J. Ehrel, R. Glowinski, and J. Périaux (1993), A time dependent approach to the solution of the Helmholtz equation at high wave numbers. In *Procedings of Sixth SIAM Conference on Parallel Processing for Scientific Computing*, R.F. Sincovec, D.E. Keyes, M.R. Leuze, L.R. Petzold, and D.A. Reed, eds., SIAM, Philadelphia, PA, 113–127.

Bristeau, M.O., V. Girault, R. Glowinski, T.W. Pan, J. Périaux, and Y. Xiang (1997), On a fictitious domain method for flow and wave problem. In *Domain Decomposition Methods in Science and Engineering*, R. Glowinski, J. Périaux, Z.C. Shi, and O. Widlund, eds., J. Wiley, Chichester, 361–386.

Bristeau, M.O., R. Glowinski, B. Mantel, J. Périaux, and M. Sefrioui (1999), Genetic algorithms for electro-magnetic backscattering multi-objective optimization. In *Electro-Magnetic Optimization by Genetic Algorithms*, Y. Rahmat-Samii and E. Michielssen, eds., J. Wiley, New York, NY, 399–434.

Bristeau, M.O., R. Glowinski, and J. Périaux (1993a), Scattering waves using exact controllability methods, *31st AIAA Aerospace Science Meeting, Reno, Nevada*, AIAA Paper 93–0460.

Bristeau, M.O., R. Glowinski, and J. Périaux (1993b), Using exact controllability to solve the Helmholtz equation at high wave numbers. In *Mathematical and Numerical Aspects of Wave Propagation*, R. Kleinman, Th. Angell,

D. Colton, F. Santosa, and I. Stackgold, eds., SIAM, Philadelphia, PA, 113–127.

Bristeau, M.O., R. Glowinski, and J. Périaux (1993c), Numerical simulation of high frequency scattering waves using exact controllability methods. In *Nonlinear Hyperbolic Problems: Theoretical, Applied, and Computational Aspects*, A. Donato and F. Oliveri, eds., Notes in Numerical Fluid Mechanics, Vol. 43, Vieweg, Branschweig, 86–108.

Bristeau, M.O., R. Glowinski, and J. Périaux (1994), Exact controllability to solve the Helmholtz equation with absorbing boundary conditions. In *Finite Element Methods*, D. Krizek, P. Neittaanmaki, and R. Sternberg, eds., Marcel Dekker, New York, NY, 79–93.

Bristeau, M.O., R. Glowinski, and J. Périaux (1998), Controllability methods for the computation of time-periodic solutions; application to scattering, *J. Comp. Phys.*, **147**, 265–292.

Brumer, P. and M. Shapiro (1995), Laser control of chemical reactions, *Scientific American*, **272**(3), 34–39.

Burns, J.A. and S. Kang (1991a), A stabilization problem for Burgers' equation with unbounded control and observation, *International Series of Numerical Mathematics*, Vol. 100, Birkhäuser, Basel.

Burns, J.A. and S. Kang (1991b), A control problem for Burgers' equation with bounded input/output, *Nonlinear Dynam.*, **2**(4), 235–262.

Burns, J.A. and H. Marrekchi (1993), Optimal fixed-finite dimensional compensator for Burgers' equation with unbounded input/output operators, ICASE Report 93-19, ICASE, Norfolk, VA.

Burq, N. (1993), Contrôle de l'équation des plaques en présence d'obstacles strictement convexes, *Mem. Soc. Math. France, Série 2*, **55**, 3–126.

Buschnell, D.M. and J.N. Hefner (eds.) (1990), *Viscous Drag Reduction in Boundary Layers*, American Institute of Aeronautics and Astronautics, Washington, DC.

Carthel, C. (1994), *Numerical Methods for the Boundary Controllability of the Heat Equation*, PhD Thesis dissertation, Department of Mathematics, University of Houston, Houston, TX.

Carthel, C., R. Glowinski, and J.L. Lions (1994), On exact and approximate boundary controllability for the heat equation: a numerical approach, *J. Optim. Theory Appl.*, **82**(3), 42–484.

Carlsson, H.M. (1991), *On the Motion of Elastic Beams: Simulation, Stabilization and Control*, Master Thesis dissertation, Department of Mechanical Engineering, University of Houston, Houston, Texas.

Chan, T.F. and H.B. Keller (1982), Arc-length continuation and multigrid techniques for nonlinear eigenvalue problems, *SIAM J. Scient. Statist. Comp.*, **3**, 173–194.

Chen, H.Q., R. Glowinski, J. Périaux, and J. Toivanen (2005), Domain embedding/controllability methods for the conjugate gradient solution of wave propagation problems. In *Domain Decomposition Methods in Science and Engineering*, R. Kornhuber, R. Hoppe, J. Périaux, O. Pironneau, O. Widlund, and J. Xu, eds., Lecture Notes in Computational Science and Engineering, Vol. 40, Springer-Verlag, Berlin, 537–546.

Chewning, W.C. (1976), Controllability of the nonlinear wave equation in several space variables, *SIAM J. Control Opt.*, **14**, 19–25.

Chiara, A. (1993), Equation des ondes et régularité sur un ouvert Lipschitzien, *C. R. Acad. Sci. Paris, Série I, Math.*, **316**, 33–36.

Ciarlet, P.G. (1978), *The Finite Element Method for Elliptic Problems*, North-Holland, Amsterdam (reprinted as Vol. 40, SIAM Classics in Applied Mathematics, SIAM, Philadelphia, PA, 2002).

Ciarlet, P.G. (1989), *Introduction to Numerical Linear Algebra and Optimization*, Cambridge University Press, Cambridge.

Ciarlet, P.G. (1990a), A new class of variational problems arising in the modeling of elastic multi-structures, *Numer. Math.*, **57**, 547–560.

Ciarlet, P.G. (1990b), *Plates and Junctions in Elastic Multi-Structures: An Asymptotic Analysis*, Masson, Paris.

Ciarlet, P.G. (1991), Basic error estimates for elliptic problems. In *Handbook of Numerical Analysis*, Vol. II, P.G. Ciarlet and J.L. Lions, eds., North-Holland, Amsterdam, 17–351.

Ciarlet, P.G., H. Le Dret, and R. Nzengwa (1989), Junctions between three-dimensional and two-dimensional linearly elastic structures, *J. Math. Pures et Appl.*, **68**, 261–295.

Cirina, M.A. (1969), Boundary controllability of nonlinear hyperbolic systems, *SIAM J. Control Opt.*, **7**, 198–212.

Collet, M., F. Bourquin, M. Joly, and L. Ratier (2004), An efficient feedback control algorithm for beams: experimental investigations, *J. Sound and Vibrations*, **278**, 181–206.

Collis, C.S., K. Ghayour, and M. Heikenschloss (2003), Optimal transpiration boundary control for aero-acoustics, *AIAA J.*, **41**(7), 1257–1270.

Collis, C.S., K. Ghayour, M. Heikenschloss, M. Ulbrich, and S. Ulbrich (2002), Optimal control for unsteady compressible viscous flow, *Int. J. Num. Meth. Fluids*, **40**, 1401–1429.

Coron, J.M. (1996a), On the controllability of 2-D incompressible perfect fluids, *J. Math. Pures Appl.*, **75**(2), 155–188.

Coron, J.M. (1996b), On the controllability of the 2-D incompressible Navier–Stokes equations with the Navier slip boundary conditions, *ESAIM: COCV*, **1**, 35–75.

Crandall, M.G. and P.L. Lions (1985), Hamilton–Jacobi equations in infinite dimensions, Part I, *J. Funct. Anal.*, **62**, 379–396.

Crandall, M.G. and P.L. Lions (1986a), Hamilton–Jacobi equations in infinite dimensions, Part II, *J. Funct. Anal.*, **65**, 368–405.

Crandall, M.G. and P.L. Lions (1986b), Hamilton–Jacobi equations in infinite dimensions, Part III, *J. Funct. Anal.*, **68**, 214–247.

Crandall, M.G. and P.L. Lions (1990), Hamilton–Jacobi equations in infinite dimensions, Part IV, *J. Funct. Anal.*, **90**, 237–283.

Crandall, M.G. and P.L. Lions (1991), Hamilton–Jacobi equations in infinite dimensions, Part V, *J. Funct. Anal.*, **97**, 417–465.

Crépeau, E. (2003), Exact controllability of the Boussinesq equation on a bounded domain, *Diff. Int. Equ.*, **16**(3), 303–326.

Daniel, J. (1970), *The Approximate Minimization of Functionals*, Prentice Hall, Englewood Cliffs, NJ.

Dean, E.J., R. Glowinski, and C.H. Li (1989), Supercomputer solution of partial differential equation problems in computational fluid dynamics and in control, *Comp. Phys. Commun.*, **53**, 401–439.

Dean, E.J., R. Glowinski and D. Trevas (1996), An approximate factorization/least squares solution method for a finite element approximation of the Cahn-Hilliard equation, *Jap. J. Ind. Appl. Math.*, **13**(3), 495–517.

Dean, E.J. and P. Gubernatis (1991), Pointwise control of Burgers' equation: a numerical approach, *Comput. Math. Appl.*, **22**, 93–100.

Decker, D.W. and H.B. Keller (1980), Multiple limit point bifurcation, *J. Math. Anal. Appl.*, **75**, 417–430.

Dehman, B., G. Lebeau, and E. Zuazua (2003), Stabilization and control for the sub-critical semi-linear wave equation, *Annales Ecole Normale Supérieure de Paris*, **36**(4), 525–551.

Dennis, J.E. and R.B. Schnabel (1983), *Numerical Methods for Unconstrained Optimization and Nonlinear Problems*, Prentice Hall, Englewood Cliffs, NJ.

Dennis, J.E. and R.B. Schnabel (1996), *Numerical Methods for Unconstrained Optimization and Nonlinear Problems*, SIAM, Philadelphia, PA.

Diaz, J.I. (1991), Sobre la controlabilidad approximada de problemas no lineales disipativos. In the proceedings of *Jornadas Hispano-Francesas Sobre Control de Sistemas Distribuidos*, Unversidad de Malaga, 41–48.

Diaz, J.I. (1994), On the controllability of simple climate models. In *Environment, Economics and their Mathematical Models*, J.I. Diaz and J.L. Lions, eds., Masson, Paris, 29–43.

Diaz, J.I. (1996), Obstruction and some approximate controllability results: Burgers equation and related problems. In *Control of Partial Differential Equations and Applications*, Lecture Notes in Pure and Applied Mathematics, **174**, E. Casas, ed., Marcel Dekker, New York, NY, 63–76.

Diaz, J.I. and A.V. Fursikov (1997), Approximate controllability of the Stokes system on cylinders by external unidirectional forces, *J. Math. Pures et Appl.*, **76**(4), 353–375.

Diaz, J.I. and A.M. Ramos (1997), Approximate controllability and obstruction phenomena for quasi-linear diffusion equations. In *Computational Science for the 21st Century*, M.O. Bristeau, G. Etgen, W. Fitzgibbon, J.L. Lions, J. Périaux, and M.F. Wheeler, eds., J. Wiley, Chichester, 698–707.

Downer, J.D., K.C. Park, and J.C. Chiou (1992), Dynamics of flexible beams for multi-body systems: a computational procedure, *Comput. Method Appl. M.*, **96**, 373–408.

Dupont, T., R. Glowinski, W. Kinton, and M.F. Wheeler (1992), Mixed finite element methods for time dependent problems: applications to control. In *Finite Element in Fluids*, Vol. 8, T. Cheung, ed., Hemisphere, Washington, DC, 137–163.

Ekeland, I. and R. Temam (1974), *Analyse Convexe et Problèmes Variationnels*, Dunod, Paris.

Eller, M., I. Lasiecka and R. Triggiani (2000), Simultaneous exact/approximate boundary controllability of thermo-elastic plates with variable transmission coefficients and moment control, *J. Math. Anal. Appl.*, **251**(2), 452–478.

Engquist, B., B. Gustafsson, and J. Vreeburg (1978), Numerical solution of a PDE system describing a catalytic converter, *J. Comput. Phys.*, **27**, 295–314.

Fabre, C., J.P. Puel, and E. Zuazua (1993), Contrôlabilité approchée de l'équation de la chaleur linéaire avec des contrôles de norme L^∞ minimale, *C. R. Acad. Sci. Paris, Série I, Math.*, **316**, 679–684.

Fattorini, H.O. (1975), Local controllability of a nonlinear wave equation, *Math. Systems Theory*, **9**, 363–366.

Faurre, P. (1971), *Navigation Inertielle Optimale et Filtrage Statistique*, Dunod, Paris.

Fernandez-Cara, E. (1999), On the approximate and null-controllability of the Navier–Stokes equations, *SIAM Rev.*, **41**(2), 269–277.

Fernandez-Cara, E., S. Guerrero, O.Y. Imanuvilov and, J.P. Puel (2004), Local exact controllability of the Navier–Stokes system, *J. Math. Pures Appl.*, **83**(12), 1501–1542.

Fornberg, B. (1980), A numerical study of steady flow past a circular cylinder, *J. Fluid Mech.*, **98**, 819–855.

Fortin, M. and R. Glowinski (1983), *Augmented Lagrangians: Application to the Numerical Solution of Boundary Value Problems*, North-Holland, Amsterdam 1983.

Foss, F. (2006), *On the Exact Point-Wise Interior Controllability of the Scalar Wave Equation and Solution of Nonlinear Elliptic Eigenproblems*, PhD dissertation, Department of Mathematics, University of Houston, Houston, Texas.

Friedman, A. (1988), Modeling catalytic converter performance. In *Mathematics in Industrial Problems*, Part IV, Chapter 7, A. Fiedman, ed., Springer-Verlag, New York, NY, 70–77.

Friend, C.M. (1993), Catalysis and surfaces, *Scientific American*, April, 74–79.

Fujita, H. and T. Suzuki (1991), Evolution problems. In *Handbook of Numerical Analysis*, Vol. II, P.G. Ciarlet and J.L. Lions, eds., North-Holland, Amsterdam, 789–928.

Fursikov, A.V. (1992), Lagrange principle for problems of optimal control of ill-posed or singular distributed systems, *J. Math. Pures et Appl.*, **71**(2), 139–194.

Fursikov, A.V. (1995), Exact boundary zero-controllability of three-dimensional Navier–Stokes equations, *J. Dynam. Contr. Systems*, **1**(3), 325–350.

Fursikov, A.V. and O.Y. Imanuvilov (1994), On exact boundary zero-controllability of two-dimensional Navier–Stokes equations, *Acta Applicandae Mathematicae*, **37**, 67–76.

Fursikov, A.V. and O.Y. Imanuvilov (1998), Local exact controllability of the Boussinesq equations, *SIAM J. Control Opt.*, **36**, 391–421.

Gabay, D. (1982), Application de la méthode des multiplicateurs aux inequations variationnelles. In *Méthodes de Lagrangiens Augmentés: Application à la Résolution Numérique des Problèmes aux Limites*, M. Fortin and R. Glowinski, eds., Dunod, Paris, 279–307.

Gabay, D. (1983), Application of the method of multipliers to variational inequalities. In *Augmented Lagrangian Methods: Application to the Numerical Solution of*

Boundary Value Problems, M. Fortin and R. Glowinski, eds., North-Holland, Amsterdam, 299–331.

Gad-el-Hak, M. (1989), Flow control, *Applied Mech. Rev.*, **42**, 261–292.

Gama, S., U. Frisch, and H. Scholl (1991), The two-dimensional Navier–Stokes equations with a large scale instability of the Kuramoto-Sivashinsky type: numerical exploration on the Connection Machine, *J. Scient. Comp.*, **6**(4), 425–452.

George, J.A. (1971), *Computer Implementation of the Finite Element Method*, PhD Thesis dissertation, Computer Sciences Department, Stanford University, Stanford, CA.

Ghattas, O. and J.H. Bark (1997), Optimal control of two and three dimensional incompressible Navier–Stokes flow, *J. Comp. Phys.*, **136**(2), 231–244.

Girault, V. and P.A. Raviart (1986), *Finite Element Methods for Navier–Stokes Equations: Theory and Algorithms*, Springer-Verlag, Berlin.

Glowinski, R. (1984), *Numerical Methods for Nonlinear Variational Problems*, Springer-Verlag, New York, NY.

Glowinski, R. (1991), Finite element methods for the numerical simulation of incompressible viscous flow. Introduction to the control of the Navier–Stokes equations. In *Vortex Dynamics and Vortex Methods*, C.R. Anderson and C. Greengard, eds., Lectures in Applied Mathematics, Vol. 28, American Mathematical Society, Providence, RI, 219–301.

Glowinski, R. (1992a), Ensuring well-posedness by analogy: Stokes problem and boundary control for the wave equation, *J. Comp. Phys.*, **103**, 189–221.

Glowinski, R. (1992b), Boundary controllability problems for the wave and heat equations. In *Boundary Control and Boundary Variations*, J.P. Zolezio, ed., Lecture Notes in Control and Information Sciences, Vol. 178, Springer-Verlag, Berlin, 221–237.

Glowinski, R. (2003), Finite Element Methods for Incompressible Viscous Flow. In: *Handbook of Numerical Analysis*, Vol. IX, P.G. Ciarlet and J.L. Lions, eds., North-Holland, Amsterdam, 3–1176.

Glowinski, R. and J.W. He (1998), On shape optimization and related issues. In *Computational Methods for Optimal Design and Control*, J. Borggaard, J. Burns, E. Cliff, and S. Schreck, eds., Birkhäuser, Boston, MA, 151–179.

Glowinski, R., J.W. He, and J.L. Lions (2002), On the controllability of wave models with variable coefficients: a numerical investigation, *Computational and Applied Mathematics*, **21**(1), 191–225.

Glowinski, R., H.B. Keller, and L. Reinhart (1985), Continuation-conjugate gradient methods for the least-squares solution of nonlinear boundary value problems, *SIAM J. Scient. Statist. Comp.*, **6**, 793–832.

Glowinski, R., W. Kinton, and M.F. Wheeler (1989), A mixed finite element formulation for the boundary controllability of the wave equation, *Int. J. Numer. Meth. Engrg.*, **27**, 623–635.

Glowinski, R. and P. Le Tallec (1989), *Augmented Lagrangian and Operator-Splitting Methods in Nonlinear Mechanics*, SIAM, Philadelphia, PA.

Glowinski, R. and C.H. Li (1990), On the exact Neumann boundary controllability of the wave equation. *C. R. Acad. Sci., Paris, Série I, Math.*, **311**, 135–142.

Glowinski, R. and C.H. Li (1991), On the numerical implementation of the Hilbert uniqueness method for the exact boundary controllability of the wave equation. In *Mathematical and Numerical Aspects of Wave Propagation Phenomena*, G. Cohen, L. Halpern and P. Joly, eds., SIAM, Philadelphia, PA, 15–24.

Glowinski, R., C.H. Li, and J.L. Lions (1990), A numerical approach to the exact boundary controllability of the wave equation (I) Dirichlet controls: description of the numerical methods, *Japan J. Appl. Math.*, **7**, 1–76.

Glowinski, R. and J.L. Lions (1994), Exact and approximate controllability of distributed parameter systems, Part I, *Acta Numerica 1994*, Cambridge University Press, Cambridge, 269–378.

Glowinski, R. and J.L. Lions (1995), Exact and approximate controllability of distributed parameter systems, Part II, *Acta Numerica 1995*, Cambridge University Press, Cambridge, 159–333.

Glowinski, R., J.L. Lions, and R. Trémolières (1976), *Analyse Numérique des Inéquations Variationnelles*, Dunod, Paris.

Glowinski, R., J.L. Lions, and R. Trémolières (1981), *Numerical Analysis of Variational Inequalities*, North-Holland, Amsterdam.

Glowinski, R., J. Périaux, and J. Toivanen (2003), Time-periodic solutions of wave equations via controllability and fictitious domain methods. In *Mathematical and Numerical Aspects of Wave Propagation (WAVES 2003)*, G.C. Cohen, E. Heikkola, P. Joly, and P. Neittaanmaki, eds., Springer-Verlag, Berlin, 805–810.

Glowinski, R. and A.M. Ramos (2002), A numerical approach to the Neumann control of the Cahn-Hilliard equation. In *Computational Methods for Control Applications*, R. Glowinski, H. Kawarada, and J. Périaux, eds., Gakkotosho Co., Tokyo, 111–115.

Glowinski, R. and T. Rossi (2006), A mixed formulation and exact controllability approach for the computation of the periodic solutions of the scalar wave equation. (I): Controllability problem formulation and related iterative solution, *C. R. Acad. Sci. Paris, Série I, Math.*, **343**, 493–498.

Gorman, M., M. El-Hamdi, and K.A. Robbins (1994), Experimental observations of ordered cellular flames, *Combust. Sci. and Tech.*, **98**, 37–45.

Gresho, P.M. and R.L. Sani (1998), *Incompressible Flow and the Finite Element Method: Advection-Diffusion Equations and Isothermal Laminar Flow*, J. Wiley, Chichester.

Griewank, A. (1992), Achieving logarithmic growth of temporal and spatial complexity in reverse automatic differentiation, *Optimization Methods and Software*, **1**, 35–54.

Gunzburger, M.D. (1989), *Finite Element Method for Viscous Incompressible Flows*, Academic Press, Boston, MA.

Gunzburger, M.D. (ed.) (1995), *Flow Control*, **68**, IMA Volumes in *Mathematics and its Applications*, Springer-Verlag, New York, NY.

Gunzburger, M.D. (1999), Navier–Stokes equations for incompressible flows: finite element methods. In *Handbook of Computational Fluid Mechanics*, R. Peyret, ed., Academic Press, Boston, 99–157.

Gunzburger, M.D. (2000), Adjoint equation-based methods for control problems in viscous, incompressible flows, Flow, *Turbulence and Combustion*, **65**, 249–272.

Gunzburger, M.D., A. Fursikov, and L. Hou (1998), Boundary value problems and optimal boundary control for the Navier–Stokes system: The two-dimensional case, *SIAM J. Control Opt.*, **36**, 852–894.

Gunzburger, M.D., A. Fursikov and L. Hou (2005), Optimal boundary control for the evolutionary Navier–Stokes system: The three-dimensional case, *SIAM J. Control Opt.*, **43**, 2191–2232.

Gunzburger, M.D., L. Hou, S. Manservisi, and Y. Yan (1998), Computations of optimal control for incompressible flows, *Int. J. Comp. Fluid Dynamics*, **11**, 181–191.

Gunzburger, M.D., L. Hou, and T. Svobodny (1989), Numerical approximation of an optimal control problem associated with the Navier–Stokes equations, *Appl. Math. Lett.*, **2**, 29–31.

Gunzburger, M.D., L. Hou, and T. Svobodny (1992), Boundary velocity control of incompressible flow with an application to viscous drag reduction, *SIAM J. Control Opt.*, **30**, 167–181.

Gunzburger, M.D. and H.C. Lee (1996), Feedback control of Kármán vortex shedding, *J. Appl. Mech.*, **63**, 828–835.

Gunzburger, M.D. and S. Manservisi (1999), The velocity tracking problem for Navier–Stokes flows with bounded distributed controls, *SIAM J. Control Opt.*, **37**, 1913–1945.

Gunzburger, M.D. and S. Manservisi (2000a), The velocity tracking problem for Navier–Stokes flows with boundary controls, *SIAM J. Control Opt.*, **39**, 594–634.

Gunzburger, M.D. and S. Manservisi (2000b), Analysis and approximation of the velocity tracking problem for Navier–Stokes flows with distributed controls, *SIAM J. Num. Anal.*, **37**, 1481–1512.

Hackbush, W. (1985), *Multigrid Methods and Applications*, Springer-Verlag, Berlin.

Hansen, S.W. and B.Y. Zhang (1997), Boundary control of a linear thermo-elastic beam, *J. Math. Anal. Appl.*, **210**(1), 182–205.

Haraux, A. (1989), Séries lacunaires et contrôle semi-interne des vibrations d'une plaque rectangulaire, *J. Math. Pures Appl.*, **68**(4), 457–465.

Haraux, A. and S. Jaffard (1991), Pointwise and spectral control of plate vibrations, *Revista Matematica Iberoamericana*, **7**, 1–24.

He, J.W., M. Chevalier, R. Glowinski, R. Metcalfe, A. Nordlander, and J. Périaux (2000), Drag reduction by active control for flow past cylinders. In *Computational Mathematics Driven by Industry*, V. Capasso, H. Engl, and J. Périaux, eds., Lecture Notes in Mathematics, Vol. 1739, Springer-Verlag, Berlin, 287–363.

He, J.W. and R. Glowinski (1998), Neumann control of unstable parabolic systems: numerical approach, *J. Optim. Theory Appl.*, **96** (1), 1–55.

He, J.W., R. Glowinski, M. Gorman and J. Périaux (1998), Some results on the controllability and stabilization of the Kuramoto-Sivashinsky equation. In *Equations aux Dérivées Partielles et Applications. Articles dédiés à Jacques-Louis Lions*, Gauthier-Villars/Elsevier, Paris, 571–590.

He, J.W., R. Glowinski, R. Metcalfe, A. Nordlander, and J. Périaux (2000), Active control and drag reduction for flow past a circular cylinder. I. Oscillatory cylinder rotation, *J. Comp. Phys.*, **163**, 83–117.

He, J.W., R. Glowinski, R. Metcalfe, and J. Périaux (1998), A numerical approach to the control and stabilization of advection-diffusion systems: application to drag reduction, *Int. J. Comp. Fluid Dynamics*, **11**, 131–156.

He, J.W., R. Glowinski, R. Metcalfe, and J. Périaux (2002), Active control for incompressible viscous fluid flow: application to drag reduction for flow past circular cylinders. In *Computational Methods for Control Applications*, R. Glowinski, H. Kawarada and J. Périaux, eds., Gakkotosho Co, Tokyo, 233–292.

Heikkola, E., S. Monkola, A. Pennanen, and T. Rossi (2005), Solution of the Helmholtz equation with controllability and spectral element methods, *J. Structural Mech.*, **8**(3), 121–124.

Henderson, R.D. (1997), Nonlinear dynamics and patterns in turbulent wake transition, *J. Fluid Mech.*, **352**, 65–112.

Henry, J. (1978), *Contrôle d'un Réacteur Enzymatique à l'Aide de Modèles à Paramètres Distribués. Quelques Problèmes de Contrôlabilité de Systèmes Paraboliques*, Thèse d'Etat, Université P. et M. Curie, Paris.

Hintermüller, M., K. Kunisch, Y. Spasov, and S. Volkwein (2004), Dynamical system-based optimal control of incompressible fluids, *Int. J. Num. Meth. Fluids*, **46**(4), 345–359.

Hinze, M. and K. Kunisch (2001), Second order methods for optimal control of time-dependent fluid flow, *SIAM J. Control Opt.*, **40**(3), 925–946.

Hiriart-Urruty, J.B. and C. Lemarechal (1993), *Convex Analysis and Minimization Algorithms*, Springer-Verlag, Berlin.

Holmes, Ph., J.L. Lumley, and G. Berkooz (1996), *Turbulence, Coherent Stuctures, Dynamical Systems and Symmetry*, Cambridge University Press, New York, NY.

Homescu, C., I.M. Navon, and Z. Li (2002), Suppression of vortex shedding for flow around a circular cylinder using optimal control, *Int. J. Num. Meth. Fluids*, **38**, 43–69.

Hörmander, L. (1976), *Linear Partial Differential Operators*, Springer-Verlag, Berlin.

Hou, L.S. and S.S. Ravindran (1996), Computation of boundary optimal control for an electrically conducting fluid, *J. Comp. Phys.*, **128**(2), 319–330.

Hyman, J.M. and B. Nicolaenko (1986), The Kuramoto-Sivashinski equation: a bridge between PDEs and dynamical systems, *Physica-D*, **18**, 113–126.

Ingham, D.B. and T. Tang (1990), A numerical investigation into the steady flow past a rotating circular cylinder at low and intermediate Reynolds numbers, *J. Comp. Phys.*, **87**, 91–107.

Ito, K. and S.S. Ravindran (1998), A reduced-order method for simulation and control of fluid flow, *J. Comp. Phys.*, **143**(2), 403–425.

Jaffard, S. (1988), Contrôle interne exact des vibrations d'une plaque carrée, *C. R. Acad. Sc. Paris, Série I, Math.*, **307**(14), 759–762.

Joó, I. (1991) Controlabilité exacte et propriétés d'oscillations de l'équation des ondes par analyse non harmonique, *C. R. Acad. Sc. Paris, Série I, Math.*, **312**, 119–122.

Kalman, R.E. (1960), A new approach to linear filtering and prediction problems, *J. of Basic Engineering*, March 1960, 35–45.

Kalman, R.E. (1963), New methods in Wiener filtering theory. In *Proceedings of the 1st Symposium on Engineering Applications of Random Functions*, J.L. Bogdanoff and F. Korin, ed., Wiley.

Kalman, R.E. and R.S. Bucy (1961), New results in linear filtering and prediction theory, *J. of Basic Engineering*, March 1961, 95–108.

Kalman, R.E., P.L. Falb, and M.A. Arbib (1969), *Topics in Mathematical System Theory*, Mc Graw Hill, New York, NY.

Kapitanov, B. (1994), Stabilization and exact boundary controllability for Maxwell's equations, *SIAM J. Control Opt.*, **32**, 408–421.

Kapitanov, B. and G. Perla-Menzela (2003), Uniform stabilization and exact control of a multilayered piezoelectric body, *Portugaliae Mathematica*, **60**(4), 411–454.

Kapitanov, B. and M.A. Raupp (2003), Simultaneous exact control of of piezoelectric media, *Comp. Appl. Math.*, **22**(2), 249–277.

Karniadakis, G.E. and G.S. Triantafyllou (1989), Frequency selection and asymptotic states in laminar wakes, *J. Fluid Mech.*, **199**, 441–469.

Keener, J.P. and H.B. Keller (1973), Perturbed bifurcation theory, *Archives Rat. Mech. Anal.*, **50**, 159–175.

Keener, J.P. and H.B. Keller (1974), Positive solutions of convex nonlinear eigenvalue problems, *J. Diff. Equations*, **16**, 103–125.

Keller, H.B. (1977), Numerical solution of bifurcation and nonlinear eigenvalue problems. In *Applications of Bifurcation Theory*, P. Rabinowitz, ed., Academic Press, New York, NY, 359–384.

Keller, H.B. (1982), Practical procedures in path following near limit points. In *Computing Methods in Applied Sciences and Engineering (V)*, R. Glowinski and J.L. Lions, eds., North-Holland, Amsterdam, 177–183.

Keller, H.B. and W.F. Langford (1972), Iterations, perturbations and multiplicities for nonlinear bifurcation problems, *Archives Rat. Mech. Anal.*, **48**, 83–108.

Kime, K. (1995), Simultaneous control of a rod equation and of a simple Schrödinger equation, *System Control Lett.*, **24**(4), 301–306.

Krevedis, I.G., B. Nicolaenko and J.C. Scovel (1990), Back in the saddle again: a computer assisted study of the Kuramoto-Sivashinski equation, *SIAM J. Appl. Math.*, **50**(3), 760–790.

Kwon, K. and H. Choi (1996), Control of laminar vortex shedding behind a circular cylinder, *Phys. Fluids*, **8**, 479–486.

Ladyzenskaya, O.A., V.A. Solonnikov, and N.N. Ural'ceva (1968), *Linear and Quasi-Linear Equations of Parabolic Type*, American Mathematical Society, Providence, RI.

Ladyzenskaya, O.A. (1969), *The Mathematical Theory of Viscous Incompressible Flow*, Gordon and Breach, New York, NY.

Lagnese, J.E. (1989), Exact boundary controllability of Maxwell's equations in a general region, *SIAM J. Control Opt.*, **27**, 374–388.

Lagnese, J.E., G. Leugering, and G. Schmidt (1992), Modeling and controllability of networks of thin beams. In *System Modeling and Optimization*, Lecture Notes in Control and Information, Vol. 97, Springer-Verlag, Berlin, 467–480.

Lagnese, J.E., G. Leugering, and G. Schmidt (1994), *Modeling, Analysis and Control of Dynamic Elastic Multi-Link Structures*, Birkhäuser, Boston.

Lagnese, J.E. and J.L. Lions (1988), *Modeling, Analysis and Control of Thin Plates*, Masson, Paris.

Lasiecka, I. and D. Tataru (1993), Uniform boundary stabilization of semi-linear wave equations with nonlinear boundary damping, *Diff. Int. Equ.*, **6**(3), 507–533.

Lasiecka, I. and R. Triggiani (1991), Exact controllability of semi-linear abstract systems with applications to waves and plates boundary control problems, *Appl. Math. Optim.*, **23**(2), 109–154.

Lasiecka, I. and R. Triggiani (1992), Optimal regularity, exact controllability and uniform stabilization of Schrodinger equations with Dirichlet control, *Diff. Int. Equ.*, **5**(3), 521–535.

Lasiecka, I. and R. Triggiani (2000a), *Control Theory for Partial Differential Equations: Continuous and Approximation Theories. Vol. I: Abstract Parabolic Systems*, Cambridge University Press, Cambridge, UK.

Lasiecka, I. and R. Triggiani (2000b), *Control Theory for Partial Differential Equations: Continuous and Approximation Theories. Vol. II: Abstract Hyperbolic-Like Systems Over a Finite Time Horizon*, Cambridge University Press, Cambridge, UK.

Laursen, T.A. and J.C. Simo (1993), A continuum-based finite element formulation for the implicit solution of multi-body, large deformation frictional contact problems, *Int. J. Num. Meth. Engrg.*, **36**, 3451–3486.

Lax, P.D. and R.S. Phillips (1989), *Scattering Theory*, Academic Press, New York, NY.

Lebeau, G. (1992), Contrôle de l'équation de Schrödinger, *J. Math. Pures Appl.*, **71**, 267–291.

Lebeau, G. and L. Robbiano (1995), Contrôle exact de l'équation de la chaleur, Commun. in *Partial Differential Equations*, **20**(1&2), 335–356.

Lebeau, G. and E. Zuazua (1998), Null-controllability of a system of linear thermoelasticity, *Archives Rat. Mech. Anal.*, **141**(4), 297–329.

León, L. and E. Zuazua (2002), Boundary controllability of the finite difference space semi-discretization of the beam equation, *ESAIM Contrôle Optim. Calc. Variat. CV*, **8**, 827–862.

Li, B., G. Turinici, V. Ramakrishna, and H. Rabitz (2002), Optimal dynamic discrimination of similar molecules through quantum learning control, *J. Phys. Chem. B.*, **106**(33), 8125–8131.

Li, T.T. and B.P. Rao (2002), Local exact boundary controllability for a class of quasilinear hyperbolic systems, *Chinese Annals of Math. B*, **23**(2), 209–218.

Li, T.T. and B.P. Rao (2003), Exact boundary controllability for quasilinear hyperbolic systems, *SIAM J. Control Optim.*, **41**(6), 1748-1755.

Li, T.T., B.P. Rao, and J. Yi (2000), Semi-global C^1 solution and exact boundary controllability for reducible quasilinear hyperbolic systems, *Math. Model. Numer. Anal.*, **34**, 399–408.

Li, T.T., B.P. Rao, and J. Yi (2001), Solution C^1 semi-globale et contrôlabilité exacte frontière de systèmes hyperboliques quasi-linéaires, *C.R. Acad. Sci. Paris, Série I*, **333**, 219–224.

Li, T.T. and Y.L. Xu (2003), Local exact boundary controllability for nonlinear vibrating string equations, *Int. J. of Modern Physics B*, **17**(22–24), 4062–4071.

Li, T.T. and B.Y. Zhang (1998), Global exact boundary controllability of a class of quasilinear hyperbolic systems, *J. Math. Anal. Appl.*, **225**, 289–311.

Lions, J.L. (1961), *Equations Différentielles Opérationnelles et Problèmes aux Limites*, Springer-Verlag, Heidelberg.

Lions, J.L. (1968) *Contrôle Optimal des Systèmes Gouvernés par des Equations aux Dérivées Partielles*, Dunod, Paris.

Lions, J.L. (1969), *Quelques Méthodes de Résolution des Problèmes aux Limites Non Linéaires*, Dunod, Paris.

Lions, J.L. (1971), *Optimal Control of Systems Governed by Partial Differential Equations*, Springer-Verlag, New York, NY.

Lions, J.L. (1983), *Contrôle des Systèmes Distribués Singuliers*, Gauthier-Villars, Paris (English translation: Gauthier-Villars, Paris, 1985).

Lions, J.L. (1986), Contrôllabilité exacte des systèmes distribués, *C. R. Acad. Sci., Paris, Série I, Math.*, **302**, 471–475.

Lions, J.L. (1988a), Exact controllability, stabilization and perturbation for distributed systems, *SIAM Rev.*, **30**, 1–68.

Lions, J.L. (1988b), *Contrôlabilité Exacte, Perturbation et Stabilisation des Systèmes Distribués*, Vols. 1 and 2, Masson, Paris.

Lions, J.L. (1989), Sentinels for periodic distributed systems, *Chinese Annals of Mathematics*, **10B**(3), 285–291.

Lions, J.L. (1990), *El Planeta Tierra*, Instituto de España, Espasa Calpe, S.A., Madrid.

Lions, J.L. (1991a), Exact controllability for distributed systems: some trends and some problems. In *Applied and Industrial Mathematics*, R. Spigler, ed., Kluwer, Dordrecht, 59–84.

Lions, J.L. (1991b), Approximate controllability for parabolic systems, Harvey Lectures, Israel Institute of Technology (Technion), Haiffa.

Lions, J.L. (1992a), *Sentinelles pour les systèmes distribués à données incomplètes*, Masson, Paris.

Lions, J.L. (1992b), On some hyperbolic equation with a pressure term. In *Partial Differential Equations and Related Subjects*, M. Miranda, ed., Longman Scientific and Technical, Harlow, UK, 196–208.

Lions, J.L. (1993), Quelques remarques sur la contrôlabilité en liaison avec des questions d'environnement. In *Les Grands Systèmes des Sciences et de la Technologie*, J. Horowitz and J.L. Lions, eds., Masson, Paris, 240–264.

Lions, J.L. (1997a), On the approximate controllability with global state constraints. In *Computational Science for the 21st Century*, M.O. Bristeau, G.

Etgen, W. Fitzgibbon, J.L. Lions, J. Périaux and M.F. Wheeler, eds., J. Wiley, Chichester, 718–727.

Lions, J.L. (1997b), Very rapid oscillations and control. In *HERMIS '96, Proceedings of the Third Hellenic-European Conference on Mathematics and Informatics*, E.A. Lipitakis, ed., LEA Publisher, Athens, 1–10.

Lions, J.L. (2002), *Quelques Méthodes de Résolution des Problèmes aux Limites Non Linéaires*, Dunod, Paris.

Lions, J.L. (2003), *Oeuvres Choisies*, Vol. III, SMAI/EDP Sciences, Paris.

Lions, J.L. and E. Magenes (1968), *Problèmes aux Limites Non Homogènes*, Vol. I, Dunod, Paris.

Lions, J.L. and E. Zuazua (1996), Approximate controllability of a hydro-elastic system, *ESAIM: COCV*, **1**, 1–15.

Lions, J.L. and E. Zuazua (1997), The cost of controlling unstable systems: time irreversible systems, *Revista Matematica Complutense*, **10**(2), 481–523.

Lions, J.L. and E. Zuazua (1998), Exact boundary controllability of Galerkin's approximation of Navier–Stokes equations, *Ann. Scuola Norm. Sup. Pisa*, **XXVI**(98), 605–621.

Lions, P.L. (1996), *Mathematical Topics in Fluid Mechanics. Vol. I: Incompressible Models*, Oxford University Press, Oxford.

Lions, P.L. and B. Mercier (1979), Splitting algorithms for the sum of two nonlinear operators, *SIAM J. Numer. Anal.*, **16**, 964–979.

Littman, W. and L. Markus (1988), Exact boundary controllability of a hybrid system of elasticity, *Arch. Rat. Mech. Anal.*, **103**, 193–236.

Liu, D.C. and J. Nocedal (1989), On the limited memory BFGS method for large scale optimization, *Math. Program.*, **45**, 503–528.

Machtyngier, E. (1990), Contrôlabilité exacte et stabilisation frontière de l'équation de Schrödinger, *C.R . Acad. Sci. Paris, Série I*, **310**(12), 801–806.

Machtyngier, E. (1994), Exact controllability for the Schrödinger equation, *SIAM J. Control Opt.*, **32**(1), 24–34.

Machtyngier, E. and E. Zuazua (1994), Stabilization of the Schrödinger equation, *Portugaliae Matematica*, **51**(2), 243–256.

Maday, Y. and G. Turinici (2003), New formulas of monotonically convergent quantum control algorithms, *J. Chem. Phys.*, **118**, 8191–8196.

Marchuck, G.I. (1990), Splitting and alternating direction methods. In *Handbook of Numerical Analysis*, Vol. I, P.G. Ciarlet and J.L. Lions, eds., North-Holland, Amsterdam, 197–462.

Marini, G., P. Testa, and V. Valente (1994), Exact controllability of a spherical cap: numerical implementation of HUM, *J. Opt. Theo. Appl.*, **81**(2), 329–341.

Marsden, J.E. and T.J.R. Hughes (1983), *Mathematical Foundations of Elasticity*, Prentice Hall, Englewood Cliffs, NJ.

Marsden, J.E. and T.J.R. Hughes (1994), *Mathematical Foundations of Elasticity*, Dover, New York, NY.

Masserey, A. (2003), *Optimisation et Simulation Numérique du Chauffage par Induction pour le Procédé du Thixoformage*, PhD dissertation, Department of Mathematics, Federal Institute of Technology, Lausanne, Switzerland.

McManus, K., Th. Poinsot, and S. Candel (1993), Review of active control of combustion instabilities, *Progr. Energy and Combustion Sci.*, **19**, 1–29.

Meyer, Y. (1989), Private communication.

Mizohata, S. (1958), Unicité du prolongement des solutions pour quelques opérateurs différentiels paraboliques, *Mem. Coll. Sci. Univ. Kyoto*, **A 31**, 219–239.

Moin, P. and T. Bewley (1994), Feedback control of turbulence, *Appl. Mech. Rev.*, **47**(6), S3–S13.

Moubachir, M. (2002), *Mathematical and Numerical Analysis of Inverse and Control Problems for Fluid-Structure Interactions*, PhD dissertation, Ecole Nationale des Ponts et Chaussées, Paris, France.

Muñoz-Rivera, J.E. and M.G. Naso (2003), Exact boundary controllability in thermoelasticity with memory, *Adv. Diff. Equ.*, **8**, 471–490.

Narukawa, J. (1983), Boundary value control of thermo-elastic systems, *Hiroshima J. Math.*, **13**, 227–272.

Nečas, J. (1967), *Les Méthodes Directes en Théorie des Equations Elliptiques*, Masson, Paris.

Nedelec, J.C. (2001), *Acoustic and Electromagnetic Equations: Integral Representations for Harmonic Problems*, Springer-Verlag, New York, NY.

Neittaanmäki, P. and D. Tiba (1994), *Optimal Control of Nonlinear Parabolic Systems: Theory, Algorithms, and Applcations*, Marcel Dekker, New York, NY.

Nicaise, S. (2000), Exact boundary controllability of Maxwell's equations in heterogeneous media and an application to an inverse source problem, *SIAM J. Control Opt.*, **38**(4), 1145–1170.

Nicolas-Carrizosa, A. (1991), A factorization approach to the Kuramoto-Sivashinski equation. In *Advances in Numerical Partial Differential Equations and in Optimization*, S. Gomez, J.P. Hennart and R.A. Tapia, eds., SIAM, Philadelphia, PA, 262–272.

Nocedal, J. (1992), Theory of algorithms for unconstrained optimization, *Acta Numerica 1992*, Cambridge University Press, Cambridge, 199–242.

Nordlander, A. (1998), *Active Control and Drag Optimization for Flow Past a Circular Cylinder*, Master Thesis dissertation, Department of Mechanical Engineering, University of Houston, Houston, TX.

Ott, E., T. Sauer and J.A. Yorke (eds.) (1994), *Coping with Chaos*, J. Wiley, New York, NY.

Park, D.S., D.M. Ladd, and E.W. Hendricks (1994), Feedback control of Von Kármán vortex shedding behind a circular cylinder at low Reynolds numbers, *Phys. Fluids*, **6**(7), 2390–2405.

Park, K.C., J.C. Chiou, and J.D. Downer (1990), Explicit-implicit staggered procedures for multi-body dynamics analysis, *J. Guidance and Control Dyn.*, **13**, 562–570.

Peaceman, D. and H. Rachford (1955), The numerical solution of parabolic and elliptic differential equations, *J. Soc. Ind. Appl. Math.*, **3**, 28–41.

Peirce, A.P., M.A. Dahleh, and H. Rabitz (1988), Optimal control of quantum-mechanical systems: existence, numerical approximation, and applications, *Physical Review A*, **37**, 4950–4967.

Périaux, J., B. Mantel, and H.Q. Chen (1997), Application of exact controllability and genetic algorithms to the problem of scattering waves. In *Computational Science for the 21st Century*, M.O. Bristeau, G. Etgen, W. Fitzgibbon, J.L. Lions, J. Périaux, and M.F. Wheeler, eds., J. Wiley, Chichester, 518–525.

Phung, K.D. (2001), Observability and control of Schrödinger equation, *SIAM J. Control Opt.*, **40**(1), 211–230.

Pironneau, O. (1989), *Finite Element Methods for Fluids*, J. Wiley, Chichester.

Polack, E. (1971), *Computational Methods in Optimization*, Academic Press, New York, NY.

Powell, M.J.D. (1976), Some convergence properties of the conjugate gradient method, *Math. Program.*, **11**, 42–49.

Prandtl, L. (1925), The Magnus effect and wind-powered ships, *Naturwissenschaften*, **13**, 93–108.

Quarteroni, A., R. Sacco and F. Saleri (2000), *Numerical Mathematics*, Springer-Verlag, New York, NY.

Rabitz, H., G. Turinici and E. Brown (2003), Control of Quantum Dynamics: Concepts, Procedures and Future Prospects. In *Handbook of Numerical Analysis*, Vol. X, Special Volume Computational Chemistry, P.G. Ciarlet and C. Le Bris, eds., North-Holland, Amsterdam, 833–887.

Ramos, A.M, R. Glowinski and J. Périaux (2002), Pointwise control of the Burgers equation and related Nash equilibrium problems: Computational approach, *J. Optim. Theory Appl.*, **112**(3), 499–516.

Rannacher, R. (2000), Finite element methods for incompressible Navier–Stokes equations. In *Fundamental Directions in Mathematical Fluid Mechanics*, G.P. Galdi, J.G. Heywood, and R. Rannacher, eds., Birkhäuser, Basel, 191–293.

Raviart, P.A. and J.M. Thomas (1983), *Introduction à l'Analyse Numérique des Equations aux Dérivées Partielles*, Masson, Paris.

Raymond, J.P. (2006), Feedback boundary stabilization of the two-dimensional Navier–Stokes equations, *SIAM J. Control Opt.*, **45**(3), 790–828.

Roberts, J.E. and J.M. Thomas (1991), Mixed and hybrid methods. In *Handbook of Numerical Analysis*, Vol. II, P.G. Ciarlet and J.L. Lions, eds., North-Holland, Amsterdam, 523–639.

Rockafellar, T.R. (1970), *Convex Analysis*, Princeton University Press, Princeton, NJ.

Roshko, A. (1954), On the development of turbulent wakes from vortex streets, NACA Rep. 1191.

Roshko, A. (1955), On the wake and drag of bluff bodies, *J. Aerospace Sci.*, **22**, 124–132.

Roussopoulos, K. (1993), Feedback control of vortex shedding at low Reynolds numbers, *J. Fluid Mech.*, **248**, 267–296.

Russel, D.L. (1978), Controllability and stabilizability theory for linear partial differential equations. Recent progress and open questions, *SIAM Rev.*, **20**, 639–739.

Samaniego, J.M., B. Yip, Th. Poinsot, and S. Candel (1993), Low frequency combustion instability mechanism inside dump combustor, *Combustion and Flame*, **94**, 363–381.

Samarskii, A.A., V.A. Galaktionov, S.P. Kurdyumov, and A.P. Mikhailov (1995), *Blow-Up in Quasi-Linear Parabolic Equations*, Walter de Gruyter, Berlin.

Sanchez-Hubert, J. and E. Sanchez-Palencia (1989), *Vibrations and Coupling of Continuous Systems (Asymptotic Methods)*, Springer-Verlag, Berlin.

Saut, J.C. and B. Scheurer (1987), Unique continuation for some evolution equations, *J. Diff. Equations*, **66**, 118–139.

Sellin, R.H. and T. Moses (1989), *Drag Reduction in Fluid Flow*, Ellis Horwood, Chichester.

Shapiro, M. and P. Brumer (1997), Quantum control of chemical reactions, *J. Chem. Soc. Faraday trans.*, **93**(7), 1263–1277.

Sivashinsky, G. (1977), Nonlinear analysis of hydrodynamic instability in laminar flames. (I): Derivation of basic equations, *Acta Astronautica*, **4**, 1177–1206.

Sritharan, S.S. (1991a), Dynamic programming of Navier–Stokes equations, *Syst. Control Lett.*, **16**, 299–307.

Sritharan, S.S. (1991b), An optimal control problem for exterior hydrodynamics. In *Distributed Parameter Control Systems: New Trends and Applications*, G. Chen, E.B. Lee, W. Littman, and L. Markus, eds., Marcel Dekker, New York, NY, 385–417.

Sritharan, S.S. (ed.) (1998), *Optimal Control of Viscous Flow*, SIAM, Philadelphia, PA.

Stoer, J. and R. Bulirsh (1993), *Introduction to Numerical Analysis*, Springer-Verlag, New York, NY.

Tabata, M. and K.A. Itakura (1998), A precise computation of drag coefficients of a sphere, *Int. J. Comp. Fluid Dynamics*, **23**(9–10), 1493–1556.

Talflove, A. (1992), Re-inventing electromagnetics: supercomputing solution of Maxwell's equations via direct time integration on space grids, *30th AIAA Aerospace Sciences Meeting, Reno, Nevada*, AIAA Paper 92-0333.

Tartar, L. (1978), *Topics in Nonlinear Analysis*, Publications Mathématiques d'Orsay, Université Paris-Sud, Département de Mathématiques, Orsay, France.

Tartar, L. (2006), *An Introduction to Navier–Stokes Equations and Oceanography*, Springer-Verlag, Berlin.

Temam, R. (1977), *Theory and Numerical Analysis of the Navier–Stokes Equations*, North-Holland, Amsterdam.

Temam, R. (1988), *Infinite-Dimensional Dynamical Systems in Mechanics and Physics*, Springer-Verlag, New York, NY.

Temam, R. (2001), *Theory and Numerical Analysis of the Navier–Stokes Equations*, American Mathematical Society, Providence, RI.

de Teresa, L. and E. Zuazua (1996), Controllability of the linear system of thermoelastic plates, *Adv. Diff. Equ.*, **1**, 369–402.

Thomee, V. (1990), Finite difference methods for linear parabolic equations. In *Handbook of Numerical Analysis*, Vol. I, P.G. Ciarlet and J.L. Lions, eds., North-Holland, Amsterdam, 5–196.

Tokumaru, P.T. and P.E. Dimotakis (1991), Rotary oscillation control of a cylinder wake, *J. Fluid Mech.*, **224**, 77–90.

Tucsnak, M. (1996), Regularity and exact controllability for a beam with piezoelectric actuators, *SIAM J. Control Opt.*, **34**, 922–930.

Turek, S. (1999), *Efficient Solvers for Incompressible Flow Problems: An Algorithmic and Computational Approach*, Springer-Verlag, Berlin.

Turicini, G. and H. Rabitz (2001), Quantum wave function controllability, *J. Chem. Phys.*, **267**, 1–9.

Williams, D.R., H. Mansy and C. Amato (1992), The response and symmetry properties of a cylinder wake submitted to localized surface excitation, *J. Fluid Mech.*, **234**, 71–96.

Williamson, C.H.K. (1989), Oblique and parallel modes of vortex shedding in the wake of a circular cylinder at low Reynolds numbers, *J. Fluid Mech.*, **206**, 579–627.

Yserentant, H. (1993), Old and new convergence proofs for multigrid methods, *Acta Numerica 1993*, Cambridge University Press, Cambridge, 285–326.

Zhang, X. and E. Zuazua (2003), Polynomial decay and control for a 1-D model of fluid structure interaction, *C. R. Acad. Sci. Paris, Série I*, **336**, 745–750.

Zhu, W. and H. Rabitz (1998), A rapid monotonically convergent iteration algorithm for quantum optimal control over the expectation value of a positive definite operator, *J. Chem. Phys.*, **109**, 385–391.

Zuazua, E. (1988), Contrôlabilité exacte en un temps arbitrairement petit de quelques modèles de plaques. Appendix 1 of *Contrôlabilité Exacte, Perturbation et Stabilisation des Systèmes Distribués*, Vol. 1, by J.L. Lions, Masson, Paris, 465–491.

Zuazua, E. (1990a), Exact controllability for the semi-linear wave equation, *J. Math. Pures Appl.*, **69**, 1–32.

Zuazua, E. (1990b), Contrôlabilité exacte d'une équation des ondes surlinéaire à une dimension d'espace, *C. R. Acad. Sci. Paris, Série I*, **311**, 285–290.

Zuazua, E. (1993a), Exact controllability for the semi-linear wave equation in one space dimension, *Ann. Inst. H. Poincaré (C) Analyse Non Linéaire*, **10**(1), 109–129.

Zuazua, E. (1993b), Contrôlabilité du système de la thermo-élasticité, *C. R. Aad. Sci. Paris, Série I, Math.*, **317**, 371–376.

Zuazua, E. (1995), Controllability of the linear system of thermo-elasticity, *J. Math. Pures Appl.*, **74**(4), 291–315.

Zuazua, E. (1996), Controllability of the linear system of thermo-elastic plates, *Adv. Diff. Equ.*, **1**(3), 369–402.

Zuazua, E. (2002), Controllability of partial differential equations and its semi-discrete approximation, *Discrete and Continuous Dynamical Systems*, **8**(2), 469–513.

Zuazua, E. (2003), Remarks on the controllability of the Schrödinger equation. In *Quantum Control: Mathematical and Numerical Challenges*, A. Bandrauk, M.C. Delfour and C. Le Bris, eds., CRM Proc. Lectures Notes Series, Vol. 33, AMS Publications, Providence, RI, 181–199.

Zuazua, E. (2005), Propagation, observation, and control of waves approximated by finite difference methods, *SIAM Rev.*, **47**(2), 197–243.

Zuazua, E. (2007), Controllability and observability of partial differential equations: some results and open problems. In *Handbook of Differential Equations*, Vol. 3: *Evolutionary Equations*, Chapter 7, C.M. Dafermos and E. Feireisl, eds., Elsevier, Amsterdam, 527–621.

Index of names

Index of subjects